CONCEPTS AND METHODS IN MODERN THEORETICAL CHEMISTRY

ELECTRONIC STRUCTURE AND REACTIVITY

ATOMS, MOLECULES, AND CLUSTERS
Structure, Reactivity, and Dynamics
Series Editor: Pratim Kumar Chattaraj

Aromaticity and Metal Clusters
Edited by Pratim Kumar Chattaraj

Concepts and Methods in Modern Theoretical Chemistry:
Electronic Structure and Reactivity
Edited by Swapan Kumar Ghosh and Pratim Kumar Chattaraj

Concepts and Methods in Modern Theoretical Chemistry:
Statistical Mechanics
Edited by Swapan Kumar Ghosh and Pratim Kumar Chattaraj

Quantum Trajectories
Edited by Pratim Kumar Chattaraj

ATOMS, MOLECULES, AND CLUSTERS

CONCEPTS AND METHODS IN MODERN THEORETICAL CHEMISTRY

ELECTRONIC STRUCTURE AND REACTIVITY

EDITED BY

SWAPAN KUMAR GHOSH
PRATIM KUMAR CHATTARAJ

CRC Press
Taylor & Francis Group
Boca Raton London New York

CRC Press is an imprint of the
Taylor & Francis Group, an **informa** business

CRC Press
Taylor & Francis Group
6000 Broken Sound Parkway NW, Suite 300
Boca Raton, FL 33487-2742

First issued in paperback 2019

© 2013 by Taylor & Francis Group, LLC
CRC Press is an imprint of Taylor & Francis Group, an Informa business

No claim to original U.S. Government works

ISBN-13: 978-1-4665-0528-5 (hbk)
ISBN-13: 978-0-367-38032-8 (pbk)

Library of Congress Cataloging-in-Publication Data

Concepts and methods in modern theoretical chemistry : electronic structure and
    reactivity / edited by Swapan Kumar Ghosh and Pratim Kumar Chattaraj.
    pages cm. -- (Atoms, molecules, and clusters)
    Includes bibliographical references and index.
    ISBN 978-1-4665-0528-5 (hardback)
    1. Chemical reaction, Conditions and laws of. 2. Electron distribution. 3. Chemistry,
Physical and theoretical. 4. Quantum chemistry. I. Ghosh, Swapan Kumar. II. Chattaraj,
Pratim Kumar.

QD503.C665 2013
541'.28--dc23   2012044529

Visit the Taylor & Francis Web site at
http://www.taylorandfrancis.com

and the CRC Press Web site at
http://www.crcpress.com

Contents

Series Preface

ATOMS, MOLECULES, AND CLUSTERS: STRUCTURE, REACTIVITY, AND DYNAMICS

While atoms and molecules constitute the fundamental building blocks of matter, atomic and molecular clusters lie somewhere between actual atoms and molecules and extended solids. Helping to elucidate our understanding of this unique area with its abundance of valuable applications, this series includes volumes that investigate the structure, property, reactivity, and dynamics of atoms, molecules, and clusters.

The scope of the series encompasses all things related to atoms, molecules, and clusters including both experimental and theoretical aspects. The major emphasis of the series is to analyze these aspects under two broad categories: approaches and applications. The *approaches* category includes different levels of quantum mechanical theory with various computational tools augmented by available interpretive methods, as well as state-of-the-art experimental techniques for unraveling the characteristics of these systems including ultrafast dynamics strategies. Various simulation and quantitative structure–activity relationship (QSAR) protocols will also be included in the area of approaches.

The *applications* category includes topics like membranes, proteins, enzymes, drugs, biological systems, atmospheric and interstellar chemistry, solutions, zeolites, catalysis, aromatic systems, materials, and weakly bonded systems. Various devices exploiting electrical, mechanical, optical, electronic, thermal, piezoelectric, and magnetic properties of those systems also come under this purview.

The first two books in the series are (a) *Aromaticity and Metal Clusters* and (b) *Quantum Trajectories*. A two-book set on *Concepts and Methods in Modern Theoretical Chemistry*, edited by Swapan Kumar Ghosh and Pratim Kumar Chattaraj, is the new addition to this series. The first book focuses on the electronic structure and reactivity of many-electron systems and the second book deals with the statistical mechanical treatment of collections of such systems.

Pratim Kumar Chattaraj
Series Editor

Foreword

A certain age comes when it is no longer unseemly to reflect on one's contribution to the world and, in the case of a scientist, the mark one has left on one's career. Professor B. M. Deb has reached such an age and can look back with considerable satisfaction on his scientific legacy. I knew him long ago, when his career was still to come, when he was at Oxford and was forming his aspirations and skills. Now, long after, in these volumes, we are seeing where those aspirations and skills in due course led.

One of the principal contributions of theoretical chemistry to what might be called "everyday" chemistry is its development of powerful computational techniques. Once such techniques were regarded with suspicion and of little relevance. But in those days the techniques were primitive, and the hardware was barely adequate for the enormous computations that even the simplest molecules require. Then, over the decades, techniques of considerable sophistication emerged, and the hardware evolved in unimaginable ways to accommodate and inspire even more imagination and effort. Now, the computations give great insight and sometimes surpass even actual measurements.

Of these new techniques, the most intriguing, and currently one in high fashion, has been the density functional theory. That Professor Deb has contributed so much in this field is demonstrated by the number of contributions in these volumes that spring from his work. Fashions, of course, come and go, but these techniques are currently having a considerable impact on so many branches of chemistry that they are undoubtedly a good reason for Professor Deb to reflect, with characteristic but misplaced modesty, on what he has done to promote and advance the technique.

It was for me a great pleasure to know the young Professor Deb and to discern promise and to know that the contributions to these volumes show that that promise has been more than amply fulfilled in a lifetime of contributions to theoretical chemistry. Professor Deb must be enormously proud of having inspired these volumes, and justly so.

Peter Atkins
Oxford

Preface

This collection presents a glimpse of selected topics in theoretical chemistry by leading experts in the field as a tribute to Professor Bidyendu Mohan Deb in celebration of his seventieth birthday.

The research of Professor Deb has always reflected his desire to have an understanding and rationalization of the observed chemical phenomena as well as to predict new phenomena by developing concepts or performing computations with the help of available theoretical, modeling, or simulation techniques. Formulation of new and more powerful theoretical tools and modeling strategies has always formed an ongoing and integral part of his research activities. Proposing new experiments, guided by theoretical insights, has also constituted a valuable component of his research that has a fairly interdisciplinary flavor, having close interconnections with areas like physics and biology.

The concept of single-particle density has always fascinated him, perhaps starting with his work on force concept in chemistry, where the density is sufficient to obtain Hellmann–Feynman forces on the nuclei in molecules. His two reviews on "Force Concept in Chemistry" and "Role of Single Particle Density in Chemistry," published in *Reviews of Modern Physics*, have provided a scholarly exposition of the intricate concepts, inspiring tremendous interest and growth in this field. These have culminated in two edited books. The force concept provided the vehicle to go to new ways of looking at molecular shapes, the HOMO postulate being an example of his imaginative skills. The concept of forces on the nuclei was soon generalized to the concept of stress tensor within the electron cloud in molecules, the role of which in determining chemical binding and stability of molecules was also explored. Various aspects of the density functional theory (DFT) were investigated. The static aspects were soon viewed as only a special case of the corresponding dynamical theory, the so-called quantum fluid dynamics (QFD), which was developed in 3-D space and applied to study collision phenomena, response to external fields, and other related problems.

His mind has always opened new windows to bring in the fresh flavor of novel concepts for interpreting the "observed," predicting the "not yet observed," and also created tools and strategies to conquer unknown territory in the world of molecules, materials, and phenomena. "Concepts are the fragrance of science," he always emphasizes. His research has often seemed to be somewhat unconventional in the sense that he has always stressed conceptual developments that are often equally suited for practical applications as well. He has a thirst for looking into the secret of "why things are the way they are" and the mystery behind "being to becoming," focusing on the structure and dynamics of systems and phenomena, both of which have been enriched immensely by his contributions. Aptly, we have the two present books covering structure and dynamics, respectively.

The topics in *Concepts and Methods in Modern Theoretical Chemistry: Electronic Structure and Reactivity* include articles on DFT, particularly the functional and conceptual aspects, excited states, molecular electrostatic potentials, intermolecular interactions, general theoretical aspects, application to molecules, clusters and solids, electronic stress, the information theory, the virial theorem, new periodic tables, the role of the ionization potential and electron affinity difference, etc. The majority of the chapters in *Concepts and Methods in Modern Theoretical Chemistry: Statistical Mechanics* include time-dependent DFT, QFD, photodynamic control, nonlinear dynamics, molecules in laser field, charge carrier mobility, excitation energy transfer, chemical reactions, quantum Brownian motion, the third law of thermodynamics, transport properties, nucleation, etc.

In the Indian context, theoretical chemistry has experienced significant growth over the years. Professor Deb has been instrumental in catalyzing this growth by providing the seed and nurturing young talents. It is the vision and effort of Professor Deb that made it possible to inspire the younger generation to learn, teach, and practice theoretical chemistry as a discipline. In this context, it is no exaggeration to describe him as the doyen of modern theoretical chemistry in India.

Professor Deb earned a PhD with Professor Charles Coulson at the University of Oxford and then started his professional career at the Indian Institute of Technology, Bombay, in 1971. Being a scientist–humanist of the highest order, he has always demanded a high sense of integrity and a deep involvement from his research group and other students. He has never sacrificed his own human qualities and never allowed other matters to overtake the human aspects of life.

While his research has focused on conceptual simplicity, computational economy, and sound interpretive aspects, his approach to other areas of life reflects the same. We have often wondered at the expanse of his creativity, which is not restricted to science but also covers art, literature, and life in general. His passion for work has, of course, never overshadowed his warmth, affection, and helpfulness to others. He has an extraordinary ability to act as a creative and caring mentor. His vast knowledge in science, art, literature, and many other of the finer aspects of life in general, together with his boundless sources of enthusiasm, creativity, and imagination, has often made him somewhat unconventional in his thinking, research, and teaching. Designing new experiments in class and introducing new methods in teaching have also been his passion. His erudition and versatility are also reflected in his writings on diverse topics like the cinema of Satyajit Ray and lectures on this as well as various aspects of art.

We are privileged to serve as editors of these two books on Concepts and Methods in Modern Theoretical Chemistry and offer the garland of scholarly essays written by experts as a dedication to this great scientist–humanist of recent times with affection and a deep sense of respect and appreciation for all that he has done for many of us and continues to do so. We also gratefully acknowledge the overwhelming and hearty response received from the contributors, to whom we express our indebtedness.

We are grateful to all the students, associates, and collaborators of Professor B. M. Deb who spontaneously contributed to the write-up of the "Reminiscences" and, in particular, Dr. Amlan K. Roy for compiling it in a coherent manner to the present form. Finally, we are deeply indebted to Professor B. M. Deb for his kind help, guidance, and encouragement throughout our association with him.

Swapan Kumar Ghosh
Pratim Kumar Chattaraj

Reminiscences

It is indeed a great pleasure to pen this note in celebration of Professor B. M. Deb's seventieth birthday. For many of us, he is a mentor, confidante, and adviser. Many others look at him as an extraordinary teacher; a patient, encouraging, and motivating guide; a warm and caring human being; and a connoisseur of literature, art, and so on. His dedication and passion for science is infectious.

Many of us have been fortunate to attend his lectures on quantum chemistry, structure, bonding, symmetry, and group theory, which were all about the interlinking of abstract concepts that are often sparsely scattered. After trudging along a series of lectures, one is rewarded with the eventual conclusion that all chemical bonds are mere manifestations of a single phenomenon, namely, the redistribution of electron density. Often, he would explain physics from real-life analogies rather than try to baffle and intimidate audiences with lots of mathematics—a popular trick often used in the community. Just paying attention in his class gives one enough confidence to tackle the most challenging problems in quantum chemistry. His recent endeavor to initiate a course on Indian heritage has been highly appreciated. It is not a history class, as the title may imply to some people, but rather a scientific evaluation of the Indian past. Taking examples from our glorious past, the course differentiates between easy and right about scientific ethics and logically establishes the path one should follow for uplifting individual souls and society as a whole. Although a theoretician, his enthusiasm and excitement for practical applications of science is no less. The experiments on beating hearts and chemical oscillations are among the most popular in the class.

His books *The Force Concept in Chemistry* and *The Single-Particle Density in Physics and Chemistry* were hugely influential among those who sought, in quantum chemistry, not just a computational tool for the calculation of molecular properties, but a fundamental understanding of the physics of chemical bonding and molecular reactivity. The application of the Hellmann–Feynman theorem to provide qualitative insights into chemical binding in molecules as well as molecular shapes caught the interest of even R. P. Feynman. As a research student, his communication with Professor Feynman was a matter of great amazement, motivation, and pride for many of his early PhD students, as Dr. Anjuli S. Bamzai recalls. Despite his considerable work in density functional theory (DFT), he held an agnostic attitude toward it, in the sense that he did not regard the search for a functional as the holy grail of DFT or see DFT as being somehow in opposition to wave function–based theories. He was also not against approximations and freely employed them wherever useful. But he was convinced that the electron density held the key to a deeper understanding of the chemical phenomena. Thus, in a way, he was willing to entertain the need for considering the phase in addition to density to achieve a consistent treatment of excited states and time-dependent phenomena.

To have worked with him has been a major turning point in our lives. We discover him as a scientist with high morality and professional ethics. It is not only learning the concepts in theoretical chemistry but also a more holistic approach toward research, learning, and science itself. While scrupulously fair, he expected his students to be conscientious. He gave his all to his students and to his research. Reasonably enough, he expected no less from his students and from his colleagues, a favorite expression being that he wanted the students "to go flat out" on their prospective research problems. The amount of hard work that he put, propelled by tiny seeds of imagination and analytical logic, always inspired us. But while the force of his scientific conviction was strong, he was always open to arguments and discussion. Even in turbulent times and under less-than-ideal conditions, he was not willing to compromise on his scientific standards or integrity. He had a knack for choosing and working on problems that were emerging frontiers of theoretical chemistry. That was because of his intuition to choose research projects for us so that we could contribute to the field effectively, despite the fact that all his research works were done in India in relative isolation. Although much of his research career spanned the overlap between physics and chemistry, he had no sympathy for those who would regard chemistry as inferior to physics. When a physicist, after hearing Professor Deb speak about his current research, praised him with the words, "You are almost doing physics," he rejoined with a wry smile, "No, I am doing good chemistry." With this statement, even his detractors would agree!

It feels amazing that we have learned as much from anecdotal informal interaction with him as from the research experience. What added to the pleasure of working with him were discussions about science and nonscientific matters. It was fascinating to listen to him talk about poetry, literature, movies, food, art, and cultures across the world. We would occasionally visit his residence and spend time with him at the dining table discussing the progress of our projects while partaking of delicious snacks and meals prepared by Mrs. Deb. For many of us, it was something like a home away from home, and we soon learned that a combination of food and food for thought goes well together. The amazement of such an experience is narrated here by Dr. Bamzai. Their home was decorated with the works of some of the greatest artists of all time. Often one would come across a discussion about Leonardo da Vinci's *The Last Supper* or Picasso's *Guernica* and how the artist, through his work, had conveyed the tragedies of war and its horrific impact on innocent civilians. At other times, he would discuss how M. C. Escher's art effectively conveys important concepts such as symmetry and transformations in crystallography. He has serious concern also about science, culture, and heritage. He constantly engages into the popularization of science as well as the improvement of the education system in India. It is surprising how he was able to impart knowledge on such a diverse array of topics.

Given his varied interests and the positive energy that he imbues into his surroundings, we know that he will never stop being an academic. Despite his own and Mrs. Deb's deteriorating health, they have stood beside their students and colleagues with constant support and encouragement. Many of us remember the act of good Samaritan-ship by Professor Deb and his family toward his colleagues. One such act is vividly recollected here by Professor Harjinder Singh, whose daughter was struggling in an intensive care unit at that time. They needed to stay at a place close to the hospital. Deb's family extended their wholehearted support during that crisis, not minding any inconvenience caused to them, especially when the city of Chandigarh was going through the political turmoil of a full-blown secessionist movement, regular terrorist threats, shootings, bus bombings, and assassinations.

A lesson we learned from Professor Deb that we have carried throughout our life was his admonition: "Beware of the fourth rater who calls the third rater good." It was a call and a challenge to aspire to the highest standards of excellence in life, and it is the pursuit of this gold standard that he strived to inculcate in us, despite potential temptations to discard it so often! We consider ourselves very fortunate to have Professor Deb as our teacher, philosopher, and guide. His work and work ethic will continue to influence and nurture future generations via many students and postdocs he has taught and guided. He remains a source of inspiration to all who wish to be an ideal teacher, a thorough researcher, and, above all, a decent human being. We feel privileged to be a part of his extended family and take this opportunity to express our sincere gratitude to him for his support, kindness, and patience. We are indebted to him and send our best wishes to his family.

Anjuli S. Bamzai
Pratim K. Chattaraj
Mukunda Prasad Das
Swapan K. Ghosh
Neetu Gupta
Geeta Mahajan
Smita Rani Mishra
Amitabh Mukherjee
Aniket Patra
Amlan K. Roy
Mainak Sadhukhan
R. P. Semwal
Harjinder Singh
Ranbir Singh
Nagamani Sukumar
Vikas
Amita Wadehra

Editors

Swapan Kumar Ghosh earned a BSc (Honors) and an MSc from the University of Burdwan, Bardhaman, India, and a PhD from the Indian Institute of Technology, Bombay, India. He did postdoctoral research at the University of North Carolina, Chapel Hill. He is currently a senior scientist with the Bhabha Atomic Research Centre (BARC), Mumbai, India, and head of its theoretical chemistry section. He is also a senior professor and dean-academic (Chemical Sciences, BARC) of the Homi Bhabha National Institute, Department of Atomic Energy (DAE), India, and an adjunct professor with the University of Mumbai–DAE Centre of Excellence in Basic Sciences, India.

He is a fellow of the Indian Academy of Sciences, Bangalore; Indian National Science Academy, New Delhi; National Academy of Sciences, India, Allahabad; Third World Academy of Sciences (TWAS), Trieste, Italy (currently known as the Academy of Sciences for the Developing World); and Maharashtra Academy of Sciences. He is a recipient of the TWAS prize in chemistry; silver medal of the Chemical Research Society of India (CRSI); the Jagdish Shankar Memorial Lecture Award of the Indian National Science Academy; the A. V. Rama Rao Prize of Jawarharlal Nehru Centre for Advanced Scientific Research, Bangalore, India; and the J. C. Bose Fellowship of the Department of Science and Technology, India. He is currently also one of the vice presidents of CRSI.

His research interests are theoretical chemistry, computational materials science, and soft condensed matter physics. He has been involved in teaching and other educational activities including the Chemistry Olympiad Program. He has twice been the mentor and delegation leader of the Indian National Chemistry Olympiad Team participating in the International Chemistry Olympiad at Athens (Greece) and Kiel (Germany).

Pratim Kumar Chattaraj earned a BSc (Honors) and an MSc from Burdwan University and a PhD from the Indian Institute of Technology (IIT), Bombay, India, and then joined the faculty of the IIT, Kharagpur, India. He is now a professor with the Department of Chemistry and also the convener of the Center for Theoretical Studies there. In the meantime, he visited the University of North Carolina, Chapel Hill, as a postdoctoral research associate and several other universities throughout the world as a visiting professor. Apart from teaching, Professor Chattaraj is involved in research on density functional theory, the theory of chemical reactivity, aromaticity in metal clusters, *ab initio* calculations, quantum trajectories, and nonlinear dynamics. He has been invited to deliver special lectures at several international conferences and to contribute chapters to many edited volumes.

Professor Chattaraj is a member of the editorial board of *J. Mol. Struct. Theochem* (currently *Comp. Theo. Chem.*), *J. Chem. Sci., Ind. J. Chem.-A, Nature Collections Chemistry*, among others. He has edited three books and special issues of different journals. He was the head of the Department of Chemistry, IIT, Kharagpur, and a council member of the Chemical Research Society of India. He is a recipient of the University Gold Medal, Bardhaman Sammilani Medal, INSA Young Scientist Medal, B. C. Deb Memorial Award, B. M. Birla Science Prize, and CRSI Medal. He was an associate of the Indian Academy of Sciences. He is a fellow of the Indian Academy of Sciences (Bangalore), the Indian National Science Academy (New Delhi), the National Academy of Sciences, India (Allahabad), and the West Bengal Academy of Science and Technology. He is also a J. C. Bose National Fellow and a member of the Fonds Wetenschappelijk Onderzoek (FWO), Belgium.

Contributors

Rodrigo J. Alvarez-Mendez
Departamento de Química
Cinvestav
Mexico City, Mexico

G. G. N. Angilella
Dipartimento di Fisica e Astronomia
Università di Catania
and
Scuola Superiore di Catania
Università di Catania
and
CNISM
UdR di Catania
and
INFN
Sezione di Catania
Catania, Italy

Paul W. Ayers
McMaster University
Hamilton, Ontario, Canada

S. Bhattacharya
Department of Materials Science
Indian Association for the Cultivation
   of Science
Jadavpur, Kolkata, India

S. P. Bhattacharyya
Department of Physical Chemistry
Indian Association for the Cultivation
   of Science
Jadavpur, Kolkata, India

Ramon Carbó-Dorca
Institut de Química Computacional
Universitat de Girona
Girona, Catalonia, Spain

Carlos Cárdenas
Departamento de Física
Facultad de Ciencias
Universidad de Chile
and
Centro para el Desarrollo de las
   Nanociencias y Nanotecnología
CEDENNA
Santiago, Chile

Debajit Chakraborty
McMaster University
Hamilton, Ontario, Canada

K. R. S. Chandrakumar
Theoretical Chemistry Section
Bhabha Atomic Research Center
Mumbai, India

Lan Cheng
Department of Chemistry and
   Biochemistry
University of Texas
Austin, Texas

G. P. Das
Department of Materials Science
Indian Association for the Cultivation
   of Science
Jadavpur, Kolkata, India

Aurélien de la Lande
Laboratoire de Chimie Physique
Université Paris Sud
Orsay, France

Jorge M. del Campo
Departamento de Física y Química
   Teórica
Facultad de Química
Universidad Nacional Autónoma de
   México
Mexico City, Mexico

Patricio Fuentealba
Departamento de Física
Facultad de Ciencias
Universidad de Chile
and
Centro para el Desarrollo de las
   Nanociencias y Nanotecnología
CEDENNA
Santiago, Chile

José L. Gázquez
Departamento de Química
Universidad Autónoma
   Metropolitana-Iztapalapa
Mexico City, Mexico

M. K. Harbola
Department of Physics
Indian Institute of Technology
Kanpur, India

M. Hemanadhan
Department of Physics
Indian Institute of Technology
Kanpur, India

Rahul Kar
Department of Chemistry
Dibrugarh University
Dibrugarh, Assam, India

Susmita Kar
Department of Physical Chemistry
Indian Association for the Cultivation
   of Science
Jadavpur, Kolkata, India

Andreas M. Köster
Departamento de Química
Cinvestav
Avenida Instituto Politécnico Nacional
   2508
Mexico, Mexico

R. Mahesh Kumar
Chemical Laboratory
CSIR-Central Leather Research
   Institute
Chennai, India

Priya Mahadevan
S. N. Bose National Center for Basic
   Sciences
JD-Block, Sector III
Salt Lake, Kolkata, India

N. H. March
Oxford University
Oxford, United Kingdom

and

Department of Physics
University of Antwerp
Antwerp, Belgium

Debashis Mukherjee
Raman Center for Atomic, Molecular
   and Optical Sciences
Indian Association for the Cultivation
   of Science
Kolkata, India

Jane S. Murray
CleveTheoComp
Cleveland, Ohio

and

Department of Chemistry
University of New Orleans
New Orleans, Los Angeles

Á. Nagy
Department of Theoretical Physics
University of Debrecen
Debrecen, Hungary

Roman F. Nalewajski
Department of Theoretical Chemistry
Jagiellonian University
Cracow, Poland

Ashis Kumar Nandy
S. N. Bose National Center for Basic
    Sciences
JD-Block, Sector III
Salt Lake, Kolkata, India

and

Indian Association for the Cultivation
    of Sciences
Jadavpur, Kolkata, India

Sourav Pal
Physical Chemistry Division
National Chemical Laboratory
Pune, India

Rudolph Pariser
Chapel Hill, North Carolina

Robert G. Parr
Department of Chemistry
University of North Carolina
Chapel Hill, North Carolina

S. H. Patil
Department of Physics
Indian Institute of Technology
Mumbai, India

John P. Perdew
Department of Physics and Engineering
    Physics
Tulane University
New Orleans, Louisiana

Peter Politzer
CleveTheoComp
Cleveland, Ohio

and

Department of Chemistry
University of New Orleans
New Orleans, Los Angeles

Elisa Rebolini
Laboratoire de Chimie Théorique
Université Pierre et Marie Curie and
    CNRS
Paris, France

Dennis R. Salahub
Department of Chemistry and IBI—
    Institute for Biocomplexity and
    Informatics
University of Calgary
Calgary, Alberta, Canada

P. Samal
School of Physical Sciences
National Institute of Science Education
    and Research
Bhubaneswar, India

D. D. Sarma
Solid State and Structural Chemistry
    Unit
Indian Institute of Science
Bangalore, India

and

Council of Scientific and Industrial
    Research-Network of Institutes for
    Solar Energy (CSIR-NISE)
New Delhi, India

G. Narahari Sastry
Molecular Modeling Group
CSIR-Indian Institute of Chemical
Technology
Hyderabad, India

Andreas Savin
Laboratoire de Chimie Théorique
Université Pierre et Marie Curie and
CNRS
Paris, France

Avijit Sen
Raman Center for Atomic, Molecular
and Optical Sciences
Indian Association for the Cultivation
of Science
Kolkata, India

K. D. Sen
School of Chemistry
University of Hyderabad
Hyderabad, India

Md. Shamim
Department of Physics
Indian Institute of Technology
Kanpur, India

V. Subramanian
Chemical Laboratory
CSIR-Central Leather Research
Institute
Chennai, India

Akitomo Tachibana
Department of Micro Engineering
Kyoto University
Kyoto, Japan

Andreas K. Theophilou
Democritos National Center for
Scientific Research
Athens, Greece

Julien Toulouse
Laboratoire de Chimie Théorique
Université Pierre et Marie Curie and
CNRS
Paris, France

Samuel B. Trickey
Quantum Theory Project
Department of Physics and Department
of Chemistry
University of Florida
Gainesville, Florida

Alberto Vela
Departamento de Química
Cinvestav
Mexico City, Mexico

Dolly Vijay
Chemical Laboratory
CSIR-Central Leather Research
Institute
Chennai, India

An Interview with B. M. Deb

(This interview was conducted by Richa Malhotra for the journal *Current Science*. An edited version of the interview was published in *Current Science* on January 25, 2012, and an expanded version appears in the present book. Courtesy of *Current Science*.)

- **How has the field of theoretical and computational chemistry evolved over the years?**

   One has to write a book to answer this! It is similar to answering how science has evolved in the last century and the present century so far. Let me try to explain how the broad contours of theoretical and computational chemistry have developed over many years.

   Theoretical chemistry has been operating at the interface between chemistry, physics, biology, mathematics, and computational science. It deals with *systems* and *phenomena* concerning these large subjects. The *systems* are microscopic, mesoscopic, and macroscopic, *viz.*, atoms, molecules, clusters, and other nanosystems, soft and hard condensed matter. The *phenomena* involve a holistic combination of *structure, dynamics,* and *function. Structure* concerns itself with geometry, where both Group Theory and Topology (especially, Graph Theory) come in. *Dynamics* deal with evolution in time; structure is a consequence of dynamics, and *vice versa. Function* implies all kinds of properties, *viz.,* electrical, magnetic, optical, chemical, biological, and even mechanical properties. Let me show this by a triangular SDF figure, which is actually *valid for every field of human endeavor,* including literature and arts.

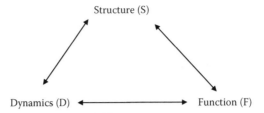

The disciplines which study all of these are quantum chemistry (both nonrelativistic and relativistic), quantum biology and biochemistry, quantum pharmacology, spectroscopy, molecular reaction dynamics, equilibrium and nonequilibrium statistical mechanics, equilibrium and nonequilibrium thermodynamics, nonlinear dynamics, mathematical methods of chemistry, etc., with various subdisciplines. It is interesting to note that Graph Theory, which is a branch of mathematics and is also used by chemists, physicists, biologists, sociologists, electrical engineers, neural and other network scientists, had drawn primary inspiration in the 1870s from structural formulas in chemistry which denote *connectivity*. As you see, at this level, it is really not possible to distinguish between theoretical chemistry and theoretical physics or, for that matter, theoretical biology. Atomic and molecular physics, polymer and condensed matter physics—bringing in materials science—and even certain issues of structure and interaction in nuclear physics are of interest to theoretical chemists. Present-day mathematical chemistry, which uses topology—though not necessarily in conjunction with quantum mechanics—to develop quantitative structure–activity relations for drug design, hazard chemicals assessment, etc., is another aspect of theoretical chemistry.

Computational chemistry has been primarily concerned with the development and application of computer software, using theoretical chemistry methodologies, utilizing numerical methods and computer programming in a significant way. Nowadays, not all theoretical chemists and computational chemists develop their own codes. Only some do, if necessary, whereas others employ standard and/or commercially available software packages for performing computations on electronic structure; geometry; various chemical, physical, and biological properties; as well as various kinds of classical and semiclassical simulations of structures and dynamics of large molecular systems such as proteins, polymers, and liquids. Since the 1930s, theoretical and computational chemists have been a major driving force behind developments in computational sciences, including both computer hardware and software development, especially number crunching and graphics. Graphics are particularly important because chemists find it difficult to work without detailed visualization. Also, representing millions of computed numbers in terms of colorful pictures, which could undulate in time as well, greatly enhances our insight into the phenomenon being studied. Presently, we feel that any equation which cannot be solved analytically but the solution exists can be solved numerically with an accuracy which goes beyond experimental accuracy as long as the variables are not too many in number.

Because of their subject's complex multidisciplinary nature, theoretical chemists have been somewhat like orphans! You can find them everywhere, in chemistry, biochemistry, physics, mathematics, computer science, chemical engineering, materials science, as well as in industries across the world, though I believe very few, if any at all, in Indian industries. Since earth scientists are currently using theoretical chemistry computations for interpretations of their data, perhaps we will soon have a theoretical chemist in an earth science department!

Historically, mathematicians, physicists, chemists, computer scientists, and even economists have contributed to theoretical chemistry. Apart from the mathematician J. Sylvester's realization in the 1870s that structural formulas in chemistry have a hidden algebraic structure, I think Gibbs's development of thermodynamics, followed by the axiomatic development of the subject by Carathéodory and Born, Debye–Hückel–Onsager theory of strong electrolytes, Lewis's electronic theory of valence, the vector atom model, etc., are some of the important landmarks in the early days of theoretical chemistry. Once quantum mechanics came into being, there resulted an explosive growth in the areas that I have mentioned earlier.

- **What have been your key contributions and areas of interest in chemistry?**

It is embarrassing to talk about "key contributions" of a mediocre scientist. I always believed that (1) theory should not only explain current experiment but should also make predictions for future experiment and that (2) concepts are the fragrance of science. Therefore, along with my research students, I have been struggling to develop rigorous concepts in chemistry which can lead to deep insights, as well as accurate results which are amenable to physical and pictorial interpretation. Whatever we have been able to do has been possible only because of my courageous students.

Because of my persistent interest in geometry, our first work in India was to develop a purely qualitative and general molecular–orbital approach (without computation and by using group theory extensively), leading to a force model for explaining and predicting various features of molecular geometry of small- and medium-sized molecules based on the electron density in the highest occupied molecular orbital. This was followed by semi-empirical computations of electronic structure and geometry of quite a few unknown molecules, predicting that they are capable of independent existence. Along with these, we wrote an article entitled "the force concept in chemistry." The responses to this article changed the course of our research, especially Fano's suggestion that we think of how the electron density can be calculated without the wave function and Feynman's suggestion that we look into internal stresses in molecules. Even though we did not know then of Hohenberg–Kohn theorems and Kohn–Sham equations, we were already completely convinced of the fundamental significance of electron density in three-dimensional space and strongly felt that, through the electron density, nonrelativistic quantum phenomena can have "classical" interpretations, which are necessary for pictorial understanding. Based on Feynman's suggestion, we defined an electrostatic stress tensor using the electron density and showed that this has the same form as Maxwell's stress tensor for classical electromagnetic fields. Furthermore, along the bond direction in a diatomic molecule, the appropriate component of the stress tensor shows an extremum at the equilibrium internuclear distance. In trying to understand why stress tensor should be such an important entity, we realized that we have to go to classical fluid dynamics. The fluid dynamical interpretation of one-electron systems was already in existence but taking it to many-electron systems was rather difficult. We therefore developed what we called a quantum fluid dynamical interpretation of many-electron systems in terms of the electron density and defined a comprehensive stress tensor for such a system in terms of the full nonrelativistic Hamiltonian, i.e., by incorporating kinetic, electrostatic, exchange, and correlation terms. This still had the same form as Maxwell's tensor. We then defined a general criterion for the stability of matter, *viz.*, the force density obtained from this stress tensor must vanish at every point in three-dimensional space. The stress tensor, however, did not yield a deterministic equation for the electron density which has to incorporate both *space and time*.

Time was of the essence in our struggle. The two interlinked bottlenecks in the electron density approach were time dependence and excited states. We first developed a rigorous time-dependent density functional theory for a certain class of potentials by utilizing QFD. Since this version of density functional theory was not exact for all potentials, we also developed a similar approach in terms of natural orbitals which are exact in principle. This approach yielded an equation for the ground state density whose accuracy was very good. Using this, we calculated the frequency-dependent multipole (2^l-pole, l = 1, 2, 3, 4) polarizabilities of atoms. Some of these computed numbers still await experimental verification.

Efforts continued to generate more accurate equations for directly determining density by a single equation no matter how many electrons are there in the system. One such effort yielded a quadratic—rather than a differential—time-independent equation whose density yielded many interesting results. This equation led to an effort to justify the existence of empirically finite atomic, ionic, and Van der Waals radii—even though quantum mechanically these radii are infinite—by adopting a conjecture that such finite radii are decided by a single universal value of the electron density in space. Finally, we were able to obtain a fascinating nonrelativistic nonlinear single partial differential equation for the direct determination of electron density and properties of many-electron systems. Besides applying this equation to the ground state and time-dependent situations, application was also made to proton-atom high-energy scattering; it was possible to identify *approach, encounter, and departure regimes,* which should be helpful in chemical reactions. This equation has a number of interesting mathematical properties, some of which we have examined while others remain unexplored. A relativistic quantum fluid dynamical density approach was also developed. Additionally, we have written quite a bit in trying to emphasize the fundamental significance of electron density in understanding structures, dynamics, and properties.

An important job of theory is to explain and predict phenomena. Two decades ago, we became interested in atoms and molecules under extreme conditions such as intense laser and strong magnetic fields. We pushed our time-dependent density equation into these difficult situations. With lasers, quite exciting results and insights were obtained into various multiphoton processes such as spatial shifting of density in both femtoseconds and attoseconds, photoionization spectra, high-order harmonics generation (its implication is the creation of attosecond and X-ray lasers), suppression of ionization under a superintense laser, Coulomb explosion in molecular dissociation, etc. The mechanism of shortening of bond length in a diatomic molecule under strong magnetic fields was also studied. We have predicted that, if an oscillating strong magnetic field of an appropriate frequency interacts with a hydrogen atom, coherent radiation should be emitted. This remains to be experimentally verified. We have also proposed a new dynamical signature of quantum chaos and demonstrated it with strong magnetic fields.

We now come to excited states. Using a hybrid wave function-density approach and an interpretation of exchange proposed by other workers, we have been able to calculate excited state energies and densities of several hundred excited states of various atoms. These were singly, doubly, and triply excited states, autoionizing states, satellite states, etc., and involved both small and large energy differences.

- **You have worked mainly outside the boundaries of "chemistry." What are the interdisciplinary areas that you have worked on?**

    I do not think I have worked outside the boundaries of chemistry, which are actually limitless. An interdisciplinary research area that I have pursued is the quantum theory of structures, dynamics, and properties of atoms and molecules. I have had great pleasure in devoting some time over the years for designing exciting and colorful chemical experiments, based on research literature, for undergraduate and postgraduate teaching laboratories. Each of these brings as many sciences as possible on the common platform of one single experiment. This was to partly satisfy my hunger for experimental chemistry! Also, writing on integrated learning in sciences, designing new curricula and developing new courses have been something of a passion. I have had a life-long interest in science, mathematics, literature, and art in Ancient and Medieval India, on which I am writing a book for the last five years. The idea of holism of Ancient Indians that everything is connected to everything else has always fascinated me because this is the essence of multidisciplinarity.

- **What do you see as things that have changed in the field of chemistry, especially theoretical chemistry and computational chemistry?**

    I see considerable development in the interfaces between chemistry and biology, as well as between chemistry and materials science. Some development has also occurred in the interface between chemistry and earth science, as well as between chemistry and archaeology (e.g., archaeometry). With the advent of improved computer hardware and software, the way chemistry used to be done has changed, in the way data are recorded and analyzed. Computational chemistry software are being used almost routinely by many experimental chemists. Computational chemists are themselves using standard software packages to tackle more and more exciting and challenging problems. Two- and three-dimensional visualizations (graphics) are increasingly being employed. Experimentally, attempts are being made to probe single molecules rather than molecules in an assembly. Combinatorial chemistry, as well as green chemistry, has been in existence for some time. A synthesis protocol using artificial intelligence also exists. Attosecond (10^{-18} s) phenomena, concerned with electronic motions, have emerged very recently. Overall, I sense a great churning taking place in chemistry.

- **What do you think lies in the future of theoretical chemistry and computational chemistry?**

  If I am not wrong, of the total global population of theoretical and computational chemists, 90% or even more are computational chemists. Two things ought to be noted here. Software packages represent the technology of theoretical chemistry, and they employ existing theories which cannot be regarded as "perfect." Everybody knows that "all exact sciences are dominated by approximations." Chemical systems being highly complex, it would be rather unrealistic to play with toy models which admit analytical solutions. Therefore, the need for developing new concepts for improving existing theories would remain strong because this is an open-ended quest. Needless to say, software packages should not be used as "black boxes."

  I have a feeling that the number of theoretical chemists who can traverse the whole gamut of theoretical chemistry, *viz.*, generation of concepts, formalisms, algorithms, computer codes, and new ways of interpreting computed numbers, is decreasing steadily all over the world. Urgent replenishments are needed through the induction of bright, imaginative, and capable young chemists. In a way, theoretical chemists are akin to poets, admittedly with a practical bent of mind. We need to ensure that poetry, imagination, and the fun of making predictions do not disappear from chemistry.

- **Where do you think physical chemistry stands relative to other areas like inorganic and organic chemistry? (in terms of number of researchers, publications, Nobel Prizes, etc.)**

  Since my undergraduate days, I have been acutely uncomfortable with the attitude that chemistry can be completely classified into inorganic, organic, and physical chemistry. These are artificial intellectual barriers. The numbers of researchers and publications in certain areas of chemistry have been steadily increasing and will continue to do so. In terms of the number of researchers in various areas, there has been a seriously lopsided development in some countries because of the tripartite classification. One even hears of cases where there is a large number of Ph.D. students with just one Supervisor. I hope the situation will improve and a balanced development will take place. Until 1960s, successive Nobel Committees apparently did not find theoretical chemists worthy of the Nobel Prize, although the latter had enormous impact on the whole of chemistry. That also changed from the 1960s. Of late, even a theoretical condensed matter physicist has received the Nobel Prize in chemistry. So, the earlier we teach ourselves to climb over these barriers, the better for the growth of chemistry.

- **How are Physical Chemistry and Chemical Physics different from each other?**

Since both the terms refer to the interface between chemistry and physics, they should have the same meaning. However, in usage, this is not so. The term "chemical physics" was coined in the postquantum mechanical era and popularized by journals in chemical physics. One might simplistically say that, if, in the chemistry–physics interface, one is veering more toward chemistry, then one is doing physical chemistry, whereas, if one veers more toward physics, one is doing chemical physics. Alternatively, since science develops by progressive quantification, one might say that chemical physics is the modern more quantified version of physical chemistry. But I think all such distinctions are somewhat contrived. However, chemical physics has certainly been enriched by contributions from many physicists who probably felt more comfortable with this term than "physical chemistry." It may be worth noting here that an "overzealous" scientist had once defined physical chemistry as "the study of everything that is interesting"!

- **How has computation changed the way research in chemistry is carried out?**

Over the years, there has been a sea of change in the attitude of chemists. Earlier, any theoretical method and the numbers computed from it had to be justified by comparing with experimental results. This has drastically changed because of two reasons: First, the sophistication in theory, algorithms, and computer codes is now so good that these frequently deliver computed numbers much beyond present-day experimental accuracy. Second, there are many situations in which it is extremely difficult to perform an experiment, e.g., to study a very short-lived molecule or study a phenomenon like the folding of a protein in a biological environment. Theoretical and computational chemistry could be the only route to take in such cases.

Let me give you an example of accuracy of a theoretical method. In the last five years, it has been possible to numerically solve the Schrödinger equation for some systems with a precision of forty significant figures! While I do not understand the experimental significance of numbers beyond a certain significant figure or what we can do with such precise numbers, the fact remains that such computational accuracy is now deliverable and it challenges experiment. This is definitely good for overall development. Another recent development is the experimental tomographical picture of the highest occupied molecular orbital of the N_2 molecule, which proved the physical existence of a wave function.

With the availability of dependable and commercially available software packages developed by theoretical and computational chemists, in collaboration with experts in numerical methods, an interesting situation has come about. The synthesis and structure of a new molecule discovered in the chemical laboratory is nowadays justified by experimentalists by doing a geometry optimization according to a good software package. On the other hand, the experimental determination of structure generally resorts to a combination of methods.

However, we should never forget that experiment and theory are the two wings of a bird named science. It cannot fly on only one wing.

- **On one hand, the boundaries between chemistry and other sciences are blurring, and on the other, chemistry is branching out into specialized courses/fields. How do you think this is making a difference? Is the effect getting balanced out in some way?**

I look at it differently, instead of a balance or an imbalance. I like to spell "Chemistry" as "Chemistree." The Tree of Chemistry is large. It has deep roots and spreads in all directions. It has many branches and subbranches. New branches, subbranches, and leaves sprout in the course of time. As chemists, we are like birds living on this tree. A group of birds might nest in a small subbranch. There is no harm in that as long as the birds leave their nest once in a while and become aware of the large tree.

I believe there is a network of sciences, humanities, and social sciences with chemistry as a central science. I think future teaching and research in chemistry might develop along such a network.

- **What significance did the International Year of Chemistry (2011) have for you? What would you like to see changing in the future about research in chemistry?**

Let me answer the first question first. Chemistry has always been a deeply humanistic subject. For six thousand years, chemistry has worked for the benefit of humankind. Therefore, IYC did not remind me anew of chemistry in the service of humankind. Instead, it reminded me of two individuals whom I greatly admire: Madame Marie Sklodowska Curie and Acharya (Sir) Prafulla Chandra Rây. Besides being the centenary of Madame Curie's Nobel Prize in chemistry, 2011 was also the 150th birth anniversary of Acharya Rây, the first modern chemist of India and, along with Acharya (Sir) J. C. Bose, the founder of modern scientific research in India. As a teacher, Acharya Rây had inspired Meghnad Saha (the founder of modern astrophysics), Satyendra Nath Bose (the founder of quantum statistics), Jnan Chandra Ghosh (pre-Debye–Hückel theory of strong electrolytes), and many others. Besides his own well-known researches in chemistry, he was the founder of the chemical and pharmaceutical industries in India and an indefatigable social reformer. He was one of the greatest sons and builders of modern India, greatly admired by Mahatma Gandhi and Rabindranath Tagore, as well as numerous other persons, because of his asceticism, scientific modernism, deep knowledge about classical Indian culture, and a life totally dedicated to others.

There are striking parallels in the lives of Madame Curie and Acharya Râ_y which set guidelines for other human beings: poverty, suffering, indomitable spirit which does not recognize any obstacle, pioneering works in spite of continued ill health, tremendous leadership, building of multiple institutions, as well as an ascetic life totally devoid of self and completely dedicated to the welfare of others.

Coming to the second question. Within the global scenario, I believe we are not too bad in dealing with problems of fundamental importance in chemistry. However, I would like to see much greater intensity here, in terms of issues which were not tackled before. Where I would like to see extensive leapfrogging is in the development of new and sophisticated technologies born in chemical research laboratories, in collaboration with other scientists and engineers, wherever necessary. Some examples would be attosecond and X-ray lasers, a working quantum computer, new drug molecules by drastically cutting down the laboratory-to-market time schedule through a clever but absolutely safe multidisciplinary approach, etc. The list is actually quite long. Increasingly sophisticated chemical technologies which would be inexpensive and eco-friendly and can improve the lives of common people, especially those in rural and impoverished areas all over the world, need to be developed as quickly as possible.

- **You were involved in the development of chemistry curriculum for Indian universities. What are the key aspects of a good chemistry curriculum according to you?**

A curriculum involves a combination of teaching, learning, and assessment. Irrespective of what an individual chemist may practice in his/her own research, a chemistry curriculum must not split chemistry into inorganic, organic, and physical chemistry, and there should be no specialization in any of these three up to the graduation level (pre-Ph.D.). I strongly believe that this tripartite splitting has done enormous damage to the free development of chemistry in certain countries. Throughout the undergraduate years, there should be self-exploration by the student through as many small and medium projects as possible. Science can be learnt only through a dialog with Nature, through experiments in the laboratory, and in natural environments outside the laboratory. Laboratory programs in certain countries are not in a good state. We must bring back imagination, excitement, and wonder into the laboratory courses in chemistry. This is easier said than done. Here, theoretical and computational chemists should join hands with their experimental colleagues in devising concept-oriented, technique-intensive, and generally fun experiments for students. We must also bring back experimental demonstrations during classroom lectures. Let us not forget that chemistry is a subject combining magic, logic, and aesthetics.

The life-blood of any educational program is a dedicated and conscientious band of teachers. I would request the teachers concerned that, in formulating any chemistry curriculum, they should keep in mind that chemistry is a *central science*, overlapping with practically any subject under the sun and even the processes occurring in the sun, in the earth, and elsewhere. Here, I would like to draw a connectivity network which depicts chemistry as a central science, and teachers as well as students may keep this in mind. Note that seven lines, some of which are coincident, radiate out from every subject toward other subjects. Both teaching and research in chemistry might develop in the future along this network.

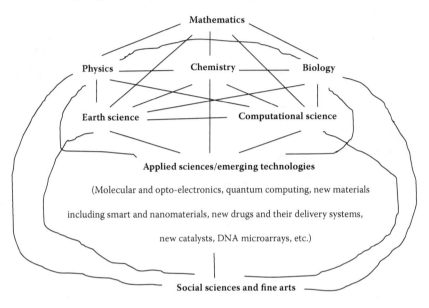

Within this pattern, a chemistry curriculum must impart both *intellectual and manual skills* to the student and try to *integrate both skills*. This is the essence of creativity.

- **What kind of prospects do young theoretical and computational chemists have?**

    I strongly feel that every chemistry department in colleges and universities should appoint at least one theoretical and computational chemist. In universities, the critical number would be three. Because of the multidisciplinary nature of the subject, a theoretical chemist can teach quite a few areas and would therefore lend strength to the teaching programs. Secondly, industries in a number of countries do not seem to have felt the need to appoint theoretical and computational chemists. All these have drastically reduced the employability of young theoretical and computational chemists, who show enormous personal courage to go into these areas. As a result, theoretical and computational chemists have found employment only in a limited number of institutions. I find this overall situation fraught with danger for the future development of chemistry.

- **What sparked your interest in Chemistry? (You had done a Ph.D. in mathematics.)**

    I drifted into chemistry. I could see myself also in literature or medicine or biology or physics. However, even though my father was a legendary teacher of mathematics, I never saw myself as a mathematician. An encounter with a highly charismatic teacher of chemistry put me into chemistry at Presidency College, Kolkata. I began to love the subject because of its all-encompassing nature. I was seriously thinking about going into experimental chemistry. It was a distinguished polymer chemist who advised me to pursue my doctoral studies with Professor C. A. Coulson. Since Professor Coulson was the Director of the Mathematical Institute of Oxford University, my D.Phil. degree was under the Mathematics Faculty at Oxford. Still, it took me six months at Oxford to firmly conclude that theoretical chemistry with its unlimited expanse will be my life, mainly because I knew that I would never be able to come to grips with it.

Looking back, I am convinced that it was my teachers right from high school to the doctoral level who were instrumental in charting my professional life.

- **Is there a particular incident from your research career, an anecdote, that you would like to share with the readers?**

I recall an incident which helped to redefine the course of my research. Around 1970, I had written an article on what I called "the force concept in chemistry" for students and teachers of chemistry. The chemistry journals I sent it to declined to consider it for publication, saying this would be beyond their readership. Exasperated, I sent it to Professor Coulson for his critical comments. Professor Coulson decided to communicate it himself to Reviews of Modern Physics (RMP). It was highly interesting that, while the referee(s) accepted the article, Professor U. Fano, the Editor of RMP, wanted me to rewrite parts of it, making an extremely important point that I comment on how the electron density can be calculated directly. I wrote whatever I could, and the article was published. I was rather unnerved but elated when I received many letters from people belonging to various disciplines, including several highly respected scientists. One of them was Professor R. P. Feynman who, besides telling me that he liked the article, suggested that I look into stresses in molecules, which he himself was interested in at one time but never published anything on it. Enclosed with his letter came the Xerox copies of relevant pages on stresses from his B.S. dissertation (under J. C. Slater) at MIT, which contained his famous work on the Hellmann–Feynman theorem. These two suggestions changed my research.

1 Kinetic Energy Functionals of Electron Density and Pair Density

Debajit Chakraborty and Paul W. Ayers

CONTENTS

1.1 ALTERNATIVES TO WAVE FUNCTION FOR DESCRIBING *N*-ELECTRON SYSTEMS

Wave function–based approaches to the electronic Schrödinger equation are computationally daunting because the dimensionality of wave function grows as the number of electrons increases. Therefore, one confronts either exponentially increasing computational costs (if one chooses to explicitly account for the increasing dimensionality) or steadily decreasing computational accuracy (if one chooses a simple, computationally tractable, model wave function with a lower effective dimension). Density-based approaches are based on the idea that there are alternative descriptors for electronic systems that are much simpler than the wave function. The most popular of such descriptors is electron density.[1-5]

Electron density is the probability of observing an electron at a specified point in space:

$$\rho(\mathbf{r}) = \left\langle \Psi \left| \sum_{i=1}^{N} \delta\left(\mathbf{r}_i - \mathbf{r}\right) \right| \Psi \right\rangle = N \sum_{\sigma_1=\alpha,\beta} \sum_{\sigma_2=\alpha,\beta} \sum_{\sigma_3=\alpha,\beta} \cdots$$
$$\sum_{\sigma_N=\alpha,\beta} \int\int \cdots \int \left| \Psi\left(\mathbf{r},\sigma_1;\mathbf{r}_2,\sigma_2;\mathbf{r}_3,\sigma_3;\ldots;\mathbf{r}_N,\sigma_N\right) \right|^2 d\mathbf{r}_2, d\mathbf{r}_3,\ldots,d\mathbf{r}_N.$$

$$(1.1)$$

Electron density is defined so that it is normalized to the number of electrons

$$N = \int \rho(\mathbf{r}) d\mathbf{r}. \tag{1.2}$$

The probability of observing an electron at a point in space is always greater than or equal to zero, so the electron density is nonnegative

$$0 \le \rho(\mathbf{r}). \tag{1.3}$$

Somewhat surprisingly, every property of an electronic system can be determined from its electron density.[6,7] In particular, ground-state energy can be determined from the variational principle for electron density

$$E_0 = \min_{\left\{\rho \left| \begin{matrix} 0 \le \rho(\mathbf{r}) \\ N = \int \rho(\mathbf{r})\, d\mathbf{r} \end{matrix}\right.\right\}} E_v[\rho]. \tag{1.4}$$

The notation indicates that one should search over all nonnegative N-electron densities and find the one with the lowest energy. That energy is the ground-state electronic energy; its associated density is the ground-state electron density. Unlike wave function–based methods, where the energy functional

$$E[\Psi] = \left\langle \Psi \left| \hat{H} \right| \Psi \right\rangle \tag{1.5}$$

is known exactly, in density-based methods, the energy functional $E_v[\rho]$ must be approximated. In density functional theory (DFT), one uses electron density as the fundamental descriptor of an electronic system.

Developing improved approximations to the energy density functional $E_v[\rho]$ is very difficult. More than 80 years after the first approximations were proposed, approximating the energy functional remains a very active area of research.[8-11] This frustratingly slow progress has motivated researchers to consider alternative descriptors that contain more information than the electron density but less information that the electronic wave function.[12] For example, electron pair density (or two-electron distribution function) can also be used as the fundamental descriptor of an electronic system.[13,14] More generally, k-density (or k-electron distribution function)

$$\rho_k\left(\mathbf{q}_1, \mathbf{q}_2, \ldots, \mathbf{q}_k\right) = \left\langle \Psi \left| \sum_{i_1=1}^{N} \sum_{\substack{i_2=1 \\ i_2 \ne i_1}}^{N} \cdots \sum_{\substack{i_k=1 \\ i_k \ne i_1 \\ i_k \ne i_2 \\ \vdots \\ i_k \ne i_{k-1}}}^{N} \delta\left(\mathbf{r}_{i_1} - \mathbf{q}_1\right) \delta\left(\mathbf{r}_{i_2} - \mathbf{q}_2\right) \cdots \delta\left(\mathbf{r}_{i_k} - \mathbf{q}_k\right) \right| \Psi \right\rangle \tag{1.6}$$

can be used as the fundamental descriptor of chemical substances.[15,16] The advantage of approaches based on the k-density, with $k \ge 2$, is that potential energy can be evaluated exactly, so only the kinetic energy portion of $E_v[\rho]$ needs to be approximated. However, there has been very little work performed on designing kinetic energy functionals for higher-order electron distribution functions, and there has been even less testing of such functionals.

Most of research studies on higher-order electron distribution functions have focused on pair density.[13,14,17] Pair density represents the probability of simultaneously observing one electron at $\mathbf{r}_1$ and another electron at $\mathbf{r}_2$. It is normalized to the number of electron pairs

$$N(N-1) = \iint \rho_2(\mathbf{r}_1, \mathbf{r}_2)\, d\mathbf{r}_1\, d\mathbf{r}_2. \tag{1.7}$$

As before, ground-state energy and pair density can be determined using a variational principle:

$$E_0 = \min_{\substack{\rho_2 \\ \left\{ \begin{array}{l} 0 \le \rho(\mathbf{r}_1, \mathbf{r}_2) \\ N(N-1) = \int \int \rho_2(\mathbf{r}_1, \mathbf{r}_2) d\mathbf{r}_1, d\mathbf{r}_2 \\ \rho 2(\mathbf{r}_1, \mathbf{r}_2) \text{ is } N\text{-representable} \end{array} \right\}}} E_v[\rho_2]. \tag{1.8}$$

Using pair density, instead of electron density, as the fundamental descriptor of the system has two disadvantages. The first disadvantage is that the pair density is a six-dimensional function, so it is a more complicated mathematical object than the three-dimensional electron density function. The more severe disadvantage is that the minimization in Equation 1.8 has a very difficult constraint, namely, the pair density must correspond to some N-electron system.[16–20] That is, there must exist some mixture of N-electron wave functions for which

$$\rho_2(\mathbf{q}_1, \mathbf{q}_2) = \sum_{k=1}^{n \text{ states}} p_k \left\langle \Psi_k \left| \sum_{i_1=1}^{N} \sum_{\substack{i_2=1 \\ i_2 \neq i_1}}^{N} \delta(\mathbf{r}_1 - \mathbf{q}_1) \delta(\mathbf{r}_2 - \mathbf{q}_2) \right| \Psi_k \right\rangle. \tag{1.9}$$

Such pair densities are said to be N-representable. Pair densities that are not N-representable do not describe N-electron systems. That is, the requirement that the variational principle for the pair density consider only N-representable functions is analogous to the requirement that the variational principle for the wave function consider only normalized antisymmetric wave functions.

Determining whether a given six-dimensional function corresponds to an N-electron system is extraordinarily difficult; thus, in practice, both the energy functional and the constraints on minimization must be approximated in pair DFT (2-DFT). These problems are exacerbated in methods based on even higher-order electron distribution functions because the computational difficulty of working with k-density and the difficulty of determining whether a given k-density is N-representable grow rapidly with k.

The difficulty of approximating the kinetic energy functional is removed if one considers, instead of the k-density, the k-electron reduced density matrices[21–24]

$$\Gamma_k(\mathbf{q}_1, \dots, \mathbf{q}_k; \mathbf{q}_1', \dots \mathbf{q}_k')$$

$$= \binom{N}{k} \int \dots \int \Gamma_N(\mathbf{q}_1, \dots, \mathbf{q}_k; \mathbf{r}_{k+1}, \dots \mathbf{r}_N; \mathbf{q}_1', \dots \mathbf{q}_k', \mathbf{r}_{k+1}, \dots \mathbf{r}_N) d\mathbf{r}_{k+1}, \dots, d\mathbf{r}_N. \tag{1.10}$$

Alternatively,

$$\Gamma_k\left(\mathbf{q}_1,\ldots,\mathbf{q}_k;\mathbf{q}_1',\ldots\mathbf{q}_k'\right)=\sum_{k=1}^{n\text{ states}}p_k\left\langle\Psi_k\left|\hat{\psi}^+\left(\mathbf{q}_1'\right)\hat{\psi}^+\left(\mathbf{q}_2'\right)\right.\right.$$
$$\left.\left.\cdots\hat{\psi}^+\left(\mathbf{q}_k'\right)\hat{\psi}\left(\mathbf{q}_k\right)\cdots\hat{\psi}\left(\mathbf{q}_2\right)\hat{\psi}\left(\mathbf{q}_1\right)\right|\Psi_k\right\rangle,$$

(1.11)

where $\hat{\psi}^+(\mathbf{q})\cdot\left(\hat{\psi}(\mathbf{q})\right)$ is the quantum mechanical field operator for creating (anni-hilating) an electron at point $\mathbf{q}$. The k-electron distribution function (k-density) is obtained by setting the primed and unprimed variables equal in the k-electron den-sity matrix (k-matrix)

$$\rho_k\left(\mathbf{q}_1,\ldots,\mathbf{q}_k\right)=\Gamma_k\left(\mathbf{q}_1,\ldots,\mathbf{q}_k;\mathbf{q}_1,\ldots,\mathbf{q}_k\right).$$

(1.12)

In density matrix functional theory, the k-matrix is used as the fundamental descriptor of an electronic system. For $k \geq 2$, the exact energy functional of the k-matrix is known explicitly, but the N-representability problem is still daunting.[19] For $k = 1$, the exact energy functional is not known, but the exact kinetic energy func-tional is known, and the N-representability problem is unproblematic.[25] By regarding the k-matrix as an integral kernel, one can define its eigenvectors and eigenvalues

$$\int\int\cdots\int\Gamma_k\left(\mathbf{q}_1,\ldots,\mathbf{q}_k;\mathbf{q}_1',\ldots\mathbf{q}_k'\right)\phi_i\left(\mathbf{q}_1',\mathbf{q}_2',\ldots,\mathbf{q}_k'\right)d\mathbf{q}_1',d\mathbf{q}_2',\ldots,d\mathbf{q}_k'=n_i\phi_i\left(\mathbf{q}_1,\mathbf{q}_2,\ldots,\mathbf{q}_k\right).$$

(1.13)

The eigenvectors of the k-matrix represent the natural k-electron states of the system, or k-bitals. The associated eigenvalues are the occupation numbers of these states, so that

$$\Gamma_k\left(\mathbf{q}_1,\ldots,\mathbf{q}_k;\mathbf{q}_1',\ldots\mathbf{q}_k'\right)=\sum_i n_i\phi_i\left(\mathbf{q}_1,\mathbf{q}_2,\ldots,\mathbf{q}_k\right)\phi_i^*\left(\mathbf{q}_1',\mathbf{q}_2',\ldots,\mathbf{q}_k'\right).$$

(1.14)

Most chemical applications using k-matrices are for $k = 1$ or $k = 2$, or both. The natural states of the two-matrix

$$\int\int\Gamma_2\left(\mathbf{q}_1,\mathbf{q}_2;\mathbf{q}_1',\mathbf{q}_2'\right)\phi_i\left(\mathbf{q}_1',\mathbf{q}_2'\right)d\mathbf{r}_1'd\mathbf{r}_2'=n_i\phi_i\left(\mathbf{q}_1,\mathbf{q}_2\right)$$

(1.15)

are called natural geminals. In order to follow the Pauli exclusion principle, no geminal can be occupied by more than $N - 1$ electrons. However, imposing con-straints on geminal occupation numbers is not enough to ensure that the two-matrix is N-representable.[26–30]

The one-matrix is so important that it is given a special notation:

$$\gamma\left(\mathbf{q},\mathbf{q}'\right)=\Gamma_1\left(\mathbf{q},\mathbf{q}'\right)=\sum_i n_i\phi_i(\mathbf{q})\phi_i^*(\mathbf{q}'). \tag{1.16}$$

If the coordinate $\mathbf{q}$ is considered to include spin, then the Pauli principle indicates that the natural orbital occupation numbers are between 0 and 1 (inclusive)

$$0 \le n_i \le 1. \tag{1.17}$$

(If spin is not included, then one can put up to two electrons in a single spatial orbital, so that $0 \le n_i \le 2$.) An N-electron Slater determinant corresponds to a one-matrix with N unit occupation numbers (corresponding to spin orbitals that appear in the Slater determinant); all the other natural orbital occupation numbers are 0. Because all the eigenvalues of a Slater determinantal one-matrix are 0 or 1, the resulting one-matrix is idempotent:

$$\int \gamma\left(\mathbf{q},\mathbf{q}'\right)\gamma\left(\mathbf{q}',\mathbf{q}''\right)d\mathbf{q}' = \gamma\left(\mathbf{q},\mathbf{q}''\right). \tag{1.18}$$

More generally, it follows from Equation 1.17 that the matrix $\gamma - \gamma^2$ is positive semidefinite:

$$0 \le \gamma(\mathbf{q},\mathbf{q}'') - \int \gamma(\mathbf{q},\mathbf{q}')\gamma(\mathbf{q}',\mathbf{q}'')d\mathbf{q}'. \tag{1.19}$$

A matrix is positive semidefinite if and only if all its eigenvalues are nonnegative. Equation 1.19 (or the corresponding bounds on the natural orbital occupation numbers, Equation 1.17) is the N-representability constraint for the one-matrix.[25] That is, every one-matrix that satisfies Equation 1.19 can be derived from an average of N-electron states, as in Equation 1.11.

1.2 DENSITY FUNCTIONAL THEORY

1.2.1 OVERVIEW

In DFT, the ground-state electron density replaces the wave function as the fundamental descriptor of an electronic system. This replacement is justified by the first Hohenberg–Kohn theorem, which indicates that the ground-state electron density determines all measurable properties of an electronic system, including the ground-state energy.[6] The second Hohenberg–Kohn theorem indicates that the ground-state energy and ground-state density can be determined through the variational principle[31]

$$E_{g,s} = \min_{\substack{\rho \mid 0 \le \rho(\mathbf{r}) \\ N = \int \rho(\mathbf{r})d\mathbf{r}}} E_v[\rho] \tag{1.20}$$

$$\rho_{\text{g.s.}}(\mathbf{r}) = \arg \min_{\left\{\rho \left| \begin{array}{l} 0 \le \rho(\mathbf{r}) \\ N = \int \rho(\mathbf{r}) d\mathbf{r} \end{array} \right. \right\}} E_v[\rho] \tag{1.21}$$

The constraints on the domain of electron densities included in the variational search ensure that the electron density is N-representable.[32,33]

The primary difficulty in DFT is that the energy functional is unknown. This can be contrasted with the situation in wave function theory, where the formula for evaluating the energy is explicit and computationally feasible but the form of the exact wave function is unknown and—because it is very complicated—probably unknowable. In DFT, the formerly intractable wave function is replaced by the mathematically simple electron density, but the energy functional is unknown and—because it is very complicated—probably unknowable.

Some progress can be made by decomposing the energy into contributions from kinetic energy, electron–nuclear attraction potential energy, and electron–electron repulsion potential energy

$$E_v[\rho] = T[\rho] + V_{ne}[\rho;v] + V_{ee}[\rho]. \tag{1.22}$$

The electron–nuclear attraction energy has a simple explicit form:

$$V_{ne}[\rho; v] = \int \rho(\mathbf{r}) v(\mathbf{r}) d\mathbf{r}, \tag{1.23}$$

where $v(\mathbf{r})$ is the external potential that binds the electrons. For a molecular system, the external potential is simply

$$v(\mathbf{r}) = -\sum_{\alpha=1}^{P} \frac{Z\alpha}{|\mathbf{r} - \mathbf{R}_\alpha|}. \tag{1.24}$$

The electron–electron repulsion contribution is often decomposed into classical electrostatic repulsion energy, plus corrections for the Pauli principle (exchange) and electron correlation

$$V_{ee}[\rho] = J[\rho] + V_x[\rho] + V_c[\rho] = J[\rho] + V_{xx}[\rho]. \tag{1.25}$$

The classical Coulomb repulsion energy is, by far, the largest contribution to the electron–electron repulsion; it is also known in an explicit form:

$$J[\rho] = \frac{1}{2} \int \int \frac{\rho(\mathbf{r})\rho(\mathbf{r}')}{|\mathbf{r} - \mathbf{r}'|} d\mathbf{r} d\mathbf{r}'. \tag{1.26}$$

The exchange and correlation contributions must be approximated, but acceptably accurate models were available even in the 1930s (e.g., the Fermi–Amaldi model and the Dirac model for exchange[34,35] and the Wigner model for correlation[36]).

The kinetic energy functional is more problematic. There is no classical kinetic energy expression analogous to Equations 1.23 and 1.26 that can be used as a building block. A semiclassical expression in powers of Planck's constant[37,38] gives, at leading order, the approximate functional of Fermi[39] and Thomas.[40] This approximation is much less accurate than the analogous expressions Equations 1.23 and 1.26 for the potential energy contributions to the energy. This is partly because, for a stable geometric arrangement of atomic nuclei, the virial theorem indicates that the kinetic energy of a molecule has the same magnitude as the electronic energy $T[\rho_{g.s.}] = -E_{g.s.}$. Merely determining the order of magnitude of chemical equilibrium constants and reaction rates requires errors of less than 0.001 atomic units (1 milliHartree) in energy; for a moderately large molecule, this means that parts-per-million accuracy in the energy is required. Kinetic energy functionals must also be accurate to within a few parts per million. This exquisite accuracy is unattainable with the Thomas–Fermi theory, where errors in the kinetic energy are about 10%.

1.2.2 KOHN–SHAM DFT

The difficulty of developing good kinetic energy functional motivated Kohn and Sham to design a model system whose kinetic energy closely resembles that of the target electronic system.[7] As a model system, they chose a system of noninteracting fermions with the same density as the target electronic system. The exact ground-state wave function for a system of noninteracting fermions is a Slater determinant, so the one-matrix of the model system is idempotent. The kinetic energy of the non-interacting system can then be computed from the Slater determinant

$$T_s[\rho] = \left\langle \Phi_s \left| \sum_{i=1}^{N} -\frac{1}{2}\nabla_i^2 \right| \Phi_s \right\rangle \tag{1.27}$$

from the one-matrix

$$T_s[\rho] = \sum_{\sigma=\alpha,\beta} \int\int \delta(\mathbf{r}-\mathbf{r}')\left(-\frac{1}{2}\nabla_{\mathbf{r}}^2\gamma^{\sigma\sigma}(\mathbf{r}-\mathbf{r}')\right)d\mathbf{r}d\mathbf{r}' \tag{1.28}$$

or from the occupied Kohn–Sham orbitals

$$T_s[\rho] = \sum_{\sigma=\alpha,\beta}\sum_{\{i\in\text{occupied orbitals}\}} \left\langle \phi_i^\sigma \left| -\frac{1}{2}\nabla^2 \right| \phi_i^\sigma \right\rangle. \tag{1.29}$$

The remaining contribution to the kinetic energy is due to electron correlation

$$T_c[\rho] = T[\rho] - T_s[\rho] \ge 0. \tag{1.30}$$

The Kohn–Sham approximation to the kinetic energy $T_s[\rho]$ is extremely accurate. The correlation kinetic energy $T_c[\rho]$ is usually smaller than the magnitude of the correlation energy.

The unknown portions of the energy functional—the correlation kinetic energy, the exchange potential energy, and the correlation potential energy—are added together to define the exchange-correlation energy

$$E_{xc}[\rho] = T_c[\rho] + V_{xc}[\rho]. \tag{1.31}$$

In Kohn–Sham DFT (KS-DFT), the only portion of the total energy

$$E_v[\rho] = T_s[\rho] + \int \rho(\mathbf{r})v(\mathbf{r})\,d\mathbf{r} + J[\rho] + E_{xc}[\rho] \tag{1.32}$$

that needs to be approximated is the exchange-correlation energy functional. This is why the pursuit for better exchange-correlation functionals is the dominant research topic in DFT.[8–11,41] Good approximations to the exchange-correlation functional exist, and it is increasingly difficult to design new exchange-correlation functionals that are significantly better than existing functionals such as[42] the ubiquitous B3LYP functional.[43–47]

KS-DFT is effective because the exchange-correlation energy is a relatively small number. For this reason, a relatively inaccurate functional for the exchange-correlation energy (with, e.g., about a 1% error) can give results that are accurate enough to be useful for computational chemistry. By contrast, a kinetic energy functional with a 1% error is worthless for quantitative chemical modeling.

Substituting Equation 1.27 into Equation 1.32 and invoking the variational principle give rise to the Kohn–Sham equation

$$\left(-\frac{1}{2}\nabla^2 + v_s^\sigma(\mathbf{r}) \right)\phi_i^\sigma(\mathbf{r}) = \varepsilon_i^\sigma \phi_i^\sigma(\mathbf{r}). \tag{1.33}$$

The potential that enters into this equation is called the Kohn–Sham potential

$$v_s^\sigma(\mathbf{r}) = v(\mathbf{r}) + v_J\left[\rho^\sigma; \mathbf{r}\right] + v_{xc}\left[\rho^\sigma; \mathbf{r}\right] \tag{1.34}$$

The Coulomb potential

$$v_J\left[\rho^\sigma; \mathbf{r}\right] = \int \frac{\rho^\alpha(\mathbf{r}') + \rho^\beta(\mathbf{r}')}{|\mathbf{r} - \mathbf{r}'|}\,d\mathbf{r}' = \frac{\delta J\left[\rho^\alpha, \rho^\beta\right]}{\delta\rho^\alpha(\mathbf{r})} \tag{1.35}$$

is the functional derivative of the classical Coulomb repulsion energy. The exchange-correlation potential is the functional derivative of the exchange-correlation energy

$$v_{xc}\left[\rho^{\sigma};\mathbf{r}\right]=\frac{\delta E_{xc}\left[\rho^{\alpha},\rho^{\beta}\right]}{\delta\rho^{\sigma}(\mathbf{r})} \qquad (1.36)$$

The energy of the noninteracting reference system is

$$E_s\left[v_s;N\right]=\sum_{\sigma=\alpha,\beta}\sum_{\{i\in\text{occupied orbitals}\}}\varepsilon_i^{\sigma}=\sum_{\sigma=\alpha,\beta}\sum_{\{i\in\text{occupied orbitals}\}}\left\langle\phi_i^{\sigma}\left|-\frac{1}{2}\nabla^2+v_s(\mathbf{r})\right|\phi_i^{\sigma}\right\rangle$$

$$=\sum_{\sigma=\alpha,\beta}T_s[\rho^{\sigma}]+\int\rho^{\sigma}(\mathbf{r})v_s^{\sigma}(\mathbf{r})d\mathbf{r}. \qquad (1.37)$$

The energy of the true system of interacting electrons is

$$E_{\text{g.s.}}=E_s\left[v_s^{\sigma};N\right]+E_{xc}\left[\rho^{\alpha},\rho^{\beta}\right]-J\left[\rho^{\alpha},\rho^{\beta}\right]-\sum_{\sigma=\alpha,\beta}\int\rho^{\sigma}(\mathbf{r})v_{xc}^{\sigma}(\mathbf{r})d\mathbf{r}. \qquad (1.38)$$

The value of other observable properties has a similar form. Because the electron densities of the noninteracting system and the interacting system are the same, the value of any property can be written as the sum of its value in the noninteracting system plus a correction that is a density functional:

$$Q\left[\rho^{\alpha},\rho^{\beta}\right]=Q_s\left[\rho^{\alpha},\rho^{\beta}\right]+Q_{\text{correction}}\left[\rho^{\alpha},\rho^{\beta}\right]$$

$$=\left\langle\Phi_s\left|\hat{Q}\right|\Phi_s\right\rangle+Q_{\text{correction}}\left[\rho^{\alpha},\rho^{\beta}\right]. \qquad (1.39)$$

The most important example in this article is the exact kinetic energy, which can be written as

$$T\left[\rho^{\alpha},\rho^{\beta}\right]=T_s\left[\rho^{\alpha},\rho^{\beta}\right]+T_c\left[\rho^{\alpha},\rho^{\beta}\right]=\left\langle\Phi_s\left|\hat{T}\right|\Phi_s\right\rangle+T_c\left[\rho^{\alpha},\rho^{\beta}\right]. \qquad (1.40)$$

The utility of KS-DFT arises because the properties of the noninteracting reference system often closely resemble the properties of the target interacting electron system.

1.2.3 ORBITAL-FREE DFT

Almost all DFT-based computations use the Kohn–Sham method. This approach, however, arguably violates the spirit of DFT. The simplicity and computational efficiency of DFT are compromised because, instead of needing to determine only one three-dimensional function (the electron density), in KS-DFT one must determine N three-dimensional functions (the Kohn–Sham spin orbitals). The conceptual beauty of DFT is compromised because, unlike the electron density, which is physically observable with a simple interpretation, the Kohn–Sham orbitals are neither physically observable nor directly interpreted. For these reasons, there is still interest in orbital-free DFT (OF-DFT), wherein the kinetic energy is written as an explicit functional of the electron density. The dominant research topic in OF-DFT is the pursuit of better kinetic energy functionals.[48–56]

Because the accuracy and reliability of present-day kinetic energy functionals are poor, OF-DFT is usually used either for specific types of systems where existing functionals are reliable (simple metals and similar materials[57–59] and warm dense matter[60]) or for applications where low-cost, low-accuracy models for the kinetic energy suffice. For example, OF-DFT is useful for massive molecular dynamics simulations of material fracture and deformation.[57,61] It is also useful for embedding high-level calculations inside lower-level calculations and for dividing systems into subsystems; in these cases, the kinetic energy functional is used to model the "kinetic energy pressure" for the surroundings that keeps the electrons confined to a particular subsystem.[51,62–69] Finally, sometimes OF-DFT is used to provide a qualitative description of a substance's electronic structure. For example, even very approximate kinetic energy functionals can be used to provide a picture of electron pairing in molecules and materials.[70–74]

1.2.4 PROPERTIES OF KINETIC ENERGY FUNCTIONAL

The conventional approach to deriving new density functionals is to first make a list of the properties of the exact functional; these properties then constrain the types of functionals that need to be considered.

Most researchers choose to model not the total kinetic energy $T[\rho]$ but only the Kohn–Sham piece $T_s[\rho]$. This is partly because accurate approximations to the correlation kinetic energy $T_c[\rho]$ are already available from KS-DFT. In particular, most OF-DFT programs use the same energy expression that is used in Kohn–Sham,

$$E_v[\rho] = \tilde{T}_s[\rho] + J[\rho] + \tilde{E}_{xc}[\rho] + \int \rho(\mathbf{r})v(\mathbf{r})\,d\mathbf{r}, \qquad (1.41)$$

so that standard approximate exchange-correlation functionals from KS-DFT can be used. In Equation 1.41, the functionals that must be approximated in practical calculations are decorated with $\tilde{a}$. When Equation 1.41 is used, the correlation kinetic energy contribution is included in the exchange-correlation energy functional.

The other reason that researchers choose to approximate the Kohn–Sham kinetic energy is that this functional has simpler mathematical properties. For example, the Kohn–Sham kinetic energy is homogeneous of degree two with respect to coordinate scaling of the electron density[75]

$$T_s[\rho_\alpha] = \alpha^2 T_s\,[\rho_{\alpha=1}] \qquad (1.42)$$

$$\rho_\alpha(x, y, z) = \alpha^3 \rho(\alpha x, \alpha y, \alpha z). \qquad (1.43)$$

From this equation, one can derive the virial equation for the Kohn–Sham kinetic energy[75–77]:

$$T_s[\rho] = \frac{1}{2} \int \rho(\mathbf{r}) \mathbf{r} \cdot \nabla v_s(\mathbf{r}) d\mathbf{r} = -\frac{1}{2} \int \rho(\mathbf{r}) \mathbf{r} \cdot \nabla \frac{\delta T_s[\rho]}{\delta \rho(\mathbf{r})} d\mathbf{r}$$

$$= -\frac{1}{2} \int v_s(\mathbf{r}) \big(3\rho(\mathbf{r}) - \mathbf{r} \cdot \nabla \rho(\mathbf{r}) \big) = \frac{1}{2} \int \frac{\delta T_s[\rho]}{\delta \rho(\mathbf{r})} \big(3\rho(\mathbf{r}) - \mathbf{r} \cdot \nabla \rho(\mathbf{r}) \big). \tag{1.44}$$

This expression is much simpler than the corresponding expression for the total kinetic energy.[78–80]

Coordinate scaling requirements seem arbitrary, but because they are equivalent to the virial relationship, functionals that violate coordinate scaling conditions often have poor variational properties.[81] Moreover, scaling the atomic charge in isoelectronic series of ions changes the density in a way that is qualitatively similar to coordinate scaling,[82] so failing to describe coordinate scaling correctly usually gives poor results for isoelectronic series.

Another important constraint related to the variational principle is the N-representability of the kinetic energy functional. The kinetic energy functional is N-representable if and only if, for every noninteracting system, the kinetic energy functional gives a higher energy compared with the kinetic energy portion of the energy. That is, if for all electron densities $\rho(\mathbf{r})$ and for all Kohn–Sham potentials $v_s(\mathbf{r})$,

$$\tilde{T}_s[\rho] \geq E_s\big[v_s;N\big] - \int \rho(\mathbf{r}) v_s(\mathbf{r}) d\mathbf{r}, \tag{1.45}$$

then the functional $\tilde{T}_s[\rho]$ is N-representable. This ensures that the energy computed with an approximated kinetic energy functional is always an upper bound to the KS-DFT energy. It also ensures that there exists an N-representable one-matrix that has this electron density and this kinetic energy:

$$\rho(\mathbf{r}) = \gamma(\mathbf{r}, \mathbf{r})$$

$$\tilde{T}_s = \iint \delta(\mathbf{r} - \mathbf{r}') \left(-\frac{1}{2} \nabla_r^2 \gamma(\mathbf{r}, \mathbf{r}') \right) d\mathbf{r} d\mathbf{r}'. \tag{1.46}$$

1.2.5 Approaches to Kinetic Energy Functional

1.2.5.1 One-Electron Density Matrix Models

In molecular quantum mechanics, the kinetic energy is usually computed from the one-matrix

$$T_s[\rho] = \iint \delta(\mathbf{r} - \mathbf{r}') \left(-\frac{1}{2} \nabla^2 \gamma_s[\rho; \mathbf{r}, \mathbf{r}'] \right) d\mathbf{r} d\mathbf{r}'. \tag{1.47}$$

For finite systems, integration by parts lets one rewrite the kinetic energy in a form where its nonnegativity is manifest

$$T_s[\rho] = \int \int \delta(\mathbf{r} - \mathbf{r}') \left(\frac{1}{2} \nabla \cdot \nabla' \gamma_s[\rho; \mathbf{r}, \mathbf{r}'] \right) d\mathbf{r} d\mathbf{r}'. \tag{1.48}$$

These two formulas give the same result for any well-behaved one-matrix.

If one can model the noninteracting one-matrix $\gamma_s(\mathbf{r}, \mathbf{r}')$ as a functional of the electron density, then, using Equation 1.47 or Equation 1.48, one can compute the kinetic energy. This is the most straightforward approach to deriving kinetic energy functionals, and the Thomas–Fermi functional and the Weizsäcker functional can both be derived in this way. (Indeed, all of the most popular functionals can be derived in several different ways.) The one-matrix can also be modeled based on weighted density approximation (WDA), which we will discuss subsequently.

1.2.5.2   Exchange Hole Models

For an idempotent density matrix, Equation 1.48 and

$$T_s[\rho] = \int \int \frac{\left| \nabla (\gamma[\rho; \mathbf{r}, \mathbf{r}'])^2 \right|^2}{8(\gamma[\rho; \mathbf{r}, \mathbf{r}'])^2} d\mathbf{r} d\mathbf{r}' \tag{1.49}$$

give identical results.[83] This expression can be rewritten in terms of the exchange hole

$$h_x^{\sigma\sigma}(\mathbf{r}, \mathbf{r}') = - \frac{\left| \gamma^{\sigma\sigma}(\mathbf{r}, \mathbf{r}') \right|^2}{\rho^\sigma(\mathbf{r}) \rho^\sigma(\mathbf{r}')}, \tag{1.50}$$

giving[18]

$$T_s[\rho^\alpha, \rho^\beta] = \sum_{\sigma = \alpha, \beta} \left[\int \frac{\left| \nabla \rho^\sigma(\mathbf{r}) \right|^2}{8\rho^\sigma(\mathbf{r})} d\mathbf{r} - \int \int \rho^\sigma(\mathbf{r}) \rho^\sigma(\mathbf{r}') \frac{\left| \nabla h_x^{\sigma\sigma}[\rho^\sigma; \mathbf{r}, \mathbf{r}'] \right|^2}{8 h_x^{\sigma\sigma}[\rho^\sigma; \mathbf{r}, \mathbf{r}']} d\mathbf{r} d\mathbf{r}' \right]. \tag{1.51}$$

Therefore, if one can model the exchange hole in terms of the electron density, one can build a density functional. This link between the exchange hole and the kinetic energy motivates the use of exchange energy expressions to derive kinetic energy functionals. When this is done, the exchange energy functional and the kinetic energy functional are said to be conjoint.[84]

1.2.5.3   Momentum Density Models

The kinetic energy is most easily computed in momentum space, where the classical expression

$$T_s[\rho] = \int \frac{1}{2} p^2 \Pi_s[\rho; \mathbf{p}] d\mathbf{p} \qquad (1.52)$$

is valid. Therefore, if one can model the momentum density $\Pi_s(\mathbf{p})$ using the electron density, one has derived a kinetic energy functional. The Thomas–Fermi functional or a generalization of the Weizsäcker functional can be derived using a model for the momentum density.[85]

1.2.5.4 Quasi-Probability Distribution Models

When approximating the potential energy portions of the energy, it was very helpful to start with the result from classical electrostatics and then to add a correction for quantum mechanics. A similar approach for the kinetic energy can be derived using the quasi-probability distribution function $F(\mathbf{r},\mathbf{p})$. In classical physics, the classical phase–space probability distribution function gives the probability of observing a particle at $\mathbf{r}$ with momentum $\mathbf{p}$. The Heisenberg uncertainty principle indicates that position and momentum cannot be observed simultaneously, so there is no unique analogue to the classical quantity. Instead, one has a quasi-probability distribution function[86,87]:

$$F_s(\mathbf{r},\mathbf{p}) = \left(\frac{1}{2\pi}\right)^6 \iiint e^{-i\tau \cdot \mathbf{p}} e^{-i\theta(\mathbf{r}-\mathbf{u})} f(\theta,\tau) \gamma_s \left(\mathbf{u} + \frac{1}{2}\tau, \mathbf{u} - \frac{1}{2}\tau\right) d\mathbf{u}\, d\theta\, d\tau, \quad (1.53)$$

where $f_1(\theta,\tau)$ is any integrable function that satisfies the constraints

$$1 = f(\mathbf{0},\,\tau) = f(\theta,\,\mathbf{0}) \qquad (1.54)$$

and

$$(f(\theta,\,\tau))^* = f(-\theta,\,-\tau). \qquad (1.55)$$

The Heisenberg uncertainty principle is captured by the fact there are many different choices for $f(\theta,\tau)$. Equivalently, the quantum-classical correspondence is not unique; this is why, for example, there are two (and even more) equivalent forms for the kinetic energy (Equations 1.47 and 1.48). All $f(\theta,\tau)$ that are consistent with Equation 1.54 provide a suitable classical correspondence; the most popular choice, $f(\theta,\tau) = 1$, corresponds to the Wigner quasi-probability distribution function.[88,89]

Given a quasi-probability distribution function, the electron density and the momentum density are given by

$$\rho(\mathbf{r}) = \int F_s(\mathbf{r},\mathbf{p}) d\mathbf{p} \qquad (1.56)$$

and

$$\Pi_s(\mathbf{p}) = \int F_s(\mathbf{r},\mathbf{p}) d\mathbf{r}. \qquad (1.57)$$

If one can model the quasi-probability distribution function using the electron density, then the kinetic energy can be expressed using[90,91]

$$T_s[\rho] = \iint \frac{1}{2} p^2 F_s[\rho; \mathbf{r}, \mathbf{p}] d\mathbf{r} d\mathbf{p}. \tag{1.58}$$

This approach to derive kinetic energy functionals was pioneered by Anderson et al.,[90] Ayers et al.,[91] Ghosh and Berkowitz,[92] and Lee and Parr.[93] The Thomas–Fermi model and generalizations thereto have been derived using this approach.

1.2.5.5   Local Kinetic Energy Models

With the quasi-probability distribution function as an intermediary, the local kinetic energy can be defined as[94]

$$t_s(\mathbf{r}) = \int \frac{1}{2} p^2 F_s[\rho; \mathbf{r}, \mathbf{p}] d\mathbf{p}. \tag{1.59}$$

Because the quasi-probability distribution is not uniquely defined, neither is the local kinetic energy.[90–93] If one can model the local kinetic energy using the electron density, then the kinetic energy functional can be written as

$$T_s[\rho] = \int t_s[\rho; \mathbf{r}] d\mathbf{r}. \tag{1.60}$$

Most kinetic energy functionals are expressions in the form of Equation 1.60.

1.2.5.6   Local Temperature Models

By analogy to the classical expression for the kinetic energy of a monatomic ideal gas, k.e. $= \frac{3}{2} k_B \Theta \rho$, Ghosh and Berkowitz[92] and Ghosh et al.[95] proposed writing the local kinetic energy as

$$t_s(\mathbf{r}) = \frac{3}{2} \Theta(\mathbf{r}) \rho(\mathbf{r}), \tag{1.61}$$

where $\Theta(\mathbf{r})$ is the local temperature.[96] Therefore, if one can build a model for the local temperature from the electron density, one can derive a kinetic energy functional,

$$T_s[\rho] = \int \frac{3}{2} \Theta[\rho; \mathbf{r}] \rho(\mathbf{r}) d\mathbf{r}. \tag{1.62}$$

This approach can also be used to derive Thomas–Fermi–like functionals.[93]

1.2.5.7 Other Approaches

There are other even less conventional approaches to deriving kinetic energy functionals. There are recent approaches using hydrodynamic tensors,[97,98] semiclassical expansions,[97,99–101] information theory,[102–108] theory of moments,[82,109–112] analysis of quantum fluctuations,[113–115] and higher-order electron distribution functions.[15,107,114,116]

1.3 ONE-POINT APPROXIMATE ORBITAL-FREE KINETIC ENERGY FUNCTIONALS

1.3.1 THOMAS–FERMI–BASED FUNCTIONALS

In Sections 1.3.2 and 1.3.3, we will briefly summarize the main families of kinetic energy functionals. Many functionals are necessarily omitted; the intention is to provide the flavor of mainstream approaches, rather than an encyclopedic list of functionals.

The first kinetic energy density functional was derived, independently, by Fermi[39] and Thomas[40] in 1928 and 1927, respectively. The Thomas–Fermi functional is the simplest local density approximation.

In a uniform electron gas (UEG) of noninteracting fermions with density ρ, the local kinetic energy per fermion is a constant:

$$\tau(\rho) = \frac{(3\pi)^{2/3}}{10} \rho^{2/3}(\mathbf{r}). \tag{1.63}$$

The local kinetic energy per unit volume is then the probability of observing a fermion at the point $\mathbf{r}$ times the local kinetic energy per fermion

$$t[\rho] = \rho(\mathbf{r})\tau(\rho). \tag{1.64}$$

Integrating the overall volume gives the Thomas–Fermi functional

$$T_s^{\text{TF}}[\rho] = C_F \int \rho^{5/3}(\mathbf{r})d\mathbf{r} \tag{1.65}$$

with

$$C_F = \frac{(3\pi)^{2/3}}{10}. \tag{1.66}$$

If one chooses a model system other than the uniform electron gas, the value of the constant C_F changes.[93] However, the exponent 5/3 is required by the coordinate scaling condition Equation 1.42 and should not be changed.

When the number of electrons per unit volume is very large, the number of occupied states becomes very large, and most of the fermions are in states with very high quantum numbers. This is the classical limit. If one neglects relativistic effects, then the relative error in the Thomas–Fermi kinetic energy of neutral atoms goes to 0 as the atomic number increases. However, the absolute error of the Thomas–Fermi kinetic energy is not small for any atom in the periodic table.[117–119]

The Thomas–Fermi model does not suffice for chemistry. In the Thomas–Fermi theory, all atomic anions are predicted to be unstable. Also, the energy of a molecule is always greater than the energy of isolated atoms, so the Thomas–Fermi theory predicts that no molecule is stable. This result is called the Teller nonbinding theorem.[117–120]

For the electron density of closed-shell atoms, the Thomas–Fermi kinetic energy is in error by about 5%. Variational minimization of the density increases this error to about 10% and leads to electron densities that diverge at the atomic nucleus.[121] Adding constraints to force the correct electron–nuclear cusp conditions on the electron density,[122] like other attempts to improve the simple Thomas–Fermi model, gives disappointing results.

1.3.2 Gradient-Corrected Thomas–Fermi Functionals

1.3.2.1 Overview

Since the Thomas–Fermi functional is exact for the uniform electron gas, its failings must arise because the electron densities of chemical substances are far from uniform. This suggests that we construct the gradient expansion about the uniform electron gas limit; such functionals will be exact for nearly uniform electron gases. An alternative perspective is to recall that the Thomas–Fermi theory is exact in the classical high-quantum number limit. The gradient expansion can be derived as a Maclaurin series in powers of $\hbar$; it adds additional quantum effects to the Thomas–Fermi model.

The second-order gradient expansion approximation is[37]

$$T_s^{GEA2}[\rho] = C_F \int \left[\rho^{5/3}(\mathbf{r}) + \frac{|\nabla\rho(\mathbf{r})|^2}{72\rho(\mathbf{r})} \right] d\mathbf{r}$$

(1.67)

$$= T_s^{TF}[\rho] + \frac{1}{9}T_w[\rho].$$

Here,

$$T_w[\rho] = \int \frac{|\nabla\rho(\mathbf{r})|^2}{8\rho(\mathbf{r})} d\mathbf{r}$$

(1.68)

is the kinetic energy functional proposed by Weizsäcker.[123] The error in $T^{GEA2}[\rho]$ is about 1% of the total energy; this error is still several thousand times larger than the minimum error acceptable for chemical applications.[52]

The gradient expansion was extended to fourth order,[124]

$$T^{\text{GEA4}}[\rho] = T^{\text{GEA2}}[\rho] + \frac{1}{540(3\pi)^{2/3}} \int \rho^{5/3}(\mathbf{r}) \left[\begin{array}{c} \left(\dfrac{\nabla^2 \rho(\mathbf{r})}{\rho^{5/3}(\mathbf{r})}\right)^2 + \dfrac{1}{3}\left(\dfrac{|\nabla\rho(\mathbf{r})|}{\rho^{4/3}(\mathbf{r})}\right)^4 \\[2em] -\dfrac{9}{8}\left(\dfrac{\nabla^2\rho(\mathbf{r})}{\rho^{5/3}(\mathbf{r})}\right)\left(\dfrac{|\nabla\rho(\mathbf{r})|}{\rho^{4/3}(\mathbf{r})}\right)^2 \end{array} \right] d\mathbf{r},$$

$$(1.69)$$

and later to sixth order.[125] The fourth-order results are generally an improvement on the second-order results, but they are still woefully inadequate for modeling chemical reaction energies.[126,127] The sixth-order functional diverges because of the cusp in the electron density at the location of atomic nuclei.[125] Molecular electron densities are too different from the uniform electron gas for the gradient expansion to converge.

The divergence of the gradient expansion motivated researchers to consider generalized gradient approximations for the kinetic energy. Based on the form of the convergent terms in the gradient expansion, it is observed that the kinetic energy starts with a Thomas–Fermi term, which is then embellished by a dimensionless enhancement factor (see Equation 1.69). That is, it is reasonable to propose kinetic energy functionals of the form

$$T_s^{\text{GGA}}[\rho] = C_F \int \rho^{5/3}(\mathbf{r}) F_{\text{enh}}[s(\rho), p(\rho)] d\mathbf{r}, \tag{1.70}$$

where

$$s[\rho; \mathbf{r}] = \frac{1}{2(3\pi^2)^{1/3}} \frac{|\nabla\rho(\mathbf{r})|}{\rho^{4/3}(\mathbf{r})} \tag{1.71}$$

is the reduced gradient and

$$p[\rho; \mathbf{r}] = \frac{1}{4(3\pi^2)^{2/3}} \frac{\nabla^2\rho(\mathbf{r})}{\rho^{5/3}(\mathbf{r})} \tag{1.72}$$

is the reduced Laplacian and, ideally,

$$F_{\text{enh}}(0, 0) = 1. \tag{1.73}$$

The functional in Equation 1.70 is exact for the uniform electron gas if and only if Equation 1.73 is satisfied. According to the conjointness hypothesis, it is reasonable to choose the same enhancement functional for kinetic energy functionals that one uses for exchange energy functionals.[84]

Many generalized gradient approximations for the kinetic energy have been proposed.[55,56,128] A few of the more popular and innovative approximations are now presented.

1.3.2.2 Linear Combinations of Thomas–Fermi and Weizsäcker Functionals

The simplest approach to improving the gradient expansion is to reweight the Weizsäcker correction to the gradient expansion; such approximations are called TF + λW approximations:

$$T_s^{\text{TF}+\lambda \text{W}}[\rho] = T_s^{\text{TF}}[\rho] + \lambda T_w[\rho]. \tag{1.74}$$

Popular choices include the gradient expansions ($\lambda = 1/9$),[37] $\lambda = 1/5$,[129] and $\lambda = 1$.[130,131] Thakkar[54] proposed choosing $\lambda = b/9$, where b is a system-dependent constant (analogous to the α parameter in $X\alpha$ theory). Pearson and Gordon[132] analyzed the gradient expansion as an asymptotic series and wrote λ as a functional of the reduced gradient, so that the gradient correction is only applied in regions where the gradient is small enough for this correction to be reliable.

1.3.2.3 N-Dependent Functionals

In this case, the enhancement factor is considered to be a simple function of the number of electrons N.[133] The most common form for the enhancement factor, based on the asymptotic expansion of atomic kinetic energies,[134–136] is

$$F_{\text{enh}}^{\text{TF-}N}(N) = 1 + \frac{a}{N^{1/3}} + \frac{b}{N^{2/3}}, \tag{1.75}$$

where a and b are parameters to be fit.

Unfortunately, N-dependent functionals are not size consistent. If one considers 10 atoms very far from each other, the energy of the system should be the sum of the isolated atomic energies. This is not true for N-dependent functionals because one uses a different functional for the isolated atoms and the supersystem (because they have different numbers of electrons).

1.3.2.4 Rational Function Approximation

Often the enhancement factor is written as a rational function; this is motivated by the observation that the gradient expansion diverges and by the knowledge that Pade approximants are often very effective approaches for resummation of divergent series. A representative functional of this form is the Depristo–Kress functional[137]

$$F_{\text{enh}}^{\text{DK}}(x) = \frac{1 + 0.95x(\mathbf{r}) + 14.28111x^2(\mathbf{r}) - 19.57962x^3(\mathbf{r}) + 26.6477x^4(\mathbf{r})}{1 - 0.05x(\mathbf{r}) + 9.99802x^2(\mathbf{r}) + 2.96085x^3(\mathbf{r})}, \tag{1.76}$$

where

$$x(\mathbf{r}) = \frac{\frac{1}{9}t_{\mathrm{w}}[\rho;\mathbf{r}]}{t_{\mathrm{s}}^{\mathrm{TF}}[\rho;\mathbf{r}]} \tag{1.77}$$

and $t_{\mathrm{w}}[\rho;\mathbf{r}]$ and $t_{\mathrm{s}}^{\mathrm{TF}}[\rho;\mathbf{r}]$ are the local kinetic energies in the Weizsäcker (Equation 1.68) and Thomas–Fermi (Equation 1.65) functionals, respectively.

1.3.2.5 Conjoint Gradient Corrected Functionals

The conjointness hypothesis is based on the similarity between kinetic energy functional and exchange energy functional.[84] This can be motivated, for example, by the form for the kinetic energy in terms of the exchange hole (Equation 1.51). A kinetic energy functional is conjoint to an exchange energy functional if it uses the same enhancement factor as the exchange functional. (Sometimes the parameters in the enhancement factor are adjusted; often they are not.) Many popular generalized gradient approximations for the kinetic energy are built using the conjointness hypothesis.[84,138–143] The PW91 functional has a typical form:

$$F_{\mathrm{enh}}^{\mathrm{PW91}}(s) = \frac{1 + 0.19645s \cdot \operatorname{arcsin} h(7.7956s) + s^2(0.2743 - 0.1508e^{-100s^2})}{1 + 0.19645s \cdot \operatorname{arcsin} h(7.7956s) + 0.004s^4}, \tag{1.78}$$

where s is the reduced gradient (cf. Equation 1.71).

The most successful generalized gradient approach (GGA) forms based on conjointness are the Perdew-Burke-Ernzerhof (PBE)-motivated forms

$$F_{\mathrm{enh}}^{\mathrm{PBE}}(s) = 1 + \sum_{k=1}^{n-1} C_k \left[\frac{s^2}{1 + a_k s^2} \right]^k \tag{1.79}$$

with $n = 2$ and $n = 4$. The popular PBE-TW functional is an $n = 2$ functional.

1.3.2.6 Functionals Consistent with Nonuniform Coordinate Scaling

The uniform density scaling condition in Equation 1.42 was generalized to nonuniform density scaling

$$\rho_{\alpha,\beta,\gamma}(x,y,z) = \alpha \cdot \beta \cdot \gamma \cdot \rho(\alpha x, \beta y, \gamma z) \tag{1.80}$$

by Ou-Yang and Levy.[144] Based on these conditions, Ou-Yang and Levy proposed two new kinetic energy functionals, OL1 and OL2. The Thakkar functional, which was fit to the kinetic energies of 77 molecules, satisfies the nonuniform scaling constraints and is reported to be one of the most accurate GGA functionals[54]:

$$F_{\text{enh}}^{\text{T92}}(s) = 1 + \frac{0.0055(2(6\pi^2)^{1/3}s)^2}{1 + 0.0253(2(6\pi^2)^{1/3}s)\operatorname{arcsinh}(2(6\pi^2)^{1/3}s)} - \frac{0.072(2(6\pi^2)^{1/3}s)}{1 + 2^{5/3}(2(6\pi^2)^{1/3}s)}. \quad (1.81)$$

1.3.3 WEIZSÄCKER-BASED FUNCTIONALS

The Thomas–Fermi kinetic energy is accurate when the electron density is almost constant and when the number of electrons per unit volume is very large (because one is approaching the classical limit). The Weizsäcker functional is accurate in complementary situations: the Weizsäcker functional is exact for one-electron systems and two-electron systems with a nondegenerate ground state; it is also accurate in regions where the electron density changes very rapidly (e.g., near the atomic nuclei) and in all regions of a system where the dominant contribution to the electron density comes from a single Kohn–Sham orbital (e.g., the asymptotic regions far from the atomic nuclei). Moreover, unlike the Thomas–Fermi functional, the Weizsäcker functional is consistent with chemical binding. This suggests that the Weizsäcker functional, instead of the Thomas–Fermi functional, might be a better starting point for chemical applications of OF-DFT.[145]

The Weizsäcker functional is a lower bound to the true kinetic energy, but it is a very weak lower bound.[15,107,146,147] It is even less accurate than the Thomas–Fermi functional.

1.3.4 CORRECTED WEIZSÄCKER-BASED FUNCTIONALS

1.3.4.1 Weizsäcker Plus Thomas–Fermi Functionals

The simplest approach to correcting the Weizsäcker functional is to add a fraction of Thomas–Fermi functional, forming a W + λTF functional,

$$T_s^{\text{W}+\lambda\text{TF}}[\rho] = T_w[\rho] + \lambda T_s^{\text{TF}}[\rho]. \quad (1.82)$$

The only choice for λ that is consistent with the uniform electron gas limit is λ = 1; unfortunately, that functional gives answers far above the true kinetic energy.[130,131,148]

This model can be improved by making the mixing factor λ system dependent. Several authors considered making λ dependent on N.[102,149] These functionals give reasonable results for atomic kinetic energies, but, unfortunately, they are not size consistent. Moreover, when they are used in variational optimization, atomic densities have no shell structure.[150] Thus,

$$\lambda(N) = \left(1 - \frac{2}{N}\right)\left(1 - \frac{A_1}{N^{1/3}} + \frac{A_2}{N^{2/3}}\right). \quad (1.83)$$

Shell structure is retained if one chooses λ, so that it varies at different points in space[151]:

$$T_s^{\text{DG}}[\rho] = T_w[\rho] + C_F \int \lambda(\mathbf{r})\rho^{5/3}(\mathbf{r})\,d\mathbf{r}. \quad (1.84)$$

1.3.4.2 Enhancement Factors from Pauli Potential

The Weizsäcker functional is exact for noninteracting bosons. The correction to the Weizsäcker functional then is solely due to the Pauli principle. The Pauli kinetic energy can then be defined as

$$T_\theta[\rho] = T_s[\rho] - T_W[\rho] \ge 0. \tag{1.85}$$

Because the Weizsäcker kinetic energy functional is a lower bound, the Pauli kinetic energy is always positive.[15,107,146,147,152,153] If one differentiates the resulting energy function, one obtains a differential equation for the square root of the electron density,[153–155]

$$\left(\frac{-1}{2}\nabla^2 + v_s(\mathbf{r}) + v_\theta(\mathbf{r})\right)\sqrt{\rho(\mathbf{r})} = \varepsilon\sqrt{\rho(\mathbf{r})}, \tag{1.86}$$

where the Pauli potential is[153]

$$v_\theta(\mathbf{r}) = \frac{\delta T_\theta[\rho]}{\delta\rho(\mathbf{r})} = \frac{1}{\rho(\mathbf{r})}\left(t_\theta(\mathbf{r}) + \sum_{\{i \in \text{occupied orbitals}\}} (\varepsilon_{\text{HOMO}} - \varepsilon_i)|\phi_i(\mathbf{r})|^2\right) \ge 0. \tag{1.87}$$

Here $t_\theta(\mathbf{r})$ is the local Pauli kinetic energy, which is defined as

$$t_\theta(\mathbf{r}) = \frac{1}{2}\left[\nabla \cdot \nabla\gamma_s(\mathbf{r},\mathbf{r}')\right]_{\mathbf{r}=\mathbf{r}'} - \frac{1}{2}\left|\nabla\sqrt{\rho(\mathbf{r})}\right|^2, \tag{1.88}$$

where $\gamma_s(\mathbf{r},\mathbf{r}')$ is the Kohn–Sham one-matrix. Beyond its conceptual utility,[156] the approach based on the Pauli potential is useful for its computational ease: solving Equation 1.86 is equivalent in difficulty to solving the Kohn–Sham equations for a one-electron system.

The nonnegativity constraints on the Pauli correction and its potential give stringent constraints on the types of functionals that can be considered. The most popular form for the kinetic energy has attempted to modify the enhancement factors from Thomas–Fermi–based kinetic energy functionals, defining[157]

$$T_\theta[\rho] = \int C_F\rho^{5/3}(\mathbf{r})F_\theta\big(s(\mathbf{r})\big)d\mathbf{r} \tag{1.89}$$

with the modified enhancement factors

$$F_\theta(s) = F_{\text{enh}}(s) - \frac{5}{3}s^2. \tag{1.90}$$

The resulting functionals, however, become negative (and even diverge) near the atomic nuclei.[158] This problem is avoided by using so-called reduced gradient approximations, where the Pauli kinetic energy has the form[56]

$$T_{\theta}[\rho] = \int C_F \rho^{5/3}(\mathbf{r}) F_{\theta}(K_2, K_4) d\mathbf{r}, \tag{1.91}$$

where

$$K_2 = s + bp \tag{1.92}$$

and

$$K_4 = s^4 + \frac{18}{13} p^2 - \frac{30}{13} s^2 p \tag{1.93}$$

are specific linear combinations of the reduced gradient s and the reduced Laplacian p (cf. Equations 1.71 and 1.72).

1.4 TWO-POINT APPROXIMATE ORBITAL-FREE KINETIC ENERGY FUNCTIONALS

1.4.1 FUNCTIONALS BASED ON NONINTERACTING RESPONSE KERNEL

The functionals considered in Section 1.3 are all semilocal: the local kinetic energy at the point $\mathbf{r}$ depends only on the electron density and its derivatives at the point $\mathbf{r}$. Improved models for the kinetic energy require considering how the electron density at other points $\mathbf{r}'$ affects the local kinetic energy at the point $\mathbf{r}$. Without including these effects, the oscillations in electron density that are essential for modeling the shell structure of atoms and differentiating between core and valence electrons in molecules cannot be recovered.

Most nonlocal functionals aim to reproduce the Kohn–Sham linear response function

$$\chi_s(\mathbf{r}, \mathbf{r}') = \left(\frac{\delta \rho(\mathbf{r})}{\delta v_s(\mathbf{r}')} \right)_{\mu}. \tag{1.94}$$

The Kohn–Sham linear response predicts long-range and short-range density oscillations in the nearly uniform electron gas; in particular, it captures how the electron density becomes nonuniform when the Kohn–Sham potential is changed by an external perturbation,

$$d\rho(\mathbf{r}) = \int x_s(\mathbf{r}, \mathbf{r}') \delta v_s(\mathbf{r}') d\mathbf{r}'. \tag{1.95}$$

For the uniform electron gas, the response function depends only on the distance between the perturbation and the point at which the density change is measured

$$d\rho(\mathbf{r}) = \int x_s(|\mathbf{r} - \mathbf{r}'|) \delta v_s(\mathbf{r}') d\mathbf{r}', \tag{1.96}$$

and the Fourier transform of the response function[159] can be determined analytically with[160]

$$x_{\text{Lind}}(\eta) = -\frac{k_F}{\pi^2}\left(\frac{1}{2} + \frac{1-\eta^2}{4\eta}\ln\left|\frac{1+\eta}{1-\eta}\right|\right). \tag{1.97}$$

Here

$$k_F = (3\pi^2\rho)^{1/3} \tag{1.98}$$

is the Fermi wave vector and

$$\eta \frac{q}{2k_F} \tag{1.99}$$

is the scaled Fermi momentum.

The response kernel provides a useful constraint on kinetic energy functionals because the second derivative of the noninteracting kinetic energy is related to the inverse of the linear response function

$$\frac{\delta^2 T_s[\rho]}{\delta\rho(\mathbf{r})\delta\rho(\mathbf{r})} = -x_s^{-1}(\mathbf{r},\mathbf{r}'). \tag{1.100}$$

Taking the Fourier transform of both sides, one may require that, in the uniform electron gas limit,

$$\hat{F}\left[\frac{\delta^2 T_s[\rho]}{\delta\rho(\mathbf{r})\delta\rho(\mathbf{r})}\right] = \frac{-1}{x_{\text{Lind}}(\eta)}. \tag{1.101}$$

This constraint cannot be satisfied by any semilocal functional. That is, a semilocal functional is incapable of describing how the kinetic energy of the uniform electron gas changes when it is simultaneously perturbed away from uniformity at two different points. The idea that imposing the correct linear response would improve the description of nonuniform electron densities emerged from the work of Herring[148] and Chacon et al.[161]

Many functionals that reproduce the Lindhard response have been proposed in the intervening years.[55,61,128,162–170] These functionals are among the most accurate functionals available, but they are difficult to formulate in real space and are still inadequate when one is far from the uniform electron gas limit. For this reason, these functionals are much more useful for solid-state physics than they are for molecular chemistry.

While there are many ways to design functionals that are consistent with the Lindhard response, most functionals are based on one of two approaches. The first approach was pioneered by Chacon et al. The Chacon-Alvarellos-Tarazona (CAT) functionals have the form[161]

$$T_s^{CAT}[\rho] = T_w[\rho] - \alpha T_s^{TF}[\rho] + (1+\alpha)\int C_F \rho(\mathbf{r})\left((\tilde{\rho}_\beta(\mathbf{r}))\right)^{2/3\beta} d\mathbf{r}, \qquad (1.102)$$

where $\tilde{\rho}_\beta(\mathbf{r})$ is an effective electron density, averaged over a nonlocal integral kernel,

$$\tilde{\rho}_\beta(\mathbf{r}) = \int \rho^\beta(\mathbf{r}')\Omega\left(k_F(\mathbf{r})|\mathbf{r}-\mathbf{r}'|\right)d\mathbf{r}'. \qquad (1.103)$$

The function $\Omega(k_F(\mathbf{r})|\mathbf{r}-\mathbf{r}'|)$ is chosen so that the Lindhard response is recovered in the uniform electron gas limit. The parameters α and β can be used to satisfy constraints and fit reference data.

There are several generalizations of this functional form; most differ on whether the Fermi wave vector in Equation 1.103 is considered to be constant,[61,166–168] to vary with $\mathbf{r}$ (as shown in Equation 1.103), or to be replaced by a symmetrized form based on the generalized p-mean[164]

$$k_F(\mathbf{r},\mathbf{r}') = \left(\frac{k_F^p(\mathbf{r}) + k_F^p(\mathbf{r}')}{2}\right)^{1/p}. \qquad (1.104)$$

The symmetrized form seems to be the best for recovering shell structure in the electron density.[164] Taking the Fermi momentum to be constant has computational advantages because the kinetic energy functional can be evaluated very quickly as a double convolution of densities with the weighting function. In this case, the form of the functional also simplifies, becoming

$$T_s^{SNDA}[\rho] = T_w[\rho] - \alpha T_s^{TF}[\rho] + (1+\alpha)(2k_F^3)\int C_F \rho^{5/3+\beta}(\mathbf{r})\omega\left(2k_F|\mathbf{r}-\mathbf{r}'|\right)\rho^\beta(\mathbf{r}')d\mathbf{r}. \qquad (1.105)$$

The weighting factor in the function is again chosen to ensure that the Lindhard response is recovered. Functionals like Equation 1.105 are called simplified nonlocal density approximations.[61,165–167]

1.4.2 WEIGHTED DENSITY APPROXIMATION

1.4.2.1 Overview
The idea behind the CAT functional and its generalizations is that if the linear response function of the uniform electron gas is correct, then at least some of the shell structure in the uniform electron gas will also be reproduced. The shell structure, however, is directly implied by the exchange hole (cf. Equation 1.50) and, therefore, also by the one-matrix. The conventional WDA is based on the desire to recover the one-matrix of the uniform electron gas perfectly.[130,171,172] The main difference between the various types of WDA functionals and the various types of CAT functionals then is that the nonlocal function that is being reproduced is the one-matrix for WDAs but the response kernel for CATs.

WDAs are especially convenient for molecular systems because it is impossible to write the Lindhard response function explicitly in real space. However, the one-matrix of the uniform electron gas has the simple real-space form

$$\gamma_{UEG}^{\sigma\sigma}\left[\rho^{\sigma};\mathbf{r},\mathbf{r}'\right]=\sqrt{\rho^{\sigma}(\mathbf{r})\rho^{\sigma}(\mathbf{r}')}g_{UEG}^{\sigma\sigma}\left[\rho^{\sigma};\mathbf{r},\mathbf{r}'\right], \tag{1.106}$$

where

$$g_{UEG}^{\sigma\sigma}\left(k_{F}^{\sigma}\left|\mathbf{r}-\mathbf{r}'\right|\right)=3\left(\frac{\sin\left(k_{F}^{\sigma}\left|\mathbf{r}-\mathbf{r}'\right|\right)-\left(k_{F}^{\sigma}\left|\mathbf{r}-\mathbf{r}'\right|\right)\cos\left(k_{F}^{\sigma}\left|\mathbf{r}-\mathbf{r}'\right|\right)}{\left(k_{F}^{\sigma}\left|\mathbf{r}-\mathbf{r}'\right|\right)^{3}}\right). \tag{1.107}$$

Another nice feature of WDAs is that after one has determined the density matrix, one automatically obtains not only an approximation to the kinetic energy

$$T_{S}^{WDA\text{-}UEG}\left[\rho^{a},\rho^{\beta}\right]=\sum_{\sigma=a,\beta}\delta\left(\mathbf{r}-\mathbf{r}'\right)\left(\frac{1}{2}\nabla.\nabla'\gamma_{UEG}^{\sigma\sigma}\left[\rho^{\sigma};\mathbf{r},\mathbf{r}'\right]\right)d\mathbf{r}d\mathbf{r}' \tag{1.108}$$

but also an approximation to the exchange energy

$$E_{\chi}^{WDA\text{-}UEG}\left[\rho^{a},\rho^{\beta}\right]=\sum_{\sigma=a,\beta}\left(\frac{-1}{2}\right)\iint\frac{\left(\gamma_{UEG}^{\sigma\sigma}\left[\rho_{\sigma};\mathbf{r},\mathbf{r}'\right]\right)^{2}}{\left|\mathbf{r}-\mathbf{r}'\right|}d\mathbf{r}d\mathbf{r}'. \tag{1.109}$$

1.4.2.2 Types of WDAs

When one makes a WDA, one must make three choices. First, one must choose the form of the model one-matrix. The one-matrix of the uniform electron gas is one choice, but certainly it is not the only choice. (For example, a Gaussian model for the density matrix is prevalent in the literature.[93,173,174])

Second, one must choose how the Fermi momentum enters into the formula. It is possible to choose k_{F}^{σ} to be a constant, equal to the average value of the Fermi wave vector. This will not be accurate for atoms and molecules, though, where the electron density varies over many orders of magnitude. Then, one can choose the Fermi wave vector to be dependent on one of the points, $k_{F}(\mathbf{r})$. The resulting one-matrix

$$\gamma_{UEG}^{\sigma\sigma}\left[\rho^{\sigma};\mathbf{r},\mathbf{r}'\right]=\sqrt{\rho^{\sigma}(\mathbf{r})\rho^{\sigma}(\mathbf{r}')}g_{UEG}^{\sigma\sigma}\left(k_{F}^{\sigma}(\mathbf{r})\left|\mathbf{r}-\mathbf{r}'\right|\right), \tag{1.110}$$

however, is not symmetric. Finally, one can force the one-matrix to be symmetric. One way to do this is to add the one-matrix to its transpose

$$\gamma_{UEG}^{\sigma\sigma}\left[\rho^{\sigma};\mathbf{r},\mathbf{r}'\right]=\sqrt{\rho^{\sigma}(\mathbf{r})\rho^{\sigma}(\mathbf{r}')}\left(g_{UEG}^{\sigma\sigma}\left(k_{F}^{\sigma}(\mathbf{r})\left|\mathbf{r}-\mathbf{r}'\right|\right)+g_{UEG}^{\sigma\sigma}\left(k_{F}^{\sigma}(\mathbf{r}')\left|\mathbf{r}-\mathbf{r}'\right|\right)\right). \tag{1.111}$$

This choice has been proposed in the literature[175,176] but seemingly never tested. Instead, one might use the generalized p-mean from Equation 1.104

$$\gamma_{\text{UEG}}^{\sigma\sigma}\left[\rho^{\sigma};\mathbf{r},\mathbf{r}'\right] = \sqrt{\rho^{\sigma}(\mathbf{r})\rho^{\sigma}(\mathbf{r}')}\,g_{\text{UEG}}^{\sigma\sigma}\left(k_{F}^{\sigma}\left(\mathbf{r},\mathbf{r}'\right)\left|\mathbf{r}-\mathbf{r}'\right|\right). \tag{1.112}$$

This form is symmetric. Note that the three different choices here correspond exactly to the three different choices that can be made when the Lindhard response function is used to define a nonlocal weighting function.

Finally, one must choose how the Fermi wave vector is to be determined. It is easiest to use the classical local-density approximation

$$k_{\text{F;LDA}}^{\sigma}(\mathbf{r}) = \left(6\pi^{2}\rho^{\sigma}(\mathbf{r})\right)^{1/3}. \tag{1.113}$$

Alternatively, one can determine the "effective value" of the Fermi wave vector by imposing a constraint. We want the one-matrix to be idempotent, so we would like to satisfy the equations

$$\left\{\int \gamma^{\sigma\sigma}\left[k_{F}^{\sigma}(\mathbf{r});\mathbf{r},\mathbf{r}'\right]\gamma^{\sigma\sigma}\left[k_{F}^{\sigma}(\mathbf{r}'');\mathbf{r}',\mathbf{r}''\right]d\mathbf{r}' = \gamma^{\sigma\sigma}\left[k_{F}^{\sigma}(\mathbf{r});\mathbf{r},\mathbf{r}''\right]\right\}_{\sigma=a,\beta}. \tag{1.114}$$

This system of nonlinear equations is underdetermined because there is an equation for every pair of points $(\mathbf{r}, \mathbf{r}'')$, but there is only one unknown at each point $k_F(\mathbf{r})$. We could try to make a more sophisticated model for the density matrix, but if we force idempotency exactly,[177] then the method is equivalent to the Kohn–Sham method and will have computational cost similar to it.

The simplest thing to do is to reduce the dimensionality of the system by setting $\mathbf{r}, \mathbf{r}''$ in the idempotency condition (Equation 1.114). One then has a set of nonlinear equations, one for each grid point

$$\left\{\int \gamma^{\sigma\sigma}\left[k_{F}^{\sigma}(\mathbf{r});\mathbf{r},\mathbf{r}'\right]\gamma^{\sigma\sigma}\left[k_{F}^{\sigma}(\mathbf{r});\mathbf{r}',\mathbf{r}\right]d\mathbf{r}' = \rho^{\sigma}(\mathbf{r})\right\}_{\sigma=a,\beta}. \tag{1.115}$$

This can be rewritten in terms of the exchange hole

$$\left\{\int \rho^{\sigma}(\mathbf{r}')h_{\chi}^{\sigma\sigma}\left[k_{F}^{\sigma}(\mathbf{r});\mathbf{r},\mathbf{r}'\right]d\mathbf{r}' = -1\right\}_{\sigma-a,\beta}. \tag{1.116}$$

This condition then ensures that the functionals are self-interaction free and that each σ-spin electron excludes another σ-spin electron from its immediate vicinity. From a computational perspective, when the nonsymmetric form of the one-matrix is used, then the nonlinear equations (Equation 1.115) are decoupled; in this case, the equations can be solved using Newton's method in one dimension. But when the symmetric form of the one-matrix (cf. Equation 1.112) is used, then the nonlinear equations in Equation 1.115 are coupled. Chakraborty et al.[178] solved these equations using a limited-memory bad Broyden method to update an approximation to the inverse Jacobian, which is the inverse of the diagonal of the Jacobian.[179] It has been shown that the cost of the symmetrized WDA one-matrix method is only one order of magnitude slower than the cost of the asymmetric WDA one-matrix.[178]

When one is considering the exchange energy, all these variants give different results; for example, if one solves for the effective Fermi vector in Equation 1.115 using the asymmetric form of the one-matrix (Equation 1.110) and then symmetrizes the one-matrix using the p-mean, the exchange energy that one computes is different from what one would have obtained without symmetrization. That is, even if one uses the same values of $k_F(\mathbf{r})$, the following formulas will give different results:

$$-\frac{1}{2}\iint \frac{\left(\sqrt{\rho(\mathbf{r})\rho(\mathbf{r}')}\, g\left(k_F(\mathbf{r})|\mathbf{r}-\mathbf{r}'|\right)\right)^2}{|\mathbf{r}-\mathbf{r}'|} d\mathbf{r}d\mathbf{r}'$$

$$\neq -\frac{1}{2}\iint \frac{\left(\sqrt{\rho(\mathbf{r})\rho(\mathbf{r}')}g\left(\left\{\frac{k_F^p(\mathbf{r})+k_F^p(\mathbf{r}')}{2}\right\}^{1/p}|\mathbf{r}-\mathbf{r}'|\right)\right)^2}{|\mathbf{r}-\mathbf{r}'|} d\mathbf{r}d\mathbf{r}'$$

(1.117)

That is, for the exchange energy, the results change if you symmetrize the one-matrix after determining $k_F(\mathbf{r})$. However, the formulas for the kinetic energy are the same

$$\iint \delta(\mathbf{r}-\mathbf{r}')\left(\frac{-1}{2}\nabla_\mathbf{r}^2\sqrt{\rho(\mathbf{r})\rho(\mathbf{r}')}g\left(k_F(\mathbf{r})|\mathbf{r}-\mathbf{r}'|\right)\right)d\mathbf{r}d\mathbf{r}'$$

$$=\iint \delta(\mathbf{r}-\mathbf{r}')\left(\frac{-1}{2}\nabla_\mathbf{r}^2\sqrt{\rho(\mathbf{r})\rho(\mathbf{r}')}g\left(\left\{\frac{k_F^p(\mathbf{r})+k_F^p(\mathbf{r}')}{2}\right\}^{1/p}|\mathbf{r}-\mathbf{r}'|\right)\right)d\mathbf{r}d\mathbf{r}'. \quad (1.118)$$

Moreover, Equation 1.118 can be simplified to a simple three-dimensional integral,[130,166]

$$T_s^{WDA}\left[\rho^\alpha,\rho^\beta\right]=\sum_{\sigma=\alpha,\beta}T_w\left[\rho^\sigma\right]-\frac{3}{2}g''(0)\int\rho^\sigma(\mathbf{r})\left(k_F^\sigma(\mathbf{r})\right)^2 d\mathbf{r}, \quad (1.119)$$

Where $g''(0)$, the curvature of the one-matrix model at the origin, equals $-1/5$ for the uniform electron gas models. Notice that this formula does not depend on the value of p in Equation 1.104.

1.4.2.3 Conclusions Made from Computations of WDA

If one chooses the formula for the Fermi vector that is appropriate for the uniform electron gas (Equation 1.113), one recovers the TF + W ($\lambda = 1$) functional. Although this functional is exact for the uniform electron gas, the kinetic energies that it assigns to atoms and molecular systems are far too high.[130,131]

The conventional WDA for the kinetic energy results when the effective Fermi vector is determined by substituting the asymmetric one-matrix (Equation 1.110) into the diagonal idempotency condition (Equation 1.115). This functional is also exact for the uniform electron gas, but the kinetic energies of atoms and molecules are still predicted to be far too high.[130,171,172] Indeed, this functional is only slightly more accurate than the Thomas–Fermi functional. This is surprising, since the WDA and the TF functional were derived from the same formula for the one-matrix, but the WDA adds an additional exact constraint.

When the effective Fermi vector is determined by substituting the symmetric one-matrix (Equation 1.112) into the diagonal idempotency condition, the results are much better. This functional is also exact for the uniform electron gas, but now the kinetic energies of atoms and molecules are only slightly too high. The performance of this symmetrized WDA is comparable to the second-order gradient expansion.[178] This seems to be a general result: the accuracy of the symmetrized WDA for exchange is comparable to conventional second-order gradient-corrected density functionals.

1.4.2.4 Other Related Approaches

By using local scaling transformation of the electron density,[180–188] one can rewrite any kinetic energy functional as a one-point nonsymmetric WDA.[180,181] Specifically, one has

$$\rho^{WDA}(\mathbf{r}) = \left(\frac{\tilde{t}\left[\rho;\mathbf{r}\right] - t_w(\mathbf{r}) + \frac{1}{4}\nabla^2\rho(\tilde{r})}{c_F\rho(\mathbf{r})} \right)^{3/2}, \qquad (1.120)$$

where $\tilde{t}[\rho;\mathbf{r}]$ is the local kinetic energy of the approximate functional and $t_w(\mathbf{r})$ is the local Weizsäcker kinetic energy $t_w(\mathbf{r}) = |\nabla\rho(\mathbf{r})|/8\rho(\mathbf{r})$. This indicates that every approximate kinetic energy functional can be reproduced by a suitable WDA or a suitable local scaling transformation. That is, both the WDA and the local scaling transformation are completely general—and, in principle, exact—approaches to developing kinetic energy functionals.

There are many other approaches that relate to the WDA; most of these approaches are designed to be more computationally efficient than the conventional WDA.[130,171,172] We briefly mention the average density approximation (ADA),[189] the modified WDA (MWDA),[190] the semilocal ADA (SADA),[161,191] the reduced WDA,[175] and the generalized WDA.[175,176]

1.5 APPROXIMATE KINETIC ENERGY FUNCTIONALS IN *k*-DENSITY FUNCTIONAL THEORY

It seems extremely difficult to approximate the kinetic energy as an explicit functional of the electron density. In wave function–based methods, when a single-particle theory fails to give satisfactory results, one has recourse to a hierarchy of increasingly complicated approaches. In particular, one builds better and better models by considering, first, two-particle correlations, then three-particle correlations, and so on. The CCSDTQ... and CISDTQ... hierarchies are of this form. For most chemical systems, these hierarchies quickly converge to the exact answer.

There is a similar hierarchy in DFT: instead of using the one-electron distribution function (the electron density), one can use the two-electron distribution function (the pair density),[13,14,192] the three-electron distribution function, and so on.[15,16,116] Using a higher-order electron distribution function as the fundamental descriptor for an electronic system should give increasingly accurate results for chemical properties, including the electronic kinetic energy.

Chemists and physicists have performed a lot of theoretical work on the pair density–based theory,[12–18,20,70,116,192–220] but only a few systematic tests of the approximate kinetic energy functionals have been proposed.[221]

The most simple and popular one is the generalized form of Weizsäcker functional for higher-order electron distribution functions[15,123,196,211,213,222]:

$$T_w[\rho_k] = \frac{(N-k)!}{2(N-1)!} \iint \cdots \int \left| \nabla_{r_1} \sqrt{\rho_k(\mathbf{r}_1,\ldots,\mathbf{r}_k)} \right|^2 d\mathbf{r}_1 \ldots d\mathbf{r}_k. \qquad (1.121)$$

This extended form in Equation 1.121 forms an increasing sequence of lower bounds to the exact kinetic energy[15]

$$T_w^{(1)} \leq T_w^{(2)} \leq T_w^{(3)} \leq \cdots \leq T_w^{(N)} = T_{exact}, \qquad (1.122)$$

with the $k = N$ formula being exact for any N-electron distribution function that arises from a real-valued wave function.

This approach has many features in common with the Weizsäcker-based approaches. For example, the k-electron distribution function can be determined by solving for the ground state of an effective k-electron Schrödinger equation:

$$\left(\sum_{i=1}^{k} \frac{-1}{2} \nabla_i^2 + v(\mathbf{r}_k) + \frac{(N-1)}{(k-1)} \sum_{j=1}^{i-1} \frac{1}{|\mathbf{r}_i - \mathbf{r}_j|} + v_\theta(\mathbf{r}_1, \mathbf{r}_2, \ldots, \mathbf{r}_k) \right)$$
$$\times \sqrt{\rho_k(\mathbf{r}_1, \mathbf{r}_2, \ldots, \mathbf{r}_k)} = \varepsilon \sqrt{\rho_k(\mathbf{r}_1, \mathbf{r}_2, \ldots, \mathbf{r}_k)}, \qquad (1.123)$$

where $v_\theta(\mathbf{r}_1, \mathbf{r}_2, \ldots, \mathbf{r}_k)$ is a potential that accounts for the effects of the Pauli exclusion principle.[211,222]

One might hope that, just as the wave function–based hierarchy rapidly converges to the exact kinetic energy, the hierarchy based on electron distribution functions would also do the same. This is not true of the Weizsäcker functional: only a very small portion of the error in the Weizsäcker kinetic energy functional is corrected when one considers the two-electron distribution function. Similarly, results for the three-electron distribution function are only slightly better than those for the two-electron distribution function. The Weizsäcker kinetic energy functional does not appear to be a good starting point for methods based on the k-electron distribution function.

Another approach is the March–Santamaria formula for the kinetic energy density functional (cf. Equation 1.49), which is a functional for the noninteracting two-fermion distribution function.[15,83] Substituting the *interacting* distribution function into this expression gives

$$
T^{MS}[\rho_k] = \iint \frac{\left| \nabla_r \left(\rho_2(\mathbf{r},\mathbf{r}') - \rho(\mathbf{r})\rho(\mathbf{r}') \right) \right|^2}{8 \left(\rho_2(\mathbf{r},\mathbf{r}') - \rho(\mathbf{r})\rho(\mathbf{r}') \right)} d\mathbf{r} d\mathbf{r}'.
\tag{1.124}
$$

Since this functional is exact for the Hartree–Fock theory, one might expect that it would give even better results when correlated pair densities are used. But unfortunately, this is not the case: the quality of Equation 1.124 actually *deteriorates* as the amount of electron correlation increases.[223]

The effectiveness of the Kohn–Sham theory arises because it imposes the Pauli principle by computing the kinetic energy from an N-electron wave function or, equivalently, an N-representable one-matrix. The previous results suggest that something similar should be performed in higher-order density functional theories. Some results of this sort already exist in the literature. Ayers and Levy[193] have proposed using a Slater determinant to approximate the kinetic energy and then computing the correlation kinetic energy by comparing the true pair density to the noninteracting model pair density. Higuchi and Higuchi[201,202,204,205,208] and Higuchi et al.[206,207,209] have proposed to compute the kinetic energy directly, using either Slater determinants or correlated wave functions. Gonis et al.[197,198] have proposed computing the kinetic energy using a model two-matrix (albeit one that is not necessarily N-representable). Approaches like these are much more computationally demanding than wave function–free approaches, such as the ones we tried, but they seem to be required.

1.6 SUMMARY

Fermi[39] and Thomas[40] derived the first approximate kinetic energy density functional in 1928 and 1927, respectively. For the next decade, researchers patiently pursued better kinetic energy density functionals, but there were no major improvements until 1965, when Kohn and Sham[7] revolutionized the field by introducing orbitals to approximate the kinetic energy. The Kohn–Sham orbital-based DFT has come to dominate applications in chemistry and physics.

However, there is still interest in old-fashioned Thomas–Fermi–like OF-DFT. Part of this interest is practical: in KS-DFT, one must determine the form of N three-dimensional functions (the occupied Kohn–Sham orbitals), but in OF-DFT, one needs to determine only one three-dimensional function (the electron density). Therefore, OF-DFT methods are computationally easier than KS-DFT. Some of the interest is certainly aesthetic: density-based methods are appealing precisely because they avoid the conceptual and computational complexity inherent in the wave function. At a practical level, however, the Kohn–Sham approach is indistinguishable from wave function–based "Hartree plus correction" and "Hartree–Fock plus correction" approaches. Finally, some of the interest is intellectual: the literature on kinetic energy density functionals is extensive and challenging; there are few facets of DFT that are as intellectually stimulating as the quixotic quest for the kinetic energy functional.

Starting around 1980, there was a surge in new approaches to the kinetic energy density functional. Much progress has been made; modern functionals are much more accurate than anything available to Kohn and Sham. A breakthrough is still needed, however: no orbital-free density functional method has ever been shown to achieve high accuracy (e.g., chemical reaction energies and reaction barriers to within 5 kcal/mol). It seems doubtful whether any known orbital-free kinetic energy functional even achieves errors of 50 kcal/mol for reaction thermochemistry and kinetics.

Most of the better approximate functionals give reasonable, albeit unspectacular, results for the kinetic energy when accurate atomic and molecular densities are used. However, when the functionals are used in the variational principle for the electron density, the results deteriorate catastrophically, giving kinetic energies, total energies, and electron densities that are qualitatively incorrect. These failings may be attributed to the N-representability problem for density functionals: the kinetic energy functional does not correspond to an acceptable N-electron system.

We have been pursuing a method that addresses this failing directly. The idea is that the Pauli principle is partly encapsulated by the requirement that each electron excludes another electron with the same spin from its immediate vicinity. Mathematically, this condition is manifest in the normalization of the exchange hole

$$-1 = \int \rho^\sigma(\mathbf{r}')h_\chi^{\sigma\sigma}(\mathbf{r},\mathbf{r}')d\mathbf{r}'. \tag{1.125}$$

Invoking the link between the exchange hole and the one-matrix

$$\left(\gamma^{\sigma\sigma}(\mathbf{r},\mathbf{r}')\right)^2 = -\rho^\sigma(\mathbf{r})h_\chi^{\sigma\sigma}(\mathbf{r},\mathbf{r}')\rho^\sigma(\mathbf{r}'), \tag{1.126}$$

and the link between the kinetic energy and the one-matrix, every possible model for the exchange hole implies a corresponding model for the kinetic energy. We made an attempt to apply this approach using the uniform electron gas model for the exchange hole.[178,224] The results are encouraging, but the error in these functionals is still far from sufficient for chemical applications.

Instead of modeling the exchange hole directly, one can use the exact exchange-correlation hole. Alternatively, one can use the pair density (two-density)

$$\rho_2^{\sigma_1\sigma_2}(\mathbf{r}_1,\mathbf{r}_2) = \rho^{\sigma_1}(\mathbf{r}_1)\rho^{\sigma_2}(\mathbf{r}_2)\left(1 + h_{\chi c}^{\sigma_1\sigma_2}(\mathbf{r}_1,\mathbf{r}_2)\right) \tag{1.127}$$

as the fundamental descriptor for the electronic system and variationally minimize the energy with respect to the two-density (rather than the electron density). This orbital-free theory is afflicted by the N-representability problem for both the two-density and the kinetic energy functional of the two-density. Several kinetic energy functionals of the two-density have been tested, but the results are extremely disappointing.[221,223] In particular, the generalized Weizsäcker functional for the two-density is not much accurate when compared with the conventional Weizsäcker density functional. Comparatively, the March–Santamaria functional is more accurate and gives exact results for Slater determinant pair densities, but it incorrectly predicts that the kinetic energy decreases, instead of increases, with increasing electron correlation.

One might expect that moving to the still higher-order electron distribution functions would remedy the situation. It seems not to. The generalized Weizsäcker kinetic energy formula (which is arguably the most logical starting point) is still extremely inaccurate for the three-density, and our preliminary investigations of the four-density were not very encouraging either. It seems that explicit orbital-free approaches using the k-density are just as challenging as the OF-DFT.

Eighty-five years after Thomas developed the first kinetic energy density functional, the quixotic quest for accurate kinetic energy functionals continues. No existing orbital-free kinetic energy functionals can compete with the robustness and accuracy of conventional KS-DFT methods. One might sensibly abandon the quest, but kinetic energy functionals are important not only for computational modeling but also for theoretical reasons. For example, the famous Lee–Yang–Parr correlation energy functional is built upon a kinetic energy functional. Orbital-free kinetic energy functionals are also useful for DFT embedding calculations, where one uses the kinetic energy functional to define the "kinetic pressure" that the electrons in a subsystem feel due to its surroundings. Finally, the kinetic energy density is useful for chemical interpretation because it can be used to locate electron pair regions within a molecular structure. Fortunately, these applications require much less accuracy than computing accurate reaction thermochemistry or kinetics. Indeed, it seems possible that there will never be kinetic energy functionals that will give results competitive (in terms of scope and accuracy) with the chemical reaction energies computed with KS-DFT.

REFERENCES

1. Parr, R. G.; Yang, W. *Density-Functional Theory of Atoms and Molecules*. Oxford University Press: New York, 1989.
2. Levine, I. N. *Quantum Chemistry*. Prentice Hall: Englewood Cliffs, NJ, 1999.
3. Cramer, C. J. *Essentials of Computational Chemistry*. Wiley: Chichester, UK, 2002.
4. Kohn, W.; Becke, A. D.; Parr, R. G. Density functional theory of electronic structure. *J. Phys. Chem.* **1996**, *100*, 12974–12980.
5. Kohn, W. Nobel lecture: electronic structure of matter: wave functions and density functionals. *Rev. Mod. Phys.* **1999**, *71*, 1253–1266.
6. Hohenberg, P.; Kohn, W. Inhomogeneous electron gas. *Phys. Rev. B* **1964**, *136*, 864–871.
7. Kohn, W.; Sham, L. J. Self-consistent equations including exchange and correlation effects. *Phys. Rev. A* **1965**, *140*, 1133–1138.

8. Perdew, J. P.; Ruzsinszky, A.; Tao, J. M.; Staroverov, V. N.; Scuseria, G. E.; Csonka, G. I. Prescription for the design and selection of density functional approximations: more constraint satisfaction with fewer fits. *J. Chem. Phys.* **2005**, *123*, 062201.

9. Cohen, A. J.; Mori-Sanchez, P.; Yang, W. T. Insights into current limitations of density functional theory. *Science* **2008**, *321*, 792–794.

10. Sousa, S. F.; Fernandes, P. A.; Ramos, M. J. General performance of density functionals. *J. Phys. Chem. A* **2007**, *111*, 10439–10452.

11. Goerigk, L.; Grimme, S. A. Thorough benchmark of density functional methods for general main group thermochemistry, kinetics, and noncovalent interactions. *Phys. Chem. Chem. Phys.* **2011**, *13*, 6670–6688.

12. Ayers, P. W.; Golden, S.; Levy, M. Generalizations of the Hohenberg–Kohn theorem: I. Legendre transform constructions of variational principles for density matrices and electron distribution functions. *J. Chem. Phys.* **2006**, *124*, 054101.

13. Ziesche, P. Pair density-functional theory—a generalized density-functional theory. *Phys. Lett. A* **1994**, *195*, 213–220.

14. Ziesche, P. Attempts toward a pair density functional theory. *Int. J. Quantum Chem.* **1996**, *60*, 1361–1374.

15. Ayers, P. W. Generalized density functional theories using the k-electron densities: development of kinetic energy functionals. *J. Math. Phys.* **2005**, *46*, 062107.

16. Ayers, P. W. Using classical many-body structure to determine electronic structure: an approach using k-electron distribution functions. *Phys. Rev. A* **2006**, *74*, 042502.

17. Ayers, P. W.; Davidson, E. R. Linear inequalities for diagonal elements of density matrices. *Adv. Chem. Phys.* **2007**, *134*, 443–483.

18. Davidson, E. R. N-Representability of the electron-pair density. *Chem. Phys. Lett.* **1995**, *246*, 209–213.

19. Garrod, C.; Percus, J. K. Reduction of the N-particle variational problem. *J. Math. Phys.* **1964**, *5*, 1756–1776.

20. Pistol, M. E. N-Representability of two-electron densities and density matrices and the application to the few-body problem. *Chem. Phys. Lett.* **2004**, *400*, 548–552.

21. Coleman, A. J.; Yukalov, V. I. *Reduced Density Matrices: Coulson's Challenge.* Springer: Berlin, Germany, 2000.

22. Davidson, E. R. *Reduced Density Matrices in Quantum Chemistry.* Academic Press: New York, 1976.

23. Lowdin, P. O. Quantum theory of many-particle systems: I. Physical interpretation by means of density matrices, natural spin-orbitals, and convergence problems in the method of configuration interaction. *Phys. Rev.* **1955**, *97*, 1474–1489.

24. Mazziotti, D. A. Quantum chemistry without wave functions: two electron reduced density matrices. *Acc. Chem. Res.* **2006**, *39*, 207–215.

25. Coleman, J. Structure of fermion density matrices. *Rev. Mod. Phys.* **1963**, *35*, 668–687.

26. Yoseloff, M. L.; Kuhn, H. W. Combinatorial approach to the N-representability of P-density matrices. *J. Math. Phys.* **1969**, *10*, 703–706.

27. Kuhn, H. W. *Proc. Symp. Appl. Math.* **1960**, *10*, 141.

28. Kummer, H. N-Representability problem for reduced density matrices. *J. Math. Phys.* **1967**, *8*, 2063–2081.

29. Davidson, E. R. Linear inequalities for density matrices. *J. Math. Phys.* **1969**, *10* (4), 725–734.

30. Mcrae, W. B.; Davidson, E. R. Linear inequalities for density matrices II. *J. Math. Phys.* **1972**, *13* (10), 1527–1538.

31. Levy, M.; Ivanov, S.; Gorling, A. Correlation energy in a high-density limit from adiabatic connection perturbation theory. In Dobson, J. F.; Vignale, G.; Das, M. P., Eds. *Electron Density Functional Theory: Recent Progress and New Directions.* Plenum Press, New York and London, **1998**, 133–147.

32. Lieb, E. H. Density functionals for Coulomb systems. *Int. J. Quantum Chem.* **1983**, *24*, 243–277.

33. Gilbert, T. L. Hohenberg–Kohn theorem for nonlocal external potentials. *Phys. Rev. B* **1975**, *12*, 2111–2120.

34. Dirac, P. A. M. Note on exchange phenomena in the Thomas atom. *Proc. Cambridge Philos. Soc.* **1930**, *26*, 376–385.

35. Fermi, E.; Amaldi, E. Le Orbite Oos Degli Elementi. *Accad. Ital. Rome* **1934**, *6*, 117–149.

36. Wigner, E. On the interaction of electrons in metals. *Phys. Rev.* **1934**, *46*, 1002–1011.

37. Kirzhnits, D. A. Quantum corrections to the Thomas–Fermi equation. *Sov. Phys.* **1957**, *5*, 64–71.

38. Parr, R. G.; Yang, W. *Density-Functional Theory of Atoms and Molecules.* Oxford University Press: New York, 1989.

39. Fermi, E. A statistical method for the determination of some atomic properties and the application of this method to the theory of the periodic system of elements. *Z. Phys.* **1928**, *48*, 73–79.

40. Thomas, L. H. The calculation of atomic fields. *Proc. Cambridge Philos. Soc.* **1927**, *23*, 542–548.

41. Cramer, C. J.; Truhlar, D. G. Density functional theory for transition metals and transition metal chemistry. *Phys. Chem. Chem. Phys.* **2009**, *11*, 10757–10816.

42. Gill, P. M. W. Obituary: density functional theory (1927–1993). *Aust. J. Chem.* **2001**, *54*, 661–662.

43. Becke, A. D. Density-functional exchange-energy approximation with correct asymptotic-behavior. *Phys. Rev. A* **1988**, *38*, 3098–3100.

44. Lee, C.; Yang, W.; Parr, R. G. Development of the Colle–Salvetti correlation-energy formula into a functional of the electron density. *Phys. Rev. B* **1988**, *37*, 785–789.

45. Miehlich, B.; Savin, A.; Stoll, H.; Preuss, H. Results obtained with the correlation-energy density functionals of Becke and Lee, Yang and Parr. *Chem. Phys. Lett.* **1989**, *157* (3), 200–206.

46. Becke, A. D. Density-functional thermochemistry: 3. The role of exact exchange. *J. Chem. Phys.* **1993**, *98*, 5648–5652.

47. Becke, A. D. A new mixing of Hartree–Fock and local density-functional theories. *J. Chem. Phys.* **1993**, *98*, 1372–1377.

48. Wang, Y. A.; Carter, E. A.; Schwartz, S. D. Orbital-free kinetic-energy density functional theory. In *Theoretical Methods in Condensed Phase Chemistry.* Kluwer: Dordrecht, The Netherlands, 2000, pp. 117–184.

49. Chen, H. J.; Zhou, A. H. Orbital-free density functional theory for molecular structure calculations. *Numer. Math. Theor. Methods Appl.* **2008**, *1*, 1–28.

50. Garcia-Aldea, D.; Alvarellos, J. E. A study of kinetic energy density functionals: a new proposal. In Simos, T.; Maroulis, G., Eds. *Advances in Computational Methods in Sciences and Engineering*, Springer, vols. 4A and 4B. 2005, pp. 1462–1466.

51. Wesolowski, T. A. Quantum chemistry 'without orbitals'—an old idea and recent developments. *Chimia* **2004**, *58* (5), 311–315.

52. Iyengar, S. S.; Ernzerhof, M.; Maximoff, S. N.; Scuseria, G. E. Challenge of creating accurate and effective kinetic-energy functionals. *Phys. Rev. A* **2001**, *63* (5), Art-052508.

53. Chan, G. K. L.; Handy, N. C. An extensive study of gradient approximations to the exchange-correlation and kinetic energy functionals. *J. Chem. Phys.* **2000**, *112* (13), 5639–5653.

54. Thakkar, A. J. Comparison of kinetic-energy density functionals. *Phys. Rev. A* **1992**, *46*, 6920–6924.

55. Garcia-Aldea, D.; Alvarellos, J. E. Kinetic energy density study of some representative semilocal kinetic energy functionals. *J. Chem. Phys.* **2007**, *127*, 144109.

56. Karasiev, V. V.; Jones, R. S.; Trickey, S. B.; Harris, F. E. Recent advances in developing orbital-free kinetic energy functionals. In Paz, J. L.; Hernandez, A. J., Eds. *New Developments in Quantum Chemistry*. Transworld Research Network: Kerala, India, **2009**, 25–54.

57. Ho, G. S.; Ligneres, V. L.; Carter, E. A. Introducing PROFESS: a new program for orbital-free density functional theory calculations. *Comput. Phys. Commun.* **2008**, *179*, 839–854.

58. Chai, J. D.; Weeks, J. D. Orbital-free density functional theory: kinetic potentials and ab initio local pseudopotentials. *Phys. Rev. B* **2007**, *75*, 205122.

59. Ho, G. S.; Huang, C.; Carter, E. A. Describing metal surfaces and nanostructures with orbital-free density functional theory. *Curr. Opin. Solid State Mater. Sci.* **2007**, *11*, 57–61.

60. Dufty, J. W.; Trickey, S. B. Scaling, bounds, and inequalities for the noninteracting density functionals at finite temperature. *Phys. Rev. B* **2011**, *84*, 125118.

61. Smargiassi, E.; Madden, P. A. Orbital-free kinetic-energy functionals for 1st-principles molecular-dynamics. *Phys. Rev. B* **1994**, *49*, 5220–5226.

62. Govind, N.; Wang, Y. A.; Carter, E. A. Electronic-structure calculations by first-principles density-based embedding of explicitly correlated systems. *J. Chem. Phys.* **1999**, *110* (16), 7677–7688.

63. Cortona, P. Self-consistently determined properties of solids without band-structure calculations. *Phys. Rev. B* **1991**, *44* (16), 8454–8458.

64. Vaidehi, N.; Wesolowski, T. A.; Warshel, A. Quantum-mechanical calculations of solvation free-energies—a combined ab initio pseudopotential free-energy perturbation approach. *J. Chem. Phys.* **1992**, *97* (6), 4264–4271.

65. Wesolowski, T.; Warshel, A. Ab-initio free-energy perturbation calculations of solvation free-energy using the frozen density-functional approach. *J. Phys. Chem.* **1994**, *98*, 5183–5187.

66. Wesolowski, T. A.; Warshel, A. Frozen density-functional approach for ab-initio calculations of solvated molecules. *J. Phys. Chem.* **1993**, *97* (30), 8050–8053.

67. Wesolowski, T. A.; Weber, J. Kohn–Sham equations with constrained electron density: an iterative evaluation of the ground-state electron density of interacting molecules. *Chem. Phys. Lett.* **1996**, *248*, 71–76.

68. Wesolowski, T. A.; Weber, J. Kohn–Sham equations with constrained electron density: the effect of various kinetic energy functional parameterizations on the ground-state molecular properties. *Int. J. Quantum Chem.* **1997**, *61*, 303–311.

69. Wesolowski, T. A. One-electron equations for embedded electron density: challenge for theory and practical payoffs in multi-scale modelling of complex polyatomic molecules. In Leszczynski, J., Ed. *Computation Chemistry: Reviews of Current Trends*. World Scientific: Singapore, **2006**, 1–82.

70. Ayers, P. W.; Levy, M. Generalized density-functional theory: conquering the *N*-representability problem with exact functionals for the electron pair density and the second-order reduced density matrix. *J. Chem. Sci.* **2005**, *117*, 507–514.

71. Stash, A. I.; Tsirelson, V. G. Modern possibilities for calculating some properties of molecules and crystals from the experimental electron density. *Crystallogr. Rep.* **2005**, *50*, 177–184.

72. Tsirelson, V.; Stash, A. Analyzing experimental electron density with the localized-orbital locator. *Acta Crystallogr. Sect. B Struct. Sci.* **2002**, *58*, 780–785.

73. Tsirelson, V.; Stash, A. Determination of the electron localization function from electron density. *Chem. Phys. Lett.* **2002**, *351*, 142–148.

74. Tsirelson, V. G. The mapping of electronic energy distributions using experimental electron density. *Acta Crystallogr. Sect. B Struct. Sci.* **2002**, *58*, 632–639.

75. Levy, M.; Perdew, J. P. Hellmann–Feynman, virial, and scaling requisites for the exact universal density functionals. Shape of the correlation potential and diamagnetic susceptibility for atoms. *Phys. Rev. A* **1985**, *32*, 2010–2021.

76. Ghosh, S. K.; Parr, R. G. Density-determined orthonormal orbital approach to atomic energy functionals. *J. Chem. Phys.* **1985**, *82*, 3307–3315.

77. Levy, M. Electron densities in search of Hamiltonians. *Phys. Rev. A* **1982**, *26*, 1200–1208.

78. Levy, M.; Yang, W.; Parr, R. G. A new functional with homogeneous coordinate scaling in density functional theory: *F*[rho,lambda]. *J. Chem. Phys.* **1985**, *83* (5), 2334–2336.

79. Wang, Y. A. Coordinate scaling and adiabatic-connection formulation in density-functional theory. *Phys. Rev. A* **1997**, *56* (2), 1646–1649.

80. Wang, Y. A.; Liu, S. B.; Parr, R. G. Laurent series expansions in density functional theory. *Chem. Phys. Lett.* **1997**, *267* (1–2), 14–22.

81. Lowdin, P. O. Correlation problem in many-electron quantum mechanics: I. Review of different approaches and discussion of some current ideas. *Adv. Chem. Phys.* **1959**, *2*, 207–322.

82. Ayers, P. W.; Lucks, J. B.; Parr, R. G. Constructing exact density functionals from the moments of the electron density. *Acta Chim. Phys. Debricina* **2002**, *34–35*, 223–248.

83. March, N. H.; Santamaria, R. Nonlocal relation between kinetic and exchange energy densities in Hartree–Fock theory. *Int. J. Quantum Chem.* **1991**, *39* (4), 585–592.

84. Lee, H.; Lee, C.; Parr, R. G. Conjoint gradient correction to the Hartree–Fock kinetic- and exchange-energy density functionals. *Phys. Rev. A* **1991**, *44* (1), 768–771.

85. Chakraborty, D.; Ayers, P. W. Derivation of generalized Von-Weizsäcker kinetic energies from quasi-probability distribution functions. In *Statistical Complexity*. Springer Science, 2011.

86. Cohen, L. Can quantum mechanics be formulated as classical probability theory. *Philos. Sci.* **1966**, *33*, 317–322.

87. Cohen, L. Generalized phase–space distribution functions. *J. Math. Phys.* **1966**, *7*, 781–786.

88. Weyl, H. Quantenmechanik Und Gruppentheorie. *Z. Phys.* **1927**, *46*, 1–46.

89. Wigner, E. On the quantum correction for thermodynamic equilibrium. *Phys. Rev.* **1932**, *40*, 749–759.

90. Anderson, J. S. M.; Ayers, P. W.; Hernandez, J. I. R. How ambiguous is the local kinetic energy? *J. Phys. Chem.* **2010**, *114*, 8884–8895.

91. Ayers, P. W.; Parr, R. G.; Nagy, A. Local kinetic energy and local temperature in the density-functional theory of electronic structure. *Int. J. Quantum Chem.* **2002**, *90*, 309–326.

92. Ghosh, S. K.; Berkowitz, M. A. Classical fluid-like approach to the density-functional formalism of many-electron systems. *J. Chem. Phys.* **1985**, *83*, 2976–2983.

93. Lee, C.; Parr, R. G. Gaussian and other approximations to the first-order density matrix of electronic systems, and the derivation of various local-density–functional theories. *Phys. Rev. A* **1987**, *35*, 2377–2383.

94. Cohen, L. Local kinetic energy in quantum mechanics. *J. Chem. Phys.* **1979**, *70*, 788–789.

95. Ghosh, S. K.; Berkowitz, M.; Parr, R. G. Transcription of ground-state density-functional theory into a local thermodynamics. *Proc. Natl. Acad. Sci.* **1984**, *81*, 8028–8031.

96. Mazo, R. M.; Kirkwood, J. G. Statistical thermodynamics of quantum fluids. *J. Chem. Phys.* **1958**, *28* (4), 644–647.

97. Ovchinnikov, I. V.; Neuhauser, D. Orbital-free tensor density functional theory. *J. Chem. Phys.* **2006**, *124* (2), 024105(5).

98. Ovchinnikov, I. V.; Bartell, L. A.; Neuhauser, D. Hydrodynamic tensor density functional theory with correct susceptibility. *J. Chem. Phys.* **2007**, *126* (13), 134101(9).

99. Cangi, A.; Lee, D.; Elliott, P.; Burke, K. Leading corrections to local approximations. *Phys. Rev. B* **2010**, *81*, 235128.

100. Cangi, A.; Lee, D.; Elliott, P.; Burke, K.; Gross, E. K. U. Electronic structure via potential functional approximations. *Phys. Rev. Lett.* **2011**, *106*, 236404(4).

101. Elliott, P.; Lee, D. H.; Cangi, A.; Burke, K. Semiclassical origins of density functionals. *Phys. Rev. Lett.* **2008**, *100*, 256406.

102. Acharya, P. K.; Bartolotti, L. J.; Sears, S. B.; Parr, R. G. An atomic kinetic energy functional with full Weizsäcker correction. *Proc. Natl. Acad. Sci.* **1980**, *77*, 6978–6982.

103. Ghiringhelli, L. M.; Le Site, L.; Mosna, R. A.; Hamilton, I. P. Information-theoretic approach to kinetic-energy functionals: the nearly uniform electron gas. *J. Math. Chem.* **2010**, *48*, 78–82.

104. Ghiringhelli, L. M.; Hamilton, I. P.; Le Site, L. Interacting electrons, spin statistics, and information theory. *J. Chem. Phys.* **2010**, *132*, 014106.

105. Le Site, L. On the scaling properties of the correlation term of the electron kinetic functional and its relation to the Shannon measure. *Europhys. Lett.* **2009**, *86* (4), 40004 (5).

106. Le Site, L. Erratum on "On the scaling properties of the correlation term of the electron kinetic functional and its relation to the Shannon measure." *Europhys. Lett.* **2009**, *88*(1),19901(2).

107. Sears, S. B.; Parr, R. G.; Dinur, U. On the quantum-mechanical kinetic energy as a measure of the information in a distribution. *Isr. J. Chem.* **1980**, *19* (1–4), 165–173.

108. Trickey, S. B.; Karasiev, V. V.; Vela, A. Positivity constraints and information-theoretical kinetic energy functionals. *Phys. Rev. B* **2011**, *84*, 075146.

109. Ayers, P. W.; Rodriguez, J. I. Out of one, many—using moment expansions of the virial relation to deduce universal density functionals from a single system. *Can. J. Chem.* **2009**, *87*, 1540–1545.

110. Nagy, A. Hierarchy of equations for the energy functional of the density-functional theory. *Phys. Rev. A* **1993**, *47*, 2715–2719.

111. Nagy, A.; Liu, S. B.; Parr, R. G. Density-functional formulas for atomic electronic energy components in terms of moments of the electron density. *Phys. Rev. A* **1999**, *59* (5), 3349–3354.

112. Nagy, A. Hierarchy of equations in the generalized density functional theory. *Int. J. Quantum Chem.* **2006**, *106*, 1043–1051.

113. Hamilton, I. P.; Mosna, R. A.; Le Site, L. Classical kinetic energy, quantum fluctuation terms and kinetic-energy functionals. *Theor. Chem. Acc.* **2007**, *118*, 407–415.

114. Mosna, R. A.; Hamilton, I. P.; Le Site, L. Quantum-classical correspondence via a deformed kinetic operator. *J. Phys. A Math. Gen.* **2005**, *38*, 3869–3878.

115. Mosna, R. A.; Hamilton, I. P.; Le Site, L. Variational approach to dequantization. *J. Phys. A Math. Gen.* **2006**, *39*, L229–L235.

116. Ayers, P. W. Constraints for hierarchies of many electron distribution functions. *J. Math. Chem.* **2008**, *44*, 311–323.

117. Lieb, E. H.; Simon, B. Thomas–Fermi theory revisited. *Phys. Rev. Lett.* **1973**, *31*, 681–683.

118. Lieb, E. H.; Simon, B. The Thomas–Fermi theory of atoms, molecules, and solids. *Adv. Math.* **1977**, *23*, 22–116.

119. Lieb, E. H. Thomas–Fermi and related theories of atoms and molecules. *Rev. Mod. Phys.* **1981**, *53*, 603–641.

120. Teller, E. On the stability of molecules in the Thomas–Fermi theory. *Rev. Mod. Phys.* **1962**, *34*, 627–631.

121. March, N. H. *Electron Density Theory of Atoms and Molecules*. Academic Press: New York, 1992.

122. Parr, R. G.; Ghosh, S. K. Thomas–Fermi theory for atomic systems. *Proc. Natl. Acad. Sci.* **1986**, *83*, 3577–3479.

123. Weizsäcker, C. F. Zur Theorie Dier Kernmassen. *Z. Phys.* **1935**, *96*, 431–458.

124. Hodges, C. H. Quantum corrections to the Thomas–Fermi approximation: the Kirzhnits method. *Can. J. Phys.* **1973**, *51*, 1428–1437.

125. Murphy, D. R. 6th-order term of the gradient expansion of the kinetic-energy density functional. *Phys. Rev. A* **1981**, *24* (4), 1682–1688.

126. Lee, C.; Ghosh, S. K. Density-gradient expansion of the kinetic-energy functional for molecules. *Phys. Rev. A* **1986**, *33*, 3506–3507.

127. Murphy, D. R.; Wang, W. P. Comparative-study of the gradient expansion of the atomic kinetic-energy functional–neutral atoms. *J. Chem. Phys.* **1980**, *72* (1), 429–433.

128. Garcia-Aldea, D.; Alvarellos, J. E. Kinetic-energy density functionals with nonlocal terms with the structure of the Thomas–Fermi functional. *Phys. Rev. A* **2007**, *76*, 052504.
129. Berk, A. Lower-bound energy functionals and their application to diatomic systems. *Phys. Rev. A* **1983**, *28*, 1908–1923.
130. Alonso, J. A.; Girifalco, L. A. Nonlocal approximation to exchange potential and kinetic energy of an inhomogeneous electron-gas. *Phys. Rev. B* **1978**, *17* (10), 3735–3743.
131. Katsumi, Y. Energy levels for an extended Thomas–Fermi–Dirac potential. *J. Phys. Soc. Jpn.* **1967**, *22*, 1127–1132.
132. Pearson, E. W.; Gordon, R. G. Local asymptotic gradient corrections to the energy-functionals of an electron-gas. *J. Chem. Phys.* **1985**, *82*, 881–889.
133. Thakkar, A. J.; Pedersen, W. A. Local density functional approximations and conjectured bounds for momentum moments. *Int. J. Quantum Chem.* **1990**, 327–338.
134. March, N. H.; White, R. J. Non-relativistic theory of atomic and ionic binding energies for large atomic number. *J. Phys. B* **1972**, *5*, 466–475.
135. March, N. H.; Parr, R. G. Chemical-potential, Teller's theorem, and the scaling of atomic and molecular energies. *Proc. Natl. Acad. Sci.* **1980**, *77*, 6285–6288.
136. Tal, Y.; Bartolotti, L. J. On the $Z-1$ and $N-1/3$ expansions of Hartree–Fock atomic energies. *J. Chem. Phys.* **1982**, *76*, 4056–4062.
137. Depristo, A. E.; Kress, J. D. Kinetic-energy functionals via Pade approximations. *Phys. Rev. A* **1987**, *35*, 438–441.
138. Lacks, D. J.; Gordon, R. G. Tests Of nonlocal kinetic-energy functionals. *J. Chem. Phys.* **1994**, *100*, 4446–4452.
139. Becke, A. D. On the large-gradient behavior of the density functional exchange energy. *J. Chem. Phys.* **1986**, *85* (12), 7184–7187.
140. Becke, A. D. Density functional calculations of molecular-bond energies. *J. Chem. Phys.* **1986**, *84*, 4524–4529.
141. Depristo, A. E.; Kress, J. D. Rational-function representation for accurate exchange energy functionals. *J. Chem. Phys.* **1987**, *86*, 1425–1428.
142. Perdew, J. P.; Yue, W. Accurate and simple density functional for the electronic exchange energy: generalized gradient approximation. *Phys. Rev. B* **1986**, *33*, 8800–8802.
143. Perdew, J. P.; Chevary, J. A.; Vosko, S. H.; Jackson, K. A.; Pederson, M. R.; Singh, D. J.; Fiolhais, C. Atoms, molecules, solids, and surfaces—applications of the generalized gradient approximation for exchange and correlation. *Phys. Rev. B* **1992**, *46*, 6671–6687.
144. Ou-Yang, H., Levy, M. Nonuniform coordinate scaling requirements in density-functional theory. *Phys. Rev. A* **1990**, *42* (1), 155–160.
145. Goodisman, J. Modified Weizsäcker corrections in Thomas–Fermi theories. *Phys. Rev. A* **1970**, *1*, 1574–1576.
146. Hoffmann-Ostenhof, M.; Hoffmann-Ostenhof, T. "Schrodinger inequalities" and asymptotic behavior of the electron density of atoms and molecules. *Phys. Rev. A* **1977**, *16*, 1782–1785.
147. Sagvolden, E.; Perdew, J. P. Discontinuity of the exchange-correlation potential: support for assumptions used to find it. *Phys. Rev. A* **2008**, *77*, 012517.
148. Herring, C. Explicit estimation of ground-state kinetic energies from electron-densities. *Phys. Rev. A* **1986**, *34*, 2614–2631.
149. Gazquez, J. L.; Robles, J. On the atomic kinetic-energy functionals with full Weizsäcker correction. *J. Chem. Phys.* **1982**, *76*, 1467–1472.
150. Bartolotti, L. J.; Acharya, P. K. On the functional derivative of the kinetic-energy density functional. *J. Chem. Phys.* **1982**, *77*, 4576–4585.
151. Deb, B. M.; Ghosh, S. K. New method for the direct calculation of electron-density in many-electron systems: 1. Application to closed-shell atoms. *Int. J. Quantum Chem.* **1983**, *23* (1), 1–26.

152. Kurth, S.; Perdew, J. P.; Blaha, P. Molecular and solid-state tests of density functional approximations: LSD, GGAs, and meta-GGAs. *Int. J. Quantum Chem.* **1999**, *75*, 889–909.

153. Levy, M.; Perdew, J. P.; Sahni, V. Exact differential-equation for the density and ionization-energy of a many-particle system. *Phys. Rev. A* **1984**, *30*, 2745–2748.

154. March, N. H. The local potential determining the square root of the ground-state electron-density of atoms and molecules from the Schrodinger-equation. *Phys. Lett. A* **1986**, *113* (9), 476–478.

155. March, N. H. Differential-equation for the electron-density in large molecules. *Int. J. Quantum Chem.* **1986**, *13*, 3–8.

156. Liu, S. B. Steric effect: a quantitative description from density functional theory. *J. Chem. Phys.* **2007**, *126*, 244103.

157. Ludena, E. V.; Karasiev, V.; Sen, K. D. *Reviews of Quantum Chemistry: A Celebration of the Contributions of Robert G. Parr.* World Scientific: Singapore, 2002, pp. 612–656.

158. Karasiev, V. V.; Trickey, S. B.; Harris, F. E. Born–Oppenheimer interatomic forces from simple, local kinetic energy density functionals. *J. Comput.-Aided Mater. Des.* **2006**, *13*, 111–129.

159. Pick, R. M.; Cohen, M. H.; Martin, R. M. Microscopic theory of force constants in the adiabatic approximation. *Phys. Rev. B* **1970**, *1*, 910.

160. Lindhard, J. On the properties of a gas of charged particles. *K. Dan. Vidensk. Selsk. Mat.-Fys. Medd.* **1954**, *28*, 8.

161. Chacon, E.; Alvarellos, J. E.; Tarazona, P. Nonlocal kinetic-energy functional for non-homogeneous electron-systems. *Phys. Rev. B* **1985**, *32*, 7868–7877.

162. Garcia-Aldea, D.; Alvarellos, J. E. Approach to kinetic energy density functionals: nonlocal terms with the structure of the von Weizsäcker functional. *Phys. Rev. A* **2008**, *77*, 022502.

163. Garciagonzalez, P.; Alvarellos, J. E.; Chacon, E. Nonlocal kinetic-energy-density functionals. *Phys. Rev. B* **1996**, *53*, 9509–9512.

164. Garciagonzalez, P.; Alvarellos, J. E.; Chacon, E. Kinetic-energy density functional: atoms and shell structure. *Phys. Rev. A* **1996**, *54*, 1897–1905.

165. Huang, C.; Carter, E. A. Nonlocal orbital-free kinetic energy density functional for semiconductors. *Phys. Rev. B* **2010**, *81*, 045206.

166. Perrot, F. Hydrogen–hydrogen interaction in an electron-gas. *J. Phys. Condens. Matter* **1994**, *6*, 431–446.

167. Wang, L. W.; Teter, M. P. Kinetic-energy functional of the electron-density. *Phys. Rev. B* **1992**, *45*, 13196–13220.

168. Wang, Y. A.; Govind, N.; Carter, E. A. orbital-free kinetic-energy functionals for the nearly free electron gas. *Phys. Rev.* **1998**, *58*, 13465–13471.

169. Wang, Y. A.; Govind, N.; Carter, E. A. Orbital-free kinetic-energy density functionals with a density-dependent kernel. *Phys. Rev. B* **1999**, *60*, 16350–16358.

170. Zhou, B. J.; Ligneres, V. L.; Carter, E. A. Improving the orbital-free density functional theory description of covalent materials. *J. Chem. Phys.* **2005**, *122*, 044103.

171. Alonso, J. A.; Girifalco, L. A. Nonlocal approximation to exchange energy of non-homogeneous electron-gas. *Solid State Commun.* **1977**, *24*, 135–138.

172. Gunnarsson, O.; Jonson, M.; Lundqvist, B. I. Exchange and correlation in inhomogeneous electron-systems. *Solid State Commun.* **1977**, *24* (11), 765–768.

173. Sadd, M.; Teter, M. P. Weighted density approximation applied to diatomic molecules. *Phys. Rev. B* **1996**, *54*, 13643–13648.

174. Sadd, M.; Teter, M. P. A. Weighted spin density approximation for chemistry: importance of shell partitioning. *THEOCHEM* **2000**, *501*, 147–152.

175. Wang, Y. A. Natural variables for density functionals. *Phys. Rev. A* **1997**, *55* (6), 4589–4592.

176. Wu, Z. G.; Cohen, R. E.; Singh, D. J. Comparing the weighted density approximation with the LDA and GGA for ground-state properties of ferroelectric perovskites. *Phys. Rev. B* **2004**, *70*, 104112.

177. Cuevas-Saavedra, R.; Chakraborty, D.; Rabi, S.; Cardenas, C.; Ayers, P. W. Unpublished work, 2012.
178. Chakraborty, D.; Cuevas-Saavedra, R.; Ayers, P. W. Two-point weighted density approximations for the kinetic energy density functional. *J. Chem. Phys.* **2012** (under review).
179. Matthew, C; Cuevas-Saavedra, R.; Chakraborty, D.; Ayers, P. W. Unpublished work, 2012.
180. Diez, R. P.; Karasiev, V. V. A relationship between the weighted density approximation and the local-scaling transformation version of density functional theory. *J. Phys. B.* **2003**, *36*, 2881–2890.
181. Karasiev, V. V.; Ludena, E. V.; Artemyev, A. N. Electronic-structure kinetic-energy functional based on atomic local-scaling transformations. *Phys. Rev. A* **2000**, *62*, 062510.
182. Lopezboada, R.; Pino, R.; Ludena, E. V. Explicit expressions for $T - 8[rho]$ and $E - X[rho]$ by means of Pade approximants to local-scaling transformations. *Int. J. Quantum Chem.* **1997**, *63*, 1025–1035.
183. Ludena, E. V.; Lopez-Boada, R.; Maldonado, J. E.; Valderrama, E.; Kryachko, E. S.; Koga, T.; Hinze, J. Local-scaling transformation version of density-functional theory. *Int. J. Quantum Chem.* **1995**, *56*, 285–301.
184. Ludena, E. V.; Lopezboada, R.; Pino, R. Approximate kinetic energy density functionals generated by local-scaling transformations. *Can. J. Chem.* **1996**, *74*, 1097–1105.
185. Ludena, E. V.; Lopez-Boada, R. *Density Funct. Theory I* **1996**, *180*, 169.
186. Ludena, E. V.; Lopezboada, R.; Pino, R.; Karasiev, V. Density functionals generated by local-scaling transformations. *Abstr. Pap. Am. Chem. Soc.* **1997**, *213*, 81-COMP.
187. Ludena, E. V.; Karasiev, V. V.; Echevarria, L. Realizations of the noninteracting kinetic energy functional enhancement factor through local-scaling transformations: atoms. *Int. J. Quantum Chem.* **2003**, *91*, 94–104.
188. Ludena, E. V.; Karasiev, V. V.; Nieto, P. Exact and approximate forms of the kinetic energy functional $T - S[Rho]$ for molecules obtained via local-scaling transformations. *Theor. Chem. Acc.* **2003**, *110*, 395–402.
189. Gunnarsson, O.; Jonson, M.; Lundqvist, B. I. Exchange and correlation in atoms, molecules and clusters. *Phys. Lett. A* **1976**, *59* (3), 177–179.
190. Gunnarsson, O.; Jones, R. O. Density functional calculations for atoms, molecules and clusters. *Phys. Scr.* **1980**, *21*, 394–401.
191. Garcia-Gonzalez, P.; Alvarellos, J. E.; Chacon, E. Kinetic-energy density functionals based on the homogeneous response function applied to one-dimensional fermion systems. *Phys. Rev. A* **1998**, *57* (6), 4192–4200.
192. Levy, M.; Ziesche, P. The pair density functional of the kinetic energy and its simple scaling property. *J. Chem. Phys.* **2001**, *115*, 9110–9112.
193. Ayers, P. W.; Levy, M. Using the Kohn–Sham formalism in pair density-functional theories. *Chem. Phys. Lett.* **2005**, *416*, 211–216.
194. Ayers, P. W.; Davidson, E. R. Necessary conditions for the N-representability of pair distribution functions. *Int. J. Quantum Chem.* **2006**, *106*, 1487–1498.
195. Cuevas-Saavedra, R.; Ayers, P. W. Coordinate scaling of the kinetic energy in pair density functional theory: a Legendre transform approach. *Int. J. Quantum Chem.* **2009**, *109*, 1699–1705.
196. Furche, F. Towards a practical pair density-functional theory for many-electron systems. *Phys. Rev. A* **2004**, *70*, 022514.
197. Gonis, A.; Schulthess, T. C.; Vanek, J.; Turchi, P. E. A. A general minimum principle for correlated densities in quantum many-particle systems. *Phys. Rev. Lett.* **1996**, *77*, 2981–2984.
198. Gonis, A.; Schulthess, T. C.; Turchi, P. E. A.; Vanek, J. Treatment of electron–electron correlations in electronic structure calculations. *Phys. Rev. B* **1997**, *56*, 9335–9351.
199. Higuchi, K.; Higuchi, M. Arbitrary choice of basic variables in density functional theory: II. Illustrative applications. *Phys. Rev. B* **2004**, *69*, 165118.

200. Higuchi, K.; Higuchi, M. Density functional theory with arbitrary basic variables. *J. Magn. Magn. Mater.* **2004**, *272–276*, 659–661.
201. Higuchi, K.; Higuchi, M. Computational schemes for the ground-state pair density. *J. Phys* **2009**, *21*, 064206.
202. Higuchi, K.; Higuchi, M. Computational pair density functional theory: a proposal for the kinetic energy functional. *Phys. Rev. B* **2010**, *82*, 155135.
203. Higuchi, M.; Higuchi, K. Arbitrary choice of basic variables in density functional theory: formalism. *Phys. Rev. B* **2004**, *69*, 035113.
204. Higuchi, M.; Higuchi, K. A proposal of the approximate scheme for calculating the pair density. *Phys. Rev. B Condens. Matter* **2007**, *387*, 117–121.
205. Higuchi, M.; Higuchi, K. Pair density-functional theory by means of the correlated wave function. *Phys. Rev. A* **2007**, *75*, 042510.
206. Higuchi, M.; Miyasita, M.; Kodera, M.; Higuchi, K. Simultaneous equations for calculating the pair density. *J. Phys., Condens. Matter* **2007**, *19*, 365219.
207. Higuchi, M.; Miyasita, M.; Kodera, M.; Higuchi, K. Density functional scheme for calculating the ground-state pair density. *J. Magn. Magn. Mater.* **2007**, *310*, 990–992.
208. Higuchi, M.; Higuchi, K. Pair density functional theory utilizing the noninteracting reference system: an effective initial theory. *Phys. Rev. B* **2008**, *78*, 125101.
209. Higuchi, M.; Miyasita, M.; Higuchi, K. A restrictive condition on approximate forms of the kinetic energy functional of the pair density functional theory. *Int. J. Quantum Chem.* **2010**, *110*, 2283–2285.
210. Kristyan, S. Properties of the multi-electron densities "between" the Hohenberg–Kohn theorems and variational principle. *THEOCHEM* **2008**, *858*, 1–11.
211. Nagy, A. Density-matrix functional theory. *Phys. Rev. A* **2002**, *66*, 022505.
212. Nagy, A.; Amovilli, C. Effective potential in density matrix functional theory. *J. Chem. Phys.* **2004**, *121*, 6640–6648.
213. Nagy, A. Spherically and system-averaged pair density functional theory. *J. Chem. Phys.* **2006**, *125*, 184104.
214. Nagy, A.; Romera, E. Relation between Fisher measures of information coming from pair distribution functions. *Chem. Phys. Lett.* **2010**, *490*, 242–244.
215. Pistol, M. E. Characterization of N-representable n-particle densities when N is infinite. *Chem. Phys. Lett.* **2006**, *417*, 521–523.
216. Pistol, M. E. Relations between N-representable N-particle densities. *Chem. Phys. Lett.* **2006**, *422*, 363–366.
217. Pistol, M. E. N-Representable distance densities have positive Fourier transforms. *Chem. Phys. Lett.* **2006**, *431*, 216–218.
218. Pistol, M. E. Investigations of random pair densities and the application to the N-representability problem. **2007**, *449*, 208–211.
219. Pistol, M. E.; Almbladh, C. O. Adiabatic connections and properties of coupling-integrated exchange-correlation holes and pair densities in density functional theory. *Chem. Phys. Lett.* **2009**, *480*, 136–139.
220. Samvelyan, S. K. N-Representability of diagonal elements of second-order reduced density matrices. *Int. J. Quantum Chem.* **1997**, *65* (2), 127–142.
221. Chakraborty, D.; Ayers, P. W. Failure of the Weizsäcker kinetic energy functional for one-, two- and three-electron distribution functions. *J. Math. Chem.* **2011**, *49* (8), 1810.
222. Nagy, A.; Amovilli, C. Electron–electron cusp condition and asymptotic behavior for the Pauli potential in pair density functional theory. *J. Chem. Phys.* **2008**, *128*, 114115.
223. Chakraborty, D.; Ayers, P. W. Failure of the March–Santamaria kinetic energy functional for two-electron distribution functions. *J. Math. Chem.* **2011**, *49* (8), 1822.
224. Cuevas-Saavedra, R.; Chakraborty, D.; Ayers, P. W. A symmetric two-point weighted density approximation for exchange energies. *Phys. Rev. A* **2012**.

2 Quantum Adiabatic Switching and Supersymmetric Approach to Excited States of Nonlinear Oscillators

Susmita Kar and S. P. Bhattacharyya

CONTENTS

2.1 INTRODUCTION

Nonlinear oscillators (NLOs) have been extensively used as realistic models for chemical bonds (Merzbacher 1970; Morse 1929), especially for describing the bond-breaking process (dissociation), simulating vibrational spectra of molecules (Child and Lawton 1981; Lehmann 1992), and modeling the nonlinear optical responses of several classes of molecules (Kirtman 1992; Takahashi and Mukamel 1994), catalytic bond activation, and dissociation processes (McCoy 1984). The nonlinear (anharmonic) oscillator, defined by the equation

$$\left(-\frac{d^2}{dx^2} + \frac{1}{4}x^2 + \lambda x^4 \right)\phi(x) = \varepsilon(\lambda)\phi(x), \tag{2.1}$$

with the boundary condition $\phi(x) = 0$ at $x = \pm\infty$, has drawn considerable attention because the perturbation series even for the ground-state energy diverges (Bender and Wu 1969). Bender and Wu (1969) have shown that the levels of the NLO, as a function of λ, have an infinite number of branch points with a limit point at $\lambda = 0$, and level crossings occur at each branch point. These results have important ramifications for ϕ^4 field theories.

The eigenspectra of NLOs may be "expanded" or "contracted" with reference to the completely equispaced energy levels of the linear oscillator. The Pöschl–Teller potential, for example, has an "expanded" spectrum, with the gap between consecutive energy levels increasing with quantum number (Pöschl and Teller 1933). The Morse oscillator, which has been the cornerstone of vibrational spectroscopy, has a "contracted" spectrum, with energy levels coming closer as n increases (Morse 1929). Unlike in the two preceding examples, the energy levels of NLOs are not generally known in the form of closed analytic expressions. The analysis of their spectra therefore depends on the availability of accurate numerical estimates of excited-state wave functions and energies. Accurate calculations of excited-state wave functions and energies are, however, far more difficult than the task of computing the ground-state eigenfunction and energy of the same system within, for example, the variational framework. Unless linear variational ansatz is invoked, the upper boundedness of the approximate excited-state energy values can be ensured only by enforcing the trial wave function to be explicitly orthogonal to all $(n - 1)$ lower (exact) eigenstates of the system. If exact lower eigenstates of H are unavailable, a decoupling constraint, along with the constraint of orthogonality to the $(n - 1)$ lower (may not be exact) (Hendeković 1982) eigenstates of H, is to be explicitly enforced, making such calculations difficult and unpopular. It would be profitable to have a method in which one works only with the computation of the ground-state eigenfunctions of a sequence of n appropriately constructed Hamiltonians $(H_1 \ldots H_n)$ such that the ground state of $H_i (i = 1,\ldots,n)$ matches exactly the $(i + 1)$th state or the ith excited state of the parent Hamiltonian (H_0). To elaborate, we assume that the ground state $\left| \psi_0 \right\rangle$ of H_0 is known $\left(H_0 \left| \psi_0 \right\rangle = E_0 \left| \psi_0 \right\rangle \right)$ and that $P_0 = \left| \psi_0 \right\rangle \left\langle \psi_0 \right|$ is the projector for the ground state. The Hamiltonian H_1 from which the ground state of H_0 has been projected out is $H_1 = (1 - P_0)^\dagger H_0 (1 - P_0)$. The ground state of H_1 is then the first excited state $\left| \psi_1 \right\rangle$ of H_0. Once $\left| \psi_0 \right\rangle$ and $\left| \psi_1 \right\rangle$ are known, we can construct a new Hamiltonian, $H_2 = \sum_{i=0}^{1} (1 - P_i)^\dagger H_0 (1 - P_i)$, the ground state of which is the second excited state of H_0. The hierarchy can be continued until we have exhausted all the eigenfunctions of H_0 without ever having to go beyond the recipes for ground-state calculation. We have recently explored the idea in the context of computing eigenvalues and eigenvectors of real symmetric matrices using a genetic algorithm (Nandy et al. 2002, 2011). The essence of the approach has been to construct a sequence of Hamiltonians such that the nth Hamiltonian has its ground state completely overlapping with the nth excited state of the original Hamiltonian (H_0). The connection of this idea with supersymmetric (SUSY) quantum mechanics (QM) is almost immediate.

SUSY QM (Cooper et al. 1995, 2001; Gol'fand and Likhtman 1971) provides an elegant means of handling one-dimensional energy eigenvalue problems. The central idea has been to factorize the Schrödinger Hamiltonian operator into a pair of charge operators, Q^+ and Q^-, thereby reducing the problem to solving a pair of first-order linear differential equations. Kouri et al. (2009), for example, recently demonstrated how SUSY QM can be exploited to achieve higher accuracy and faster convergence in Rayleigh–Ritz variational calculations of the excited states of one-dimensional systems by reducing the problem to the calculation of ground states of progressively higher sector SUSY Hamiltonians. Starting with an arbitrary Schrödinger Hamiltonian (H_0), Bittner et al. (2009) invoked the quantum Monte Carlo (QMC) method on a series of one-dimensional SUSY partner Hamiltonians to generate the full energy spectrum and the associated eigenfunctions of the Hamiltonian H_0. With the ground states of each of the higher sector partner Hamiltonians being nodeless, the calculations were free of the "sign" or "node" problem that has been known to plague QMC calculations. Kouri et al. (2010) also succeeded in generalizing SUSY QM to handle multidimensional and multiparticle systems. Their success has aroused interest in the potency of SUSY QM in electronic structure calculations. A time-dependent route to the different eigenstates of SUSY partner Hamiltonians has been explored by Kar and Bhattacharyya (2012) very recently. The central idea was to switch across partner potentials by exploiting quantum adiabatic switching algorithms. The initial success with simple systems prompted us to explore if the time-dependent route could be successfully used for handling NLOs in general.

   We present the methodology, analyze its powers and pitfalls, and report the results of calculations on the excited states of two NLOs. The numerical accuracy and stability of the proposed algorithm are tested.

2.2 METHODOLOGY

Let the one-dimensional NLO be described by the Hamiltonian

$$H = \frac{p^2}{2} + V(x),$$ (2.2)

where $V(x)$ is the NLO potential (with $m = e - \hbar = 1$ in atomic units).

   Let the ground-state wave function and energy be ψ_0 and E_0, respectively. We may now define the superpotential $\varphi_0(x)$ as

$$\varphi_0(x) = (-) \frac{\psi_0^{(0)'}(x)}{\psi_0^{(0)}(x)},$$ (2.3)

where $\psi_0^{(0)}(x)$ is the ground eigenfunction of the Hamiltonian $H^{(0)}$, and $\psi_0^{(0)}(x)$ is the first derivative of $\psi_0^{(0)}(x)$ with respect to x, where

$$H^{(0)} = H - E_0 = \frac{p^2}{2} + (V - E_0) = \frac{p^2}{2} + V^{(0)}(x).$$ (2.4)

   The ground eigenvalue $E_0^{(0)}$ of $H^{(0)}$ is zero because of energy shift. We can now define the pair of charge operators Q_0^+ and Q_0^- of SUSY QM as follows:

$$Q_0^+ = \frac{1}{\sqrt{2}} \left\{ -\frac{d}{dx} + \varphi_0(x) \right\} \tag{2.5}$$

and

$$Q_0^- = \frac{1}{\sqrt{2}} \left\{ \frac{d}{dx} + \varphi_0(x) \right\}. \tag{2.6}$$

The charge operators immediately lead to a pair of SUSY partner Hamiltonians $H^{(0)}$ and $H^{(1)}$,

$$H^{(0)} = Q_0^+ Q_0^- \tag{2.7}$$

and

$$H^{(1)} = Q_0^- Q_0^+, \tag{2.8}$$

where

$$H^{(0)} = \frac{p^2}{2} + V^{(0)} = \frac{p^2}{2} + \frac{1}{2} \left\{ (\varphi_0)^2 - \frac{d}{dx} (\varphi_0(x)) \right\} \tag{2.9}$$

and

$$H^{(1)} = \frac{p^2}{2} + V^{(1)} = \frac{p^2}{2} + \frac{1}{2} \left\{ (\varphi_0)^2 + \frac{d}{dx} (\varphi_0(x)) \right\}, \tag{2.10}$$

with $V^{(0)}$ and $V^{(1)}$ being called the SUSY partner potentials.

It is straightforward to see that the superpotential $\varphi_0(x)$ satisfies a nonlinear first-order differential equation, a Riccati equation,

$$\varphi_0'(x) - \{\varphi_0(x)\}^2 + 2V^{(0)}(x) = 0. \tag{2.11}$$

The solution to the Riccati equation provides not only $\varphi_0(x)$ but also the SUSY partner Hamiltonians $H^{(0)}$ and $H^{(1)}$, along with the ground-state eigenfunction $\psi_0^{(0)}(x)$ of $H^{(0)}$, as

$$\psi_0^{(0)}(x) = N \exp \left(-\int_0^x \varphi(x')\,dx' \right), \tag{2.12}$$

with N being the normalization constant.

We may now look into the possibility of adiabatically switching from one of the eigenstates of $H^{(0)}$ to the relevant eigenstate of $H^{(1)}$, the SUSY partner of $H^{(0)}$.

Quantum adiabatic theorem (QAT) has been extensively explored since the birth of QM (Born and Fock 1928; Messiah 1962). The essence of QAT is as follows. Let H_i be the initial Hamiltonian at time $t = 0$, and let H_f be the final Hamiltonian after the lapse of a long time $t = T$. We can construct a time-dependent Hamiltonian $H(t)$ that continuously and infinitesimally slowly interpolates between H_i and H_f in the time interval $0 \leq t \leq T$:

$$H(t) = H_i + \lambda(t)\{H_f - H_i\}, \tag{2.13}$$

where the switching function $\lambda(t) = 0$ at $t = 0$ and $\lambda(t) = 1$ at $t = T$. QAT asserts that if $\lambda(t)$ satisfies Kato conditions (Kato 1950), $\dfrac{d\lambda(t)}{dt}$ is sufficiently small, and the continuum is not excited, the switching will lead to the ground state of the final Hamiltonian H_f if one starts with the ground state of H_i, the initial Hamiltonian. Let us now construct a time-dependent Hamiltonian $H(t)$ using the pair of SUSY partner Hamiltonians $H^{(0)}$ and $H^{(1)}$

$$H(t) = H^{(0)} + \lambda(t)\{H^{(1)} - H^{(0)}\}, \tag{2.14}$$

where $\lambda(t)$ is the switching function satisfying Kato conditions, and T is the total switching time. In addition to $\dfrac{d\lambda(t)}{dt}$ being sufficiently small, $\lambda(t)$ is chosen to be continuous and to have continuous derivatives all along the switching path. It is now possible to solve the time-dependent Schrödinger equation (TDSE) for the Hamiltonian $H(t)$ as

$$i\frac{\partial \psi(t)}{\partial t} = H(t)\psi(t) \tag{2.15}$$

by invoking the time-dependent Fourier grid Hamiltonian (TDFGH) method (Adhikari and Bhattacharyya 1992; Adhikari et al. 1992), with $\psi(t = 0)$ chosen to be the ground state $\psi_0^{(0)}$ of $H^{(0)}$ constructed from the superpotential $\varphi_0(x)$ on a pre-defined coordinate grid. According to QAT, we would land with $\psi_0^{(1)}(x)$, the ground state of $H^{(1)}$, upon the conclusion of switching at $t = T$ represented on the same grid because we started with the ground state of $H^{(0)}$ at $t = 0$. Because $H^{(1)}$ is isospectral with $H^{(0)}$, except for the ground state (Cooper et al. 1995, 2001), the energy eigenvalues of the state reached by switching are just the first excited-state energy $E_1^{(0)}$ of $H^{(0)}$ (Kar and Bhattacharyya 2012). We can recover the first excited state of $H^{(0)}$, $\psi_1^{(0)}(x)$, by the action of the charge operator Q_0^+ on $\psi_0^{(1)}(x)$ as

$$Q_0^+ \psi_0^{(1)}(x) = \psi_1^{(0)}(x). \tag{2.16}$$

Having found the first excited state of NLO, we proceed to construct a shifted potential $V_s^{(1)} = V^{(1)} - E_1^{(0)}$. $V_s^{(1)}$ then generates a shifted Hamiltonian

$$H_s^{(1)} = \frac{p^2}{2} + V_s^{(1)}, \tag{2.17}$$

which is associated with a superpotential $\varphi_1(x)$ that satisfies the following Riccati equation:

$$\varphi_1'(x) + 2V_s^{(1)} - \left\{\varphi_1(x)\right\}^2 = 0. \tag{2.18}$$

Solving the Riccati equation for $\varphi_1(x)$, we can construct $V^{(2)}(x)$, the SUSY partner potential of $V_s^{(1)}$, and thus generate $H^{(2)}$, the SUSY partner of $H_s^{(1)}$. Once $H_s^{(1)}$ and $H^{(2)}$ are known, we can construct a time-dependent, adiabatically evolving Hamiltonian $\tilde{H}(t)$ as

$$\tilde{H}(t) = H_s^{(1)} + \lambda(t)\left\{H^{(2)} - H_s^{(1)}\right\}, \tag{2.19}$$

which obeys the TDSE

$$i\frac{\partial\tilde{\psi}(t)}{\partial t} = \tilde{H}(t)\tilde{\psi}(t). \tag{2.20}$$

We can solve the TDSE as before after setting $\tilde{\psi}(t = 0) = \psi_0^{(1)}(x)$ and adiabatically pass it on to the ground state $\psi_0^{(2)}(x)$ of $H^{(2)}$ upon the conclusion of switching. SUSY QM then tells us that (Cooper et al. 1995, 2001)

$$Q_1^+\psi_0^{(2)}(x) = \psi_1^{(1)}(x) \tag{2.21}$$

and

$$Q_0^+\psi_1^{(1)}(x) = \psi_2^{(0)}(x). \tag{2.22}$$

Thus, we have generated the first excited state of $H^{(1)}$ and the second excited state of $H^{(0)}$, respectively, by the action of appropriate charge operators. By repeating the same sequence of steps, we can generate the ground states of $H^{(0)}$, $H^{(1)}$, $H^{(2)}$, and so on, from which the ground, first excited, and second excited states of the NLO represented by H can be recovered by the action of appropriate charge operators. Thus, sequential quantum adiabatic switching through $H^{(0)} \rightarrow H^{(1)} \rightarrow H^{(2)} \ldots \rightarrow H^{(n-1)}$ leads us to the sequence of ground states $\psi_0^{(1)}, \psi_0^{(2)}, \ldots, \psi_0^{(n-1)}$ of progressively higher sector Hamiltonians from which the entire sequence of the n-eigenstates of the NLO H or $(H^{(0)})$ can be recovered through the action of an appropriate sequence of charge operators.

2.3   RESULTS AND DISCUSSION

2.3.1   POLYNOMIAL SUPERPOTENTIAL

We consider a particular class of superpotentials $\varphi(x)$ that support bound states in the domain $-\infty < x < \infty$, which has been analyzed by Kouri et al. (2009), and choose

$$\varphi_{(0)}(x) = x^3 + 2x, \tag{2.23}$$

which defines a pair of NLO Hamiltonians

$$
\begin{aligned}
H^{(0)} &= \frac{1}{2}\left\{ -\frac{d}{dx} + \varphi^{(0)} \right\}\left\{ \frac{d}{dx} + \varphi^{(0)} \right\} \\
&= \frac{p^2}{2} + \frac{1}{2}\left(x^6 + 4x^4 + x^2 - 2 \right) \\
&= \frac{p^2}{2} + V^{(0)}(x).
\end{aligned} \tag{2.24}
$$

$H^{(0)}$ has the analytic ground-state wave function $\psi_0^{(0)}(x) = N\exp\left(-\left(\frac{x^4}{4} + x^2 \right) \right)$ with $H^{(0)}\psi_0^{(0)}(x) = 0$. The second sector Hamiltonian $H^{(1)}$ then turns out to be

$$
\begin{aligned}
H^{(1)} &= \frac{p^2}{2} + \frac{1}{2}\left(x^6 + 4x^4 + 7x^2 + 2 \right) \\
&= \frac{p^2}{2} + V^{(1)}(x).
\end{aligned} \tag{2.25}
$$

Both $H^{(0)}$ and $H^{(1)}$ have been constructed in matrix form following the TDFGH method (Balint-Kurti and Ward 1991; Marston and Balint-Kurti 1989). The coordinate grid used has a length of 10 a.u., and 91 numbers of grid points have been used in the calculation. The time-dependent Hamiltonian that would adiabatically switch from an eigenstate of $H^{(0)}$ to the corresponding eigenstate of $H^{(1)}$ is

$$
\begin{aligned}
H(t) &= H^{(0)} + \lambda(t)\left\{ H^{(1)} - H^{(0)} \right\} \\
&= H^{(0)} + \left\{ 3x^2 + 2 \right\}\lambda(t),
\end{aligned} \tag{2.26}
$$

where

$$\lambda(t) = \frac{1}{\ln 2}\left(\ln\left(1 + \left(\frac{t}{T} \right)^2 \right) \right). \tag{2.27}$$

This switching function has been previously shown to perform well in the calculation of impurity-perturbed quantum dots (Hazra et al. 2008). We have used the

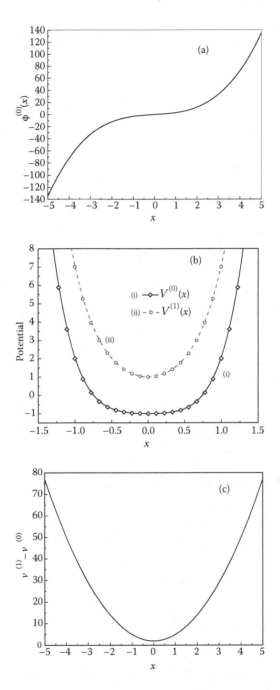

FIGURE 2.1 (a) Polynomial superpotential of the anharmonic oscillator. (b) SUSY partner potentials $V^{(0)}(x)$ and $V^{(1)}(x)$ are plotted. (c) Difference potential $(V^{(1)}(x) - V^{(0)}(x))$ is plotted as function of x.

sixth-order Runge–Kutta method for time integration with time-step size $\Delta t =$ 0.00001 a.u.

The chosen superpotential $\varphi_{(0)}(x)$ is displayed in Figure 2.1a, whereas the SUSY partner potentials derived from it are shown in Figure 2.1b. The difference potential $(V^{(1)} - V^{(0)})$ that drives the switching is shown in Figure 2.1c.

The time development of the switching function [$\lambda(t)$ of Equation 2.27, with $T = 2 \times 10^5$ a.u.] is displayed in Figure 2.2a, whereas the energy of the evolving state $E(t)$ along the switching path from $t = 0$ to $t = T$ is reported in Figure 2.2b.

The initial wave function $\psi(t = 0)$ is nodeless (Figure 2.3a), and the final wave function $\psi(t = T)$ is also nodeless, as expected (see Figure 2.3b).

The first excited-state wave function $\psi_1^{(0)}(x)$ of $H^{(0)}$ was generated by forming the function $Q_0^+ \psi_0^{(1)}$ by numerical differentiation using the central difference method. The function so generated has a single node and is displayed in Figure 2.4a. To monitor the adiabaticity of the switching process, we have computed the mean square energy fluctuation $(\langle H^2(t) \rangle - \langle H(t)^2 \rangle)$ of the system from the initiation to the conclusion of

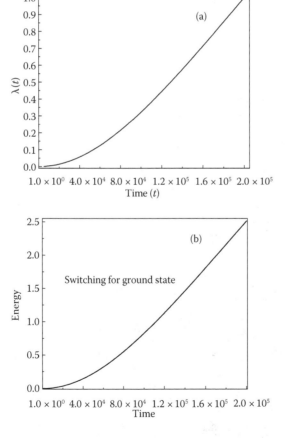

FIGURE 2.2 (a) Time profile of the logarithmic switching function is displayed. (b) Evolution of energy during switching from the ground state of $H^{(0)}$ to that of $H^{(1)}$ is shown.

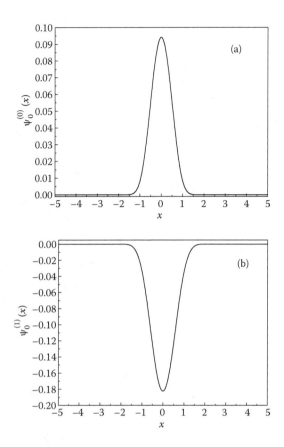

FIGURE 2.3 (a) Ground state $\psi_0^{(0)}(x)$ of $H^{(0)}$, obtained from the superpotential, is shown. (b) Ground state $\psi_0^{(1)}(x)$ of $H^{(1)}$, obtained through adiabatic passage from $\psi_0^{(0)}(x)$, is shown.

switching (Figure 2.4b). Fluctuations are small throughout, signifying that the system has remained in the instantaneous eigenstate of $H(t)$ all along.

The superpotential $(\varphi^{(1)}(x))$ for the next sector (sector 2), as computed from $\psi_1^{(0)}(x)$, is displayed in Figure 2.5a. The corresponding partner potentials $V^{(1)}(x)$ and $V^{(2)}(x)$ are shown in Figure 2.5b. The difference potential $\{V^{(2)}(x) - V^{(1)}(x)\}$ turns out to be parabolic and is displayed in Figure 2.5c.

The evolution of energy during the switching from $H^{(1)}$ to $H^{(2)}$ is shown in Figure 2.6a. The ground state of the sector 2 Hamiltonian $H^{(2)}$ is Gaussian, as can be seen from Figure 2.6b.

The first excited state of $H^{(1)}$, as constructed by forming $Q_1^+\psi_0^{(2)} = \psi_1^{(1)}(x)$ and displayed in Figure 2.7a, has a single node structure as expected. The targeted second excited state of $H^{(0)}$ has been constructed by forming $Q_0^+\psi_1^{(1)}(x)$ and has the expected two-node structure (see Figure 2.7b). The computed energy of this state $E_2^{(0)} = 5.8484094$ a.u. compares well with the value obtained by a direct diagonalization of $H^{(0)}$. One can continue with the process and compute all the excited bound states of $H^{(0)}$.

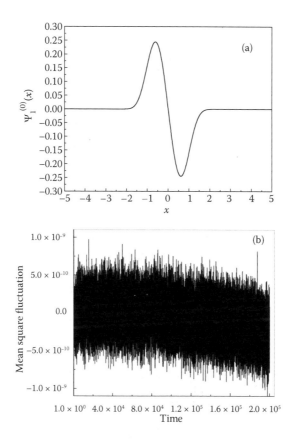

FIGURE 2.4 (a) First excited state $\psi_1^{(0)}(x)$ of $H^{(0)}$, obtained after the operation of Q_0^+ on $\psi_0^{(1)}(x)$, is plotted. (b) Mean square fluctuation in energy as a function of time during adiabatic passage from the ground state of $H^{(0)}$ to that of $H^{(1)}$.

The switching function has some effect on the final eigenvalue, which was obtained at the conclusion of the switching. Thus, with $\lambda(t) = \left(\dfrac{t}{T}\right)^4$, the first excited-state energy of $H^{(0)}$ obtained is $E_1^{(0)} = 2.5122163$ a.u., whereas with $\lambda(t)$ of Equation 2.27, the same state has the energy $E_1^{(0)} = 2.5122231$ a.u., although the same value of T is used ($E_1^{(0)}$ obtained by direct diagonalization is 2.51222394 a.u. on the same grid). Although the analysis reported here is not exhaustive, the logarithmic switching function of Equation 2.27 appears to work very well.

2.3.2 EXPONENTIAL SUPERPOTENTIAL

For the superpotential (Kouri et al. 2009)

$$\varphi_0(x) = x^3 + x + e^x, \tag{2.28}$$

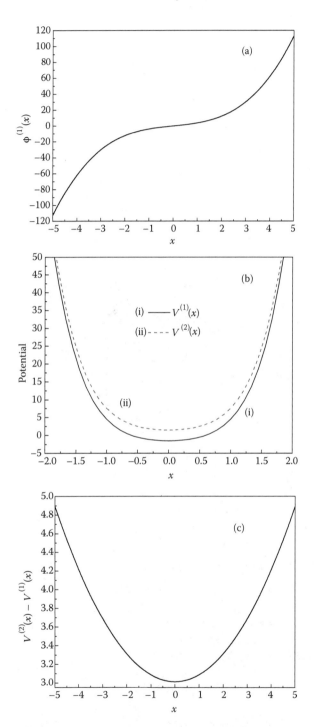

FIGURE 2.5 (a) Sector 2 superpotential $\varphi^{(1)}(x)$. (b) SUSY partner potentials for sector 2. (c) Difference potential for sector 2.

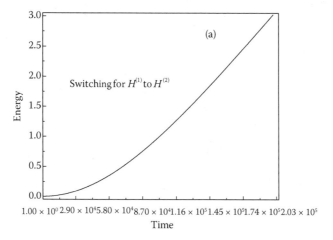

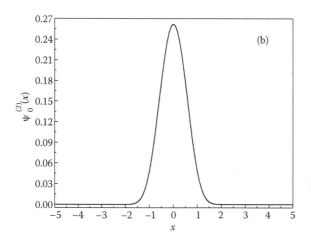

FIGURE 2.6   (a) Energy evolution for adiabatic passage in sector 2 is shown. (b) Ground state $\psi_0^{(2)}(x)$ of $H^{(2)}$, obtained through adiabatic passage from $\psi_0^{(1)}(x)$, is displayed.

the partner Hamiltonians $H^{(0)}$ and $H^{(1)}$ are easily constructed by noting that

$$V^{(0)} = \frac{1}{2}\left(x^6 + 2x^4 + 2x^3 e^x + 2xe^x + x^2 + e^{2x} - 3x^2 - e^x - 1\right) \qquad (2.29)$$

and

$$V^{(1)} = \frac{1}{2}\left(x^6 + 2x^4 + 2x^3 e^x + 2xe^x + x^2 + e^{2x} + 3x^2 + e^x + 1\right). \qquad (2.30)$$

Using the appropriate charge operators, we can write

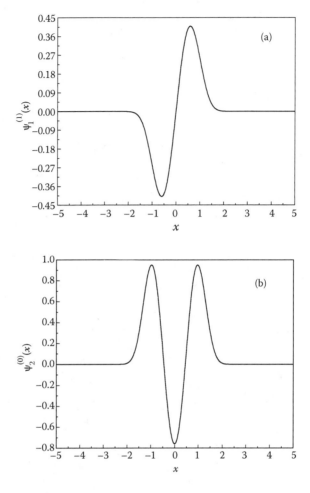

FIGURE 2.7 (a) First excited state $\psi_1^{(1)}(x)$ of $H^{(1)}$, obtained from the operation of Q_1^+ on $\psi_0^{(2)}(x)$, is shown. (b) Second excited state $\psi_2^{(0)}(x)$ of $H^{(0)}$, obtained from the operation of Q_0^+ on $\psi_1^{(1)}(x)$, is shown.

$$H^{(0)} = \frac{1}{2}\left\{-\frac{d}{dx} + x^3 + x + e^x\right\}\left\{\frac{d}{dx} + x^3 + x + e^x\right\} \tag{2.31}$$

and

$$H^{(1)} = \frac{1}{2}\left\{\frac{d}{dx} + x^3 + x + e^x\right\}\left\{-\frac{d}{dx} + x^3 + x + e^x\right\}. \tag{2.32}$$

The relevant superpotential and the SUSY partner potentials are displayed in Figure 2.8a and b, respectively.

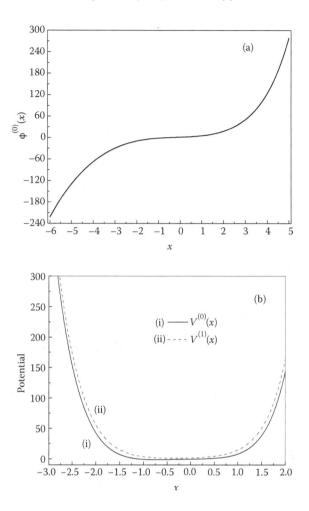

FIGURE 2.8 (a) Exponential superpotential of the chosen anharmonic oscillator is shown. (b) Corresponding SUSY partner potentials $V^{(0)}(x)$ and $V^{(1)}(x)$ are displayed.

The adiabatic switching from the ground state of $H^{(0)}$ to that of $H^{(1)}$, performed with $\lambda(t)$ of Equation 2.27, ran into numerical difficulties when the sixth-order Runge–Kutta method was used to integrate the TDFGH evolution equations into the grid-point amplitudes [the norm of $\psi(t)$ diverged as $t \to T$]. The problem could be cured by integrating the same equations with the third-order Adams–Moulton predictor–corrector method, with time-step size $\Delta t = 0.000001$ a.u.

With the Adams–Moulton predictor–corrector method as an integrator, the energy evolution profile along the switching path ($E(t)$) is shown in Figure 2.9a, whereas the mean square fluctuation in energy is displayed in Figure 2.9b. The energy $E(t)$ smoothly interpolates between the initial eigenstate and the final eigenstate. The mean square fluctuations in $E(t)$, however, oscillate around the zero value with very

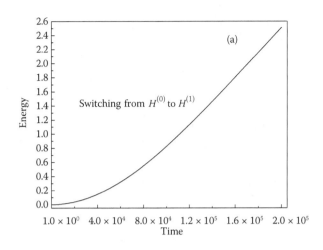

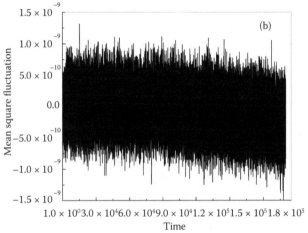

FIGURE 2.9 (a) Energy evolution profile for adiabatic passage from the eigenstate of $H^{(0)}$ to that of $H^{(1)}$. (b) Mean square fluctuations in energy along the switching path as the system adiabatically passes from the ground state of $H^{(0)}$ to that of $H^{(1)}$.

small (near-zero) amplitudes. The adiabaticity of the transition is therefore ensured, albeit at higher computational cost.

The wave functions at $t = 0(\psi(0))$ and $t = T(1 \times 10^5 \text{ a.u})(\psi(T))$ are also displayed in Figure 2.10a and b, respectively. They represent the ground-state wave functions of $H^{(0)}$ and $H^{(1)}$. The first excited-state wave function $\psi_1^{(0)}$ of $H^{(0)}$ was constructed by forming $Q_0^+\psi_0^{(1)}$ and is displayed in Figure 2.10c. The corresponding energy values $E_0^{(0)} = -3.67 \times 10^{-13}$ a.u. (0.0 a.u.) and $E_1^{(0)} = 2.63153681$ a.u. (2.63153663 a.u.) are quite accurate when compared with the corresponding values obtained by diagonalization (the values reported in parentheses).

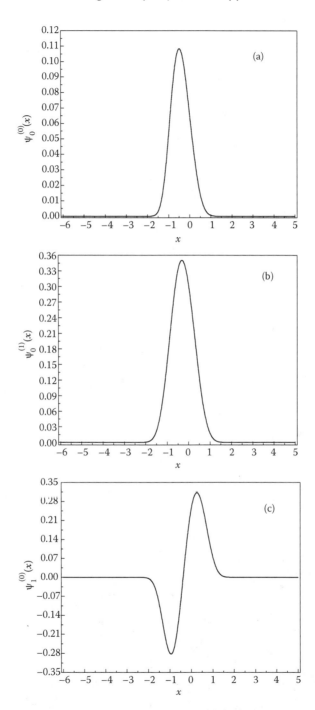

FIGURE 2.10 (a) Ground state $\psi_0^{(0)}(x)$ of $H^{(0)}$ along the coordinate axis. (b) Ground state $\psi_0^{(1)}(x)$ of $H^{(1)}$ obtained through adiabatic switching. (c) First excited state $\psi_1^{(0)}(x)$ of $H^{(0)}$.

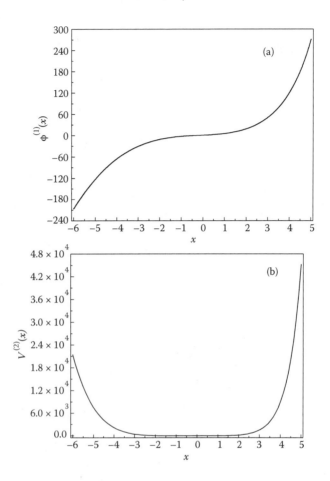

FIGURE 2.11 (a) Exponential superpotential for sector 2 is plotted. (b) SUSY partner potential $V^{(2)}(x)$ for sector 2 is displayed.

For the next sector, the computed superpotential $\varphi^{(1)}(x)$ is depicted in Figure 2.11a, and the partner potential $V^{(2)}$ is depicted in Figure 2.11b. Adiabatic switching from the ground state of $H^{(1)}$ to that of $H^{(2)}$ generates $\psi_0^{(2)}(x)$ of Figure 2.12a, whereas the second excited state $\psi_2^{(0)}(x)$ of $H^{(0)}$ obtained by the action of charge operators Q_0^+ and Q_1^+ on $\psi_0^{(2)}(x) \left(\left| \psi_2^{(0)}(x) \right\rangle = Q_0^+ Q_1^+ \left| \psi_0^{(2)}(x) \right\rangle \right)$ is depicted in Figure 2.12b.

From our experience with the two NLOs, we find that the choice of the scheme for numerical integration and differentiation is critical for the proposed method to succeed, and more work is needed to perfect the numerical implementation of the sequential scheme involving time-dependent adiabatic switching.

Although the applications presented here are simple, they demonstrate the workability of a general principle and a numerical strategy that should be extendable to higher dimensions or multiparticle systems. Andrianov et al. (1993, 1995) and Cannata et al. (2002) experimented with higher-order charge operators, whereas Kravchenko (2005) exploited Clifford algebra for developing SUSY QM in more

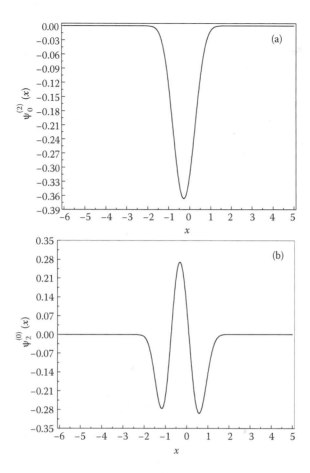

FIGURE 2.12 (a) Ground state $\psi_0^{(2)}(x)$ of $H^{(2)}$, obtained through adiabatic passage from $H^{(1)}$, laid along the x-axis. (b) Second excited state $\psi_2^{(0)}(x)$ of $H^{(0)}$, obtained from sequential operation of Q_1^+ and Q_0^+ on $\psi_0^{(2)}(x)$, is shown along the x-axis.

than one dimension. Kouri et al. (2010) suggested a generalized technique that leads one to work with an alternating series of scalar and tensor Hamiltonians but enables one to extend SUSY QM to arbitrary dimensionality or number of distinguishable particles. To our mind, it appears to be the most promising among the multidimensional SUSY methods. We would remain curious to see if Kouri et al.'s work can be interfaced with the strategy proposed here so that multidimensional systems can be handled successfully in a time-dependent framework. Very recently, Chou et al. (2012) has applied the time-dependent quantum adiabatic switching method to multidimensional systems and it appears quite encouraging. We are presently working on a time-independent adiabatic switching method for generating the excited eigenstates of $H^{(0)}$ and hope to return to the problem elsewhere.

ACKNOWLEDGMENT

S.K. would like to thank the Council of Scientific and Industrial Research, Government of India, for a senior research fellowship.

REFERENCES

Adhikari, S., and S. P. Bhattacharyya. 1992. Dissociation dynamics of a model diatomic species in an intense pulsed laser field: a time dependent Fourier grid Hamiltonian approach. *Physics Letters A* 172:155.

Adhikari, S., P. Dutta, and S. P. Bhattacharyya. 1992. A time-dependent Fourier grid Hamiltonian method: formulation and application to the multiphoton dissociation of a diatomic molecule in intense laser field. *Chemical Physics Letters* 199:574.

Andrianov, A. A., M. V. Ioffe, and V. P. Spiridonov. 1993. Higher-derivative supersymmetry and the Witten index. *Physics Letters A* 174:273.

Andrianov, A. A., M. V. Ioffe, and D. N. Nishnianidze. 1995. Polynomial SUSY in quantum mechanics and second derivative Darboux transformations. *Physics Letters A* 201:103.

Balint-Kurti, G. G., and C. L. Ward. 1991. Two computer programs for solving the Schrödinger equation for bound-state eigenvalues and eigenfunctions using the Fourier grid Hamiltonian method. *Computer Physics Communications* 67:285.

Bender, C. M., and T. T. Wu. 1969. Anharmonic oscillator. *Physical Review* 184 (5):1231.

Bittner, E. R., J. B. Maddox, and D. J. Kouri. 2009. Supersymmetric approach to excited states. *Journal of Physical Chemistry A* 113 (52):15276.

Born, V. M., and V. Fock. 1928. Beweis des Adiabatensatzes. *Zeitschrift fuer Physik* 51 (3):165.

Cannata, F., M. V. Ioffe, and D. N. Nishnianidze. 2002. New methods for the two-dimensional Schrödinger equation: SUSY—separation of variables and shape invariance. *Journal of Physics A: Mathematical and General* 35:1389.

Child, M. S., and R. T. Lawton. 1981. Local and normal vibrational states: a harmonically coupled anharmonic-oscillator model. *Faraday Discussions of the Chemical Society* 71:273.

Chou, C.-C., T. Markovich, and D. J. Kouri. 2012. Adiabatic switching approach to multidimensional supersymmetric quantum mechanics for several excited states. *Molecular Physics: An International Journal at the Interface between Chemistry and Physics* 110 (23):2977–2986.

Cooper, F., A. Khare, and U. P. Sukhatme. 1995. Supersymmetry and quantum mechanics. *Physics Reports* 251:267.

Cooper, F., A. Khare, and U. P. Sukhatme 2001. *Supersymmetry and Quantum Mechanics*. Singapore: World Scientific.

Gol'fand, Y. A., and E. P. Likhtman. 1971. Extension of the algebra of Poincare group generators and violation of *P* invariance. *Journal of Experimental and Theoretical Physics Letters* 13:323.

Hazra, R. K., M. Ghosh, and S. P. Bhattacharyya. 2008. Quantum adiabatic switching route to the impurity modulated states of 2-D quantum dots with different switching functions. *International Journal of Quantum Chemistry* 108 (4):719.

Hendeković, J. 1982. On the energy variation method. *Chemical Physics Letters* 90 (3):198.

Kar, S., and S. P. Bhattacharyya. 2012. Exploring sequential quantum adiabatic switching across supersymmetric partners for finding the eigenstates of a system. *International Journal of Quantum Chemistry* 112:2463–2474.

Kato, T. 1950. On the adiabatic theorem of quantum mechanics. *Journal of the Physical Society of Japan* 5:435.

Kirtman, B. 1992. Nonlinear optical properties of conjugated polymers from ab initio finite oligomer calculations. *International Journal of Quantum Chemistry* 43 (1):147.

Kouri, D. J., T. Markovich, N. Maxwell, and E. R. Bittner. 2009. Supersymmetric quantum mechanics, excited state energies and wave functions, and the Rayleigh–Ritz variational principle: a proof of principle study. *Journal of Physical Chemistry A* 113 (52):15257.

Kouri, D. J., K. Maji, T. Markovich, and E. R. Bittner. 2010. New generalization of supersymmetric quantum mechanics to arbitrary dimensionality or number of distinguishable particles. *Journal of Physical Chemistry A* 114 (32):8202.

Kravchenko, V. V. 2005. On the reduction of the multidimensional stationary Schrödinger equation to a first-order equation and its relation to the pseudoanalytic function theory. *Journal of Physics A: Mathematical and General* 38:851.

Lehmann, K. K. 1992. Harmonically coupled, anharmonic oscillator model for the bending modes of acetylene. *Journal of Chemical Physics* 96:8117.

Marston, C. C., and G. G. Balint-Kurti. 1989. The Fourier grid Hamiltonian method for bound state eigenvalues and eigenfunctions. *Journal of Chemical Physics* 91 (6):3571.

McCoy, B. J. 1984. Compensation effect in the dissociation of anharmonic oscillators: a model for catalysis. *Journal of Chemical Physics* 80:3629.

Merzbacher, E. 1970. *Quantum Mechanics*, 2nd edn. New York: Wiley.

Messiah, A. 1962. *Quantum Mechanics*, vol. II. Amsterdam: North-Holland Publishing Company.

Morse, P. M. 1929. Diatomic molecules according to the wave mechanics: II. Vibrational levels. *Physical Review* 34:57.

Nandy, S., P. Chaudhury, and S. P. Bhattacharyya. 2002. Stochastic diagonalization of Hamiltonian: a genetic algorithm–based approach. *International Journal of Quantum Chemistry* 90:188.

Nandy, S., R. Sharma, and S. P. Bhattacharyya. 2011. Solving symmetric eigenvalue problem via genetic algorithms: serial versus parallel implementation. *Applied Soft Computing* 11 (5):3946.

Pöschl, G., and E. Teller. 1933. Bemerkungen zur quantenmechanik des anharmonischen oszillators. *Zeitschrift fuer Physik* 83 (3–4):143.

Takahashi, A., and S. Mukamel. 1994. Anharmonic oscillator modeling of nonlinear susceptibilities and its application to conjugated polymers. *Journal of Chemical Physics* 100:2366.

3 Isomorphic Local Hardness and Possible Local Version of Hard–Soft Acids–Bases Principle

Carlos Cárdenas and Patricio Fuentealba

CONTENTS

3.1 INTRODUCTION

The concept of chemical hardness was introduced in the 1960s by Pearson to explain reactivity preferences in certain acid–base reactions. He noted that, besides the principle establishing that strong acids–bases prefer binding to strong bases–acids, there should be a second tiebreaker principle explaining the reactivity preferences of acids and bases of similar strength. Pearson then postulated the hard–soft acids–bases (HSAB) principle, which states that when the strengths of acids and bases (Lewis acids and bases) involved a reaction are similar, hard acids prefer binding to hard bases, and soft acids prefer binding to soft bases. In his original works, the hardness of a molecule was not sharply defined, but the principal characteristics of hard and soft species were outlined.[1–6] It was observed that the reactive site of hard species tends to be small, highly charged, and barely polarizable, whereas the reactive site of a soft reagent tends be large, less charged, and highly polarizable. At that time, no operative (mathematical) definition or numerical scale was given for hardness. However, in the late 1970s, Parr and Yang,[7] Chermette,[8] Ayers et al.,[9] Chattaraj ·

65

et al.,[10] Gazquez,[11] and Liu[12] began a search for a common physical and mathematical basis for basic concepts in chemistry within the framework of density functional theory. The first achievement in this direction was the association of the negative of electronegativity χ[13] with Lagrange's multiplier in the Euler equation of the density functional theory μ[14]

$$\mu = \left(\frac{\partial E[N, v(\mathbf{r})]}{\partial N} \right)_{v(r)} = -\chi, \tag{3.1}$$

where $E[N,v(r)]$ is the energy functional, and N and $v(r)$ stand for the number of electrons and the external potential, respectively. The chemical potential μ measures the tendency of electrons to escape from a molecule.

It was not until 5 years later that Parr and Pearson[15] defined absolute hardness η as the second derivative of energy with respect to the number of electrons

$$\eta = \left(\frac{\partial^2 E}{\partial N^2} \right)_{v(r)} = \left(\frac{\partial \mu}{\partial N} \right)_{v(r)} = \frac{1}{S}, \tag{3.2}$$

where the inverse of hardness is softness S. In chemistry, η is the resistance of electronegativity to change when electron transfer occurs. Comparison with experimental data reveals that Equation 3.2 coincides with experimental scales of hardness.[15–17] This definition of hardness has brought substantial comprehension of the HSAB principle from a theoretical point of view.[18–22]

Note that absolute hardness is a global (number) property of molecules. It says "nothing" with respect to the identity of an active site. A natural next question would be whether there exists a local version (position dependent) of the HSAB principle, that is, can one associate values of softness and hardness with different sites of the molecule? If so, can one postulate that the hard–soft site of an acid prefers binding to the hard–soft site of a base? There certainly seems to be such a property called *local hardness*. A clear example of this is the case of ambidentate molecules such as the anion SCN$^-$. This ligand can bind to a metal center with either nitrogen or sulfur atoms depending on the hardness of the metal center: it binds to hard cations (Mg^{2+}, Al^{3+}, etc.) with the nitrogen atom and to soft cations (Br$^+$, Ag$^+$, etc.) with the sulfur atom. Therefore, a definition of local hardness consistent with the HSAB principle must predict that N is harder than S. Although much has been done from a theoretical point of view,[23–29] the existence of such a local HSAB principle remains unclear and debatable.[30–32]

3.2 DIFFICULTIES IN FINDING SUITABLE DEFINITION FOR LOCAL HARDNESS

Any theoretical approach to the local HSAB principle needs a mathematical definition of local hardness and local softness consistent with the principle itself. The definition of local softness seems not to be a problem. Yang and Parr[33] defined local softness as

$$s(\mathbf{r}) = \left(\frac{\partial_{\rho}(\mathbf{r})}{\partial N} \right)_{v(r)} = -\left(\frac{\delta N}{\delta v(\mathbf{r})} \right)_{\mu} = sf(\mathbf{r}), \tag{3.3}$$

where $f(\mathbf{r})$ is the Fukui function.[34–36] Using this definition, Berkowitz[37] showed that soft–soft interactions ease electron transfer and are more significant when hardness is small. In other words, soft–soft interactions are indicative of covalent bonding.

In analogy to the inverse relationship between hardness and softness ($\eta s = 1$), it is desirable that local hardness and local softness be related following a certain inverse rule. Local hardness should be defined such that

$$\int s(\mathbf{r})\eta(\mathbf{r})d\mathbf{r} = 1 \tag{3.4}$$

holds. However, all attempts to define local hardness that fulfill Equation 3.4 yield ambiguous or ill-conditioned definitions.

Berkowitz[37] proposed the first definition of local hardness to be

$$\eta(\mathbf{r}) = \left(\frac{\delta\mu}{\delta\rho(\mathbf{r})} \right)_{v(r)}. \tag{3.5}$$

Then Harbola et al.[38] and Ghosh[39] gave a more extensive definition based on the hardness kernel $\eta(\mathbf{r}, \mathbf{r}')$

$$\eta(\mathbf{r}) = \int \lambda(\mathbf{r}')\eta(\mathbf{r},\mathbf{r}')d\mathbf{r}' = \int \lambda(\mathbf{r}') \frac{\delta^2 F[\rho(\mathbf{r}')]}{\delta\rho(\mathbf{r})\delta\rho(\mathbf{r}')} d\mathbf{r}'. \tag{3.6}$$

The last term in this equation defines the hardness kernel

$$\eta(\mathbf{r},\mathbf{r}') = \frac{\delta^2 F[\rho(\mathbf{r})]}{\delta\rho(\mathbf{r})\delta\rho(\mathbf{r}')} = \frac{\delta(\mu - v(\mathbf{r}))}{\delta\rho(\mathbf{r})} = \frac{\delta(u(\mathbf{r}))}{\delta\rho(\mathbf{r})}, \tag{3.7}$$

where $u(\mathbf{r})$ is the generalized potential. Both definitions (Equations 3.5 and 3.6) are ambiguous. The first is ambiguous because the restriction on the derivative limits $\delta\rho(\mathbf{r})$ only to variations in density that change N. In other words, the Hohenberg–Kohn theorem says that is not possible to have a general variation in density if the external potential is kept unchanged. Similarly, the definition of Equation 3.6 is ambiguous because it holds for any function $\lambda(\mathbf{r})$ such that

$$\int \lambda(\mathbf{r})d\mathbf{r} = 1. \tag{3.8}$$

Arguments have been given to prefer some choices to others.[26–29,40] Ayers and Parr proposed to remove the constraint in Equation 3.5 to solve the ambiguity

$$\eta(\mathbf{r}) = \frac{\delta\mu}{\delta\rho(\mathbf{r})}. \tag{3.9}$$

However, it has been recently shown that this quantity is seriously ill defined.[41] It is worth mentioning that there are other useful definitions of local hardness that do not satisfy the inverse condition (Equation 3.4).[42-45] This should not be considered as a weakness of these models but as a signal that the inverse condition might not be enough to reify the chemical concept of local hardness.

3.3 LOCAL HARDNESS IN AN ISOMORPHIC ENSEMBLE

In isomorphic representation, the state of a molecular system is characterized by the variables $\sigma(\mathbf{r})$ and N, where $\sigma(\mathbf{r})$ is the shape factor of the electron density $\sigma(\mathbf{r}) = \frac{\rho(\mathbf{r})}{N}$.[46-56] The mathematical properties of reactivity indicators in an isomorphic ensemble have been studied elsewhere.[24,57] We are not going to go deep into these details but will focus on a natural definition of local hardness that arises in this ensemble. In the late 1990s, De Proft et al.[58] proposed isomorphic total local hardness to be

$$\eta^\sigma(\mathbf{r}) = \left(\frac{\partial\mu(\mathbf{r})}{\partial N}\right)_\sigma = \left(\frac{\partial\mu}{\partial N}\right)_\sigma - \left(\frac{\partial v(\mathbf{r})}{\partial N}\right)_\sigma = \eta^\sigma - g(\mathbf{r}), \tag{3.10}$$

which they claimed to be not ambiguous. The function $g(\mathbf{r})$, which is the only position-dependent contribution to $\eta^\sigma(\mathbf{r})$, had already been interpreted by Baekelandt et al.[24] as a "disconnectivity" index in the sense that it measures how the point $\mathbf{r}$ is connected or disconnected to the system as N changes. In other words, if a set of points $\{\mathbf{r}\}_\alpha$ is associated with an atom α (this corresponds to atoms in molecule representation), then $g_\alpha = \int_{\mathbf{r}\in\{\mathbf{r}\}_\alpha} g(\mathbf{r})d\mathbf{r}$ measures how the atom α is pulled apart from the system when N changes. If g_α is positive, the average distance of the atom α to the rest of the atoms decreases with increasing number of electrons. Observe that this quantity is more related to what happens to the nuclei when electron transfer occurs than what happens to electron distribution. As we shall see, one can say that the local hardness defined in Equation 3.10 is linked to the idea of conceiving a chemical reaction as either electron following or electron preceding.[59-61]

The claim that isomorphic local hardness is not ambiguous is not necessary true. Note that it is very likely that the derivative $\left(\frac{\partial v(\mathbf{r})}{\partial N}\right)_{\sigma(r)}$ does not necessary exist. The cusp condition of density in the nuclei[62-64] is inherited by the shape factor.[46] Hence, if the position of atoms (external potential) changes upon electron transfer, so must the position of the cusp. Therefore, keeping the shape factor constant while changing the position of the atoms gives rise to the ambiguity. Note that, in our analysis, we assume no external potential different from that produced by the nuclei. Thus, the change in the external potential is due only to the changes in the position of the atoms while the number and identity of these remain constant. This seems to be the only chemically relevant case.

3.4   RELATIONSHIP WITH NUCLEAR REACTIVITY DESCRIPTORS IN ISOMORPHIC ENSEMBLE

In this section, we will show that isomorphic local hardness is linked to nuclear reactivity descriptors in the isomorphic ensemble. This relationship is given by[65,66]

$$\sum_{\alpha} \vec{\eta}_{\alpha}(\mathbf{r}) \cdot \left(\frac{\partial \mathbf{F}_{\alpha}}{\partial N} \right)_{\sigma} = \eta^{\sigma}(\mathbf{r}), \tag{3.11}$$

where $\vec{\eta}_{\alpha}(\mathbf{r})$ stands for the nuclear hardness kernel,[65–68] and $\mathbf{F}_{\alpha}$ is the force on the nucleus α. The proof is simple (a more detailed deduction is available in the studies of Cardenas[45] and Ayers[46]). In the isomorphic ensemble, the variation in the force due to a small change in the number of electrons and in the shape factor of the density is

$$\delta \mathbf{F}_{\alpha} = \left(\frac{\partial \mathbf{F}_{\alpha}}{\partial N} \right)_{\sigma} dN + \int \left(\frac{\delta \mathbf{F}_{\alpha}}{\delta \sigma(\mathbf{r})} \right)_{N} \delta \sigma(\mathbf{r}) d\mathbf{r}, \tag{3.12}$$

whereas the variation in the generalized potential is

$$\delta u(\mathbf{r}) = \left(\frac{\partial u(\mathbf{r})}{\partial N} \right)_{\sigma} dN + \int \left(\frac{\delta u(\mathbf{r})}{\delta \sigma(\mathbf{r}_2)} \right)_{N} \delta \sigma(\mathbf{r}_2) d\mathbf{r}_2$$

$$= \eta^{\sigma}(\mathbf{r}) dN + \int \left(\frac{\delta u(\mathbf{r})}{\delta \sigma(\mathbf{r}_2)} \right)_{N} \delta \sigma(\mathbf{r}_2) d\mathbf{r}_2. \tag{3.13}$$

The variation in $u(\mathbf{r})$ can also be written in the canonical ensemble, where the fundamental variables are N and $v(\mathbf{r})$,

$$-\delta u(\mathbf{r}) = \eta dN - \int \delta(\mathbf{r} - \mathbf{r}_2) \delta v(\mathbf{r}_2) d\mathbf{r}_2 - \int \left(\frac{\partial \rho(\mathbf{r})}{\partial N} \right)_{v(r)} \delta v(\mathbf{r}) d\mathbf{r} \tag{3.14}$$

and as a total differential

$$\delta u(\mathbf{r}) = \sum_{\alpha} \frac{\delta u(\mathbf{r})}{\delta \mathbf{F}_{\alpha}} \delta \mathbf{F}_{\alpha} = -\sum \vec{\eta}_{\alpha}(\mathbf{r}) \delta \mathbf{F}_{\alpha}, \tag{3.15}$$

where $\vec{\eta}_\alpha(\mathbf{r})$ is the so-called nuclear hardness kernel of the atom α.[67] Replacing $\delta\mathbf{F}_\alpha$ from Equation 3.10 with Equation 3.15 and comparing with Equation 3.14, one derives, among other identities,

$$\sum_\alpha \vec{\eta}_\alpha(\mathbf{r}) \cdot \left(\frac{\partial\mathbf{F}_\alpha}{\partial N}\right)_\sigma = \eta^\sigma(\mathbf{r}), \tag{3.16}$$

which is Equation 3.11. Note that $\vec{\eta}_\alpha(\mathbf{r})$ is a vector field and $\left(\dfrac{\partial\mathbf{F}_\alpha}{\partial N}\right)_\sigma$ is a vector. Hence, in this equation, the product is a scalar product.

It is possible to write down an explicit expression for Equation 3.16 in terms of the properties of the potential energy surface and the shape factor. As might be expected, this expression is rather complicated to perform calculations easily from it. However, for molecules in their stable structure, evaluating $\left(\dfrac{\partial\mathbf{F}_\alpha}{\partial N}\right)_\sigma$ is rather simple.

If one wants to calculate local hardness from Equation 3.11, one should first find the nuclear hardness kernel

$$\vec{\eta}_\alpha(\mathbf{r}) = -\frac{\delta u(\mathbf{r})}{\delta\mathbf{F}_\alpha} = \frac{\delta\mu}{\delta\mathbf{F}_\alpha} - \frac{\delta v(\mathbf{r})}{\delta\mathbf{F}_\alpha}. \tag{3.17}$$

The derivative $\dfrac{\delta}{\delta\mathbf{F}_\alpha}$ here has the meaning of a gradient

$$\frac{\delta}{\delta\mathbf{F}_\alpha} = \frac{\partial}{\partial f_{\alpha x}}, \frac{\partial}{\partial f_{\alpha y}}, \frac{\partial}{\partial f_{\alpha y}}, \tag{3.18}$$

where F_α has been written in terms of its component in the x, y, and z directions. The term $\dfrac{\delta\mu}{\delta\mathbf{F}_\alpha}$ is an unconstrained derivative that can be easily evaluated numerically if one keeps N constant. Besides, the i-esime component of $\dfrac{\delta v(\mathbf{r})}{\delta\mathbf{F}_\alpha}$ can be simplified using the chain rule

$$\frac{\partial v(\mathbf{r})}{\partial f_{\alpha i}} = \sum_\beta^{N \text{ atoms}} \sum_{j=x,y,z} \frac{\partial v(\mathbf{r})}{\partial R_{\beta j}}\frac{\partial R_{\beta j}}{\partial f_{\alpha i}}, \tag{3.19}$$

where $\mathbf{R}_\alpha$ is the position of the nucleus α. Note that $\dfrac{\partial R_{\beta i}}{\partial f_{\alpha j}}$ is nothing but the negative of the inverse of the element $(i,j;\alpha,\beta)$ of the Hessian matrix $k_{ji}^{\alpha\beta}$

$$\frac{\partial R_{\beta i}}{\partial f_{\alpha j}} = \frac{1}{\dfrac{\partial f_{\alpha j}}{\partial R_{\beta j}}} = -\frac{1}{k_{ji}^{\alpha\beta}}. \tag{3.20}$$

Hence,

$$\frac{\partial v(\mathbf{r})}{\partial f_{\alpha i}} = \sum_{\beta} \sum_{j=x,y,z} -\frac{\partial v(\mathbf{r})}{\partial R_{\beta j}} \frac{1}{k_{ji}^{\alpha \beta}} \cdot \tag{3.21}$$

The sum must exclude the elements of the Hessian that correspond to translation and rotation, which are zero.

It is convenient to write the last equation in matrix notation, which reads

$$\frac{\partial v(\mathbf{r})}{\partial \mathbf{F}_{\alpha}} = -\sum_{\beta} \frac{\partial v(\mathbf{r})}{\partial R_{\beta}} K_{\alpha,\beta}^{-1}, \tag{3.22}$$

where $\mathbf{K}_{\alpha\beta}^{-1}$ is the matrix with elements $1 / k_{ji}^{\alpha\beta}$

$$\mathbf{K}_{\alpha\beta}^{-1} = \begin{pmatrix} k_{\alpha\beta,xx}^{-1} & k_{\alpha\beta,xy}^{-1} & k_{\alpha\beta,xz}^{-1} \\ k_{\alpha\beta,yx}^{-1} & k_{\alpha\beta,yy}^{-1} & k_{\alpha\beta,yz}^{-1} \\ k_{\alpha\beta,zx}^{-1} & k_{\alpha\beta,zy}^{-1} & k_{\alpha\beta,zz}^{-1} \end{pmatrix}, \tag{3.23}$$

and $\dfrac{\partial v(\mathbf{r})}{\partial \mathbf{R}_{\alpha}}$ is just the negative of the electric field of the nuclei β

$$\frac{\partial v(\mathbf{r})}{\partial \mathbf{R}_{\beta}} = -\frac{\partial}{\partial \mathbf{R}_{\beta}} \sum_{\gamma} \frac{Z\gamma}{|\mathbf{r} - \mathbf{R}_{\gamma}|} = -\frac{Z_{\beta}(\mathbf{r} - \mathbf{R}_{\beta})}{|\mathbf{r} - \mathbf{R}_{\beta}|^3} \cdot \tag{3.24}$$

Assembling Equations 3.22 through 3.24 all together, the nuclear hardness kernel is written explicitly as

$$\vec{\eta}_{\alpha}(\mathbf{r}) = \frac{\delta \mu}{\delta \mathbf{F}_{\alpha}} - \sum_{\beta} \frac{Z_{\beta}(\mathbf{r} - \mathbf{R}_{\beta})}{|\mathbf{r} - \mathbf{R}_{\beta}|^3} K_{\alpha\beta}^{-1}. \tag{3.25}$$

To calculate $\eta^{\sigma}(\mathbf{r})$, we still have to deal with $\left(\dfrac{\partial \mathbf{F}_{\alpha}}{\partial N}\right)_{\sigma}$. The dependence of the force on N is given by the Hellmann–Feynman equation[69]

$$F_{\alpha} = -Z_{\alpha} N \int \frac{\sigma(\mathbf{r})(\mathbf{r} - \mathbf{R}_{\alpha})}{|\mathbf{r} - \mathbf{R}_{\alpha}|^3} d\mathbf{r} + \sum_{\beta \neq \alpha} \frac{Z_{\alpha} Z_{\beta}(\mathbf{R}_{\alpha} - \mathbf{R}_{\beta})}{|\mathbf{R}_{\alpha} - \mathbf{R}_{\beta}|^3} \cdot \tag{3.26}$$

Thus,

$$\left(\frac{\partial \mathbf{F}_{\alpha}}{\partial N}\right)_{\sigma} = -Z_{\alpha} \int \frac{\sigma(\mathbf{r})(\mathbf{r} - \mathbf{R}_{\alpha})}{|\mathbf{r} - \mathbf{R}_{\alpha}|^3} d\mathbf{r}. \tag{3.27}$$

The right side of Equation 3.27 becomes very simple if the molecule is in its equilibrium structure, that is, $\mathbf{F}_\alpha = 0$ for every atom. In such a case, the derivative of the force depends only on the distribution of the nuclei in the molecule

$$\left(\frac{\partial \mathbf{F}_\alpha}{\partial N}\right)_\sigma = -\sum_{\beta \neq \alpha} \frac{Z_\alpha Z_\beta (\mathbf{R}_\alpha - \mathbf{R}_\beta)}{N|\mathbf{R}_\alpha - \mathbf{R}_\beta|^3}. \tag{3.28}$$

With the use of Equations 3.11, 3.25, and 3.28, the isomorphic local hardness for molecules in their equilibrium geometry is

$$\eta^\sigma(\mathbf{r}) = -\sum_\alpha \sum_{\beta \neq \alpha} \frac{\delta \mu}{\delta \mathbf{F}_\alpha} \frac{Z_\alpha Z_\beta (\mathbf{R}_\alpha - \mathbf{R}_\beta)}{N|\mathbf{R}_\alpha - \mathbf{R}_\beta|^3}$$
$$+ \sum_\alpha \sum_{\beta \neq \alpha} \left(\sum_\gamma \frac{Z_\gamma (\mathbf{r} - \mathbf{R}_\gamma)}{|\mathbf{r} - \mathbf{R}_\gamma|^3} K_{\alpha\gamma}^{-1} \right) \frac{Z_\alpha Z_\beta (\mathbf{R}_\alpha - \mathbf{R}_\beta)^T}{N|\mathbf{R}_\alpha - \mathbf{R}_\beta|^3}. \tag{3.29}$$

It is surprising that one can arrive to an expression for local hardness that has no explicit information on the resistance of density to change, that is, the only dependence on the position is given by the electric field attributable to the nuclei. Of course, the position of the nuclei dictates all the properties of the system. However, all "local electronic properties" are hidden in Equation 3.29. Indeed, besides the nature and position of the nuclei, the only "electronic parameters" are the chemical potential, the number of electrons, and the Hessian. It is then expected that isomorphic local hardness will be another failed attempt to define local hardness consistent with a local HSAB principle. To get more insight on the meaning of $\eta^\sigma(\mathbf{r})$, we have computed it for a few cases of diatomic molecules. The Hessian and the chemical potential have been evaluated at the B3LYP/6-311++G(d,p) level. $\dfrac{\delta\mu}{\delta\mathbf{F}_\alpha}$ has been obtained from a Taylor series of the function $\mu(f_{\alpha,x}, f_{\alpha,y}, f_{\alpha,z})$ using 11 points very close to the geometry of equilibrium to evaluate the forces.

3.5 RESULTS AND DISCUSSION

Figure 3.1 shows contour plots of $\eta^\sigma(\mathbf{r})$ on a plane 2.0 a.u. above the molecular plane of diatomic molecules H_2, HF, CO, and BeO. The first thing to note is that heteronuclear molecules have significant regions with negative values $\eta^\sigma(\mathbf{r})$. Global hardness is always positive, and all local hardness models we are aware of are positive semidefinite. The negativeness of $\eta^\sigma(\mathbf{r})$ can easily yield negative values when one tries to recover global hardness from the inverse relation (Equation 3.4). As the Fukui function is positive in most places,[70–72] the regions where the product $\eta^\sigma(\mathbf{r})f(\mathbf{r})$ is negative could dominate the integral, resulting in a negative value for global isomorphic hardness. Another important detail to note is that $\eta^\sigma(\mathbf{r})$ is less positive in regions between the atoms. This is closely related to Berlin's binding regions.[65,73–79] Berlin[73] proposed that a molecule can be divided into binding and antibinding regions. Binding regions

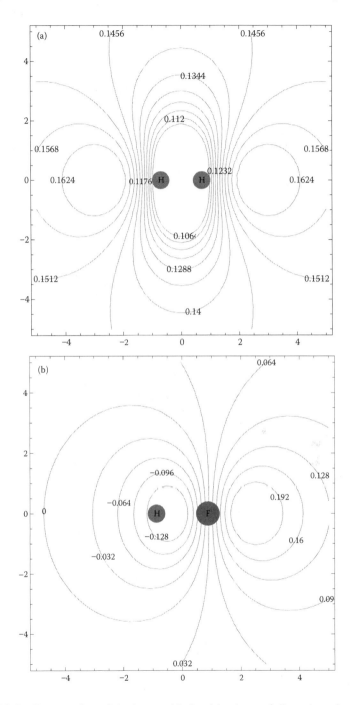

FIGURE 3.1   Contour plots of the isomorphic local hardness of diatomic molecules. The contour plane is 2.0 a.u. above the molecular plane. (a) H_2, (b) HF, (c) CO, and (d) BeO. See the text for details of calculations. All values are expressed in atomic units.

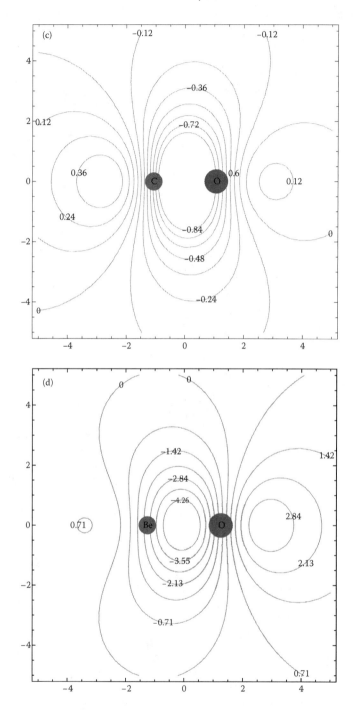

FIGURE 3.1 (*continued*) Contour plots of the isomorphic local hardness of diatomic molecules. The contour plane is 2.0 a.u. above the molecular plane. (a) H_2, (b) HF, (c) CO, and (d) BeO. See the text for details of calculations. All values are expressed in atomic units.

correspond to those places where adding an electron decreases the average distance of atoms from their center of charge. In diatomic molecules, this means that if an electron is added to or removed from a binding/antibinding region, the bond distance decreases. The same picture can be recovered from $\eta^\sigma(\mathbf{r})$. If electrons are transferred to regions where $\left(\dfrac{\partial v(\mathbf{r})}{\partial N}\right)_\sigma$ is positive, the external potential will decrease in those regions. This then implies that $v(\mathbf{r})$ decreases more in the regions where $\eta^\sigma(\mathbf{r})$ becomes less positive (more negative). That $v(\mathbf{r})$ decreases in a given region means that the nuclei tend to move to that region because $v(\mathbf{r})$ is an attractive potential. Therefore, one could say that the regions where $\eta^\sigma(\mathbf{r})$ is more negative are binding regions and vice versa.

In heteronuclear molecules, there could be regions behind the lightest atom with a slight tendency of $\eta^\sigma(\mathbf{r})$ to decrease. This effect is more pronounced when the difference between atomic numbers is large. In terms of binding and antibinding regions, $\eta^\sigma(\mathbf{r})$ shows that, in all cases, "internuclear" regions correspond to binding regions, in agreement with chemical intuition. However, the region behind H in HF is predicted to have a small tendency for binding. This same behavior is observed and much more pronounced in Berlin's function. Finally, it is well known that Berlin's binding function is not invariable to the position of origin. It is possible, for instance, to pick an origin so that the binding regions of diatomic molecules are always behind the nuclei. Contrarily, $\eta^\sigma(\mathbf{r})$ is invariable to the position of origin.[80]

3.6 CONCLUSIONS

In the pursuit of finding a theoretical approach to a local HSAB principle, definitions of local hardness and softness are necessary. The obvious condition is that these definitions have to be consistent with the principle itself. Besides, there should be an inverse relationship between local softness and local hardness. This inverse relationship has long been believed, as given by Equation 3.4. A suitable definition for softness is not a problem. Local softness in Equation 3.3 explains correctly the interaction between soft reactants, which is mostly covalent. On the other hand, all attempts to define nonempirical local hardness derived from the hardness kernel have been "unsuccessful" because all known expressions suffer from ambiguity or are dreadfully ill conditioned. In this work, we used the formalism of nuclear reactivity descriptors to find a closed expression for the isomorphic local hardness proposed by De Proft et al. back in the 1990s. We gave arguments contradicting the original claim that this local hardness is not ambiguous. Based on mathematical analysis and calculations, we conclude that isomorphic local hardness is far from being a satisfactory definition of local hardness consistent with a local HSAB principle. We also conclude that isomorphic local hardness is more closely related to a partition of a molecule in binding and antibinding regions, in analogy with Berlin's original ideas. In other words, this local hardness measures how the nuclei respond upon electron transfer and not how electron distribution resists deforming. Finally, isomorphic local hardness belongs to the family of local descriptors that describe chemical processes as electron following.

ACKNOWLEDGMENTS

This work received support from FONDECYT (grant 11090013) and Financiamiento basal para centros científicos y tecnológicos de excelencia. P.F. received support from Project CILIS-P10-003-F funded by Fondo de Innovacion para la Competividad del Ministerio de Economia Fomento y Turismo, Chile.

REFERENCES

1. Pearson, R. G. *J Am Chem Soc* 1963, *85*, 3533–3539.
2. Pearson, R. G. *Science* 1966, *151*, 172–177.
3. Pearson, R. G. *J Chem Educ* 1968, *45* (9), 581–587.
4. Pearson, R. G. *Chem Commun* 1968, (2), 65–77.
5. Pearson, R. G. *Chem Br* 1967, *3* (3), 103–107.
6. Pearson, R. G.; Songstad, J. *J Am Chem Soc* 1967, *89* (8), 1827–1836.
7. Parr, R. G.; Yang, W. *Density-Functional Theory of Atoms and Molecules*. Oxford University Press, New York, 1989.
8. Chermette, H. *Coord Chem Rev* 1998, *180*, 699–721.
9. Ayers, P. W.; Anderson, J. S. M.; Bartolotti, L. J. *Int J Quantum Chem* 2005, *101*, 520–534.
10. Chattaraj, P. K.; Sarkar, U.; Roy, D. R. *Chem Rev* 2006, *106*, 2065–2091.
11. Gazquez, J. *J Mex Chem Soc* 2008, *52* (1), 3–10.
12. Liu, S. B. *Acta Phys Chim Sin* 2009, *25* (03), 590–600.
13. Mulliken, R. S. *J Chem Phys* 1934, *2*, 782–793.
14. Parr, R. G.; Donnelly, R. A.; Levy, M.; Palke, W. E. *J Chem Phys* 1978, *68*, 3801–3807.
15. Parr, R. G.; Pearson, R. G. *J Am Chem Soc* 1983, *105*, 7512–7516.
16. Pearson, R. *J Chem Sci* 2005, *117* (5), 369–377.
17. Gazquez, J. L. *J Phys Chem A* 1997, *101*, 9464–9469.
18. Gazquez, J. L. *J Phys Chem A* 1997, *101*, 4657–4659.
19. Chattaraj, P. K. *J Phys Chem A* 2001, *105* (2), 511–513.
20. Ayers, P. W. *J Chem Phys* 2005, *122*, 141102.
21. Ayers, P. W.; Parr, R. G.; Pearson, R. G. *J Chem Phys* 2006, *124*, 194107.
22. Ayers, P. W. *Faraday Discuss* 2007, *135*, 161–190.
23. Berkowitz, M.; Ghosh, S. K.; Parr, R. G. *J Am Chem Soc* 1985, *107*, 6811–6814.
24. Baekelandt, B. G.; Cedillo, A.; Parr, R. G. *J Chem Phys* 1995, *103*, 8548–8556.
25. Langenaeker, W.; Deproft, F.; Geerlings, P. *J Phys Chem* 1995, *99*, 6424–6431.
26. Chattaraj, P. K.; Roy, D. R.; Geerlings, P.; Torrent-Sucarrat, M. *Theor Chem Acc* 2007, *118*, 923–930.
27. Fuentealba, P.; Chamorro, E.; Cardenas, C. *Int J Quantum Chem* 2007, *107*, 37–45.
28. Ayers, P. W.; Parr, R. G. *J Chem Phys* 2008, *128*, 184108.
29. Torrent-Sucarrat, M.; De Proft, F.; Ayers, P. W.; Geerlings, P. *Phys Chem Chem Phys* 2010, *12* (5), 1072–1080.
30. Loos, R.; Kobayashi, S.; Mayr, H. *J Am Chem Soc* 2003, *125* (46), 14126–14132.
31. Mayr, H.; Breugst, M.; Ofial, A. R. *Angew Chem Int Ed* 2011, *50* (29), 6470–6505.
32. Breugst, M.; Zipse, H.; Guthrie, J. P.; Mayr, H. *Angew Chem Int Ed* 2010, *49* (30), 5165–5169.
33. Yang, W. T.; Parr, R. G. *Proc Natl Acad Sci* 1985, *82*, 6723–6726.
34. Parr, R. G.; Yang, W. T. *J Am Chem Soc* 1984, *106*, 4049–4050.
35. Fukui, K. *Science* 1987, *218*, 747–754.
36. Ayers, P. W.; Levy, M. *Theor Chem Acc* 2000, *103*, 353–360.
37. Berkowitz, M. *J Am Chem Soc* 1987, *109*, 4823–4825.

38. Harbola, M. K.; Chattaraj, P. K.; Parr, R. G. *Isr J Chem* 1991, *31*, 395–402.
39. Ghosh, S. K. *Chem Phys Lett* 1990, *172*, 77–82.
40. Ayers, P. W.; Liu, S. B.; Li, T. L. *Phys Chem Chem Phys* 2011, *13* (10), 4427–4433.
41. Cuevas-Saavedra, R.; Rabi, N.; Ayers, P. W. *Phys Chem Chem Phys* 2011, *13* (43), 19594.
42. Meneses, L.; Tiznado, W.; Contreras, R.; Fuentealba, P. *Chem Phys Lett* 2004, *383*, 181–187.
43. Gal, T.; Geerlings, P.; De Proft, F.; Torrent-Sucarrat, M. *Phys Chem Chem Phys* 2011, *13* (33), 15003–15015.
44. Jin, P.; Murray, J.; Politzer, P. *Comput Lett* 2007, *2* (4), 373–385.
45. Cardenas, C. *Chem Phys Lett* 2011, *513*, 127–129.
46. Ayers, P. W. *Proc Natl Acad Sci* 2000, *97*, 1959–1964.
47. Borgoo, A.; Godefroid, M.; Sen, K. D.; De Proft, F.; Geerlings, P. *Chem Phys Lett* 2004, *399* (4–6), 363–367.
48. De Proft, F.; Ayers, P. W.; Sen, K. D.; Geerlings, P. *J Chem Phys* 2004, *120*, 9969–9973.
49. Nagy, A.; Amovilli, C. *J Chem Phys* 2004, *121*, 6640–6648.
50. Geerlings, P.; Boon, G.; Van Alsenoy, C.; De Proft, F. *Int J Quantum Chem* 2005, *101* (6), 722–732.
51. Sen, K. D.; De Proft, F.; Borgoo, A.; Geerlings, P. *Chem Phys Lett* 2005, *410* (1–3), 70–76.
52. Ayers, P. W. *Theor Chem Acc* 2006, *115*, 370–378.
53. Ayers, P. W. *Chem Phys Lett* 2007, *438*, 148–152.
54. Ayers, P. W.; De Proft, F.; Geerlings, P. *Phys Rev A* 2007, *75*, 012508.
55. Geerlings, P.; De Proft, F.; Ayers, P. W. *Theor Comput Chem* 2007, *19*, 1–17.
56. Ayers, P.; Cedillo, A. *The Shape Function. Chemical Reactivity Theory: A Density Functionl View*. Chattaraj, P. K., Ed. Taylor and Francis, Boca Raton, 2009, p. 269.
57. Cedillo, A. *Int J Quantum Chem* 1994, *52* (28), 231–240.
58. De Proft, F.; Liu, S. B.; Parr, R. G. *J Chem Phys* 1997, *107*, 3000–3006.
59. Nakatsuji, H. *J Am Chem Soc* 1973, *95*, 345–354.
60. Nakatsuji, H. *J Am Chem Soc* 1974, *96*, 24–30.
61. Nakatsuji, H. *J Am Chem Soc* 1974, *96*, 30–37.
62. Kato, T. *Commun Pure Appl Math* 1957, *10*, 151–177.
63. Steiner, E. *J Chem Phys* 1963, *39*, 2365–2366.
64. Pack, R. T.; Brown, W. B. *J Chem Phys* 1966, *45* (2), 556–559.
65. De Proft, F.; Liu, S. B.; Geerlings, P. *J Chem Phys* 1998, *108*, 7549–7554.
66. Cardenas, C.; Lamsabhi, A. M.; Fuentealba, P. *Chem Phys* 2006, *322*, 303–310.
67. Cohen, M. H.; Ganduglia-Pirovano, M. V. *J Chem Phys* 1994, *101*, 8988–8997.
68. Cohen, M. H.; Ganduglia-Pirovano, M. V.; Kudrnovsky, J. *J Chem Phys* 1995, *103*, 3543–3551.
69. Feynman, R. P. *Phys Rev* 1939, *56*, 340–343.
70. Bultinck, P.; Carbo-Dorca, R. *J Math Chem* 2003, *34*, 67–74.
71. Bultinck, P.; Carbo-Dorca, R.; Langenaeker, W. *J Chem Phys* 2003, *118*, 4349–4356.
72. Bultinck, P.; Clarisse, D.; Ayers, P. W.; Carbo-Dorca, R. *Phys Chem Chem Phys* 2011, *13* (13), 6110–6115.
73. Berlin, T. *J Chem Phys* 1951, *19*, 208–213.
74. Deb, B. *Rev Mod Phys* 1973, *45* (1), 22.
75. Wang, X.; Peng, Z. *Int J Quantum Chem* 1993, *47* (5), 393–404.
76. Koga, T.; Nakatsuji, H.; Yonezawa, T. *J Am Chem Soc* 1978, *100* (24), 7522–7527.
77. Balawender, R.; De Proft, F.; Geerlings, P. *J Chem Phys* 2001, *114* (10), 4441–4449.
78. Chamorro, E.; De Proft, F.; Geerlings, P. *J Chem Phys* 2005, *123*, 084104.
79. Chabraborty, D.; Cardenas, C.; Echegaray, E.; Toro-Labee, A.; Ayers, P. W. *Chem Phys Lett* 2012, *539–540*, 168–171.
80. Autschbach, J.; Schwarz, W. *J Phys Chem A* 2000, *104* (25), 6039–6046.

4 Quantum Chemistry of Highly Symmetrical Molecules and Free-Space Clusters, Plus Almost Spherical Cages of C and B Atoms

N. H. March and G. G. N. Angilella

CONTENTS

4.1   BACKGROUND AND OUTLINE

Some six decades ago, March [1] attempted to expose some regularities in the quantum-chemical properties of highly symmetrical molecules, particularly those with tetrahedral and octahedral geometries. To do so at that time, March found that two major simplifications are essential to making some largely analytical predictions. The first simplification was to employ semiclassical theory in the form of Thomas–Fermi (TF) statistical theory [3]. But the second, and much more drastic, simplification was to use a central field model. While this was appropriate for, say, GeH_4, it was clearly very poor for CCl_4, since it used spherical averaging of the outer nuclei about the "central" atoms (Ge and O, respectively). Nevertheless, as will be displayed in detail in Section 4.3, some valuable fingerprints of the scaling properties of equilibrium Ge–H and C–Cl distances, say R_e, and hence of the nuclear–nuclear electrostatic potential energy V_{nn} were revealed by this admittedly oversimplified model. Then some attention will be given to much more recent work on clusters in Section 4.4. Section 4.5 treats almost spherical C cages, particularly fullerenes C_n with $n = 50$, 60, 70, and 84. Section 4.6 deals quite briefly with B cages and emphasizes in that context the old problem raised by Thomson about the equilibrium configuration of different numbers of point charges constrained to lie on the surface of a sphere (see also Amovilli and March [4]). Finally, Section 4.7 constitutes a summary plus some brief comments on directions that should prove useful for future research.

4.2   CLASSIFICATION OF DENSITY FUNCTIONAL THEORY: POSSIBLY VARIATIONALLY VALID VERSUS HEURISTIC

Because of the widespread use of density functional theory (DFT) in treating molecules and clusters, Amovilli et al. [5] used quantum-chemical *ab initio* methods to classify four existing energy density functionals according to their possible variational validity. Because DFT itself is, of course, fundamentally a variational method, Amovilli et al. [5] proposed the classification of available energy density functionals into two groups: (1) heuristic (H) and (2) possibly variationally valid (PV). Then, quantum-chemical *ab initio* methods are employed on selected neutral and anionic atomic cases, as well as on molecules H_2O and LiOB, to achieve the proposed separation into H and PV categories in the case of exchange-correlation functionals LDA, PBE, PW91, and B3LYP.

Table 4.1 presents the ground-state energies of eight light atoms, atomic ions, and molecules, as evaluated with four DFT exchange-correlation choices, compared with their best available *ab initio* values [5]. Table 4.2 shows the desired classification into H and PV after the results of Table 4.1. As stressed by Amovilli et al. [5], PBE being variationally valid in all examples studied is the most impressive outcome. This is in marked contrast to B3LYP, for which all ground-state energies lie below known quantum-chemical values. While, needless to say, B3LYP is a big improvement on TF statistical functional [3], which has been known for decades to tend to the correct nonrelativistic ground-state energy for neutral atoms of atomic number Z (namely $-0.77Z^{7/3}$ a.u. from below a large atomic number Z [3]), B3LYP belongs to

TABLE 4.1
Ground-State Energies (in Atomic Units) for Systems Studied by Amovilli et al. [5] at Various Levels of DFT Calculations and Comparison with Best Available Variational Nonrelativistic Data

System	LDA	PBE	PW91	B3LYP	Best *Ab Initio*
H⁻	0.54335	0.52439	0.52697	0.53477	0.52775
He	2.87217	2.89293	2.90005	2.91522	2.90372
Li	7.39838	7.46216	7.47423	7.49296	7.47806
Li	7.43146	7.47926	7.49237	7.51171	7.50040
Be	14.52049	14.62993	14.64801	14.67333	14.66735
Ne	128.43480	128.86640	128.94669	128.98096	128.9376
H_2O	76.10341	76.38317	76.44145	76.46961	76.4274
LiBO	107.04496	107.53627	107.63019	107.69329	107.540

TABLE 4.2
Classification of Density Functionals Considered by Amovilli et al. [5] as H or PV, according to Results of Table 4.1

System	LDA	PBE	PW91	B3LYP
H⁻	H	PV	PV	H
He	PV	PV	PV	H
Li	PV	PV	PV	H
Li	PV	PV	PV	H
Be	PV	PV	PV	H
Ne	PV	PV	H	H
H_2O	PV	PV	H	H
LiBO	PV	PV	H	H

category H. The conclusion that B3LYP intrinsically overestimates the exchange-correlation energy is hard to avoid.

To complete this section, we also note that Amovilli et al. [5] pointed to the future importance of lower bounds in the context of DFT and gave relevant references to which the interested reader is referred.

4.3 STABLE TETRAHEDRAL AND OCTAHEDRAL MOLECULES: ESPECIALLY SCALING PROPERTIES

In an early work, March [1], motivated by the desire to study the scaling properties of a variety of tetrahedral (T) and octahedral (O) molecules, presented a simplistic self-consistent central field model using the TF statistical method [1,6–8]. This prompted, much later, the study of Mucci and March [9], who proceeded, but now by semiempirical analysis, to relate nuclear–nuclear potential energy at equilibrium V_{nn} to the total number of electrons N in the molecule. Their semiempirical result, which was, as expected, subject to some relatively small scatter, reads (see also March and Nagy [10])

$$V_{nn} = aN^{5/3}, \tag{4.1}$$

where the constant a is given by Mucci and March [9].

Based on the model in March [1], V_{nn} was related exactly to the equilibrium bond length R_e of a given T or O molecule by

$$V_{nn} = \frac{n(z+cn)e^2}{R_e}. \tag{4.2}$$

In Equation 4.2, ze is the central charge, and ne is the total positive charge of all outer nuclei: for example, for SF_6, $z = 16$ and $n = 54$ (with the total number of electrons N in neutral molecules treated throughout, being evidently $z + n$). The constant c in Equation 4.2 is given by

$$c = \frac{3\sqrt{6}}{32} \approx 0.23 \tag{4.3}$$

for T molecules and by

$$c = \frac{1+4\sqrt{2}}{24} \approx 0.28 \tag{4.4}$$

for O molecules. To avoid the simplistic model [1] (see also March and Parr [11]), we next combine Equations 4.1 and 4.2 to reach the scaling prediction for the equilibrium bond length R_e entering Equation 4.2 as

$$\frac{aR_e}{N^{1/3}} = c + (1-2c)\frac{z}{N} + (c-1)\left(\frac{z}{N}\right)^2. \tag{4.5}$$

Following March and Nagy [10], we find therefore that the maximum value for $aR_e/N^{1/3}$ as a function of z/N equals $1/(4(1-c))$, being ≈ 0.31 for T molecules and ≈ 0.35 for O molecules. Indeed, comparison with available experimental data for R_e in some T molecules shows only weak dependence of $aR_e/N^{1/3}$ on z/N [10].

We turn to relate the "breathing" force constant $k = \partial^2 E/\partial R^2\big|_{R=R_e}$ for T and O molecules to R_e. Turning back to the model [1], Bowers [12] fitted in this oversimplistic model—when heavy atoms such as Cl and Br were in the outer positions (in March [1],

Cl inner electrons were "compressed" into the nucleus)—the scaling properties for R_e given by the TF model in March [1], namely,

$$R_e = z^{-1/3} d\left(\frac{n}{z}\right),\tag{4.6}$$

where Bowers wrote the approximate form to fit the numerical results in March [1] as

$$d = \text{const}\left(\frac{n}{z}\right)^{0.6}.\tag{4.7}$$

Similarly for k, in Bowers [12], a fit was made to predictions from March [1], of the form

$$k = z^3 g\left(\frac{n}{z}\right),\tag{4.8}$$

where $g = \text{const}(n/z)^{-3.51}$. Combining these, March and Nagy [10] found $6\log R_e = \log(z/k) + \text{const}$, which is then employed to show a marked correlation between k and z/n for some T molecules.

4.4 SMALL (USUALLY METASTABLE) FREE-SPACE CLUSTERS

4.4.1 SiO$_n$, GeO$_n$, AND CO$_n$ ($n = 4$ AND 6) FREE-SPACE CLUSTERS IN RELATION TO SAME SOLID-STATE PROPERTIES

4.4.1.1 Neutral Free-Space Cluster of Si Atom with Six Surrounding O Atoms

Figure 4.1, calculated by Hartree–Fock (HF) theory, shows what Forte et al. [13] proposed as a low-lying isomer of the free-space cluster SiO$_4$. The geometry of this present proposal for the ground state is recorded in Table 4.3. There are two SiO distances that are determined to be 1.64 and 1.73 Å, respectively. The corresponding

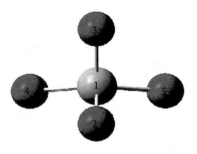

FIGURE 4.1 Optimized low-lying isomer of SiO$_4$. For precise bond lengths and angles, see Table 4.3. (Redrawn from Forte, G. et al., *Phys. Lett. A* **376**, 476, 2012.)

TABLE 4.3

Bond Lengths and Angles for Optimized Low-Lying Isomer of SiO_4, as Shown in Figure 4.1

	Distance (Å)		Angle (°)
1–2	1.729	2–1–3	82.25
1–3	1.729	2–1–4	108.5
1–4	1.639	2–1–5	108.5
1–5	1.639	4–1–5	130.14

Source: Forte, G. et al., *Phys. Lett. A* **376**, 476, 2012. With permission.

angles range from 82.3° to 130° via the (almost) tetrahedral angle of 109°. It is already relevant here to make some contact with the crystalline calculations of Saito and Ono [14] (see especially their Table II). They show that, at ambient temperature and pressure, the ground-state structure α-quartz (q-SiO_2 in their Table II) has two slightly different bond lengths, 1.60 and 1.61 Å, with angles near the tetrahedral value of ~110°. Above a pressure of 2 GPa, a rutile structure (r-SiO_2) is formed, and their Table II shows two SiO bond lengths of 1.75 and 1.79 Å, which are at least pretty near to the predicted second bond length, 1.73 Å, for the free-space SiO_4 cluster in Figure 4.1. Angles were not recorded, however, for r-SiO_2 in their Table II.

We shall return briefly to further results of Forte et al. [13] pertaining to SiO_4 clusters when we summarize below the results of the calculations on cationic free-space clusters.

4.4.1.2   Neutral Free-Space Cluster of Si Atom with Four Surrounding O Atoms

Figure 4.2a shows, first, a neutral cluster in which atom 1 is Si and tetrahedral angles for four surrounding O atoms are assumed. The energy and bond distances of Forte et al. [13] are recorded in Table 4.4, with the larger distances 1–2 and 1–3 being around 2.3 ± 0.1 Å, compared with four shorter bond lengths around 1.66 ± 0.01 Å.

(a)   (b)

FIGURE 4.2   Two low-lying isomers of SiO_6, as proposed by Forte et al. [13].

TABLE 4.4

Stability with Respect to Elements of Isomer SiO_4, as Displayed in Figure 4.1

n	$E(O)$	$E(Si)$	$E(Si + nO)$	$E(SiO_n)$	ΔE
4	−74.940	−289.017	−588.777	−589.384	−0.607

Source: Forte, G. et al., *Phys. Lett. A* **376**, 476, 2012. With permission.
Note: Energies are expressed in hartrees.

TABLE 4.5

Geometries of Low-Lying Isomers of SiO_6, as Depicted in Figure 4.2

SiO_6	Distance (Å) Figure 4.2a	Angle (°) $E = -739.4260$	Hartree
1–4	1.647	5–1–4	109.5
1–5	1.669	5–1–6	109.5
1–6	1.67	5–1–7	109.5
1–7	1.679		
1–2	2.195		
1–3	2.377		
SiO_6	Figure 4.2b	$E = -739.4343$	Hartree
1–4	1.664	5–1–4	123.07
1–5	1.664	5–1–6	84.77
1–6	1.664	5–1–7	123.07
1–7	1.664		
1–2	2.208		
1–3	2.208		

Source: Forte, G. et al., *Phys. Lett. A* **376**, 476, 2012. With permission.

Table 4.5 shows how variation in the geometry away from the tetrahedral angle of 109.5° entering Table 4.4 lowers the HF energy by ~ 0.008 hartree, with the cluster form depicted in Figure 4.2b. We note briefly that in Table II of Saito and Ono [14], the largest bond length recorded, 1.81 Å, was for r-SiO_2. Forte et al. [13] also considered the stability of the neutral cluster SiO_6 with respect to an isolated Si atom plus two ozone molecules O_3, finding a relative energy difference of $\Delta E = -0.373$ hartree.

4.4.1.3 Less Detailed Results with Si Atom Replaced by Ge in Free-Space Neutral Clusters

Prompted by the solid-state theoretical calculations presented by Saito and Ono [14], Forte et al. [13] also studied GeO_n clusters, although in somewhat less detail

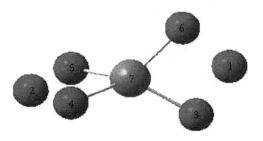

FIGURE 4.3 Low-lying isomer of GeO_6, with bond lengths and angles given in Table 4.6. (Redrawn from Forte, G. et al., *Phys. Lett. A* **376**, 476, 2012.)

TABLE 4.6

Bond Lengths and Angles for Optimized Low-Lying Isomer of GeO_6, as Shown in Figure 4.3

	Distance (Å)		Angle (°)
1–7	2.333	3–7–4	123.54
2–7	2.333	3–7–5	130.04
3–7	1.797	3–7–6	78.84
4–7	1.797		
5–7	1.796		
6–7	1.797		

Source: Forte, G. et al., *Phys. Lett. A* **376**, 476, 2012. With permission.

---

than their Si counterparts. One of the configurations of the free-space neutral cluster is depicted in Figure 4.3 for GeO_6. Their predicted bond distances and angles are collected in Table 4.6, for comparison with SiO_6 for the neutral cluster shown in Figure 4.2b.

4.4.1.4   Singly Charged Positive Ion Clusters SiO_4^+ and GeO_4^+

After treating the neutral clusters of SiO_n in some detail, Forte et al. [13] also estimated bond length distortions due to ionization for both SiO_4^+ and GeO_4^+. We note in this context that if we were dealing with the free-space T molecule silane, then we would appeal to experiment for the symmetry of the cation SiH_4^+ (see, e.g., Krishtal et al. [2]).

We therefore show in Figure 4.4 the proposed structures of Forte et al. [13] for these neutral clusters SiO_4 and cationic clusters SiO_4^+ from the present quantum-chemical technique [13]. Bond lengths and angles are correspondingly recorded in Table 4.7. Analogous results are then shown in Figure 4.5 and Table 4.8 for neutral clusters GeO_4 and cationic clusters GeO_4^+. In the light of the earlier discussion, Forte et al. [13] noted, in particular, that two bond lengths are now in evidence, with the longer one being 2.2 Å for GeO_4^+.

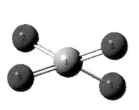

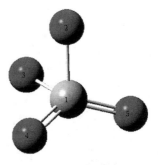

FIGURE 4.4 Low-lying isomers predicted theoretically for neutral cluster SiO_4 and cationic cluster SiO_4^+, with the total ground-state energies being $E = -589.3844$ and $E = -588.8507$ a.u., respectively. Bond lengths and angles are given in Table 4.7. (Redrawn from Forte, G. et al., *Phys. Lett. A* **376**, 476, 2012.)

TABLE 4.7

Bond Lengths and Angles for Neutral Cluster SiO_4 and Cationic Cluster SiO_4^+, as Shown in Figure 4.4

	Distance (Å)		Angle (°)
SiO_4			
1–2	1.63	2–1–4	137.12
1–3	1.63	2–1–5	137.13
1–4	1.63	2–1–3	62.24
1–5	1.63		
SiO_4^+			
1–2	1.729	2–1–4	108.5
1–3	1.729	2–1–5	108.52
1–4	1.639	2–1–3	82.25
1–5	1.639		

Source: Forte, G. et al., *Phys. Lett. A* **376**, 476, 2012. With permission.

Additionally, in Table 4.9, we record calculations [13] of Mulliken charges, which demonstrate the expected shift of electronic charge from Ge to the O nuclei owing to the high electronegativity of the latter atom.

Even though we are not dealing with stable gas-phase molecules such as SiH_4 and its cation, there are parallels with the Jahn–Teller removal of degeneracy.

The major conclusion from the theoretical study of Forte et al. [13] concerns the predicted low-lying isomers for the neutral free-space clusters SiO_n ($1 < n \leq 6$) and, in briefer fashion, for the corresponding Ge clusters. As stressed in Section 4.4.1.1, the

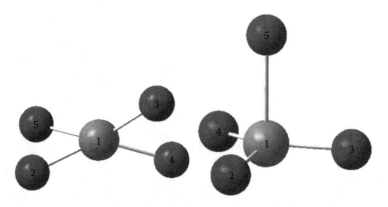

FIGURE 4.5 Low-lying isomers predicted theoretically by Forte et al. [13] for neutral cluster GeO_4 and cationic cluster GeO_4^+, with the total ground-state energies being $E = -2375.9477$ and $E = -2375.4176$ a.u., respectively. Bond lengths and angles are given in Table 4.8. (Redrawn from Forte, G. et al., *Phys. Lett. A* **376**, 476, 2012.)

immediate motivation was afforded by the recent theoretical work of Saito and Ono [14] on three different crystal structures of SiO_2 under pressure (see also an early study by Stenhouse et al. [15] on diffraction from vitreous silica). We hope that our present results will stimulate experimental studies on the free-space clusters SiO_n for small n and possibly on the comparison between electron density distribution in crystalline SiO_2 under pressure and electron density distribution in vitreous SiO_2 under compression.

TABLE 4.8
Bond Lengths and Angles for Neutral Cluster GeO_4 and Cationic Cluster GeO_4^+, as Shown in Figure 4.5

	Distance (Å)		Angle (°)
GeO_4			
1–2	1.770	2–1–4	140.31
1–3	1.770	2–1–5	57.32
1–4	1.770	2–1–3	140.35
1–5	1.770		
GeO_4^+			
1–2	1.755	2–1–4	120.58
1–3	1.755	2–1–5	95.9
1–4	1.755	2–1–3	118.13
1–5	2.252		

Source: Forte, G. et al., *Phys. Lett. A* **376**, 476, 2012. With permission.

TABLE 4.9

Summary of Mulliken Charges in Free-Space Neutral and Cationic Clusters of SiO_n (n = 4 and 6)

SiO_4		SiO_4^+		SiO_6		GeO_4		GeO_4^+	
Si	2.458	Si	2.202	Si	2.440	Ge	2.416	Ge	2.348
O	−0.615	O	−0.126	O	−0.600	O	−0.604	O	−0.420
O	−0.615	O	−0.126	O	−0.600	O	−0.604	O	−0.421
O	−0.615	O	−0.474	O	−0.020	O	−0.604	O	−0.424
O	−0.615	O	−0.474	O	−0.600	O	−0.604	O	−0.084
				O	−0.020				

Source: Forte, G. et al., *Phys. Lett. A* **376**, 476, 2012. With permission.

4.4.1.5 CO_n Clusters in Relation to Crystalline Forms of CO_2

We note here that the ordinary condensed solid phase of CO_2 is a molecular solid. At variance with silica and germania, a nonmolecular CO_2 crystalline form of carbon dioxide in which carbon atoms are tetrahedrally coordinated to oxygen atoms only exists at high pressure [16,17]. In addition, an amorphous phase of CO_2, formed by a disordered arrangement of CO_4 tetrahedra, has recently been obtained at very high pressure [18].

The CO_4 cluster has been studied in some detail by Cabria and March [19]. Free CO_4 has been experimentally detected [20]. This is a metastable species that dissociates into CO_2 and O_2. Also, as Cabria and March [19] stressed, CO_4 is an intermediate in the O exchange reaction $O_2 + CO_2$ [21]. Calculations show that free CO_4 does not have a tetrahedral structure [21,22].

Again motivated by tetrahedral coordination in CO_2 solids, we address our attention, following Cabria and March [19], to the study of CO_4 clusters with near-tetrahedral structure. Cabria and March [19] focused on three different cluster geometries. The first one is a planar structure with C_{2v} symmetry. The second is a three-dimensional structure with D_{2d} symmetry. This structure can be viewed as a highly distorted tetrahedron with two different interatomic O–O distances. Those two structures were identified in CO_4 by Elliott and Boldyrev [22]. The third one is a tetrahedron (T_d). In all cases, those geometries have been fully relaxed by Cabria and March [19] by allowing all C–O and O–O distances to change until the forces on all atoms are reduced to zero.

Quantum-chemical calculations [21,22] have shown that the ground state of CO_4 is planar, with the C atom coordinated to three O atoms. The C_{2v} cluster was detected [23] by infrared spectroscopy via its v_1 vibrational frequency of about 1941 cm^{-1}. Elliott and Boldyrev [22] also found a low-lying isomer of D_{2d} symmetry, with the C atom coordinated to four O atoms. The calculations of Cabria and March [19] agree with those results. The energy of the D_{2d} structure is 1.61 eV above the planar state (1.43 eV in the calculations of Elliott and Boldyrev [22]). Cabria and March [19] also found an isomer related to the D_{2d} structure, but with lower symmetry. In this structure, the two pairs of oxygen atoms show different O–O distances. The energy of this isomer lies 1.30 eV above the planar state (i.e., this isomer is more stable than the D_{2d} structure). Relaxation of the perfect T_d structure of CO_4 produced negligible changes in interatomic distances and angles, with the four C–O distances equal to 1.34 Å. All vibrational frequencies of this cluster were found to be positive. However, according to the PBE calculations of Cabria and March [19], the energy of the relaxed tetrahedral cluster is above the energy of the planar C_{2v} state by 3.04 eV.

From the results presented above, it is possible to establish a connection between free clusters and crystalline solids. The largest promotion energy from C_{2v} to the deformed tetrahedral and near-tetrahedral structures occurs for CO_4 rather than for SiO_4 or GeO_4 [19], a feature that correlates with the observations that CO_2 is a molecular solid at ordinary pressure and that crystalline and amorphous forms of CO_2 in which the C atom is tetrahedrally coordinated to four atoms are only produced at very high pressure.

4.5 ALMOST SPHERICAL C CAGES: π ELECTRONS IN BUCKMINSTERFULLERENE C_{60} AND C_n (n = 50, 70, AND 84)

Amovilli et al. [24] reported calculations on the π-electron eigenvalue sum of the number of atoms in almost spherical C cages. In introducing the HF calculations of Amovilli et al. [24], it will be useful to start with an admittedly oversimplistic model to gain insight. This considers free π electrons confined on the surface of a sphere [25–27]. As noted by Amovilli et al. [24], π levels, on zeroth order, are those of a rigid rotator, having degeneracy $2j + 1$ and energies $\hbar^2 j(j + 1)/2I$, with I being the moment of inertia. One then finds, with double occupancy up through the nth shell, a total energy proportional to

$$2 \sum_{j=0}^{n} j(j+1)(2j+1) = n(n+1)^2(n+2).$$ (4.9)

This result (Equation 4.9) is readily confirmed in the limit of large n by replacing the summation by an integration. Adopting the identification of I as mR^2, we then obtain [24]

$$E_\pi = \frac{n(n+1)^2(n+2)\hbar^2}{2mR^2}.$$ (4.10)

Below, following Amovilli et al. [24], we shall insert an equilibrium cage radius. Recognizing then that the total number of electrons is $N = 2 \sum_{j=0}^{n} (2j+1) = 2(n+1)^2$, one finds

$$E_\pi = \frac{N\hbar^2}{8mk_e^2} - \frac{\hbar^2}{4mk_e^2}.$$ (4.11)

Here, the constant k_e enters the equilibrium radius R_e in this model, as presented by Amovilli et al. [24], through

$$R_e = k_e \sqrt{n}. \tag{4.12}$$

Equation 4.11 is plainly of the correct general form for a graphene sheet. A related N dependence also follows the simple Hückel tight-binding model [28], in which there is a resonance integral β between each site and its three nearest neighbors. Based on the Gerschgorin circle theorem [24], all molecular orbital (MO) eigenvalues must be bounded between $-3|\beta|$ and $+3|\beta|$, so that the total N electron ground-state energy must lie between $-3N|\beta|$ and $+3N|\beta|$. One can expect the MO levels to be not too nonuniformly distributed, so that the ground-state energy is $\sim -cN$, where $c \neq 0$ and probably similar to $\dfrac{3}{2}|\beta|$, as indicated by Gutman et al. [29]. To reach Equation 4.12, Amovilli et al. [24] used an equilibrium radius R_e, determined by the "rule" that the surface area per C atom remains constant as the cage size increases, by raising the number of C atoms N.

4.5.1   HF Results for Some C Cages

For C_{60}, it is of course well established that the lowest isomer is the European football geometry. For C_{84}, Amovilli et al. [24] used the study of Manolopoulos and Fowler [30], in which classification of a wide variety of isomers is made. The one chosen by Amovilli et al. [24] (albeit with some inevitable arbitrariness) was that of the highest symmetry. The nuclear framework is shown in Figure 4.6. For C_{70}, related symmetry considerations led to the study of the nuclear structure depicted for $N = 70$ in Figure 4.6, while the framework adopted for C_{50} is also shown in Figure 4.6 (see also Schmalz et al. [31]).

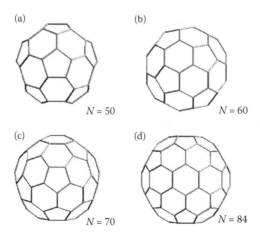

(a) $N = 50$ (b) $N = 60$ (c) $N = 70$ (d) $N = 84$

FIGURE 4.6   Structure of buckminsterfullerene (b) and the (assumed) high-symmetry structures of C_{50} (a), C_{70} (c), and C_{84} (d), according to Amovilli et al. [24].

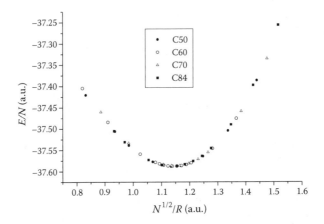

FIGURE 4.7 Total HF energy E/N plotted versus $\sqrt{N}/R$ for the four cages presented in Figure 4.6. The curves for the individual cages have been shifted by a constant to bring them into coincidence. (Redrawn from Amovilli, C. et al., *Phys. Rev. A* **66**, 013210, 2002.)

Amovilli et al. [24] then recorded HF energies for these four C cages, with their results redrawn in Figure 4.7. All the cages are presented such that each nucleus is at a common distance R from the center. While these HF calculations involved all electrons, Amovilli et al. [24] also extracted information directly connected with π electrons.

4.5.2 Total Energy Curves and Their Scaling Properties

Based on Amovilli et al. [24], the HF total energy $E(N,R)$, where R denotes the cage radius, can be well represented by the expression

$$\frac{E(N,R)}{N} = a + b\frac{N^{1/2}}{R} + c\frac{N}{R^2}. \tag{4.13}$$

This is shown in Figure 4.7, where the curves for the four individual cages have been shifted by a constant value to yield the coincidence depicted. We show in Table 4.10 both the ground-state energy E and the sum of the one-electron eigenvalues E_π (in atomic units). The equilibrium radii R_e (in angstroms) are also recorded in this table.

Figure 4.8 shows a plot of the total HF energy Evs, the eigenvalue sum denoted by E_s at the equilibrium radii R_e for C_N (where $N = 50, 60, 70,$ and 84). The dashed line represents $E = \frac{3}{2}E_s$ (see March and Plaskett [32] for atoms and Ruedenberg [33] for molecules).

To conclude this section, we point out the need to settle the question as to what are the lowest isomer structures for C_N where $N = 50, 70,$ and 84. At present, this seems to be a task for the experimentalist [34]. Current theory still seems to lack the power to decisively choose between isomer structures that will differ but little in energy.

TABLE 4.10

Ground-State Energy E and Sum of Eigenvalues for π Electrons E_π (in Atomic Units) and Equilibrium Radii R_e (in Angstroms) for Four Cages for Which HF Data Have Been Obtained

N	E/N	E_π/N	R_e	$R_e/\sqrt{N}$
50	−37.5884	−0.2232	3.278	0.463
60	−37.6029	−0.2275	3.582	0.462
70	−37.5926	−0.2234	3.865	0.462
84	−37.5874	−0.2223	4.256	0.464

Source: Amovilli, C. et al., *Phys. Rev. A* **66**, 013210, 2002. With permission.
Note: N is the number of C atoms in the cage.

Amovilli et al. [24] also made a comment on connectivity theory. They noted that, going beyond the remarkable constancy of E_π as recorded in Table 4.10, a correction of $O\left(1/\sqrt{N}\right)$ could exist. However, with the use of HF values, the behavior is not of monotonic character. For various approximations within the connectivity theory (equivalent to Hückel theory), a term of $O\left(1/\sqrt{N}\right)$ is present, for instance, in the study of Gutman and Soldatović [35].

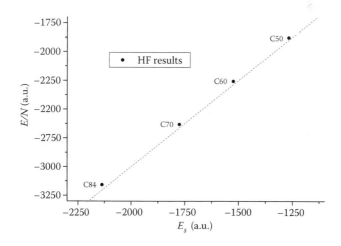

FIGURE 4.8 Total HF energy E versus eigenvalue sum E_s at the equilibrium radius R_e for C_{50}, C_{60}, C_{70}, and C_{84}. The dashed line represents $E = \dfrac{3}{2}E_s$. (Redrawn from Amovilli, C. et al., *Phys. Rev. A* **66**, 013210, 2002.)

4.6   B CAGES AND THOMSON'S PROBLEM OF POINT CHARGES ON THE SURFACE OF A SPHERE

Amovilli et al. [24] essentially replaced the use of TF statistical theory in March [1] with HF theory for the fullerenes C_{50}, C_{60}, C_{70}, and C_{84} with "almost spherical" C cages. At this point, it is relevant to make a brief digression to relate to the lower dimensionality example of planar ring clusters. The work of Amovilli and March [36] is, essentially, the two-dimensional analogue of the TF self-consistent field treatment of March [1].

Briefly then, the chemical potential μ, which is the same at every point in such a two-dimensional cluster, is in the TF method given by

$$\mu = \frac{p_F^2(\mathbf{r})}{2m} + V(\mathbf{r}), \qquad (4.14)$$

where $p_F(\mathbf{r})$ is the maximum momentum at position $\mathbf{r}$ in this semiclassical treatment, while $V(\mathbf{r})$ is the self-consistent electrostatic potential. For completion of the TF method in the present example, customary phase space arguments for two-dimensional cluster yield the ground-state density $n(\mathbf{r})$ as

$$n(\mathbf{r}) = \frac{2\pi p_F^2(\mathbf{r})}{h^2}. \qquad (4.15)$$

The next one relates n and V through the two-dimensional Poisson equation, now in linear form when Equations 4.14 and 4.15 are employed:

$$\frac{d^2V(r)}{dr^2} + \frac{1}{r}\frac{dV(r)}{dr} = 4V(r). \qquad (4.16)$$

This self-consistent equation, being linear in contrast to the three-dimensional non-linear form in March [1], can be solved analytically in terms of Bessel functions.

Amovilli and March, following March [1], compared the known results for H and C planar ring clusters. An interesting semiquantitative agreement from such a crude model, particularly for the H clusters, was found.

Amovilli and March [4] generalized the simple model in March [1] to boron cages using HF calculations. They found that the equilibrium radius of the spheroidal boron cages is proportional to $\sqrt{n}$, where n denotes the number of boron atoms in the cluster. The results of Amovilli and March [4] are therefore redrawn in Figure 4.9 to make the above comment quite concrete.

In summary, the key result for tetrahedral and octahedral classes of molecules is Equation 4.5. This shows that the quantity $aR_e/N^{1/3}$ is quadratic in the ratio z/N, where ze is the central charge, and N is the total number of electrons. As z/N tends to zero, this ratio tends to the constant c, which is ≈ 0.23 and ≈ 0.27 for T and O molecules, respectively (see Equations 4.3 and 4.4 for precise values of c). A maximum

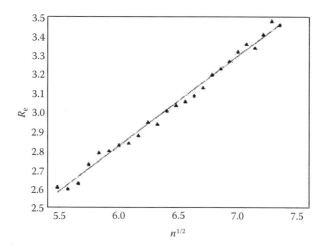

FIGURE 4.9 Equilibrium radius (in angstroms) of spheroidal boron cages against $n^{1/2}$, where n is the number of boron atoms. Triangles refer to *ab initio* computed values. (Redrawn from C. Amovilli and N. H. March, *Chem. Phys. Lett.* **347**, 459, 2001.)

occurs near $z/N = 1/3$, with the maximum value being $aR_e/N^{1/3} = 1/(4(1 - c))$. Then some scaling properties for the "breathing force" constant k are proposed for a series of five tetrachlorides, with k correlating well with a low-order polynomial in z/N.

Attention is then shifted to "almost" spherical C and B cages. As shown for C cages C_{50}, C_{60}, C_{70}, and C_{84} by Amovilli et al. [24], R_e is proportional to $\sqrt{n}$, where n is the number of C atoms. Our arguments here suggest that the force constant k is either independent of or very insensitive to the value of n. For B cages, as Amovilli and March [4] demonstrated, $R_e = 0.471\sqrt{n}\,\text{Å}$, having the same n dependence as for the four C cages discussed above.

Finally, planar ring clusters are briefly referred to again in relation to scaling properties.

4.7 SUMMARY AND FUTURE DIRECTIONS

After a brief discussion of the possible variational validity of several choices for the exchange-correlation energy in DFT, we have addressed some scaling properties in some stable "almost spherical" molecules, including GeH_4 and SF_6. Further attention is then devoted to several free-space clusters and their molecular ions, mainly on SiH_4, in relation to available experimental findings. Finally, almost spherical cages of C and B atoms are also briefly referred to.

Although the focus of the present chapter is on the quantum chemistry of highly symmetrical molecules and free-space clusters, we believe that it is of interest to summarize here an application of empirical potentials by Wales [37] to $(C_{60})_{55}$ clusters, presently beyond the practical scope of quantum mechanics. In the calculations reported by Wales [37], the intermolecular potential used was obtained by spherically averaging atom–atom Lennard–Jones terms, as previously employed in bulk simulations. As Wales emphasized, the work on $(C_{60})_{55}$ clusters aims to characterize statistically large samples of rearrangement pathways. Around 3000 pathways were calculated, and rearrangements have been analyzed by Wales [37] in terms of integrated path lengths and cooperativity indices. Wales commented on the relevance of the above studies (1) to explain the apparent instability of the bulk liquid phase of C_{60} [38–40] and (2) because the features observed for $(C_{60})_{55}$ are probably highly relevant to results obtained in studies of model metal clusters. Finally, it is evident that the cluster results for C_{60} are relevant to its vapor deposition, as it is normally produced in the Krätschmer–Huffmann synthesis [41]. For further details, we must refer the interested reader to the article by Wales [37].

REFERENCES

1. N. H. March, *Proc. Cambridge Philos. Soc.* **48**, 665 (1952). Reprinted in *Many-Body Theory of Molecules, Clusters, and Condensed Phases*, edited by N. H. March and G. G. N. Angilella (World Scientific, Singapore, 2009).
2. A. Krishtal, C. Van Alsenoy, and N. H. March, *J. Phys. Chem. A*, **115**, 12988 (2011).
3. N. H. March, *Electron Density Theory of Atoms and Molecules* (Academic Press, New York, 1992).
4. C. Amovilli and N. H. March, *Chem. Phys. Lett.* **347**, 459 (2001).
5. C. Amovilli, N. H. March, F. Bogár, and T. Gál, *Phys. Lett. A* **373**, 3158 (2009).
6. L. H. Thomas, *Proc. Cambridge Philos. Soc. Math. Phys. Sci.* **23**, 542 (1926).
7. E. Fermi, *Rend. Accad. Naz.* **6**, 602 (1927).
8. E. Fermi, *Z. Phys.* **48**, 73 (1928).
9. J. F. Mucci and N. H. March, *J. Chem. Phys.* **82**, 5099 (1985).
10. N. March and Á. Nagy, *J. Math. Chem.* **49**, 2268 (2011), 10.1007/s10910-011-9885-5.
11. N. H. March and R. G. Parr, *Proc. Natl. Acad. Sci. U. S. A.* **77**, 6285 (1980). Reprinted in *Many-Body Theory of Molecules, Clusters, and Condensed Phases*, edited by N. H. March and G. G. N. Angilella (World Scientific, Singapore, 2009).
12. W. A. Bowers, *J. Chem. Phys.* **21**, 117 (1953).
13. G. Forte, G. G. N. Angilella, V. Pittalà, N. H. March, and R. Pucci, *Phys. Lett. A* **376**, 476 (2012).
14. S. Saito and T. Ono, *Jpn. J. Appl. Phys.* **50**, 021503 (2011).
15. B. Stenhouse, P. J. Grout, and N. H. March, *Phys. Lett. A* **57**, 99 (1976).
16. V. Iota, C. S. Yoo, and H. Cynn, *Science* **283**, 1510 (1999).
17. C. S. Yoo, H. Cynn, F. Gygi, G. Galli, V. Iota, M. Nicol, S. Carlson, D. Häusermann, and C. Mailhiot, *Phys. Rev. Lett.* **83**, 5527 (1999).
18. M. Santoro, F. A. Gorelli, R. Bini, G. Ruocco, S. Scandolo, and W. A. Crichton, *Science* **441**, 857 (2006).
19. I. Cabria, M. J. López, J. A. Alonso, and N. H. March, *Eur. Phys. J. D* **66**, 105 (2012).
20. F. Cacace, G. de Petris, and A. Troiani, *Angew. Chem. Int. Ed.* **42**, 2985 (2003).
21. L. Y. Yeung, M. Okumura, J. T. Paci, G. C. Schatz, J. Zhang, and T. K. Minton, *J. Am. Chem. Soc.* **131**, 13940 (2009).
22. B. M. Elliott and A. I. Boldyrev, *J. Phys. Chem. A* **109**, 3722 (2005).
23. C. S. Jamieson, A. M. Mebel, and R. I. Kaiser, *Chem. Phys. Lett.* **440**, 105 (2007).
24. C. Amovilli, I. A. Howard, D. J. Klein, and N. H. March, *Phys. Rev. A* **66**, 013210 (2002).
25. R. C. Haddon, L. E. Brus, and K. Raghavachari, *Chem. Phys. Lett.* **125**, 459 (1986).
26. R. Saito, G. Dresselhaus, and M. S. Dresselhaus, *Phys. Rev. B* **46**, 9906 (1992).
27. M. R. Savina, L. L. Lohr, and A. H. Francis, *Chem. Phys. Lett.* **205**, 200 (1993).

28. C. A. Coulson, R. B. Mallion, and B. O'Leary, *Hückel Theory for Organic Chemists* (Academic Press, London, 1978).
29. I. Gutman, A. Nikolić, and Ž. Tomović, *Chem. Phys. Lett.* **349**, 95 (2001).
30. D. E. Manolopoulos and P. W. Fowler, *J. Chem. Phys.* **96**, 7603 (1992).
31. T. G. Schmalz, W. A. Seitz, D. J. Klein, and G. E. Hite, *J. Am. Chem. Soc.* **110**, 1113 (1988).
32. N. H. March and J. S. Plaskett, *Proc. R. Soc. A* **235**, 419 (1956). Reprinted in *Many-Body Theory of Molecules, Clusters, and Condensed Phases*, edited by N. H. March and G. G. N. Angilella (World Scientific, Singapore, 2009).
33. K. Ruedenberg, *J. Chem. Phys.* **66**, 375 (1977).
34. S. J. La Placa, P. A. Roland, and J. J. Wynne, *Chem. Phys. Lett.* **190**, 163 (1992).
35. I. Gutman and T. Soldatović, *MATCH Commun. Math. Comput. Chem.* **44**, 169 (2001).
36. C. Amovilli and N. H. March, *Phys. Rev. A* **73**, 063205 (2006).
37. D. J. Wales, *J. Chem. Phys.* **101**, 3750 (1994).
38. A. Cheng, M. L. Klein, and C. Caccamo, *Phys. Rev. Lett.* **71**, 1200 (1993).
39. M. H. J. Hagen, E. J. Meijer, G. C. A. M. Mooij, D. Frenkel, and H. N. W. Lekkerkerker, *Nature* **365**, 425 (1993).
40. N. H. March, D. Lamoen, and J. A. Alonso, *Phys. Chem. Liq.* **40**, 457 (2002).
41. W. Krätschmer, L. D. Lamb, K. Fostiropoulos, and D. R. Huffman, *Nature* **347**, 354 (1990).

5 Energy Functionals for Excited States

M. K. Harbola, M. Hemanadhan, Md. Shamim, and P. Samal

CONTENTS

5.1 INTRODUCTION

With the success of ground-state density functional theory (g-DFT) [1–6] in many-electron systems, effort has been exerted to develop a corresponding time-independent (stationary-state) excited-state density functional theory (e-DFT). The latter is envisaged to be similar to its ground-state counterpart and should enable us to calculate directly the total energy of an excited state through appropriate functionals. The difference in the energies of an excited state and the ground state would then give the corresponding transition energy. In contrast to a stationary-state theory of excited states, which is still evolving, a very popular indirect method for obtaining excited states is provided [7,8] by time-dependent density functional theory (TD-DFT) [9]. In TD-DFT, one solves for the linear response of an electronic system to a time-dependent perturbation varying sinusoidally and identifies transition energies by the poles of the response function. While the TD-DFT approach leads to accurate transition energies, the theory is not without shortcomings. For example, it is still not clear how one would obtain transition energy corresponding to two electrons being excited simultaneously (double excitation). On the other hand, in e-DFT, there is no such restriction on the number of electrons excited. Some of the earlier attempts to develop e-DFT include work by Ziegler et al. [10], Gunnarsson and Lundqvist [11], and Barth [12] on the lowest excited state of a given symmetry, as well as the ensemble-based theory introduced by Theophilou [13] and developed further by Gross et al. [14] and Oliveira et al. [15]. Ensemble-based theory was applied by Nagy [16] to calculate the transition energies of some atoms, with moderate success. Renewal of interest [17–26] in e-DFT was prompted over the past decade and a half by the work of Görling [17], Levy and Nagy [18], and Nagy and Levy [19]. Extending their work, we have focused on two important aspects of e-DFT: existence of a Kohn–Sham system for excited states [27,28] and development [21,29–32] of energy functionals for excited states. In this paper, we present our method of splitting k-space to construct excited-state energy functionals. The method is quite general in that it can be applied to obtain kinetic and exchange-correlation energy functionals. Furthermore, as our results will show, the functionals obtained through this method are very accurate. In the following, we begin with a discussion of the foundational aspects of e-DFT. We then present the idea of splitting the k-space for homogeneous electron gas (HEG) to construct excited-state energy functionals. We apply this to obtain the kinetic and exchange energy functionals and show their accuracy. Finally, we study linear response function obtained by splitting k-space for noninteracting HEG and discuss its implications for e-DFT.

Before presenting our work, we note with pleasure the extensive work [33] performed by Professor B. M. Deb and his students on excited-state calculations for atoms by employing the Harbola–Sahni exchange potential [34,35]. We feel honored to have been invited to write a chapter in a book dedicated to Professor Deb.

5.2   FOUNDATIONS OF e-DFT

The theoretical framework of e-DFT is based on the theory of ρ stationary states by Görling [17] and the bifunctional theory of Levy and Nagy [18] and Nagy and Levy [19]. Both theories stem from a constrained-search approach [36] to finding the excited-state wavefunction for a given excited-state density. According to Görling's theory, an excited-state density gives many ρ stationary states by the constrained-search procedure as follows. Given a density $\rho(\mathbf{r})$ of an excited state of an N-electron system, a ρ stationary state is the antisymmetric function $\Psi(\mathbf{r}_1, \mathbf{r}_2,\ldots, \mathbf{r}_N)$ that (1) reproduces this density and (2) simultaneously makes the expectation value $\left\langle \Psi \left| \hat{T} + \hat{V}_{ee} \right| \Psi \right\rangle$ (where $\hat{T}$ and $\hat{V}_{ee}$ are the kinetic energy and electron–electron interaction energy operators, respectively) stationary. This leads to more than one stationary state, including the true excited-state wavefunction, for an excited-state density. The true wavefunction can be identified [28] easily if we know the ground-state density $\rho_0(\mathbf{r})$. Thus, the wavefunction $\Psi(\mathbf{r}_1, \mathbf{r}_2,\ldots, \mathbf{r}_N)$ is a bifunctional $\Psi[\rho, \rho_0]$ of the excited-state density $\rho(\mathbf{r})$ and the ground-state density $\rho_0(\mathbf{r})$. Levy and Nagy proposed to obtain the wavefunction $\Psi[\rho, \rho_0]$ in a slightly different way. Their scheme is a minimization method where, for the kth excited-state density $\rho_k(\mathbf{r})$, one looks for antisymmetric function $\Psi_k(\mathbf{r}_1, \mathbf{r}_2,\ldots, \mathbf{r}_N)$ that reproduces $\rho_k(\mathbf{r})$ and is simultaneously orthogonal to the lower $(k-1)$ states determined by the ground-state density (equivalent to the external potential) $\rho_0(\mathbf{r})$ and minimizes the expectation value $\left\langle \Psi_k \left| \hat{T} + \hat{V}_{ee} \right| \Psi_k \right\rangle$. In any case, it is clear that the wavefunction corresponding to an excited-state density

$\rho(\mathbf{r})$ for a system with the ground-state density $\rho_0(\mathbf{r})$ is a bifunctional $\Psi[\rho, \rho_0]$. Thus, the total energy is given as

$$E[\rho,\rho_0] = F[\rho,\rho_0] + \int d\mathbf{r}\rho(\mathbf{r})v(\mathbf{r}), \qquad (5.1)$$

where $v(\mathbf{r})$ is the external potential and $F[\rho,\rho_0] = \left\langle \Psi[\rho,\rho_0] \left| \hat{T} + \hat{V}_{ee} \right| \Psi[\rho,\rho_0] \right\rangle$ is a bifunctional of densities $\rho(\mathbf{r})$ and $\rho_0(\mathbf{r})$. The challenge is to find the functional $F[\rho, \rho_0]$.

The Kohn–Sham system for an excited state can also be found by making the functional $\left\langle \Phi \left| \hat{T} \right| \Phi \right\rangle$, where Φ is a Slater determinant that gives the excited-state density, which is stationary with respect to variations in Φ. This also leads to a multitude of Slater determinants; the Kohn–Sham system is then identified [28] as a system with kinetic and exchange-correlation energy components closest to those of the true system. This is seen [28] to give the same configuration of the excited state as the interacting system. Thus, the existence of a Kohn–Sham system for excited states is also established.

5.3 FUNCTIONALS FOR EXCITED STATES

To apply e-DFT and make it computationally as attractive as g-DFT, we must be able to write exchange and correlation energies in terms of the excited-state density $\rho(\mathbf{r})$. Therefore, the challenging problem in e-DFT is the same as in its ground-state counterpart—construction of excited-state energy functionals. In this review, we describe our efforts in developing a systematic method (that of splitting k-space for HEG) for obtaining excited-state energy functionals. Without such functionals, there is no option but to rely on the existing ground-state energy functionals for calculations involving excited states. However, employing ground-state functionals for excited states is both qualitatively incorrect and numerically inaccurate.

The simplest and foremost approximation of the exchange-correlation energy functional for ground states is local density approximation (LDA) or its spin density counterpart, local spin density (LSD) approximation [3–5]. These approximations give reasonably accurate results [3–5] for the ground states. Furthermore, they also form the basis for obtaining better approximations, such as gradient expansion approximation (GEA) [37,38] and generalized gradient approximation (GGA) [39,40], by including the gradient of the density. For excited states, one would therefore like to start by developing the LDA or LSD functional.

In this paper, we present the idea of splitting the k-space for HEG to construct LDA functionals for excited-state energy. The method has been used [29] to construct exchange energy functionals and leads to accurate transition energies. To test the generality of the method, we have also used it to obtain [32] excited-state counterparts of Thomas–Fermi approximation [3–5,41,42] and GEA [3–5,37] for noninteracting kinetic energy. Comparison of approximate kinetic energy thus obtained with the exact kinetic energy indicates the soundness of our method. We have also analyzed [43] the response function for noninteracting HEG in excited state to learn about the gradient expansion of kinetic energy functionals. Our analysis shows that splitting k-space is the simplest way of constructing energy functionals for excited states.

5.4 SPLITTING k-SPACE TO OBTAIN EXCITED-STATE FUNCTIONALS

We start this section by examining how LDA is made for ground states. To obtain the LDA functional for kinetic energy (LDA for kinetic energy is known as the Thomas–Fermi functional) or the exchange-correlation energy of an interacting system, the orbital occupation of the corresponding Kohn–Sham system is mapped to HEG, as shown in Figure 5.1.

The density of the system at a point is given by the formula

$$\rho(\mathbf{r}) = \sum_{i=1}^{N} |\phi_i(\mathbf{r})|^2, \tag{5.2}$$

where $\phi_i(i = 1,\dots,N)$ are occupied orbitals. With this mapping, the magnitude of the space-dependent Fermi wavevector $k_F(\mathbf{r})$ for HEG is given by the formula

$$k_F(\mathbf{r}) = [3\pi^2\rho(\mathbf{r})]^{1/3}. \tag{5.3}$$

Since, for HEG, expressions for the kinetic and exchange-correlation energies per unit volume are well known in terms of k_F, the LDA for an inhomogeneous gas corresponds to writing the same expressions at each point in space and integrating the resulting energy density into the entire space. Thus, the Thomas–Fermi approximation for the kinetic energy of an inhomogeneous electronic density is given as

$$T_0[\rho] = \int \frac{k_F^5(\mathbf{r})}{10\pi^2} d\mathbf{r}. \tag{5.4}$$

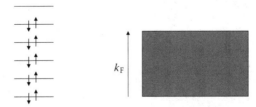

FIGURE 5.1   Orbital occupation in the ground-state configuration of an electronic system and its representation by HEG.

Similarly, the LDA for the exchange energy is given as

$$E_{\text{X}}^{\text{LDA}}[\rho] = -\int \frac{k_{\text{F}}^4(\mathbf{r})}{4\pi^3} d\mathbf{r}. \tag{5.5}$$

Expressions for correlation energy are obtained by interpolating [44–46] the results [47] of accurate Monte Carlo calculations. The question we now ask is whether a similar mapping of an excited-state configuration can be made on HEG. Furthermore, can the corresponding LDA functional be written using this mapping? The answer to both of these questions is in the affirmative, as we now discuss.

Figure 5.2 shows a schematic orbital occupation for an excited state where some lowest orbitals (core orbitals) are occupied, some orbitals (vacant orbitals) are unoccupied, and orbitals (shell orbitals) at higher energies are again occupied. The corresponding representation of this excited state by HEG is also displayed in the figure.

For the orbital occupation shown in the figure, the k-space for HEG is occupied up to wavevector k_1, vacant from k_1 to k_2, and occupied again from k_2 to k_3. These wavevectors are related to the density of the system as

$$k_1^3(\mathbf{r}) = 3\pi^2 \rho_c(\mathbf{r}), \tag{5.6}$$

$$k_2^3(\mathbf{r}) - k_1^3(\mathbf{r}) = 3\pi^2 \rho_v(\mathbf{r}), \tag{5.7}$$

and

$$k_3^3(\mathbf{r}) - k_2^3(\mathbf{r}) = 3\pi^2 \rho_s(\mathbf{r}). \tag{5.8}$$

Here, $\rho_c(\mathbf{r})$, $\rho_v(\mathbf{r})$, and $\rho_s(\mathbf{r})$ are the electron densities of the core, vacant, and shell orbitals, respectively. Thus,

$$\rho_c(\mathbf{r}) = \sum_{i=1}^{n_1} \left| \phi_i^{\text{core}}(\mathbf{r}) \right|^2, \tag{5.9}$$

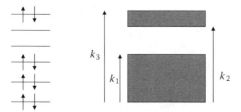

FIGURE 5.2 Orbital occupation in an excited-state configuration of an electronic system and its representation by HEG.

$$\rho_v(\mathbf{r}) = \sum_{i=n_1+1}^{n_2} \left| \phi_i^{\text{vacant}}(\mathbf{r}) \right|^2, \tag{5.10}$$

and

$$\rho_s(\mathbf{r}) = \sum_{i=n_2+1}^{n_3} \left| \phi_i^{\text{shell}}(\mathbf{r}) \right|^2, \tag{5.11}$$

where the first n_1 orbitals are occupied, orbitals from $(n_1 + 1)$ to n_2 are vacant, and orbitals from $(n_2 + 1)$ to n_3 are occupied. The total electron density $\rho(\mathbf{r})$ is given as

$$
\begin{aligned}
\rho(\mathbf{r}) &= \rho_c(\mathbf{r}) + \rho_s(\mathbf{r}) \\
&= \rho_1(\mathbf{r}) - \rho_2(\mathbf{r}) + \rho_3(\mathbf{r}),
\end{aligned}
\tag{5.12}
$$

with $\rho_1(\mathbf{r}) = \rho_c(\mathbf{r})$, $\rho_2(\mathbf{r}) = \rho_c(\mathbf{r}) + \rho_v(\mathbf{r})$, and $\rho_3(\mathbf{r}) = \rho_c(\mathbf{r}) + \rho_v(\mathbf{r}) + \rho_s(\mathbf{r})$. Having mapped the density to HEG in terms of k_1, k_2, and k_3, we find it quite easy to write the LDA for the exchange energy of the inhomogeneous system as a functional of these wavevectors or, equivalently, as a functional of $\rho_1(\mathbf{r})$, $\rho_2(\mathbf{r})$, and $\rho_3(\mathbf{r})$. This then gives the LDA for exchange energy functional for excited states with one set of vacant orbitals. We present this functional and the corresponding results in Section 5.5.

5.5   LDA FOR EXCHANGE ENERGY FUNCTIONAL FOR EXCITED STATES

The exchange energy functional corresponding to the k-space occupation shown in Figure 5.2 is given as

$$E_X^{\text{MLDA}} = E_X^{\text{core}} + E_X^{\text{shell}} + E_X^{\text{core-shell}}, \tag{5.13}$$

where MLDA stands for modified LDA. In the equation above,

$$E_X^{\text{core}}[\rho] = -\int d\mathbf{r}\, \frac{k_1^4(\mathbf{r})}{4\pi^3} \tag{5.14}$$

is the exchange energy of core electrons,

$$E_X^{\text{shell}} = -\frac{1}{8\pi^3} \int d\mathbf{r} \left[2\left(k_3^3 - k_2^3\right)(k_3 - k_2) + \left(k_3^2 - k_2^2\right)^2 \ln\left(\frac{k_3 + k_2}{k_3 - k_2}\right) \right] \tag{5.15}$$

is the exchange energy of electrons in the shell, and

$$E_X^{\text{core-shell}} = -\frac{1}{8\pi^3} \int d\mathbf{r} \left[2(k_3 - k_2)k_1^3 + 2\left(k_3^3 - k_2^3\right)k_1 + \left(k_2^2 - k_1^2\right)^2 \ln\left(\frac{k_2 + k_1}{k_2 - k_1}\right) \right.$$
$$\left. - \left(k_3^2 - k_1^2\right)^2 \ln\left(\frac{k_3 + k_1}{k_3 - k_1}\right) \right]$$

(5.16)

gives the exchange energy between the core electrons and the shell electrons. The sum of the three terms is given as

$$E_X^{\text{MLDA}} = \int d\mathbf{r}\rho\left[\varepsilon_X(k_1) - \varepsilon_X(k_2) + \varepsilon_X(k_3)\right] + \log \text{ terms}, \qquad (5.17)$$

where $\varepsilon_X(k) = -\dfrac{3k}{4\pi}$ is the exchange energy per particle for the ground state of HEG with Fermi wavevector k. Thus, we see that, except for log terms, the exchange energy per particle for the excited-state configuration of Figure 5.2 is given by combining $\varepsilon_X(k)$ corresponding to k_1, k_2, and k_3 in the same manner as is performed for the density in Equation 5.12. The MLDA functional of Equation 5.17 is easily generalized to the spin-dependent functional (modified LSD [MLSD]) through the relation

$$E_X^{\text{MLSD}}[\rho_\alpha,\rho_\beta] = \frac{1}{2} E_X^{\text{MLDA}}[2\rho_\alpha] + \frac{1}{2} E_X^{\text{MLDA}}[2\rho_\beta], \qquad (5.18)$$

where ρ_α and ρ_β represent the densities corresponding to up and down spins, respectively.

Table 5.1 shows the differences in the exchange energies of the ground state and the excited state of some atoms and ions. The first, second, and third columns give the differences obtained by Hartree–Fock (HF) theory, by employing ground-state LSD functionals, and by employing ground-state MLSD functionals, respectively.

These numbers have been calculated as follows. Results corresponding to HF theory are obtained by performing self-consistent HF calculations for the ground-state and excited-state configurations. The numbers obtained by HF theory are taken to be the exact exchange energy difference and provide the benchmark against which the accuracy of different exchange energy functionals is tested. To calculate LSD and MLSD, we perform a Kohn–Sham calculation with exchange potential $v_x(\mathbf{r}) = -\left[\dfrac{3\rho(\mathbf{r})}{\pi}\right]^{1/3}$ for the appropriate ground-state or excited-state configuration and employ the density thus obtained for calculating the exchange energies.

It is evident from Table 5.1 that the LSD functional underestimates the exact exchange energy difference; for small systems, the error is small, but for larger

TABLE 5.1

Differences in Exchange Energies [29] of Ground State and Excited States of Some Atoms and Ions, as Calculated within HF Theory and Using LSD Functional for Ground State and MLSD and MLSDSIC Functionals for Excited State

Atoms/Ions	ΔE_X^{HF}	ΔE_X^{LSD}	ΔE_X^{MLSD}	$\Delta E_X^{MLSDSIC}$
Li($2s^1\ ^2S \rightarrow 2p^1\ ^2P$)	0.0278	0.0264	0.0587	0.0282
B($2s^22p^1\ ^2P \rightarrow 2s^12p^2\ ^2D$)	0.0353	0.0319	0.0998	0.0412
C($2s^22p^2\ ^3P \rightarrow 2s^12p^3\ ^3D$)	0.0372	0.0332	0.1188	0.0454
N($2s^22p^3\ ^4S \rightarrow 2s^12p^4\ ^4P$)	0.0399	0.0353	0.1381	0.0503
O($2s^22p^4\ ^3P \rightarrow 2s^12p^5\ ^3P$)	0.1582	0.0585	0.2634	0.1624
F($2s^22p^5\ ^2P \rightarrow 2s^12p^6\ ^2S$)	0.3021	0.0891	0.3908	0.2765
Ne$^+$($2s^22p^5\ ^2P \rightarrow 2s^12p^6\ ^2S$)	0.3339	0.0722	0.4397	0.3037
S($3s^23p^4\ ^3P \rightarrow 3s^13p^5\ ^3P$)	0.1106	0.0475	0.1798	0.1252
Cl$^+$($3s^23p^4\ ^3P \rightarrow 3s^13p^5\ ^3P$)	0.1257	0.0483	0.2050	0.1441
Cl($3s^23p^5\ ^2P \rightarrow 3s^13p^6\ ^2S$)	0.2010	0.0603	0.2567	0.1969

Note: Numbers are given in atomic units.

systems, the error is quite significant. This is understood as follows. The LSD functional has been constructed for the ground state, where the overlap of orbitals is large. On the other hand, the overlap is relatively smaller for excited states. Thus, when applied to excited states, the LSD functional gives a larger magnitude of exchange energy, leading to a smaller difference with the ground-state exchange energy. The MLSD functional corrects the error of LSD in the right direction and gives exchange energy differences closer to the corresponding HF values, particularly for larger atoms. However, it tends to overcorrect for small atoms and ions; this overcorrection can be traced [29] to the self-interaction error in the LSD exchange energy for electrons involved in the transition. The MLSD functional is corrected for this self-interaction error to make the LSD functional accurate for all systems.

The self-interaction corrected functional (MLSDSIC) proposed for calculating exchange energies for excited states is given as

$$E_X^{MLSDSIC} = E_X^{MLSD} - E_X^{SIC}[\phi_{rem}] - E_X^{SIC}[\phi_{add}], \tag{5.19}$$

where $\phi_{rem}(\phi_{add})$ is the orbital where an electron is removed from (added to). Here [45]

$$E_X^{SIC}[\phi] = \frac{1}{2} \iint \frac{|\phi(\mathbf{r}_1)|^2 |\phi(\mathbf{r}_2)|^2}{|\mathbf{r}_1 - \mathbf{r}_2|} d\mathbf{r}_1 d\mathbf{r}_2 + E_X^{LSD}[\rho(\phi)], \tag{5.20}$$

where $\rho(\phi)$ is the orbital density for orbital ϕ. Results for exchange energy differences calculated by employing the MLSDSIC functional are also shown in Table 5.1.

This functional gives accurate results for all the systems shown in the table. Thus, the MLSDSIC functional can be employed for calculating the transition energies of electronic systems. In Section 5.6, we present the results for transition energies calculated by employing this functional.

5.6 TRANSITION ENERGIES

Transition energies for various atoms have been calculated using the Δ-SCF method within the exchange-only approximation as follows. HF theory results are obtained by performing self-consistent HF calculations [48] for ground-state and excited-state configurations, and the difference between the two energies gives the transition energy. Similarly, to obtain LSD transition energies, we calculate energies for the ground state and the excited state by performing a Kohn–Sham calculation with the exchange potential $v_x(\mathbf{r}) = -\left[\dfrac{3\rho(\mathbf{r})}{\pi}\right]^{1/3}$ [3–5,49] and by using the LSD functional for the exchange energies of the two states. For these calculations, the excited state chosen is such that it can be represented by a single Slater determinant since it is only for such states that the local approximation can be made [12] with reasonable accuracy. Using the same excited-state orbitals, we obtain the excited-state energy in MLSDSIC by calculating the exchange energy for the excited state using the MLSDSIC functional. The difference in energy thus obtained, with the ground-state energy calculated using the LSD functional, gives the MLSDSIC transition energies.

Table 5.2 shows the excitation energies of some atoms and ions when an electron is excited from an s orbital to a p orbital. The numbers presented are for the HF theory, the LSD functional, and the MLSDSIC functional. We consider excitations where an electron is excited from the outermost s shell to the outermost p shell ($2s \rightarrow 2p$ and $3s \rightarrow 3p$ transitions) and from a deeper s shell to the outermost p shell ($2s \rightarrow 3p$ transitions).

In all these transitions, LSD numbers are substantially in error compared to the HF theory results. On the other hand, results obtained by employing the MLSDSIC functional are highly accurate. This shows the correctness of the MLSDSIC functional for excited states. We also compare these numbers with those obtained by applying TD-DFT within a single-pole approximation. The Δ-SCF numbers are better than the TD-DFT numbers.

Next we present excitation energies corresponding to an electron being excited from an s orbital to a d orbital. These numbers are shown in Table 5.3.

We again see that the MLSDSIC functional gives transition energies that are very close to the corresponding HF values. On the other hand, single-pole approximation TD-DFT in this case does not improve the LSD numbers by any significant amount.

A problem more difficult than the single-electron excitations considered so far is the excitation of two electrons. We have also applied the MLSDSIC functional to calculate the transition energies for two-electron excitations. These results are given in Table 5.4.

It is evident that the MLSDSIC functional leads to highly accurate excitation energies. Note that there are no TD-DFT numbers given here because TD-DFT cannot be applied to calculate energies for double excitations.

TABLE 5.2
Transition Energies [29] Corresponding to Excited States of Some Atoms and Ions, as Calculated within HF Theory and Using LSD and MLSDSIC Functionals for Excited State: Electron Transfer from s Orbital to p Orbital

Atoms/Ions	ΔE^{HF}	ΔE^{LSD}	$\Delta E^{MLSDSIC}$	ΔE^{TDDFT}
$Li(2s^1\ ^2S \rightarrow 2p^1\ ^2P)$	0.068	0.065	0.067	0.072
$B(2s^22p^1\ ^2P \rightarrow 2s^12p^2\ ^2D)$	0.217	0.199	0.206	0.217
$C(2s^22p^2\ ^3P \rightarrow 2s^12p^3\ ^3D)$	0.294	0.289	0.297	0.309
$N(2s^22p^3\ ^4S \rightarrow 2s^12p^4\ ^4P)$	0.413	0.391	0.401	0.415
$O(2s^22p^4\ ^3P \rightarrow 2s^12p^5\ ^3P)$	0.626	0.524	0. 621	0.591
$F^+(2s^22p^4\ ^3P \rightarrow 2s^12p^5\ ^3P)$	0.799	0.679	0.801	0.765
$F(2s^22p^5\ ^2P \rightarrow 2s^12p^6\ ^2S)$	0.878	0.667	0.857	0.766
$Ne^+(2s^22p^5\ ^2P \rightarrow 2s^12p^6{}^2S)$	1.083	0.833	1.061	0.955
$S(3s^23p^4\ ^3P \rightarrow 3s^13p^5\ ^3P)$	0.426	0.362	0.434	0.412
$Cl^+(3s^23p^4\ ^3P \rightarrow 3s^13p^5\ ^3P)$	0.526	0.448	0.540	0.511
$Cl(3s^23p^5\ ^2P \rightarrow 3s^13p^6\ ^2S)$	0.565	0.430	0.563	0.500
$Ar^+(3s^23p^5\ ^2P \rightarrow 3s^13p^6\ ^2S)$	0.677	0.517	0.677	0.601
$P(2s^23p^3\ ^4S \rightarrow 2s^13p^4\ ^4P)$	6.882	6.419	6.956	6.157
$S(2s^23p^4\ ^3P \rightarrow 2s^13p^5\ ^3P)$	8.246	7.734	8.327	7.453
$Cl(2s^23p^5\ ^2P \rightarrow 2s^13p^6\ ^2S)$	9.714	9.165	9.817	8.869
$Ar^+(2s^23p^5\ ^2P \rightarrow 2s^13p^6\ ^2S)$	11.393	10.801	11.506	10.490

Note: The last column presents energies obtained using TD-DFT within a single-pole approximation. Numbers are given in atomic units.

TABLE 5.3
Transition Energies [29] Corresponding to Excited States of Some Atoms and Ions, as Calculated within HF Theory and Using LSD and MLSDSIC Functionals for Excited State: Electron Transfer from s Orbital to d Orbital

Atoms/Ions	ΔE^{HF}	ΔE^{LSD}	$\Delta E^{MLSDSIC}$	ΔE^{TD-DFT}
$Sc(3s^23d^1\ ^2D \rightarrow 3s^13d^2\ ^2G)$	2.156	1.858	2.122	1.865
$Ti(3s^23d^2\ ^3F \rightarrow 3s^13d^3\ ^3H)$	2.386	2.083	2.365	2.095
$V(3s^23d^3\ ^4F \rightarrow 3s^13d^4\ ^4H)$	2.610	2.311	2.611	2.327
$Mn(3s^23d^5\ ^6S \rightarrow 3s^13d^6\ ^6D)$	3.133	2.786	3.120	2.806
$Fe(3s^23d^6\ ^5D \rightarrow 3s^13d^7\ ^5F)$	3.419	3.048	3.453	3.076
$Co(3s^23d^7\ ^4F \rightarrow 3s^13d^8\ ^4F)$	3.762	3.318	3.796	3.352
$Ni(3s^23d^8\ ^3F \rightarrow 3s^13d^9\ ^3D)$	4.120	3.595	4.148	3.635

Note: The last column presents energies obtained using TD-DFT within a single-pole approximation. Numbers are given in atomic units.

TABLE 5.4

Transition Energies [29] When Two Electrons from s Orbital Are Excited to Corresponding p Orbital

Atoms/Ions	ΔE^{HF}	ΔE^{LSD}	$\Delta E^{MLSDSIC}$
Be($2s^2$ ^{1}S → $2p^2$ ^{1}D)	0.272	0.254	0.266
B($2s^22p^1$ ^{2}P → $2p^3$ ^{2}D)	0.470	0.412	0.480
C$^+$($2s^22p^1$ ^{2}P → $2p^3$ ^{2}D)	0.697	0.621	0.718
C($2s^22p^2$ ^{3}P → $2p^4$ ^{3}P)	0.743	0.595	0.731
N$^+$($2s^22p^2$ ^{3}P → $2p^4$ ^{3}P)	1.023	0.837	1.014
N($2s^22p^3$ ^{4}S → $2p^5$ ^{2}P)	1.179	0.944	1.179
O$^+$($2s^22p^3$ ^{4}S → $2p^5$ ^{2}P)	1.544	1.255	1.548
O($2s^22p^4$ ^{3}P → $2p^6$ ^{1}S)	1.503	1.133	1.474
F$^+$($2s^22p^4$ ^{3}P → $2p^6$ ^{1}S)	1.898	1.438	1.849
Mg($3s^2$ ^{1}S → $3p^2$ ^{1}D)	0.258	0.256	0.265
S($3s^23p^4$ ^{3}P → $3p^6$ ^{1}S)	1.027	0.781	1.027
P($3s^23p^3$ ^{4}S → $3p^5$ ^{2}P)	0.854	0.693	0.868

Note: Numbers are given in atomic units.

We have also applied [50] the MLSDSIC functional to calculate band gaps in a wide variety of semiconductors by exciting an electron from the top of the valence band to the bottom of the conduction band. These calculations have been performed with the linearized muffin tin orbital (LMTO) and give highly accurate band gaps. Note that band gaps calculated within the LDA underestimate the gap by about 50%.

In the above, we have presented results obtained by applying the MLSDSIC functional to excited states with one set of vacant orbitals. The question that arises is whether our method is general enough to construct functionals for other kinds of excited states. The method of splitting k-space to construct functionals for excited states is quite general and therefore easily extendable to other excited states. We have used the method to obtain [31] exchange energy functionals for excited states with two sets of vacant orbitals. Such an excited state is shown schematically in Figure 5.3.

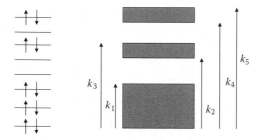

FIGURE 5.3 Orbital occupation in a two-gap excited-state configuration of an electronic system and its representation by HEG.

The expression for the functional is given by Shamim and Harbola [31]. Here we present only the results for the transition energy from single-electron excitations to excited-state configurations represented by a single Slater determinant. These are given in Table 5.5.

It is evident from the results that for these excited states also, the MLSDSIC functional leads to accurate results. For more results involving excitations of two or more electrons, we refer the reader to Shamim and Harbola [31].

So far, we have focused strictly on transition energies and demonstrated that they are given accurately by the MLSDSIC functional. This implies that the error in the ground-state energy given by the LSD functional and the error in the excited-state energy given by the MLSDSIC functional are of roughly the same magnitude. However, functionals [38–40,51] that correct the LSD functional by incorporating gradients of the density lead to highly accurate total energies for ground states. The question is whether the MLSDSIC functional can also be made accurate to the same order by including gradient corrections in it. Our ongoing work [52] indicates that, indeed, adding gradient corrections to the MLSDSIC functional leads to highly accurate total energies. In these calculations, we have performed a Kohn–Sham cal-

culation with the exchange potential $v_x(\mathbf{r}) = -\left[\dfrac{3\rho(\mathbf{r})}{\pi}\right]^{1/3}$ and calculated the total

energy with the MLSDSIC exchange energy functional with Becke [40] and Perdew–Wang [53] GGA corrections. Results thus obtained for some of the atomic excited states are shown in Table 5.6, where we show the numbers for HF energy and those calculated by using the Becke and Perdew–Wang corrections with the MLSDSIC exchange energy functional.

It is clear from the table that the total excited-state energies calculated by including the GGA correction are quite close to the HF energies. As a result, transition

---

TABLE 5.5
Transition Energies [31] for Excited-State Configurations with Two Gaps

Atoms/Ions	ΔE^{HS}	ΔE^{LSD}	$\Delta E^{MLSDSIC}$
$B(2s^12p^13s^1; M_L = 1, M_S = 1/2)$	0.406	0.380	0.437
$C(2s^12p^23s^1; M_L = 1, M_S = 1)$	0.658	0.575	0.706
$O(2s^12p^43p^1; M_L = 2, M_S = 1)$	1.230	1.044	1.313
$F(2s^12p^53p^1; M_L = 2, M_S = 1/2)$	1.463	1.259	1.556
$Ne(2s^12p^64s^1; M_L = 0, M_S = 0)$	1.796	1.538	1.837
$Mg(2s^12p^63s^24s^1; M_L = 0, M_S = 0)$	3.490	3.170	3.568
$Si(2s^12p^63s^23p^24s^1; M_L = 1, M_S = 1)$	5.785	5.385	5.914
$P(2s^12p^63s^23p^34s^1; M_L = 0, M_S = 3/2)$	7.095	6.656	7.186
$S(2s^12p^63s^23p^44s^1; M_L = 1, M_S = 1)$	8.547	8.069	8.648
$Cl(2s^12p^63s^23p^54s^1; M_L = 1, M_S = 1/2)$	10.114	9.548	10.224

Note: The near–exact exchange-only numbers are calculated using the Harbola–Sahni exchange potential [34,35]. Numbers are given in atomic units.

TABLE 5.6

Total Energies [52] of Excited States of Some Atoms, as Calculated within HF Theory $\left(E_{HF}^*\right)$ and Using MLSDSIC Functional with Becke $\left(E_{Becke}^*\right)$ and Perdew–Wang $\left(E_{PW}^*\right)$ GGA Corrections

Atoms/Ions	$-E_{HF}^*$	$-E_{Becke}^*$	$-E_{PW}^*$	ΔE^{HF}	ΔE^{Becke}	ΔE^{PW}
Li($2s^1\ ^2S \rightarrow 2p^1\ ^2P$)	7.365	7.357	7.366	0.068	0.070	0.075
B($2s^22p^1\ ^2P \rightarrow 2s^12p^2\ ^2D$)	24.312	24.304	24.330	0.217	0.211	0.220
C($2s^22p^2\ ^3P \rightarrow 2s^12p^3\ ^3D$)	37.395	37.376	37.409	0.294	0.303	0.315
N($2s^22p^3\ ^4S \rightarrow 2s^12p^4\ ^4P$)	53.988	53.989	54.026	0.413	0.409	0.423
O($2s^22p^4\ ^3P \rightarrow 2s^12p^5\ ^3P$)	74.184	74.181	74.227	0.626	0.631	0. 652
F$^+$($2s^22p^4\ ^3P \rightarrow 2s^12p^5\ ^3P$)	98.033	98.006	98.051	0.799	0.802	0.840
F($2s^22p^5\ ^2P \rightarrow 2s^12p^6\ ^2S$)	98.531	98.561	98.610	0.878	0.869	0.857
Ne$^+$($2s^22p^5\ ^2P \rightarrow 2s^12p^6\ ^2S$)	126.861	126.745	126.791	1.083	1.074	1.111
S($3s^23p^4\ ^3P \rightarrow 3s^13p^5\ ^3P$)	397.079	397.053	397.165	0.426	0.436	0.454
Cl$^+$($3s^23p^4\ ^3P \rightarrow 3s^13p^5\ ^3P$)	458.523	458.484	458.596	0.526	0.544	0.564
Cl($3s^23p^5\ ^2P \rightarrow 3s^13p^6\ ^2S$)	458.917	458.900	459.015	0.565	0.567	0.585
Ar$^+$($3s^23p^5\ ^2P \rightarrow 3s^13p^6\ ^2S$)	525.598	525.574	525.686	0.677	0.671	0.703

Note: Atoms are shown with their ground-state and excited-state configurations. The corresponding transition energies (ΔE^{HF}, ΔE^{Becke}, and ΔE^{PW}) are also given. Numbers are given in atomic units.

energies calculated by subtracting the ground-state energies obtained with the same functionals also come out to be accurate. The Becke correction usually leads to slightly more accurate results than the Perdew–Wang correction.

After constructing and testing the exchange energy functional for excited states, we now wish to understand if the method of splitting k-space has a sound conceptual foundation or it just happens to work in the case of exchange energy. For this purpose, we apply the method to produce Thomas–Fermi–like and gradient-corrected functionals for the noninteracting kinetic energy of excited states. We show that the resulting functionals give equally impressive results for kinetic energy, thereby implying the correctness of the physics invoked to construct the functionals. We also perform a response function analysis of the kinetic energy functional to demonstrate why the method of splitting k-space could be the method of choice for constructing excited-state energy functionals.

5.7 KINETIC ENERGY FUNCTIONAL IN TERMS OF DENSITY

As mentioned above, in this section, we apply the method of splitting k-space to write the kinetic energy of an excited state in terms of the densities $\rho_1(\mathbf{r})$, $\rho_2(\mathbf{r})$, and $\rho_3(\mathbf{r})$ and to study its accuracy. We note, however, that in essentially all practical applications of DFT, the noninteracting kinetic energy is calculated exactly using Kohn–Sham orbitals. The purpose of this study is therefore to test the conceptual correctness of the method proposed for obtaining excited-state energy functionals.

It is very easy to write the noninteracting kinetic energy for the split k-space. In the zeroth order (akin to the Thomas–Fermi functional), it is given by the formula [32]

$$T_0^*[\rho] = \int \left(\frac{k_1^5(\mathbf{r})}{10\pi^2} - \frac{k_2^5(\mathbf{r})}{10\pi^2} + \frac{k_3^5(\mathbf{r})}{10\pi^2} \right) d\mathbf{r} \qquad (5.21)$$

for excited states with one gap (see Figure 5.2). Here, the asterisk (*) indicates that the functional is for an excited state. The functional will be referred to as modified Thomas–Fermi functional. We point out that the kinetic energy for an excited state with one gap is obtained by combining the kinetic energy densities corresponding to $\rho_1(\mathbf{r})$, $\rho_2(\mathbf{r})$, and $\rho_3(\mathbf{r})$ in exactly the same manner as is performed for the densities in Equation 5.12. The kinetic energy formula in terms of spin densities is easily written as

$$T_0^*[\rho_\alpha, \rho_\beta] = \frac{1}{2}T_0^*[2\rho_\alpha] + \frac{1}{2}T_0^*[2\rho_\beta]. \qquad (5.22)$$

We now test the accuracy of the functional above by applying it to calculate the approximate kinetic energy for some excited states and by comparing the results with the exact noninteracting kinetic energies of a Kohn–Sham calculation. For a proper perspective, we also calculate the approximate kinetic energy by employing the traditional Thomas–Fermi functional of Equation 5.4 with spin densities (generalization similar to Equation 5.22 above), with $k_F = [3\pi^2\rho(\mathbf{r})]^{1/3}$ for excited-state density $\rho(\mathbf{r})$.

Table 5.7 shows the exact kinetic energies obtained by performing a Kohn–Sham calculation with an excited-state configuration of different atoms. The

TABLE 5.7
Comparison of Exact Kohn–Sham Kinetic Energies with Kinetic Energies Obtained Using Ground-State Functionals of Equations 5.4 and 5.21 and Functionals of Equations 5.23 and 5.24

Atom	T_0	$T_0 + T_2$	T_0^*	$T_0^* + T_2^*$	T_{exact}
Be($1s^22s^02p^03s^2$)	12.278	13.768	12.459	13.945	13.768
O($1s^22s^02p^6$)	64.154	70.068	67.545	73.704	73.094
O($1s^22s^02p^03s^23p^4$)	56.967	63.516	59.834	66.291	65.764
O($1s^22s^02p^03s^03p^6$)	56.344	62.815	59.885	66.313	65.506
Ne($1s^22s^02p^63s^2$)	109.521	118.891	116.152	125.947	124.508
Ne($1s^22s^02p^03s^23p^6$)	93.430	103.889	99.675	109.920	109.241
Mg($1s^22s^02p^63s^23p^2$)	169.083	182.740	180.095	194.392	191.942
Ar($1s^22s^02p^63s^23p^64s^2$)	443.200	474.770	474.671	507.648	501.507

Note: Numbers are given in atomic units.

exchange-correlation potential used is that given by Gunnarsson–Lundqvist param-eterization [44]. The exact kinetic energies are shown in the last column. In the sec-ond and fourth columns, we show the approximate kinetic energies obtained by the functionals of Equations 5.4 and 5.21, respectively, employed in their spin-density form given by Equation 5.22.

These results show that while the traditional Thomas–Fermi functional underes-timates the exact kinetic energy, the modified Thomas–Fermi functional gives num-bers closer to the exact values. This indicates the correctness of splitting k-space to obtain functionals for excited states.

For the ground-state theory, the Thomas–Fermi functional can be made more accurate by adding the gradient correction to it. For ground states, the gradient cor-rection up to the second order is given as [3–5]

$$T_2[\rho] = \frac{1}{72} \int \frac{|\nabla \rho(\mathbf{r})|^2}{\rho(\mathbf{r})} d\mathbf{r}. \tag{5.23}$$

Its generalization to the split k-space is the modified gradient correction [32]

$$T_2^*[\rho] = \frac{1}{72} \int \frac{|\nabla \rho_1(\mathbf{r})|^2}{\rho_1(\mathbf{r})} d\mathbf{r} - \frac{1}{72} \int \frac{|\nabla \rho_2(\mathbf{r})|^2}{\rho_2(\mathbf{r})} d\mathbf{r} + \frac{1}{72} \int \frac{|\nabla \rho_3(\mathbf{r})|^2}{\rho_3(\mathbf{r})} d\mathbf{r}. \tag{5.24}$$

Notice again that, in Equation 5.24, the gradient corrections to the kinetic energy densities corresponding to $\rho_1(\mathbf{r})$, $\rho_2(\mathbf{r})$, and $\rho_3(\mathbf{r})$ have been combined in exactly the same manner as is performed for the densities in Equation 5.12. Approximate kinetic energies obtained by adding the gradient correction to the corresponding zeroth-order functionals are also shown in Table 5.7 in the third and fifth columns, respec-tively. As is evident from these results, functional $T_0^* + T_2^*$ gives numbers that are significantly better than those given by $T_0 + T_2$. We have also studied the effect of including the fourth-order correction in the kinetic energy functional and found that it yields numbers not very different from those obtained by including corrections up to the second order. Finally, we add that the functionals for one-gap excited states can be easily generalized to those with two gaps or those with only shell electrons. The results for these systems are as accurate as those for one-gap excited states. Thus, we conclude that splitting k-space gives a conceptually correct way of con-structing energy functionals for excited states.

The gradient correction to the kinetic energy of the ground state can be derived by writing the noninteracting response function and expanding it up to the second order in the wavevector of the applied perturbation. We now wish to do the same for HEG with split k-space. In Section 5.8, we write the noninteracting response func-tion for HEG with split k-space and analyze the kinetic energy functionals discussed in this section by expanding the response function. The analysis explains why split-ting k-space is a good way of designing energy functionals for excited states.

5.8 RESPONSE FUNCTION FOR HEG WITH SPLIT k-SPACE

The noninteracting response function $\chi_0^*(k_1,k_2,k_3;q)$ for split k-space is given by combining the response function $\chi_0(k_F;\ q)$ for the ground state of HEG with Fermi wavevector k_F. This is done in a manner similar to that for density and kinetic energy density given by Equations 5.12 and 5.21, respectively. Thus [43],

$$\chi_0^*(k_1,k_2,k_3;q) = \chi_0(k_1;q) - \chi_0(k_2;q) + \chi_0(k_3;q). \tag{5.25}$$

We now expand the response function above up to order q^2 to analyze the gradient expansion of the kinetic energy, as is performed for the ground states [54]. Since [55]

$$\chi_0(k;q) = -\frac{1}{\pi^2 q}\left[\frac{qk}{2} + \left(\frac{k^2}{2} - \frac{q^2}{8}\right)\ln\left|\frac{q+2k}{q-2k}\right|\right], \tag{5.26}$$

expansion of $\chi_0^*(k_1,k_2,k_3;q)$ up to the second order in q gives

$$\chi_0^*(k_1,k_2,k_3;q) \approx -\frac{1}{\pi^2}\left[k_1 - k_2 + k_3 - \frac{q^2}{12}\left(\frac{1}{k_1} - \frac{1}{k_2} + \frac{1}{k_3}\right)\right]. \tag{5.27}$$

Connection of the response function above with the kinetic energy is as follows. For small perturbations applied to HEG, the kinetic energy that is correct up to the second order in the density change $\Delta\rho(\mathbf{r})$ can be written as [1,54]

$$T[\rho] = T_0[\rho] + \frac{1}{2}\iint K(\mathbf{r}-\mathbf{r}')\Delta\rho(\mathbf{r})\Delta\rho(\mathbf{r}')\,d\mathbf{r}d\mathbf{r}'. \tag{5.28}$$

This can also be written, in terms of the Fourier transforms $\rho(\mathbf{q}) = \int \Delta\rho(\mathbf{r})e^{-i\mathbf{q}\cdot\mathbf{r}}d\mathbf{r}$ of density change $\Delta\rho(\mathbf{r})$ and $K(\mathbf{q})$ of kernel $K(|\mathbf{r}-\mathbf{r}'|)$, as

$$T = T_0 + \frac{1}{2}\iint K(\mathbf{q})\left|\rho(\mathbf{q})\right|^2 d\mathbf{q}. \tag{5.29}$$

By making the total energy stationary with respect to density fluctuations, it is easily shown [1,43,53] that $K(\mathbf{q}) = -\dfrac{1}{\chi_0(\mathbf{q})}$ for the ground state and $K(\mathbf{q}) = -\dfrac{1}{\chi_0^*(\mathbf{q})}$ for excited states. By expanding $K(\mathbf{q})$ up to the second order in $\mathbf{q}$, we then get the expansion of the kinetic energy up to the second order in $|\nabla\rho|$ using the fact that $\nabla\rho$ $\propto q\rho(q)$. Employing the expansion of Equation 5.27, we obtain [43]

$$T[\rho] = T_0[\rho] + \frac{\pi^2}{2} \int \frac{|\Delta\rho(\mathbf{r})|^2}{(k_1 - k_2 + k_3)} d\mathbf{r} + \frac{\pi^2}{24} \int \frac{|\nabla\rho(\mathbf{r})|^2}{(k_1 - k_2 + k_3)^2} \left(\frac{1}{k_1} - \frac{1}{k_2} + \frac{1}{k_3} \right) d\mathbf{r}.$$

(5.30)

In the expression above, if we let $k_1 = k_2$ and $k_3 = k_F$, then it corresponds to the ground state with Fermi wavevector k_F. In that case, the second term on the right-hand side gives the second-order functional derivative $\frac{\delta^2 T_0}{\delta\rho(\mathbf{r})\delta\rho(\mathbf{r}')}$ of the Thomas–Fermi functional of Equation 5.4, and the last term gives the second-order correction of Equation 5.23. On the other hand, for excited states, the second term cannot be written in terms of the density. Furthermore, in many cases, the last term increases exponentially in regions asymptotically far from the system, as discussed below.

We have shown [56] in the past that in atomic excited states, the density in asymptotic regions $(r \to \infty)$ far from the nucleus varies as $\exp\left(-2\sqrt{-2\varepsilon_{\max}}\, r\right)$, where $\varepsilon_{\max}$ is the highest occupied orbital energy. Hence, $k_i (i = 1,2,3) \sim \exp\left[-\frac{2}{3}\sqrt{-2\varepsilon_i}\, r \right]$, where $\varepsilon_i (i = 1,2,3)$ are the highest eigenvalues for region 1 (corresponding to the core orbitals), region 2 (corresponding to the core + vacant orbitals), and region 3 (corresponding to the core + vacant + shell orbitals). Thus, for excited states, $k_1 < k_2 < k_3$; therefore, the gradient term in the asymptotic limit is proportional to $\frac{|\nabla\rho_3|}{k_1 k_3^2} \sim \frac{k_3^4}{k_1}$. In terms of orbital energies ε_1, ε_2, and ε_3, this expression is $\exp\left[-\frac{2}{3}\left(\sqrt{2c_1} - 4\sqrt{2c_3} \right) r \right]$. Since $|\varepsilon_1| > |\varepsilon_2| > |\varepsilon_3|$, the lowest-order gradient term diverges in the asymptotic limit for $|\varepsilon_1| > 16|\varepsilon_3|$. Thus, if we try to expand the kinetic energy functional in terms of the density, we end up facing divergent integrands. Furthermore, as pointed out above, the second term representing $\frac{\delta^2 T_0}{\delta\rho(\mathbf{r})\delta\rho(\mathbf{r}')}$ also cannot be written in terms of the density alone. In writing excited-state functionals, it is therefore better to work in terms of densities $\rho_1(\mathbf{r})$, $\rho_2(\mathbf{r})$, and $\rho_3(\mathbf{r})$ right from the beginning.

5.9   CONCLUDING REMARKS AND FUTURE WORK

In the above discussion, we have generalized LDA to the exchange energy of ground states to calculate exchange energies for excited states, too. This has been performed by splitting the k-space for HEG. The generalizability of this method to different kinds of excited states has also been demonstrated. Furthermore, the correctness of the conceptual foundations of this method is shown by the accuracy of the modified

Thomas–Fermi and gradient-corrected kinetic energy functionals. Response function analysis of HEG with split k-space shows that, to develop excited-state energy functionals, it is a good idea to work with densities corresponding to different regions of split k-space rather than working with the total density.

In this review, we have discussed only the functionals and have not paid attention to their functional derivative. The latter gives the corresponding exchange potential. Taking a direct functional derivative of the functional proposed is not possible. We have therefore tried to model the potential in various ways: (1) by taking the potential to be the HF potential corresponding to wavevector k_3—this gives the exchange potential

$$v_x(\mathbf{r}) = -\left[\frac{3\rho_3(\mathbf{r})}{\pi}\right]^{1/3}\left[1 + x_1 - x_2 - \frac{1}{2}\left(1 - x_1^2\right)\ln\left|\frac{1+x_1}{1-x_1}\right| + \frac{1}{2}\left(1 - x_2^2\right)\ln\left|\frac{1+x_2}{1-x_2}\right|\right]$$

where

$$x_1 = \frac{k_1}{k_3}, x_2 = \frac{k_2}{k_3};$$

(2) by taking the functional derivative of the MLSD functional with respect to $\rho_3(\mathbf{r})$—this makes the potential the same as in (1); and (3) by modeling the potential with carefully chosen parameters. All the three potentials yield encouraging results in terms of satisfying the theorems of Janak [57] and Levy–Perdew [58].

We are now applying the method to study the excited states of molecular systems. Work on the development of correlation energy functional is also in progress. Inclusion of gradient correction in the MLSDSIC functional has yielded very accurate total energies for atoms, and this work is being extended to calculate the total energies of molecules. It is hoped that the ideas on e-DFT presented here will make calculations for excited states as easy as the original DFT has done for ground states.

ACKNOWLEDGMENTS

We thank the editors, Dr. P. K. Chattaraj and Dr. S. K. Ghosh, for inviting us to write this chapter. M. Hemanadhan thanks the Council of Scientific and Industrial Research (CSIR) for financial support in the form of CSIR Junior Research Fellowship.

REFERENCES

1. Hohenberg, P.; Kohn, W. *Phys. Rev.* **1964**, 136, B864–B871.
2. Kohn, W.; Sham, L. J. *Phys. Rev.* **1965**, 140, A1133–A1138.
3. Parr, R. G.; Yang, W. *Density Functional Theory of Atoms and Molecules*, Oxford University Press, New York, **1989**.
4. Dreizler, R. M.; Gross, E. K. U. *Density Functional Theory*, Springer-Verlag, Berlin, **1990**.

5. March, N. H. *Electron Density Theory of Atoms and Molecules*, Academic Press, London, **1992**.
6. Harbola, M. K.; Banerjee, A. *J. Theor. Comput. Chem.* **2003**, 2, 301–322.
7. Casida, M. E. *Recent Advances in Density Functional Methods, Part I*; Chong, D. P.; Ed.; World Scientific, Singapore, **1995**, pp. 155–192.
8. Petersilka, M.; Gossmann, U. J.; Gross, E. K. U. *Phys. Rev. Lett.* **1996**, 76, 1212–1215.
9. Runge, E.; Gross, E. K. U. *Phys. Rev. Lett.* **1984**, 52, 997–1000.
10. Ziegler, T.; Rauk, A.; Baerends, E. J. *Theor. Chim. Acta* **1977**, 43, 261–271.
11. Gunnarsson, O.; Lundqvist, B. I. *Phys. Rev. B* **1976**, 13, 4274–4298.
12. Barth, U. V. *Phys. Rev. A* **1979**, 20, 1693–1703.
13. Theophilou, A. K. *J. Phys. C* **1979**, 12, 5419–5430.
14. Gross, E. K. U.; Oliveira, L. N.; Kohn, W. *Phys. Rev. A* **1988**, 37, 2809–2820.
15. Oliveira, L. N.; Gross, E. K. U.; Kohn, W. *Phys. Rev. A* **1988**, 37, 2821–2833.
16. Nagy, Á. *Phys. Rev. A* **1996**, 53, 3660–3663.
17. Görling, A. *Phys. Rev. A* **1999**, 59, 3359–3374.
18. Levy, M.; Nagy, Á. *Phys. Rev. Lett.* **1999**, 83, 4361–4364.
19. Nagy, Á.; Levy, M. *Phys. Rev. A* **2001**, 63, 052502(1)–052502(6).
20. Harbola, M. K. *Phys. Rev. A* **2002**, 65, 052504(1)–052504(6).
21. Harbola, M. K. *Phys. Rev. A* **2004**, 69, 042512(1)–042512(5).
22. Harbola, M. K.; Shamim, Md.; Samal, P.; Rahman, M.; Ganguly, S.; Mookerjee, A. *AIP Conf. Proc.* **2009**, 1108, 54–70.
23. Ziegler, T.; Seth, M.; Krykunov, M.; Autschbach, J.; Wang, F. *J. Chem. Phys.* **2009**, 130, 154102(1)–154102(8).
24. Cullen, J.; Krykunov, M.; Ziegler, T. *Chem. Phys.* **2011**, 391, 11–18.
25. Nagy, Á.; Levy, M.; Ayers, P. *Chemical Reactivity Theory: A Density Functional View*; Chattaraj, P; Ed.; Taylor & Francis, London, **2008**, pp. 121–136.
26. Ayers, P. W.; Levy, M. *Phys. Rev. A* **2009**, 80, 012508(1)–012508(16).
27. Samal, P.; Harbola, M. K.; Holas, A. *Chem. Phys. Lett.* **2006**, 419, 217–222; erratum in *Chem. Phys. Lett.* **2006**, 422, 586–586.
28. Samal, P.; Harbola, M. K. *J. Phys. B At. Mol. Opt. Phys.* **2006**, 39, 4065–4080.
29. Samal, P.; Harbola, M. K. *J. Phys. B At. Mol. Opt. Phys.* **2005**, 38, 3765–3777.
30. Harbola, M. K.; Samal, P. *J. Phys. B At. Mol. Opt. Phys.* **2009**, 42, 015003.
31. Shamim, Md.; Harbola, M. K. *J. Phys. B At. Mol. Opt. Phys.* **2010**, 43, 215002.
32. Hemanadhan, M.; Harbola, M. K. *J. Mol. Struct. (Theochem)* **2010**, 943, 152–157.
33. Singh, R.; Deb, B. M. *Phys. Rep.* **1999**, 311, 47–94.
34. Harbola, M. K.; Sahni, V. *Phys. Rev. Lett.* **1989**, 62, 489–492.
35. Harbola, M. K. *Chemical Reactivity Theory: A Density Functional View*; Chattaraj, P.; Ed.; Taylor & Francis, London, **2008**, pp. 83–103.
36. Levy, M. *Proc. Natl. Acad. Sci. U. S. A.* **1979**, 76, 6062–6065.
37. Weizsacker, C. F. V. *Z. Phys.* **1935**, 96, 431–458.
38. Herman, F.; Van Dyke, J. P.; Ortenburger, I. B. *Phys. Rev. Lett.* **1969**, 22, 807–811.
39. Perdew, J. P.; *Phys. Rev. Lett.* **1985**, 55, 1665–1668.
40. Becke, A. D. *Phys. Rev. A* **1988**, 38, 3098–3100.
41. Thomas, L. H. *Proc. Cambridge Philos. Soc.* **1927**, 23, 542–548.
42. Fermi, E. *Rend. Accad. Naz. Lincei* **1927**, 6, 602–607.
43. Hemanadhan, M.; Harbola, M. K. *Eur. Phys. J. D* **2012**, 66, 57.
44. Gunnarsson, O.; Lundqvist, B. I. *Phys. Rev. B* **1976**, 13, 4274–4298.
45. Perdew, J. P.; Zunger, A. *Phys. Rev. B* **1981**, 23, 5048–5079.
46. Vosko, S. H.; Wilk, L.; Nusair, M. *Can. J. Phys.* **1980**, 58, 1200–1211.
47. Ceperley, D. M.; Alder, B. J. *Phys. Rev. Lett.* **1980**, 45, 566–569.
48. Froese Fischer, C.; Brage, T.; Johnsson, P. *Computational Atomic Structure: An MCHF Approach*, 1st ed., Institute of physics publishing, Bristol and Philadelphia, **1997**, 35–66.

49. Gaspar, R. *Acta Phys. Acad. Sci. Hung,* **1954**, 3, 263–286.
50. Rahaman, M.; Ganguly, S.; Samal, P.; Harbola, M. K.; Saha-Dasgupta, T.; Mookerjee, A. *Physica B* **2009**, 404, 1137–1142.
51. Tau, J.; Perdew, J. P.; Staroverov, V. N.; Scuseria, E. *Phys. Rev. Lett.* **2003**, 91, 146401(1)–146401(4).
52. Shamim, Md.; Harbola, M. K. arXiv:1201.5970[physics.chem-ph].
53. Perdew, J. P.; Wang, Y. *Phys. Rev. B* **1986**, 33, 8800–8802; erratum in *Phys. Rev. B* **1989**, 40, 3399–3399.
54. Jones, R. O.; Gunnarsson, O. *Rev. Mod. Phys.* **1989**, 61, 689–746.
55. Ziman, J. M. *Principles of the Theory of Solids*, Cambridge University Press, Cambridge, **1979**.
56. Shamim, Md.; Harbola, M. K. *Chem. Phys. Lett.* **2008**, 464, 135–138.
57. Janak, J. F. *Phys. Rev. B* **1978**, 18, 7165–7168.
58. Levy, M.; Perdew, J. P. *Phys. Rev. A* **1985**, 32, 2010–2021.

6 Benchmark Studies of Spectroscopic Parameters for Hydrogen Halide Series via Scalar Relativistic State-Specific Multireference Perturbation Theory*

Avijit Sen, Lan Cheng, and Debashis Mukherjee

CONTENTS

6.1 INTRODUCTION

Both relativistic[1] and electron correlation[2,3] effects have been demonstrated to be essential to an accurate computation of heavy element–containing systems. In view of the tremendous development of relativistic methodologies[4-7] and correlation theories,[8-10] it might be envisaged that their combination will provide powerful general tools for chemical applications of the entire periodic table.

For potential energy surfaces (PES) involving bond-breaking, there is a combined interplay of three physical effects: (1) steady increase in nondynamical correlation as

---

* We dedicate this paper to Professor Bidyendu Mohan Deb on the happy occasion of his reaching 70. We take this opportunity to wish him good health and many more years of creative pursuit.

119

a bond is stretched, leading all the way to dissociation; (2) contribution of dynamical correlation, which is important throughout PES, although it essentially becomes the correlation of fragments at the dissociation limit; and (3) increasing importance of relativistic effects as the atoms constituting the molecule become heavier. Their effects are not additive, making the study of bond-breaking such a challenging proposition. For a balanced treatment of both dynamical and nondynamical corrections, it is necessary to use a correlation theory that is of potentially uniform precision throughout PES to as much extent as is possible. One important property that ensures this is maintaining size extensivity[11] in the entire range of PES. A multireference-based perturbation theory has the potential to treat large systems and those with large basis sets and will presumably be a promising method of choice for combining twin advantages—ease of implementation and simultaneous treatment of dynamical and nondynamical correlation effects. Among extant multireference perturbation theories, those based on effective Hamiltonians suffer from intruders[12,13] and, hence, are not uniformly precise in PES studies. State-specific multireference perturbation theories (SSMRPT), targeting one specific state of interest, will obviate this difficulty. Of all such theories in vogue, the one developed by Mahapatra et al.,[14,15] Ghosh et al.,[16] and Evangelista et al.,[17] known as state-specific multireference second-order perturbation theory (SSMRPT2; also called MkMRPT2), is the only known rigorously size-extensive theory and will be our method of choice. Since we will be using a spin-free (SF) Hamiltonian in our implementations, we will employ the recent version, spin-free SSMRPT2 (SF-SSMRPT2),[18,19] to dovetail the SF Hamiltonian and an SF wavefunction.

Concomitant to the simultaneous interplay of these three effects mentioned above, there is a need to use optimized orbitals in a multiconfigurational scenario to capture nondynamical correlations throughout PES at a minimalist conceptual description, usually at the complete active space self-consistent field (CASSCF) level. The complete active space (CAS) should of course contain configurations essential to describe bond-breaking. Unfortunately, the use of rigorous Dirac–Coulomb (DC) Hamiltonian at the CASSCF level[20,21] is rather demanding in comparison to nonrelativistic approaches because of spin symmetry breaking and the presence of small components. More specifically, spin symmetry breaking essentially doubles the number of active orbitals and consequently leads to formidable overhead for configuration interaction computations within the CAS, while the need to handle small-component contributions greatly complicates the orbital optimization procedure. Consequently, the use of DC CASSCF is still somewhat limited. The complication caused by spin symmetry breaking can be obviated by splitting the DC Hamiltonian into SF and spin-dependent terms and then starting out with the SF portion. This SFDC approach[22] contains full scalar relativistic effects and preserves spin symmetry. Therefore, the corresponding SFDC CASSCF approach has the same dimension of CAS as that in nonrelativistic CASSCF. Spin–orbit effects, if deemed necessary, can be included later via perturbation theory by modifying the many-body effective operator in the active space, which is diagonalized to obtain the energy of the state of interest at various geometries.[23] The complication caused by small components, which can be removed by resorting to two-component theories, still exists in the SFDC approach. In various SF two-component theories,[24–31] one converts the SF Dirac operator into a block-diagonal form with two noninteracting sectors of positive and negative energy states. The transformed positive energy block has an exact equivalence in the eigenspectra of the positive energy portion of the SF Dirac Hamiltonian. This "electrons-only" block of the transformed Hamiltonian is then combined with untransformed Coulomb interaction in subsequent many-electron treatments. We note that equivalence at the one-electron level is lost for the many-electron case since the "picture change" in the one-electron operator is not carried out in two-electron interactions. Fortunately, neglect of scalar two-electron picture change effects introduces only small errors for valence shells.[32–35] The virtue of SF two-component approaches is their support for the direct use of nonrelativistic machinery for many-electron treatments with simple modifications of one-electron Hamiltonian, including the straightforward application of the existing nonrelativistic CASSCF code. Among the varieties of SF two-component methods, the recently established SF exact two-component (X2c) theory[29–31,36–47] appears particularly promising. The SFX2c-1e scheme employs a direct block diagonalization of the SF Dirac Hamiltonian[22] in its matrix representation, where "1e" denotes the use of untransformed two-electron interactions. The simple structure of the SFX2c formulation should be emphasized in comparison to other infinite-order relativistic approaches employing sequential transformation schemes.[48–51] In particular, analytic energy gradients, which are critical to chemical applications, have been shown to be easy to implement in the framework of SFX2c-1e.[52–54] Therefore, we might argue that SFX2c-1e will become a favorable option for treating scalar relativistic effects in future chemical applications.

In this article, we present pilot studies of the PES of a prototype hydrogen halide series (HF, HCl, HBr, and HI) using SF-SSMRPT2,[18,19,55–57] combined with the SFX2c-1e scheme. The aim is to provide a consistent description of the above three effects for these single bond-breaking processes. Since this is the first ever implementation of SFX2c-1e SF-SSMRPT2 formalism, our studies also provide benchmark results in this sense.

6.2 THEORY

6.2.1 SF-SSMRPT(MP) Theory

In this subsection, we sketch the essence of SSMRPT2,[14,17–19,55–57] which is derived from the quasi-linearized form of the parent state-specific multireference coupled cluster (SSMRCC) theory (also known as MkMRCC theory in the literature).[58–60] The Jeziorski–Monkhorst (JM) ansatz[61] for the wavefunction is adopted in the parent SSMRCC theory:

$$\Psi = \sum_\mu e^{T^\mu} \phi_\mu c_\mu, \qquad (6.1)$$

where $\{|\phi_\mu\rangle\}$ is a set of model functions that span a CAS. T_μ's are cluster operators that excite model functions $\{\phi_\mu\}$ to virtual functions $\{\chi_l\}$. Since χ_l, in general, can be reached from more than one ϕ_μ, there exists an inherent redundancy of cluster amplitudes. Mahapatra et al.[58] exploited a suitable sufficiency condition to resolve this redundancy and obtained an SSMRCC formulation that is both size-extensive and intruder-free. SSMRCC working equations can be written as

$$\left\langle \chi_l \left| \bar{H}_\mu \right| \phi_\mu \right\rangle + \sum_v \left\langle \chi_l \left| e^{-T^\mu} e^{T^v} \right| \phi_\mu \right\rangle \left\langle \phi_\mu \left| \bar{H}_v \right| \phi_v \right\rangle c_v / c_\mu = 0, \; \forall \; l, \mu, \tag{6.2}$$

where the similarity-transformed Hamiltonian $\bar{H}_\mu$ is given by

$$\bar{H}_\mu = e^{-T^\mu} H e^{T^\mu}. \tag{6.3}$$

The first term in Equation 6.2 is a "direct" term similar to the matrix elements of similarity-transformed Hamiltonians in single reference coupled cluster theory, while the second term is a "coupling" term that describes the coupling between model functions. It has been shown that the structure of the coupling term is crucial for ensuring that the theory is both size-extensive and intruder-free.

A perturbative approximation to SSMRCC that is suitable for applications to large systems can be derived by writing Equation 6.2 in its quasi-linearized form and then effecting an order-by-order expansion. We partition H into H_0 and V and keep terms up to first order:

$$\left\langle \chi_l \left| H + \left[H_0, T_\mu^{(1)} \right] \right| \phi_\mu \right\rangle + \sum_v \left\langle \chi_l \left| T_v^{(1)} - T_\mu^{(1)} \right| \phi_\mu \right\rangle W_{\mu v} c_v / c_\mu = 0, \tag{6.4}$$

where $T_\mu^{(1)}$ are the first-order cluster amplitudes. In the strict perturbative sense, one should use the same partitioning for the direct and coupling terms and keep terms up to first order. But this means that we cannot use the unperturbed Hamiltonian as diagonal or as just a one-body operator if we have to maintain an intruder-free denominator. Moreover, we will lose the simple perturbative structure if two-body components of H are included in H_0. These complications are avoided by treating the "direct" term in strict perturbative sense and using the full active portion of H for the "coupling term."[16,18,19] The structure of $W_{\mu v}$ depends on the type of perturbation used. In Rayleigh–Schrödinger (RS)–type SF-SSMRPT2, $W_{\mu v}$ is the bare Hamiltonian matrix element $H_{\mu v}$. In previous papers,[18,19] we have given a detailed account of SF-SSMRPT2 and a comparative study of Møller–Plesset (MP) and Epstein–Nesbet (EN) partitioning schemes in both RS and Brillouin–Wigner (BW) versions of the formulation. It has been shown that MP partitioning generally works well. EN partitioning sometimes exhibits erratic behaviors, although, in principle, it includes more physics and its absolute error with respect to full configuration interaction is generally less. Further studies are thus required for EN partitioning. The RS version of SF-SSMRPT2 works as well as the more expensive BW version. Therefore, we have adopted RS-type SF-SSMRPT2(MP) in the present computations.

In SF-SSMRPT2 formulation, the set of model functions $\{\phi_\mu\}$ comprises configuration state functions of the Gel'fand type. c_μ's appearing in the working equation are kept fixed in CASSCF coefficients and will be denoted by c_μ^0. In MP partitioning, we employ a diagonal one body H_0. The resulting first-order equation can be written as

$$t_\mu^{l(1)}(\mu) = \frac{H_{l\mu} + \displaystyle\sum_{\nu\neq\mu} t_\mu^{l(1)}(\nu) H_{\mu\nu} c_\nu^{(0)} / c_\mu^{(0)}}{E_{CAS} - H_{\mu\mu} + E_{0,\mu\mu} - E_{0,ll}}, \tag{6.5}$$

where E_{CAS} and E_0 are, respectively, the CASSCF energy and the expectation value of H_0 with respect to a model or virtual function. Here, $t_\mu^{l(1)}$ is the first-order amplitude $\langle \chi_l | T_\mu^{(1)} | \phi_\mu \rangle$, and $t_\nu^{l(1)}$ is the amplitude $\langle \chi_l | T_\nu^{(1)} | \phi_\mu \rangle$.

In SF-SSMRPT2(MP), we adopt a multipartitioning strategy[62] in the sense that the unperturbed H_0 is chosen according to the ϕ_μ that it acts upon:

$$H_0 = \sum_\mu H_{0\mu} |\phi_\mu\rangle\langle\phi_\mu|, \tag{6.6}$$

where we take $H_{0\mu}$ to be the diagonal part "d" of the generalized Fock operator F_μ given by

$$F_\mu^{d} = \sum_p \left[f_{c_p}^p + \sum_{u\in\phi_\mu} \left(v_{pu}^{pu} - \frac{1}{2} v_{pu}^{up} \right) D_{uu}^\mu \right] = \epsilon_{\mu p}, \tag{6.7}$$

where $f_{c_p}^p$ is the core Fock operator, p is a general label for any orbital, u denotes active orbitals occupied in ϕ_μ, and D_{uu}^μ is the density (occupancy) of orbitals u in ϕ_μ.

In general, $T_\mu^{(1)} = T_{\mu 1}^{(1)} + T_{\mu 2}^{(1)}$ with

$$T_{\mu 1}^{(1)} = \sum_{A,I} t_{\mu I}^{(1)A}\{E_I^A\}, \quad T_{\mu 2}^{(1)} = \frac{1}{2}\sum_{A,B,I,J} t_{\mu IJ}^{(1)AB}\{E_{IJ}^{AB}\}, \tag{6.8}$$

where A and B are generalized particles (union of virtual and active orbital sets), and I, J, etc., are generalized holes (union of holes and active orbital sets). The upper and lower labels cannot all be active, and active spectator excitations of the type $t_{\mu Iu}^{(1)Au}$ are to be excluded. We note that $T_\mu^{(1)}$ is written in terms of SF unitary generators $\{E\}$ and in normal order with respect to the core taken as vacuum.

After solving for $T_\mu^{(1)}$, we obtain the "second-order" energy by diagonalizing the second-order matrix $\tilde{H}_{\mu\nu}^{[2]}$:

$$\sum_\nu \tilde{H}_{\mu\nu}^{[2]} c_\nu^{(1)} = E_r^{[2]} c_\mu^{(1)} \tag{6.9}$$

and

$$\tilde{H}^{[2]}_{\mu\nu} = \left\langle \phi_\mu \left| \overline{H}_\nu \right| \phi_\nu \right\rangle = H_{\mu\nu} + \sum_l H_{\mu l} t^{l^{(1)}}_\nu. \tag{6.10}$$

The energy $E^{(2)}_r$ obtained from Equation 6.9 corresponds to the value incorporating the relaxation of coefficient c_μ from $c^{(0)}_\mu$ into relaxed $c^{(1)}_\mu$. This energy is correct up to the second order but contains higher-order contributions also because of the diagonalization of the matrix $\tilde{H}^{[2]}_{\mu\nu}$. Hence, although determination of the cluster amplitude is performed with unperturbed (unrelaxed) configuration interaction (CI) coefficient $c^{(0)}_\mu$ (Equation 6.5), ultimately, the perturbed energy $E^{[2]}_r$ is obtained from Equation 6.9 in a relaxed manner via diagonalization. In case we do not wish to relax the CI coefficient, the energy $E^{[2]}_{ur}$ can be computed from

$$E^{[2]}_{ur} = \sum_{\mu\nu} c^{(0)}_\mu \tilde{H}^{[2]}_{\mu\nu} c^{(0)}_\nu. \tag{6.11}$$

We will refer to the relaxed treatment as MP and to the unrelaxed treatment as ⟨MP⟩. For details about the various choices of H_0 and the technique to accelerate the convergence of solving Equation 6.5, we refer the readers to our previous publications.[18,19]

6.2.2 SFX2c-1e Scheme

In quantum-chemical applications, the Dirac equation

$$\hat{D} \begin{pmatrix} \phi^L \\ \phi^S \end{pmatrix} = E \begin{pmatrix} \phi^L \\ \phi^S \end{pmatrix}, \quad \text{where } \hat{D} = \begin{pmatrix} \hat{V} & c(\vec{\sigma}\cdot\vec{p}) \\ c(\vec{\sigma}\cdot\vec{p}) & \hat{V} - 2c^2 \end{pmatrix}, \tag{6.12}$$

is usually discretized in terms of kinetically balanced basis[63]

$$\phi^L_p = \sum_\mu c^L_{\mu p} f_\mu, \quad \phi^S_p = \sum_\mu c^S_{\mu p} \frac{\sigma p}{2c} f_\mu \tag{6.13}$$

to ensure a proper description of relativistic kinetic energy in a finite basis expansion. The resulting matrix equation can be written as

$$D \begin{pmatrix} C^L \\ C^S \end{pmatrix} = \begin{pmatrix} S & 0 \\ 0 & \dfrac{1}{2c^2} T \end{pmatrix} \begin{pmatrix} C^L \\ C^S \end{pmatrix} E, \quad D = \begin{pmatrix} V & T \\ T & \dfrac{1}{4c^2} W - T \end{pmatrix}, \tag{6.14}$$

with corresponding matrix elements given by

$$S_{\mu\nu} = \langle f_\mu | f_\nu \rangle, \quad T_{\mu\nu} = \left\langle f_\mu \left| \frac{p^2}{2} \right| f_\nu \right\rangle, \quad V_{\mu\nu} = \left\langle f_\mu \left| \hat{V} \right| f_\nu \right\rangle, \quad W_{\mu\nu} = \left\langle f_\mu \left| (\vec{\sigma} \cdot \vec{p}) \hat{V} (\vec{\sigma} \cdot \vec{p}) \right| f_\nu \right\rangle.$$

The only spin-dependent term in Equation 6.14 $W_{\mu\nu}$ can be separated into an SF term $W_{\mu\nu}^{SF}$ and a spin–orbit term $W_{\mu\nu}^{SO}$ by applying the Dirac identity

$$W_{\mu\nu} = W_{\mu\nu}^{SO} + W_{\mu\nu}^{SF}, \quad W_{\mu\nu}^{SO} = \left\langle f_\mu \left| i\vec{\sigma} \cdot (\vec{p} \times \hat{V}\vec{p}) \right| f_\nu \right\rangle, \quad W_{\mu\nu}^{SF} = \left\langle f_\mu \left| \vec{p} \cdot (\hat{V}\vec{p}) \right| f_\nu \right\rangle.$$

One obtains the SF Dirac equation[22] when taking only the SF terms.

Four-component methods are computationally expensive since one has to deal with small-component integrals.[64] Therefore, various two-component methods in which small-component degrees of freedom are removed have flourished in the literature. We focus the present discussion on the X2c theory at one-electron level (X2c-1e).[38,42] The X2c-1e scheme consists of a one-step block diagonalization of the Dirac Hamiltonian in its matrix representation via a Foldy–Wouthuysen–type matrix unitary transformation[65]

$$U^\dagger D U = \begin{pmatrix} h_+^{FW} & 0 \\ 0 & h_-^{FW} \end{pmatrix}, \quad U = \begin{pmatrix} 1 & -X^\dagger \\ X & 1 \end{pmatrix} \begin{pmatrix} R & 0 \\ 0 & R' \end{pmatrix} \qquad (6.15)$$

and the use of the electronic block h_+^{FW} together with untransformed two-electron interactions in subsequent many-electron treatments. The "electrons-only" two-component Hamiltonian is given by

$$h_+^{FW} = R^\dagger L^{NESC} R, \qquad (6.16)$$

where

$$L^{NESC} = TX + X^\dagger T - X^\dagger TX + V + \frac{1}{4c^2} X^\dagger WX. \qquad (6.17)$$

The **X** matrix, which relates the large-component and small-component coefficients as $C^S = XC^L$, is determined by directly solving the four-component equation. The renormalization matrix **R** has been shown to take the following form:[41]

$$\mathbf{R} = \left(\tilde{S}^{-1} S \right)^{1/2}, \quad \tilde{S} = S + \frac{1}{2c^2} X^\dagger TX, \quad \mathbf{R} = S^{-1/2} \left(S^{-1/2} \tilde{S} S^{-1/2} \right)^{-1/2} S^{1/2}. \quad (6.18)$$

SFX2c-1e is obtained by using SF Dirac equation at the beginning of the above procedure. The underlying approximation of SFX2c-1e is the neglect of scalar two-electron picture change effect, which appears to be small for valence shells.[32–35] Since two-electron interactions are not transformed, the SFX2c-1e scheme can use available nonrelativistic implementation of any theory (such as orbital optimization in a CASSCF setting) in a straightforward manner. This convenience is essential to the present CASSCF and SF-SSMRPT2(MP) computations.

6.3 COMPUTATIONAL DETAILS

We have carried out computations of the PES of the hydrogen halide series with the SF-SSMRPT2 theory using both nonrelativistic and SFX2c-1e approaches. The nonrelativistic SF-SSMRPT2 computations have been undertaken to assess the importance of the effect of scalar relativistic contribution. In the present computations, we have used SFX2c-1e implementation[66] within a local version of the US-GAMESS program package[67] for CASSCF computations and generation of the one-electron and two-electron integrals in the molecular orbitals (MO) basis. The SF-SSMRPT2 program then reads these integrals and performs the RS-type SF-SSMRPT2 computations using the MP partition in the multipartitioning strategy described in Section 6.2. Both MP and $\langle$MP$\rangle$ results are reported. The atomic orbital type basis including relativistic and core correlation effects (ANO-RCC)[68] has been used in an uncontracted manner for all molecules. In all our computations, we have used pseudo-canonical CASSCF orbitals.

Spectroscopic parameters are obtained by fitting the PES to the Murrel–Sorbie (MS) function[69,70]

$$V(\rho) = -D_e(1 + a_1\rho + a_2\rho^2 + a_3\rho^3) \exp(-a_1\rho), \tag{6.19}$$

where $\rho = R - R_{eq}$, where R_{eq} is the equilibrium bond distance. The dissociation energy D_e and coefficients a_1, a_2, and a_3 are determined by fitting the PES to this analytic function. The quadratic, cubic, and quartic force constants are then calculated using the parameters in the MS potential function as

$$f_2 = D_e\left(a_1^2 - 2a_2\right), \quad f_3 = 6D_e\left(a_1a_2 - a_3 - \frac{a_1^3}{3}\right), \quad f_4 = D_ea_1^4 - 6f_2a_1^2 - 4f_3a_1. \tag{6.20}$$

Finally, the rotational constant B_e, the harmonic vibrational frequency ω_e, and the anharmonicity parameter $\omega_e\chi_e$ are computed as

$$B_e = \frac{h}{8\pi^2 c\mu R_e^2}, \quad \omega_e = \sqrt{\frac{f_2}{4\pi^2 mc^2}}, \quad \alpha_e = -\frac{6B_e^2}{\omega_e}\left(\frac{f_3R_e}{3f_2} + 1\right), \tag{6.21}$$

$$\omega_e\chi_e = \frac{B_e}{8}\left[-\frac{f_4R_e^2}{f_2} + 15\left(1 + \frac{\omega_e\alpha_e}{6B_e^2}\right)^2\right]. \tag{6.22}$$

6.4 RESULTS AND DISCUSSION

Tables 6.1 through 6.4 summarize the spectroscopic parameters for HF, HCl, HBr, and HI molecules calculated at CASSCF and SF-SSMRPT2 levels using the non-relativistic and SFX2c-1e schemes. Dynamical correlation effects are demonstrated as the differences between CASSCF and SSMRPT2 results, while scalar relativistic effects can be obtained by taking differences between SFX2c-1e and nonrelativistic computations.

Electron correlation generally plays an important role in computing the molecular properties of these molecules. CASSCF computations typically underestimate dissociation energies and harmonic frequencies. For example, dynamical correlation is particularly strong in HF molecules because of the high electronegativity of the fluorine atom. The dynamical correlation effect amounts to around 45 mH, more than 20% of the total correlation value. The second-order perturbative scheme in SSMRPT2 provides a moderately accurate description of dynamical correlation at

TABLE 6.1
HF Molecule with Uncontracted ANO-RCC

	Bond Distance r_e (Bohr)	Dissociation Energy D_e (mH)	Harmonic Frequency ω_e (cm^{-1})	Anharmonic Constant $\omega_e\chi_e$ (cm^{-1})
NR-CASSCF	1.728	182.918	4129	95.085
SFX2c-CASSCF	1.728	182.565	4127	95.261
NR–unrelaxed–SSMRPT	1.744	225.766	4140	84.686
SFX2c–unrelaxed–SSMRPT	1.744	225.470	4139	84.786
NR-SSMRPT	1.750	227.750	4177	97.929
SFX2c-SSMRPT	1.750	227.460	4176	97.968
Experimental[72]	1.7325	225.016	4138	89.88

TABLE 6.2
HCl Molecule with Uncontracted ANO-RCC

	Bond Distance r_e (Bohr)	Dissociation Energy D_e (mH)	Harmonic Frequency ω_e (cm^{-1})	Anharmonic Constant $\omega_e\chi_e$ (cm^{-1})
NR-CASSCF	2.430	144.780	2900	49.822
SFX2c-CASSCF	2.430	144.332	2895	49.808
NR–unrelaxed–SSMRPT	2.426	165.174	2976	49.668
SFX2c–unrelaxed–SSMRPT	2.426	164.868	2973	49.525
NR-SSMRPT	2.428	165.855	2981	52.812
SFX2c-SSMRPT	2.428	165.553	2977	52.669
Experimental[72]	2.4086	169.719	2991	52.05

TABLE 6.3

HBr Molecule with Uncontracted ANO-RCC

	Bond Distance r_e (Bohr)	Dissociation Energy D_e (mH)	Harmonic Frequency ω_e (cm^{-1})	Anharmonic Constant $\omega_e\chi_e$ (cm^{-1})
NR-CASSCF	2.706	126.179	2601	47.554
SFX2c-CASSCF	2.702	125.071	2590	47.532
NR–unrelaxed–SSMRPT	2.686	145.572	2677	44.404
SFX2c–unrelaxed–SSMRPT	2.680	144.992	2671	44.462
NR-SSMRPT	2.688	146.051	2665	44.885
SFX2c-SSMRPT	2.682	145.488	2659	44.975
Experimental[72]	2.6720	144.060	2649	45.21

TABLE 6.4

HI Molecule with Uncontracted ANO-RCC

	Bond Distance r_e (Bohr)	Dissociation Energy D_e (mH)	Harmonic Frequency ω_e (cm^{-1})	Anharmonic Constant $\omega_e\chi_e$ (cm^{-1})
NR-CASSCF	3.088	109.476	2285	41.942
SFX2c-CASSCF	3.078	107.318	2260	41.389
NR–unrelaxed–SSMRPT	3.058	123.532	2359	39.337
SFX2c–unrelaxed–SSMRPT	3.040	122.584	2353	39.718
NR-SSMRPT	3.060	124.093	2350	39.385
SFX2c-SSMRPT	3.040	123.175	2350	40.341
Experimental[72]	3.0406	117.290	2309	39.73

low computational cost for HF. Such strong dynamical correlation effects are not expected for other hydrogen halide systems.

Scalar relativistic effects increase steadily when one goes down the periodic table. They are essentially negligible for HF but become moderately important for HBr and more so for HI. Since the H–X bond is composed of a $1s$ orbital of hydrogen atom and s–p hybridized orbitals of halogen atom, scalar relativistic effects generally induce a bond contraction due to the relativistic spatial contraction of s-type and p-type orbitals.[1] The relativistic shortening of the bond distance becomes manifest for HBr and amounts to around 0.006 Bohr. For HI, it results in a 0.02 Bohr shortening of bond. The scalar relativistic effects are relatively small for other parameters (e.g., they amount to a few parts per million for harmonic frequencies).

We have observed errors with respect to experimental values of around 0.1–0.2 Bohr for equilibrium distances, 2–6 mH for dissociation energies, 10–50 wave numbers for frequencies, and 2–8 wave numbers for anharmonic constants. The present SFX2c-1e/SF-SSMRPT2(MP) model treats scalar relativistic, nondynamical correlation, and leading-order dynamical correlation effects consistently throughout the PES on the same footing. Since we have employed uncontracted ANO-RCC, core–valence correlation effects have also been taken into account quite adequately. The discrepancies with respect to the experimental values originate from the perturbative treatment of dynamical correlation and the neglect of spin–orbit effects. For example, as has been mentioned earlier, the HF molecule has particularly strong dynamical correlation effects and exhibits a relatively large error for almost all parameters. For HBr, spin–orbit effects have been shown to be around 6 mH for dissociation energy and several parts per million for harmonic frequencies.[71]

Figures 6.1 through 6.4 show the PES of the hydrogen halide series for both relaxed and unrelaxed SSMRPT2 in the relativistic case. To give an idea of the extent of changes in PES around equilibrium geometry, we have shown as inset an enlarged portion around r_e. SF-SSMRPT2 is not invariant with respect to the transformation among active orbitals. Thus, the results depend on whether pseudo-canonical or natural orbitals are used. As mentioned earlier, we have used pseudo-canonical CASSCF orbitals throughout our computations. Pseudo-canonical orbitals lead to quite large contributions of both Hartree–Fock–like and singly excited configuration state functions (CSFs) after the intermediate range of HX separation and their relatively important changes at intermediate distances, close to the region where PES flatten out. This avoided crossing leads to appearance of a plateau which is a consequence of the importance of singles for the ground state with pseudo-canonical orbitals. This trend prevails in all our computations. We have shown in the second

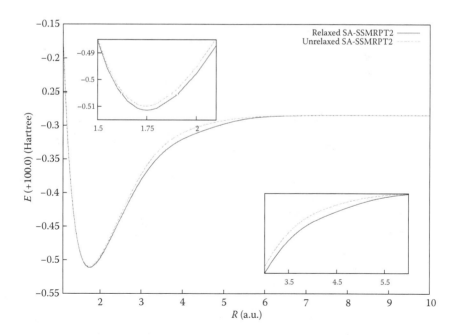

FIGURE 6.1 HF molecule with SFX2c-1e using the SF-SSMRPT2/ANO-RCC (uncontracted) theory in unrelaxed and relaxed versions.

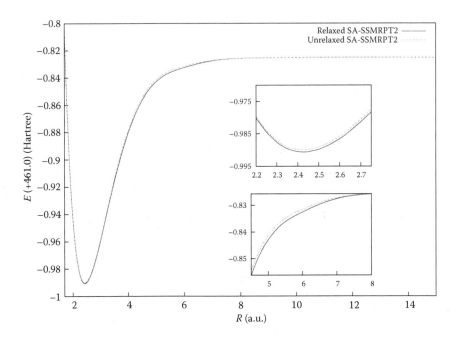

FIGURE 6.2   HCl molecule with SFX2c-1e using the SF-SSMRPT2/ANO-RCC (uncontracted) theory in unrelaxed and relaxed versions.

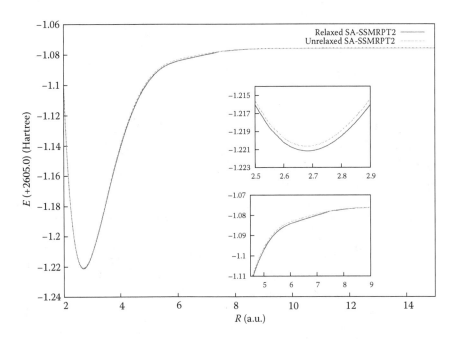

FIGURE 6.3   HBr molecule with SFX2c-1e using the SF-SSMRPT2/ANO-RCC (uncontracted) theory in unrelaxed and relaxed versions.

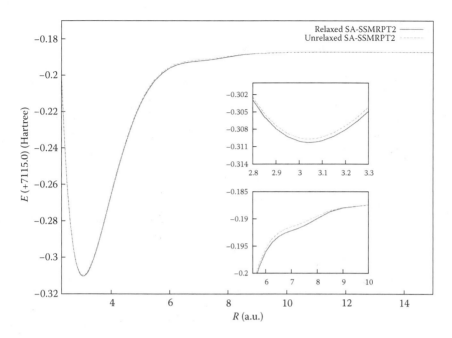

FIGURE 6.4   HI molecule with SFX2c-1e using the SF-SSMRPT2/ANO-RCC (uncontracted) theory in unrelaxed and relaxed versions.

inset a blown-up portion of the appropriate region. Although the extent of "softening" in the plateau differs for the relaxed and unrelaxed descriptions, the plateau nevertheless shows up in both versions. This trend is very much similar in the nonrelativistic computations as well, although we have not shown the corresponding plots so as not to proliferate the number of graphs. This difficulty of plateau may be avoided by using natural orbitals, although, in our opinion, the predominance of singles in CAS at longer bond lengths provides a physically more appealing picture. This is realized only if pseudo-canonical orbitals are used. Fortunately, the plateau appears in rather long bond distances and thus does not affect the spectroscopic parameters presented in this work.

ACKNOWLEDGMENTS

A.S. thanks the Council of Scientific and Industrial Research India for his research fellowship. L.C. thanks Jürgen Gauss (Mainz) and John F. Stanton (Austin) for helpful discussions and support. D.M. thanks the Alexander von Humboldt Foundation for the Humboldt Research Award and the Department of Science and Technology India for conferring on him the J. C. Bose National Fellowship. D.M. also thanks the India–European Union Science and Technology Cooperation Agreement for the MONAMI Project, the Indo-Swedish Bilateral Project, and the Indo-French CEFIPRA project for financial support. We also express our gratitude to Professors Swapan Ghosh and Pratim Chattaraj for kindly inviting us to contribute an article to this volume and for their kind patience as our article got hopelessly behind schedule.

REFERENCES

1. P. Pyykkö, *Chem. Rev.* **88**, 563 (1988).
2. K. P. Lawley, editor. *Advances in Chemical Physics: Ab Initio Methods in Quantum Chemistry, Part I*, vol. 67 (John Wiley, Chichester, England, 1987).
3. K. P. Lawley, editor. *Advances in Chemical Physics: Ab Initio Methods in Quantum Chemistry, Part II*, vol. 69 (John Wiley, Chichester, England, 1987).
4. K. G. Dyall, K. Fægri, *Relativistic Quantum Chemistry, Part III* (Oxford University Press, New York, 2007).
5. W. Liu, *Mol. Phys.* **108**, 1679 (2010).
6. W. Kutzelnigg, *Chem. Phys.* **395**, 16 (2012).
7. T. Saue, *ChemPhysChem* **12**, 3077 (2011).
8. T. Helgakar, P. Jørgensen, J. Olsen, *Molecular Electronic-Structure Theory* (Wiley, Chichester, 2000).
9. I. Shavitt, R. J. Bartlett, *Many-Body Methods in Chemistry and Physics: MBPT and Coupled-Cluster Theory* (Cambridge University Press, Cambridge, 2009).
10. W. Kutzelnigg, P. V. Herigonte, *Adv. Quantum Chem.* **36**, 186 (1999).
11. R. J. Bartlett, *Annu. Rev. Phys. Chem.* **32**, 359 (1981).
12. T. H. Schucan, H. W. Widenmueller, *Ann. Phys.* **73**, 108 (1972).
13. T. H. Schucan, H. W. Widenmueller, *Ann. Phys.* **76**, 483 (1973).
14. U. S. Mahapatra, B. Datta, D. Mukherjee, *Chem. Phys. Lett.* **299**, 42 (1999).
15. U. S. Mahapatra, B. Datta, D. Mukherjee, *J. Phys. Chem. A* **103**, 1822 (1999).
16. P. Ghosh, S. Chattopadhyay, D. Jana, D. Mukherjee, *Int. J. Mol. Sci.* **3** 733 (2002).
17. F. A. Evangelista, A. C. Simmonett, H. F. Schaefer III, D. Mukherjee, W. D. Allen, *Phys. Chem. Chem. Phys.* **11**, 4728 (2009).
18. S. Mao, L. Cheng, W. Liu, D. Mukherjee, *J. Chem. Phys.* **136**, 024105 (2012).
19. S. Mao, L. Cheng, W. Liu, D. Mukherjee, *J. Chem. Phys.* **136**, 024106 (2012).
20. H. J. Aa. Jensen, K. G. Dyall, T. Saue, K. Faegri Jr., *J. Chem. Phys.* **104**, 4083 (1996).
21. J. Thyssen, T. Fleig, H. J. Aa. Jensen, *J. Chem. Phys.* **129**, 034109 (2008).
22. K. G. Dyall, *J. Chem. Phys.* **100**, 2118 (1994).
23. A. Berning, M. Schweizer, H.-J. Werner, P. J. Knowles, P. Palmieri, *Mol. Phys.* **98**, 1823 (2000).
24. M. Douglas, N. M. Kroll, *Ann. Phys.* **82**, 89 (1974).
25. B. A. Hess, *Phys. Rev. A* **33**, 3742 (1986).
26. Ch. Chang, M. Pelissier, P. Durand, *Phys. Scr.* **34**, 394 (1986).
27. E. van Lenthe, E. J. Baerends, J. G. Snijders, *J. Chem. Phys.* **99**, 4597 (1993).
28. M. Barysz, A. J. Sadlej, J. G. Snijders, *Int. J. Quantum Chem.* **65**, 225 (1997).
29. K. G. Dyall, *J. Chem. Phys.* **106**, 9618 (1997).
30. W. Kutzelnigg, *Chem. Phys.* **225**, 203 (1997).
31. W. Kutzelnigg, W. Liu, *J. Chem. Phys.* **123**, 241102 (2005).
32. P. Knappe, N. Rösch, *J. Chem. Phys.* **92**, 1153 (1990).
33. R. Samzow, B. A. Hess, G. Jansen, *J. Chem. Phys.* **96**, 1227 (1992).
34. C. Park, J. E. Almlöf, *Chem. Phys. Lett.* **231**, 269 (1994).
35. C. van Wüllen, C. Michauk, *J. Chem. Phys.* **123**, 204113 (2005).
36. K. G. Dyall, *J. Chem. Phys.* **109**, 4201 (1998).
37. K. G. Dyall, T. Enevoldsen, *J. Chem. Phys.* **111**, 10000 (1999).
38. K. G. Dyall, *J. Chem. Phys.* **115**, 9136 (2001).
39. W. Kutzelnigg, W. Liu, *Mol. Phys.* **104**, 2225 (2006).
40. W. Liu, W. Kutzelnigg, *J. Chem. Phys.* **126**, 114107 (2007).
41. W. Liu, D. Peng, *J. Chem. Phys.* **131**, 031104 (2009).
42. M. Iliaš, T. Saue, *J. Chem. Phys.* **126**, 064102 (2007).

43. H. J. Aa. Jensen, Lecture given at the International Conference on Relativistic Effects in Heavy Element Chemistry and Physics (*REHE*; 2005).
44. W. Liu, D. Peng, *J. Chem. Phys.* **125**, 044102 (2006).
45. J. Sikkema, L. Visscher, T. Saue, M. Iliaš, *J. Chem. Phys.* **131**, 124116 (2009).
46. D. Cremer, E. Kraka, M. Filatov, *ChemPhysChem* **9**, 2510 (2008).
47. D. Peng, W. Liu, Y. Xiao, L. Cheng, *J. Chem. Phys.* **127**, 104106 (2007).
48. M. Reiher, A. Wolf, *J. Chem. Phys.* **121**, 10945 (2004).
49. D. Peng, K. Hirao, *J. Chem. Phys.* **130**, 044102 (2009).
50. M. Barysz, A. J. Sadlej, *J. Chem. Phys.* **116**, 2696 (2002).
51. M. Iliaš, H. J. Aa. Jensen, V. Kellö, B. O. Roos, M. Urban, *Chem. Phys. Lett.* **408**, 210 (2005).
52. L. Cheng, J. Gauss, *J. Chem. Phys.* **135**, 084114 (2011).
53. W. Zou, M. Filatov, D. Cremer, *J. Chem. Phys.* **134**, 244117 (2011).
54. L. Cheng, J. Gauss, *J. Chem. Phys.* **135**, 244104 (2011).
55. U. S. Mahapatra, S. Chattopadhyay, R. K. Chaudhuri, *J. Chem. Phys.* **129**, 024108 (2008).
56. U. S. Mahapatra, S. Chattopadhyay, R. K. Chaudhuri, *J. Chem. Phys.* **130**, 014101 (2009).
57. S. Chattopadhyay, U. S. Mahapatra, R. K. Chaudhuri, *Theor. Chem. Acc.* **131**, 1213 (2012).
58. U. S. Mahapatra, B. Datta, D. Mukherjee, *J. Chem. Phys.* **110**, 6171 (1999).
59. U. S. Mahapatra, B. Datta, B. Bandyopadhyay, D. Mukherjee, *Adv. Quantum. Chem.* **30**, 163 (1998).
60. U. S. Mahapatra, B. Datta, D. Mukherjee, *Mol. Phys.* **94**, 157 (1998).
61. B. Jeziorski, H. J. Monkhorst, *Phys. Rev. A* **24**, 1668 (1981).
62. A. Zaitsevskii, J. P. Malrieu, *Chem. Phys. Lett.* **250**, 366 (1996).
63. R. E. Stanton, S. Havriliak, *J. Chem. Phys.* **81**, 1910 (1984).
64. L. Visscher, *J. Comput. Chem.* **23**, 759 (2003).
65. L. L. Foldy, S. A. Wouthuysen, *Phys. Rev.* **78**, 29 (1950).
66. S. Pathak, L. Cheng, J. Gauss, D. Mukherjee, unpublished.
67. M. W. Schmidt, K. K. Baldridge, J. A. Boatz, S. T. Elbert, M. S. Gordon, J. H. Jensen, S. Koseki, N. Matsunaga, K. A. Nguyen, S. Su, T. L. Windus, M. Dupuis, J. A. Montgomery Jr., *J. Comput. Chem.* **14**, 1347 (1993).
68. B. O. Roos, R. Lindh, P. Malmqvist, V. Veryazov, P. Widmark, *J. Phys. Chem. A* **108**, 2851 (2004).
69. J. N. Murrel, K. S. Sorbie, *J. Chem. Soc. Faraday Trans.* **2**, 1552 (1974).
70. K. S. Sorbie, J. N. Murrel, *Mol. Phys.* **29**, 138 (1975).
71. T. Fleig, L. K. Sørensen, J. Olsen, *Theor. Chem. Acc.* **118**, 347 (2007).
72. K. P. Huber, G. Herzberg, *Constants of Diatomic Molecules* (Van Nostrand Reinhold, New York, 1979).

7 Local Virial Theorem for Ensembles of Excited States*

Á. Nagy

CONTENTS

7.1 INTRODUCTION

Virial theorem has an important role in quantum mechanics. It proved to be also very useful in density functional theory (see e.g. Parr and Yang, 1989). Several forms of virial theorem have been proposed in ground-state density functional theory, for example, local virial theorem (Nagy and Parr 1990), differential virial theorem (Holas and March 1995), regional virial theorem (Nagy 1992), and spin virial theorem (Nagy 1994b). In this chapter, local and differential virial theorems are extended to excited states in the frame of density functional theory.

The density functional theory was originally formalized for the ground state (Hohenberg and Kohn 1964). Gunnarsson and Lundqvist (1976) and Gunnarsson et al. (1979), von Barth, (1979) soon noticed that the original theory can also be applied to the lowest excited states with different symmetries. Earlier, Slater (1974) introduced the so-called transition-state method to calculate excitation energies. It proved to be a very efficient and simple approach and was used to solve a large variety of problems. The density functional theory was first rigorously generalized to excited states by Theophilou (1978). Formalisms for excited states have also been developed by Fritsche (1986, 1987) and English et al. (1988). The density functional theory has

---

* It is a great pleasure to dedicate this chapter to Professor Deb on his 70th birthday. Professor Deb has made very important contributions to quantum fluid dynamics. He introduced stress tensors to density functional theory. This chapter can be considered as an extension of his ideas to excited states.

also been extended by Gross et al. (1988a,b) and Oliveira et al. (1988) to ensembles of unequally weighted states.

The exact form of exchange-correlation energy functional is unknown both in the ground-state and in ensemble theories; therefore, approximations have to be applied. First, a quasi-local density approximation was proposed by Kohn (1986), and excitation energies of He atom were calculated to illustrate the method of Gross et al. Later, excitation energies of several atoms (Nagy 1990, 1991, 1995b, 1997b) were determined with several ground-state approximate functionals (Nagy and Andrejkovics 1994, 1998). Then, the optimized potential method (OPM) (Aashamar et al. 1978; Sharp and Horton 1953; Talman and Shadwick 1976) and the Krieger–Lee–Iafrate (KLI) method (Krieger et al. 1992a,b; Nagy 1997a) were generalized to ensembles of excited states. The first generalization was based on the ensemble Hartree–Fock method (Nagy 1998a) using the original ensemble theory of Gross et al. (1988a,b) and Oliveira et al. (1988). Then another extension was proposed based on a different partition of ensemble energy (Gidopoulos et al. 2001; Nagy 2001). A combination of this method and the self-interaction–free Perdew–Zunger approximation was applied to determine ensemble energies and excitation energies (Tasnádi and Nagy 2003a,b,c).

Important theorems (e.g., the adiabatic connection formula for ensemble exchange-correlation energy) (Nagy 1995a,c) were also derived in the ensemble theory (Nagy 2002a,b). Reactivity concepts, such as ensemble Kohn–Sham hardness, were introduced, and even the first excitation energy was proposed to substitute for hardness as a reactivity index (Nagy 2005). The relativistic generalization of ensemble formalism was also performed (Nagy 1994a). As an interesting example for the ensemble theory, two harmonically interacting electrons with antiparallel spins under isotropic harmonic confinement were studied (Nagy et al. 2005). [For reviews of excited-state theory, see Nagy (1997b) and Singh and Deb (1999).]

The virial theorem was also derived for ensembles of excited states (Nagy 2002a). In the ground-state theory, several forms of the virial theorem were derived. The local and differential forms proved to be especially useful. In this chapter, the local virial theorem is derived for ensembles of excited states. In Section 7.2, the ensemble theory of excited states is summarized. The ensemble local virial theorem is derived in Section 7.3. Extension of the differential virial theorem of Holas and March (1995) to ensembles is presented in Section 7.4. Finally, Section 7.5 is devoted to discussion.

7.2   ENSEMBLES OF EXCITED STATES IN DENSITY FUNCTIONAL THEORY

Consider the eigenvalue problem of the Hamiltonian

$$\hat{H} \, \Psi_k = E_k \Psi_k \quad (k = 1, \ldots, M),$$ (7.1)

where

$$E_1 \leq E_2 \leq \ldots \leq E_M.$$ (7.2)

Ensemble energy and density are defined as

$$\mathcal{E} = \sum_{k=1}^{M} w_k E_k \tag{7.3}$$

and

$$n = \sum_{k=1}^{M} w_k n_k, \tag{7.4}$$

where n_k is the electron density obtained from the wave function Ψ_k and

$$0 \le w_M \dots \le w_2 \le w_1. \tag{7.5}$$

Generalized Hohenberg–Kohn theorems are valid for the ensembles (Gross et al. 1988a,b; Oliveira et al. 1988). External potential is determined within a trivial additive constant by ensemble density. There exists a generalized Rayleigh–Ritz variational principle for ensemble energy (Gross 1988a): the ensemble energy functional takes its minimum at the correct ensemble density n. Kohn–Sham equations have the same form as in the ground-state theory

$$\left(-\frac{1}{2} \nabla^2 + v_{KS} \right) u_i = \epsilon_i u_i, \tag{7.6}$$

where u_i and ϵ_i are the ensemble orbitals and energies, respectively. The ensemble Kohn–Sham potential v_{KS} is different from the ground-state Kohn–Sham potential. The ensemble density can be expressed with the orbitals

$$n = \sum_{i=1}^{K} \lambda_i |u_i|^2, \tag{7.7}$$

where λ_i occupation numbers are not all integers. K denotes the number of orbitals with nonzero occupation numbers. The Euler equation can also be derived for the ensemble

$$\frac{\delta T_s}{\delta n} + v_{KS} = \mu, \tag{7.8}$$

where Lagrangian μ derives from keeping the total number of electrons N fixed:

$$\int n \, dr = N. \tag{7.9}$$

The noninteracting ensemble kinetic energy can be expressed with the ensemble orbitals

$$T_s = \sum_{i=1}^{K} \lambda_i \int u_i \left(-\frac{1}{2} \nabla^2 \right) u_i. \tag{7.10}$$

7.3 LOCAL VIRIAL THEOREM FOR ENSEMBLES OF EXCITED STATES

Here, real orbitals are considered following Deb and Ghosh (1979) in their derivation of the local virial theorem. Divide first the Kohn–Sham equations by the orbital u_i, take the gradient of the equations, multiply with the orbitals and occupation numbers, and sum for all i:

$$\text{div } \hat{\sigma}_s = n \nabla v_{KS}, \tag{7.11}$$

where stress tensor is defined as

$$\sigma_{s\,\mu\nu} = -\sum_{i=1}^{K} \lambda_i \frac{\partial u_i}{\partial x_\mu} \frac{\partial u_i}{\partial x_\nu} + \frac{1}{4} \nabla^2 n \, \delta_{\mu\nu}. \tag{7.12}$$

The stress tensor cannot be uniquely defined. Only the div of the stress tensor is unique (i.e., any tensor with the same div can also be used). Other definitions for the ground state can be found in Deb and Ghosh (1979) and Nagy and March (1997). Similar definitions are possible here for the ensemble case, too. Define the ensemble pressure as

$$p = -\frac{1}{3} tr \hat{\sigma}. \tag{7.13}$$

The ensemble pressure connected with noninteracting kinetic energy takes the form

$$p_s = \frac{2}{3} t_s, \tag{7.14}$$

where the noninteracting ensemble kinetic energy density is given by

$$t_s = -\frac{3}{8} \nabla^2 n + \frac{1}{2} \sum_{i=1}^{K} \lambda_i |\nabla u_i|^2. \tag{7.15}$$

Again, the kinetic energy density is not unique. Any function that integrates into the same kinetic energy as t_s can, in principle, be used. The t_s given here corresponds to the stress tensor $\hat{\sigma}_s$ above. The Kohn–Sham potential can be partitioned as

$$v_{KS} = v^* + v_{xc}, \tag{7.16}$$

where v^* is the total ensemble classical electrostatic potential, and v_{xc} is the ensemble exchange-correlation potential. The ensemble force equation has the form

$$\text{div } \hat{\sigma} = n\nabla v^*. \tag{7.17}$$

For the ground state, a similar force equation was introduced by Bartolotti and Parr (1980). The ensemble exchange-correlation stress tensor $\hat{\sigma}_{xc}$ can be defined as

$$\text{div } \hat{\sigma}_{xc} = n\nabla v_{xc}, \tag{7.18}$$

and the total ensemble stress tensor can be given by

$$\hat{\sigma} = \hat{\sigma}_s + \hat{\sigma}_{xc}. \tag{7.19}$$

Defining the ensemble exchange-correlation pressure as

$$p_{xc} = -\frac{1}{3} tr\hat{\sigma}_{xc} \tag{7.20}$$

leads to the ensemble local virial theorem

$$2t_s = 3(p - p_{xc}). \tag{7.21}$$

It has the same form as the ground-state local virial theorem (Nagy and Parr 1990).

7.4 ENSEMBLE DIFFERENTIAL VIRIAL THEOREM

The ensemble force equation derived above is the ensemble differential theorem. It can, however, be rewritten in the form of the differential theorem of Holas and March (1995)

$$2 \text{ div } \sigma_s^{HM} - \frac{1}{4}\nabla^2\nabla n = -n\nabla v_{KS}, \tag{7.22}$$

where

$$\sigma_{s\,\mu\nu}^{HM} = \frac{1}{2}\sum_{i=1}^{K}\lambda_i\frac{\partial u_i}{\partial x_\mu}\frac{\partial u_i}{\partial x_\nu}. \tag{7.23}$$

That is, the difference between the force equation and the differential virial theorem lies only in the definition of the stress tensor. The relationship between them is

$$2 \operatorname{div} \sigma_s^{HM} - \frac{1}{4} \nabla^2 \nabla n = -\operatorname{div} \hat{\sigma}_s. \tag{7.24}$$

The ensemble differential virial theorem for spherically symmetric systems (Nagy 2011) has been recently derived:

$$\tau' = -\frac{1}{8} \rho''' - \frac{1}{2} \rho v'_{KS} + \frac{q'}{r^2} - \frac{q}{r^3}, \tag{7.25}$$

where the ensemble radial kinetic energy has the form

$$\tau(r) = -\frac{1}{2} \sum_{i=1}^{K} \tilde{w}_i P_i'' P_i + \frac{q}{r^2}, \tag{7.26}$$

where P_i are the radial orbitals, ρ is the radial density

$$\rho = \sum_{i=1}^{K} \tilde{w}_i P_i^2, \tag{7.27}$$

and

$$q = \frac{1}{2} \sum_{i=1}^{K} \tilde{w}_i l_i (l_i + 1) P_i^2. \tag{7.28}$$

If the factors $\tilde{w}_i$ are equal to the weighting factors λ_i, the ensemble differential virial theorem for spherically symmetric systems is obtained. A similar theorem is valid for the ground state (Nagy and March 1989).

7.5 DISCUSSION

The ensemble virial theorems derived here have the same form as the ground-state virial theorems. The only difference is that they include quantities defined for the ensembles rather than for ground-state quantities. The reason for the similarity is that the ensemble Kohn–Sham equations have the same form as the ground-state equations.

It has been mentioned in passing that excited states can be treated not only in ensemble formalism but also in the density functional theory. There are theories for a single excited state (Levy and Nagy 1999; Nagy 1998b; Nagy and Levy 2001), and the integral form of the virial theorem was derived (Nagy 2002b).

The ensemble differential virial theorem for spherically symmetric systems (Nagy 2011) has recently turned out to be very useful in the solution of the orbital-free problem. A first-order differential equation has been derived for the functional derivative of the ensemble noninteracting kinetic energy functional:

$$\frac{1}{2} \rho \left(\frac{\delta T}{\delta n} \right)' + \rho' \frac{\delta T}{\delta n} = f, \tag{7.29}$$

where $'$ denotes the derivative with respect to the radial distance r and

$$f = -\frac{1}{8} \rho''' - \frac{\partial}{\partial r}\left(\frac{\partial \rho}{\partial \beta} \right) + \mu \rho' - \frac{1}{2r^2} \frac{\partial}{\partial r}\left(\frac{\partial \rho}{\partial \gamma} \right) + \frac{1}{2r^3} \frac{\partial \rho}{\partial \gamma}. \tag{7.30}$$

Here the factors $\tilde{w}_i$ are selected as

$$\tilde{w}_i = \lambda_i e^{\beta \epsilon_i - \gamma l_i (l_i + 1)}, \tag{7.31}$$

where β and γ are parameters. That is, the ensemble density depends not only on the radial distance r but also on the parameters β and γ: $\rho = \rho(r;\beta,\gamma)$. Instead of the original orbital-free problem, an enlarged task is solved, namely, the differential equation for the functional derivative of the ensemble noninteracting kinetic energy functional. The solution for the case $\beta = 0$ and $\gamma = 0$ provides the solution for the original orbital-free problem.

The virial theorems derived above are exact (i.e., exact functionals satisfy these equations). However, these equations do not generally hold for approximate functionals. Therefore, they can be used to judge the quality of the approximation. Exact theorems or equations can be very useful in constructing approximate functionals, too.

ACKNOWLEDGMENTS

This work was supported by projects TAMOP 4.2.1/B-09/1/KONV-2010-0007 and TAMOP 4.2.2/B-10/1-2010-0024. The project was cofinanced by the European Union and the European Social Fund. Grant OTKA No. K 100590 is also gratefully acknowledged.

REFERENCES

Aashamar, K., Luke, T. M. and Talman, J. D. 1978. *At. Data Nucl. Data Tables* 22: 443.
Bartolotti, I. and Parr, R. G. 1980. *J. Chem. Phys.* 72: 1593.
Deb, B. M. and Ghosh, S. K. 1979. *J. Phys. B* 12: 3857.
English, H., Fieseler, H. and Haufe, A. 1988. *Phys. Rev. A* 37: 4570.
Fritsche, L. 1986. *Phys. Rev. B* 33: 3976.
Fritsche, L. 1987. *Int. J. Quantum. Chem.* 21: 15.

Gidopoulos, N. I., Papakonstantinou, P. and Gross, E. K. U. 2001. *Phys. Rev. Lett.* 88: 033003.

Gross, E. K. U., Oliveira, L. N. and Kohn, W. 1988a. *Phys. Rev. A* 37: 2805.

Gross, E. K. U., Oliveira, L. N. and Kohn, W. 1988b. *Phys. Rev. A* 37: 2809.

Gunnarsson, O. and Lundqvist, B. I. 1976. *Phys. Rev.* 13: 4274.

Gunnarsson, O., Jonson, M. and Lundqvist, B. I. 1979. *Phys. Rev. B* 20: 3136.

Hohenberg, P. and Kohn, W. 1964. *Phys. Rev.* 136: B864.

Holas, A. and March N. H. 1995. *Phys. Rev. A* 51: 2040.

Kohn, W. 1986. *Phys. Rev. A* 34: 737.

Krieger, J. B., Li, Y. and Iafrate, G. J. 1992a. *Phys. Rev. A* 45: 101.

Krieger, J. B., Li, Y. and Iafrate, G. J. 1992b. *Phys. Rev. A* 46: 5453.

Levy, M. and Nagy, Á. 1999. *Phys. Rev. Lett.* 83: 4631.

Nagy, Á. 1990. *Phys. Rev. A* 42: 4388.

Nagy, Á. 1991. *J. Phys. B* 24: 4691.

Nagy, Á. 1992. *Phys. Rev. A* 46: 5417.

Nagy, Á. 1994a. *Phys. Rev. A* 49: 3074.

Nagy, Á. 1994b. *Int. J. Quantum Chem.* 49: 353.

Nagy, Á. 1995a. *Int. J. Quantum Chem.* 56: 225.

Nagy, Á. 1995b. *Int. J. Quantum Chem. Symp.* 29: 297.

Nagy, Á. 1995c. *J. Phys. B* 29: 389.

Nagy, Á. 1997a. *Phys. Rev. A* 55: 3465.

Nagy, Á. 1997b. *Adv. Quantum Chem.* 29: 159.

Nagy, Á. 1998a. *Int. J. Quantum Chem.* 69: 247.

Nagy, Á. 1998b. *Int. J. Quantum Chem.* 70: 681.

Nagy, Á. 2001. *J. Phys. B* 34: 2363.

Nagy, Á. 2002a. *Acta Phys. Chim. Debrecina* 34–35: 99.

Nagy, Á. 2002b. In *Recent Advances in Computational Chemistry*, Vol. 1, Part III. *Recent Advances in the Density Functional Method*, edited by V. Barone, A. Bencini and P. Fantucci, 247. Singapore: World Scientific.

Nagy, Á. 2005. *J. Chem. Sci.* 117: 437.

Nagy, Á. 2011. *J. Chem. Phys.* 135: 044106.

Nagy, Á. and Andrejkovics, I. 1994. *J. Phys. B* 27: 233.

Nagy, Á. and Andrejkovics, I. 1998. *Chem. Phys. Lett.* 296: 489.

Nagy, Á. and Levy, M. 2001. *Phys. Rev. A* 63: 2502.

Nagy, Á. and March, N. H. 1989. *Phys. Rev. A* 40: 554.

Nagy, Á. and March, N. H. 1997. *Mol. Phys.* 91: 597.

Nagy, Á. and Parr, R. G. 1990. *Phys. Rev. A* 42: 201.

Nagy, Á., Howard, I. A., March, N. H. and Jánosfalvi, Zs. 2005. *Phys. Lett. A* 335: 347.

Oliveira, L. N., Gross, E. K. U. and Kohn, W. 1988. *Phys. Rev. A* 37:2821.

Parr, R. G. and Yang, W. 1989. *Density Functional Theory of Atoms and Molecules*. New York: Oxford University Press.

Sharp, R. T. and Horton, G. K. 1953. *Phys. Rev.* 30: 317.

Singh, R. and Deb, B. D. 1999. *Phys. Rep.* 311: 47.

Slater, J. C. 1974. *Quantum Theory of Molecules and Solids*, vol. 4. New York: McGraw-Hill.

Talman, J. D. and Shadwick, W. F. 1976. *Phys. Rev. A* 14: 36.

Tasnádi, F. and Nagy, Á. 2003a. *Int. J. Quantum Chem.* 92: 234.

Tasnádi, F. and Nagy, Á. 2003b. *J. Phys. B* 36: 4073.

Tasnádi, F. and Nagy, Á. 2003c. *J. Chem. Phys.* 119: 4141.

Theophilou, A. K. 1978. *J. Phys. C* 12: 5419.

von Barth, U. 1979. *Phys. Rev. A* 20: 1693.

8 Information-Theoretic Probes of Chemical Bonds

Roman F. Nalewajski

CONTENTS

8.1 INTRODUCTION

Information theory (IT) [1–8] is one of the youngest branches of applied probability theory, in which probability ideas have been introduced into the fields of communication, control, and data processing. Its foundations were laid in the 1920s by Fisher [1] in his classical measurement theory and in the 1940s by Shannon [3] in his mathematical theory of communication. The electronic quantum state of a molecule is determined by the system wave function, the amplitude of the particle probability distribution that carries information. It is thus intriguing to explore the information content of electron probability distributions in molecules and to extract from it patterns of chemical bonds, reactivity trends, and other molecular descriptors (e.g., bond multiplicities ["orders"] and their covalent/ionic composition). Indeed, it directly follows from the Hohenberg–Kohn theorems of density functional theory (DFT) [9] that all properties of molecular systems are unique functionals of their electron density in their nondegenerate ground state. In this survey, some recent developments in IT probes of molecular electronic structure are summarized. In particular, this survey will examine information displacements due to subtle electron redistributions accompanying bond formation, explore locations of direct chemical bonds using nonadditive information contributions, and examine the pattern of entropic connectivities between atoms in molecules (AIM) resulting from molecular communications between atomic orbitals (AOs).

It has been amply demonstrated elsewhere [10–15] that many classical problems of theoretical chemistry can be approached afresh using this novel IT perspective. For example, displacements in information distribution in molecules relative to the promolecular reference consisting of nonbonded constituent atoms have been investigated [10–18], and the least biased partition of molecular electron distributions into subsystem contributions (e.g., densities of AIM) has been investigated as well [10,19–26]. The IT approach has been shown to lead to the "stockholder" molecular fragments of Hirshfeld [27]. These optimum density pieces have been derived from alternative global and local variational principles of IT.

The spatial localization and multiplicity of specific bonds—not to mention qualitative questions on their very existence (e.g., between bridgehead carbon atoms in small propellanes)—present another challenging problem to this novel treatment of molecular systems. Another diagnostic problem of molecular electronic structure deals with shell structure and electron localization in atoms and molecules. Nonadditive Fisher information in AO resolution has been recently used as the contragradience (CG) criterion for localizing bonding regions in molecules [11–15,28–30], while the related information density in molecular orbital (MO) resolution has been shown [10,31] to determine the vital ingredient of electron localization function (ELF) [32–34]. The communication theory of chemical bond has been developed using the basic entropy/information descriptors of molecular information (communication) channels in AIM, orbital, and local resolution levels of electron probability distributions [10–12,35–47]. The same entropic bond descriptors have been used to provide information-scattering perspectives on intermediate stages in electron redistribution processes [48], including atom promotion via orbital hybridization [49], and the communication theory for excited electron configurations has been developed [50]. Moreover, the phenomenological description of equilibria in molecular subsystems [10,51–53], which formally resembles that developed in ordinary thermodynamics, has been proposed.

Entropic probes of molecular electronic structure have provided attractive tools for describing the chemical bond phenomenon in information terms. It is the main purpose of this survey to summarize alternative local entropy/information probes of molecular electronic structure, explore the information origins of chemical bonds, and present recent developments in orbital communication theory (OCT) [11,12,46,54–57]. The importance of nonadditive effects in the chemical bond phenomenon will be emphasized, and the information cascade (bridge) propagation of electronic probabilities in molecular information systems, which generate indirect bond contributions due to orbital intermediaries [58–62], will be examined.

In the logarithmic measure of information, the logarithm is taken to base 2, $\log = \log_2$, which expresses information in bits (binary digits).

8.2   INFORMATION DISPLACEMENTS IN MOLECULES

Shannon entropy [3] in the (normalized) discrete probability vector $p = \{p_i\}$ or in the continuous probability density $p(r)$,

$$S(\boldsymbol{p}) = -\sum_i p_i \log p_i, \quad \sum_i p_i = 1,$$

$$\text{or } S[p] = -\int p(\boldsymbol{r}) \log p(\boldsymbol{r}) d\boldsymbol{r}, \quad \int p(\boldsymbol{r}) d\boldsymbol{r} = 1, \tag{8.1}$$

where the summation/integration extends over all elementary events determining the probability scheme in question, provides a measure of average indeterminacy in the distribution. This function/functional also measures the average amount of information obtained when uncertainty is removed by an appropriate measurement (experiment).

An important generalization of Shannon entropy, called relative (cross) entropy (also known as entropy deficiency, missing information, or directed divergence), has been proposed by Kullback and Leibler [5] and Kullback [6]. It measures the information "distance" between two (normalized) probability distributions for the same set of events:

$$\Delta S(\boldsymbol{p} \mid \boldsymbol{p}^0) = \sum_i p_i \log\left(p_i/p_i^0\right) \geq 0$$

$$\text{or } \Delta S[p \mid p^0] = \int p(\boldsymbol{r}) \log[p(\boldsymbol{r})/p^0(\boldsymbol{r})] d\boldsymbol{r} \geq 0. \tag{8.2}$$

Entropy deficiency thus provides a measure of information resemblance between two probability vectors or densities. The more the two distributions differ from each other, the larger the information distance becomes. For individual events, the logarithm of probability ratio $I_i = \log\left(p_i/p_i^0\right)$ or $I(\boldsymbol{r}) = \log[p(\boldsymbol{r})/p^0(\boldsymbol{r})]$, called *surprisal*, provides a measure of the event information in the current distribution relative to that in the reference distribution. The equality in Equation 8.2 takes place only for vanishing surprisal for all events (i.e., when the two probability distributions are identical).

The electron densities $\left\{\rho_i^0\right\}$ of separated atoms define the (isoelectronic) molecular prototype called "promolecule" [10,27], given by the sum of free-atom distributions shifted to the actual locations of AIM. The resulting electron density $\rho^0 = \sum_i \rho_i^0$ of this collection of "frozen" atomic distributions defines the initial stage in the bond formation process and determines a natural reference for extracting changes due to formation of chemical bonds. They are embodied in the density difference function $\Delta\rho = \rho - \rho^0$ between the molecular ground-state density $\rho = \sum_i \rho_i$, a collection of AIM densities $\{\rho_i\}$, and the promolecular electron distribution ρ^0, which has been widely used to probe the electronic structure of molecular systems. In this section a comparison of this "deformation—density" with selected local IT probes will be made to explore the information content of the molecular ground-state electron distribution $\rho(\boldsymbol{r})$ or its shape (probability) factor $p(\boldsymbol{r}) = \rho(\boldsymbol{r})/N$ obtained from Kohn–Sham (KS) [63] calculations in the local density approximation (LDA) of DFT.

In this section the density $\Delta s(\boldsymbol{r})$ of the Kullback–Leibler functional for the information distance between molecular and promolecular electron distributions,

$$\Delta S[\rho \mid \rho^0] = \int \rho(r) \ln[\rho(r)/\rho^0(r)]dr = \int \rho(r)I[w(r)]dr \equiv \int \Delta s(r) dr, \qquad (8.3)$$

where $w(r)$, and $I[w(r)]$ denote the density-enhancement and surprisal functions relative to the system promolecule, respectively, will be used as potential diagnostic tool of chemical bonds. The density $h_\rho(r)$ of the associated displacement in the Shannon entropy,

$$\mathcal{H}[\rho] = S[\rho] - S[\rho^0] \equiv -\int \rho(r) \ln \rho(r) dr + \int \rho^0(r) \ln \rho^0(r) dr \equiv \int h_\rho(r) dr, \qquad (8.4)$$

will also be examined by testing its performance as an alternative IT tool for diagnosing chemical bonds and monitoring the effective valence states of bonded atoms.

The molecular electron density $\rho(r) = \rho^0(r) + \Delta\rho(r)$ is, on average, only slightly modified relative to the promolecular distribution $\rho^0(r)$, $\rho(r) \approx \rho^0(r)$, or $w(r) \approx 1$. Indeed, the formation of chemical bonds involves only a minor reconstruction of electronic structure, mainly in the valence shells of constituent atoms, so that $|\Delta\rho(r)| \equiv |\rho(r) - \rho^0(r)| \ll \rho(r) \approx \rho^0(r)$; hence, the ratio $\Delta\rho(r)/\rho(r) \approx \Delta\rho(r)/\rho^0(r)$ is generally small in energetically important regions of large density values. As explicitly shown in the first column of Figure 8.1, the largest values of the density difference $\Delta\rho(r)$ are observed mainly in the bond region between the nuclei of chemically bonded atoms; the reconstruction of atomic lone pairs can also lead to an appreciable displacement in electron density. By expanding the logarithm of the molecular surprisal $I[w(r)]$ around $w(r) = 1$ to the first order in the relative displacement of electron density, one obtains the following approximate relation between the local value of the molecular surprisal and that of the density difference function:

$$I[w(r)] = \ln[\rho(r)/\rho^0(r)] = \ln\{[\rho^0(r) + \Delta\rho(r)]/\rho^0(r)\} \cong \Delta\rho(r)/\rho^0(r) \approx \Delta\rho(r)/\rho(r). \quad (8.5)$$

It provides a semiquantitative IT interpretation of relative density difference diagrams and links the local surprisal of IT to the density difference function of quantum chemistry. This equation also relates the integrand of the information distance functional with the corresponding displacement in electron density: $\Delta s(r) = \rho(r)I[w(r)] \cong \Delta\rho(r)w(r) \approx \Delta\rho(r)$. This approximate relation is numerically verified in Figure 8.1, where the contour diagram of the directed divergence density $\Delta s(r)$ (second column) is compared with the corresponding map of its first-order approximation $\Delta\rho(r)w(r)$ (third column) and the density difference itself (first column). A general similarity between the diagrams in each row confirms the semiquantitative character of this first-order expansion. Indeed, a comparison between panels of the first two columns in Figure 8.1 shows that the two displacement maps so strongly resemble each other that they are hardly distinguishable. This confirms a close relation between the local density and entropy deficiency relaxation patterns, thus attributing to the former the complementary IT interpretation of the latter. The

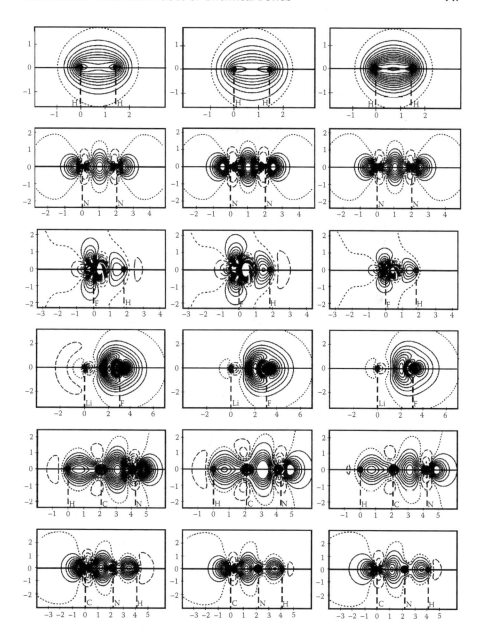

FIGURE 8.1 Contour diagrams of the molecular density difference function $\Delta\rho(r) = \rho(r) - \rho^0(r)$ (first column), the information-distance density $\Delta s(r) = \rho(r)I[w(r)]$ (second column), and its approximate, first-order expansion $\Delta s(r) \cong \Delta\rho(r)w(r)$ (third column), for selected diatomic and linear triatomic molecules: H_2, HF, LiF, HCN, and HNC. The solid, pointed, and broken lines denote the positive, zero, and negative values, respectively, of (equally spaced) contours. (From Nalewajski, R. F. et al., *Int. J. Quantum Chem.*, 2002, *87*, 198.)

density displacement and the missing information distribution can thus be viewed as practically equivalent probes of system chemical bonds.

The single covalent bond in H_2 is seen to give rise to a relative accumulation of electrons in the bond region (between the two nuclei) at the expense of the outer nonbonding regions of space. The triple-bond pattern for N_2 reflects density accumulation in the bonding region, due to both σ and π bonds, and the accompanying increase in the density of lone pairs on both nitrogen atoms, due to their expected sp hybridization. One also observes a decrease in electron density in the vicinity of the nuclei and an outflow of electrons from the $2p_\pi$ AO to their overlap area, a clear sign of orbital involvement in the formation of the double π-bond system. In triatomic molecules, one identifies a strongly covalent pattern of electron density displacements in the regions of single N–H and C–H atoms. A typical buildup of bond charge due to multiple CN bonds in the two isomers HCN and HNC can be also observed. The increase in lone-pair electron density on the terminal heavy atom (N in HCN or C in HNC) can be also detected, thus confirming the expected sp hybridization of these bonded atoms. Both heteronuclear diatomics represent partially ionic bonds between the two atoms, exhibiting small and large differences in their electronegativity (chemical hardness) descriptors, respectively. In HF one detects a common possession of valence electrons by the two atoms, giving rise to the bond charge located between them and, hence, a small H→F charge transfer. In LiF substantial Li→F electron transfer can be detected.

In Figure 8.2 the contour maps of the entropy displacement density $h_\rho(r)$ are compared with the corresponding density difference diagrams for representative linear molecules of Figure 8.1. Again, $\Delta\rho$ and h_ρ diagrams for H_2 are seen to qualitatively resemble each other and the corresponding Δs map in Figure 8.1. The main feature of h_ρ diagram—an increase in electron uncertainty between nuclei—is attributable to the inflow of electrons to this region. Again, this covalent charge/entropy accumulation reflects electron-sharing effect and delocalization of AIM valence electrons toward the bond partner.

As an additional illustration, the main results of a related analysis of the central bond problem in small [1.1.1], [2.1.1], [2.2.1], and [2.2.2] propellanes shown in Figure 8.3 are presented. The main purpose of this study [7,14] was to examine the effect of an increasing size of carbon bridges on the central C′–C′ bond between "bridgehead" carbon atoms. The contour maps in Figure 8.4 compare $\Delta\rho$, Δs, and h_ρ probes in the planes of sections shown in Figure 8.3. They have been generated using DFT-LDA calculations in the extended (DZVP) basis set.

The density difference exhibits depletion of electron density between bridgehead carbons in [1.1.1] and [2.1.1] propellanes, while larger bridges in [2.2.1] and [2.2.2] systems generate a density buildup in this region. A similar conclusion follows from the entropy displacement and entropy deficiency plots of the figure. The two entropic diagrams are again seen to be qualitatively similar to the corresponding density difference plots. As before this resemblance is seen to be particularly strong between the figure $\Delta\rho$ and Δs diagrams.

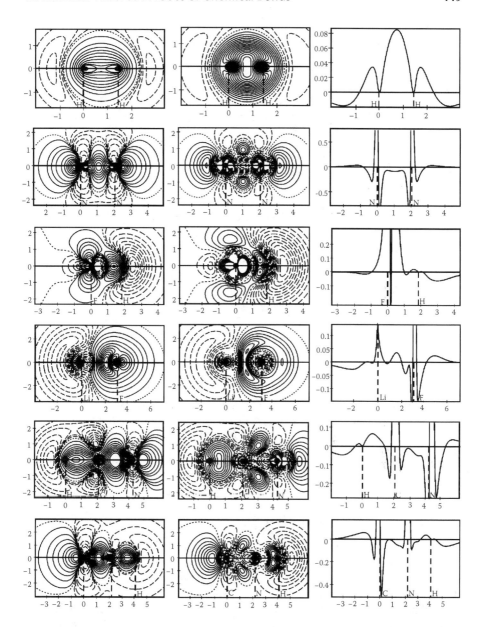

FIGURE 8.2 A comparison between diagrams of nonequidistant contours of the density difference function $\Delta\rho(\mathbf{r})$ (first column) and the entropy difference function $h_\rho(\mathbf{r})$ (second column) for the linear molecules in Figure 8.1. The corresponding profiles of $h_\rho(\mathbf{r})$ for cuts along the bond axis are shown in the third column. (Reprinted from Nalewajski, R. F. and Broniatowska, E. *J. Phys. Chem. A*, 2003, *107*, 6270.)

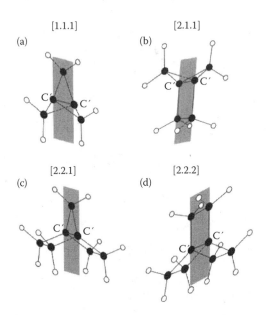

FIGURE 8.3 Propellane structures and planes of sections containing the bridge and bridge-head (C') carbon atoms identified by black circles.

8.3 NONADDITIVE FISHER INFORMATION AND BOND LOCALIZATION

Each density functional $A[\rho]$ can be regarded as the associated multicomponent functional $A[\rho] \equiv A^{total}[\rho]$ of the pieces $\rho = \{\rho_\alpha\}$ into which the electron distribution is decomposed:

$$\rho = \sum_\alpha \rho_\alpha. \qquad (8.6)$$

For example, such functionals appear in the non–Born–Oppenheimer DFT of molecular systems [64], in the partitioning of electron density into distributions assigned to AIM [10], and in fragment-embedding problems [10,65–69]. Each resolution of the molecular electron density also implies the associated division of the molecular (total) quantity $A^{total}[\rho]$ into its additive contribution $A^{add}[\rho]$ and its nonadditive contribution $A^{nadd}[\rho]$:

$$A^{add}[\rho] \equiv \sum_\alpha A[\rho_\alpha], \quad A^{nadd}[\rho] = A^{total}[\rho] - A^{add}[\rho]. \qquad (8.7)$$

For example, the Gordon–Kim–type division [65] of the kinetic energy functional defines its nonadditive contribution, which constitutes the basis of the DFT-embedding

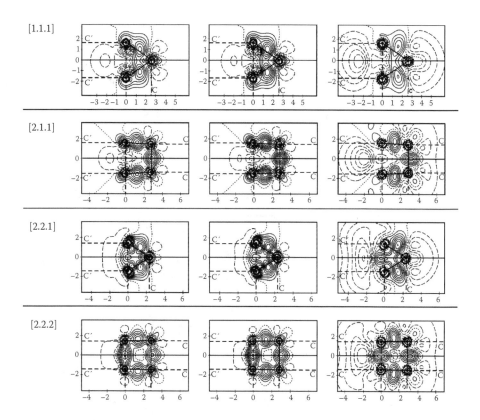

FIGURE 8.4 A comparison between maps of equidistant contours of the density difference function $\Delta\rho(r)$ (first column), the information distance density $\Delta s(r)$ (second column), and the entropy displacement density $h_\rho(r)$ (third column) for the four propellanes in Figure 8.3. (Reprinted from Nalewajski, R. F. and Broniatowska, E. *J. Phys. Chem. A*, 2003, *107*, 6270.)

concept of Cortona [66], Wesołowski and Warshel [67], and Wesołowski [68,69]. A similar partition can be used to resolve the information quantities themselves [10–11,28].

Consider the Fisher information for locality [1,2], called intrinsic accuracy, which historically predates the Shannon entropy by about 25 years being proposed at about the same time that the final form of quantum mechanics was shaped. This local measure of the information content of the continuous (normalized) classical probability density $p(r)$ reads

$$I[p] = \int p(r')[\nabla \ln p(r')]^2 \, dr' = \int [\nabla p(r)]^2 / p(r) \, dr. \tag{8.8}$$

The above functional can be simplified when expressed in terms of the associated (real) amplitude $A(r) = \sqrt{p(r)}$ of the probability distribution $p(r)$:

$$I[p] = 4 \int [\nabla A(r)]^2 \, dr \equiv I[A]. \qquad (8.9)$$

The Fisher information, reminiscent of von Weizsäcker's [70] inhomogeneity correction to electronic kinetic energy in the Thomas–Fermi theory, characterizes the compactness of the probability density. For example, the Fisher information in normal distribution measures the inverse of its variance, called invariance, while the complementary Shannon entropy is proportional to the logarithm of variance, thus monotonically increasing with the spread of Gaussian distribution. Therefore, Shannon entropy and intrinsic accuracy describe complementary facets of the probability density: the former reflects the distribution's ("spread" ("disorder", a measure of uncertainty), while the latter measures its "narrowness" ("order").

The inverse of nonadditive Fisher information in MO resolution level has been shown to define the IT-ELF concept [31] in the spirit of the original Becke and Edgecombe formulation [32], while the related quantity in the AO resolution scheme of the self-consistent field (SCF) MO theory generates the CG criterion to localize the chemical bonds in molecules [10–15,28–30]. It has been argued elsewhere [28] that the valence basins of negative CG density identify the bonding (constructive) interference of AO in a direct bonding mechanism, while the positive values of this local IT probe delineate antibonding regions in the molecule (Figure 8.5).

For the given pair $A(r)$ and $B(r)$ of $1s$ AO originating from atoms A and B, respectively, a negative contribution to the nonadditive Fisher information results when the gradient of one AO exhibits a negative projection in the direction of the gradient of the other AO, explaining the name of the CG criterion itself. The zero contour, which encloses the bonding region and separates it from the antibonding environment, is thus identified by the equation $\nabla A(r) \cdot \nabla B(r) = 0$. As shown in the qualitative diagram of Figure 8.5, for two $1s$ orbitals in H_2, this dividing surface on which the two AO gradients are mutually perpendicular defines the sphere passing through both nuclei. Integration of i^{cg} over the whole space gives the associated CG integral (a.u.)

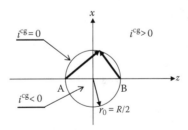

FIGURE 8.5 Circular contour of the vanishing CG integrand $i^{cg}(r) = \nabla A(r) \cdot \nabla B(r) = 0$ of two $1s$ orbitals in H_2 ($A(r)$ and $B(r)$ centered on atoms A and B, respectively) passes through both nuclei. It separates the bonding region (inside the circle), where $i^{cg}(r) < 0$, from the antibonding region of positive values, $i^{cg}(r) > 0$ (outside the circle). The heavy arrows, representing the negative gradients of the two basis functions, are mutually perpendicular on this spherical $i^{cg}(r) = 0$ surface.

$$I^{cg} = \int i^{cg}(r)\,dr = \int \nabla A(r)\cdot\nabla B(r)\,dr = -\int A(r)\Delta B(r)\,dr = 2\langle A|\hat{T}|B\rangle \equiv 2T_{A,B},$$

(8.10)

which is proportional to the coupling (off-diagonal) element $T_{A,B}$ of the electronic kinetic energy operator $\hat{T} = -1/2\Delta$. Such integrals are routinely calculated in all electronic structure calculations. The CG criterion stresses the importance of the kinetic energy of electrons in elucidating the origins of the chemical bond [71–74].

Consider a general case of the N-electron system and the AO basis set $\chi = (\chi_1,\chi_2,\ldots, \chi_m)$. Examine the ground-state configuration defined by the singly occupied subspace $\psi = \{\psi_k\}$ of N lowest spin MO, with the spatial (MO) parts $\varphi = \chi C = \{\varphi_k, k = 1, 2,\ldots, N\}$ generated either by Hartree–Fock (HF) or KS SCF calculations [63]. In these MO approximations, the nonadditive Fisher information density in the AO resolution for the ground-state electron configuration,

$$I^{nadd}[\chi] = 4\sum_{k=1}^{m}\sum_{l=1}^{m}\int \gamma_{k,l}(1-\delta_{k,l})\nabla\chi_l^*(r)\cdot\nabla\chi_k(r)\,dr \equiv 2\int f^{nadd}(r)\,dr = 8T^{nadd}[\chi],$$

(8.11)

is proportional to the nonadditive part $T^{nadd}[\chi]$ of the system average kinetic energy of electrons in AO resolution $T[\chi] = \mathrm{tr}(\gamma T)$, where $\mathbf{T} = \{T_{k,l} = \langle\chi_k|\hat{T}|\chi_l\rangle\}$ stands for the kinetic energy matrix in AO representation, and the elements of the charge and bond order (CBO) matrix

$$\gamma = \left\langle \chi\left|\left(\sum_{k=1}^{N}|\varphi_k\rangle\langle\varphi_k|\right)\right|\chi\right\rangle = \langle\chi|\varphi\rangle\langle\varphi|\chi\rangle \equiv \langle\chi|\hat{P}_\varphi|\chi\rangle = \mathbf{C}\mathbf{C}^\dagger = \{\gamma_{u,w}\} \qquad (8.12)$$

provide the AO representation of the projection operator $\hat{P}_\varphi$ onto the whole occupied MO subspace φ.

One uses the most extended ("valence") basins of the negative CG density $f^{nadd}(r) < 0$, enclosed by the corresponding $f^{nadd}(r) = 0$ surface, as locations of chemical bonds. This probe has been successfully validated in recent numerical calculations [30]. In the remaining part of this section, representative results of this study are presented. They have been obtained using standard HF SCF MO (GAMESS) calculations in the minimum (STO-3G) Gaussian basis set. The contour maps for optimized geometries are reported in atomic units, with negative CG basins identified by broken-line contours. For visualization purposes, selected perspective views of bonding CG regions are also reported.

The contour map for H_2 in Figure 8.6 confirms the qualitative prediction in Figure 8.5, of the spherical CG bonding region between the two nuclei. The accompanying increases in CG density, observed in nonbonding surroundings of this bond basin, are ultimately responsible for the net increase in the electronic kinetic energy relative to separated atom value. Figure 8.7 presents a similar analysis for HF identifying three basins of the negative CG density: a large valence region between the two atoms and two small volumes in the inner shell of the heavy atom.

In the triple bond of N_2 (Figure 8.8) the bonding basin is now distinctly extended away from the bond axis owing to the presence of the two π bonds accompanying the central σ bond. Small core polarization basins are again seen in the perspective plot. The atom promotion due to the sp hybridization in the nonbonding regions of both atoms is again evident in the accompanying contour map, and the bonding region is seen to be "squeezed" between the two atomic cores. The positive values of CG density in transverse directions on each nucleus reflect the charge displacements accompanying π-bond formation.

The chemical bonds in illustrative hydrocarbons are probed in Figures 8.9 and 8.10. These diagrams again testify to the efficiency of the CG criterion in localizing both C–C and C–H bonding regions in acetylene, butadiene, and benzene. In Figure 8.9a, the CG pattern of the triple bond between the carbon atoms in acetylene strongly resembles that observed in N_2. It also directly follows from the two perpendicular cuts in Figure 8.9b that the π bond between the neighboring peripheral carbons in butadiene is stronger than its central counterpart.

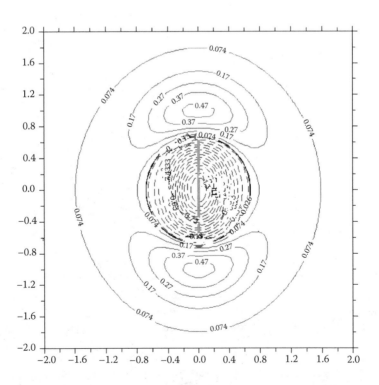

FIGURE 8.6 Contour map of CG density $f^{\text{nadd}}(r)$ for H_2.

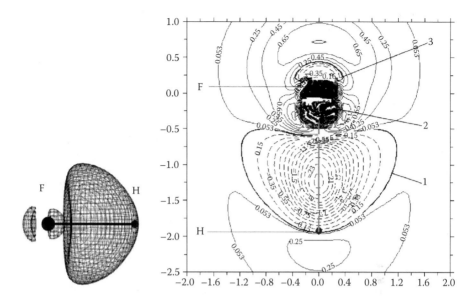

FIGURE 8.7   Perspective view of negative CG basins and contour map of $f^{nadd}(r)$ for HF.

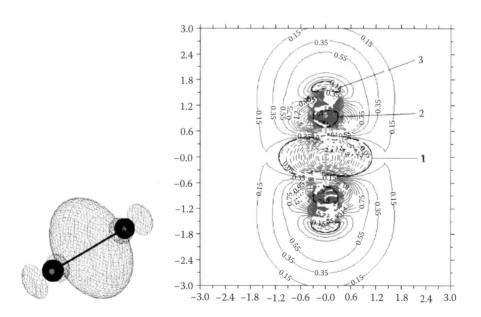

FIGURE 8.8   Same as in Figure 8.7 for N_2.

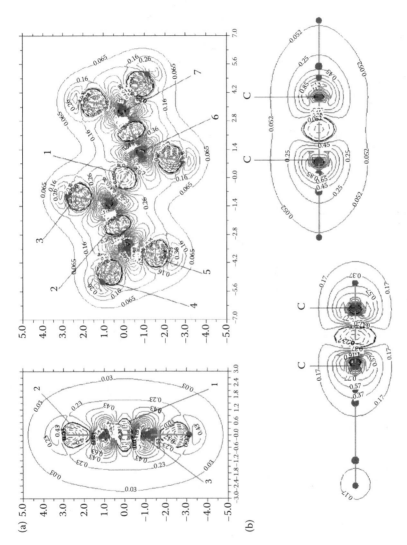

FIGURE 8.9 Contour map of the CG density $f^{\text{nadd}}(\mathbf{r})$ for acetylene (left diagram of a) and butadiene (remaining diagrams). The right diagram of a shows the contour map in the molecular plane of butadiene, reflecting only σ-bond contributions, while b reports additional perpendicular cuts for this molecule passing through the terminal (left) and central (right) C–C bonds, which additionally include π-bond contributions.

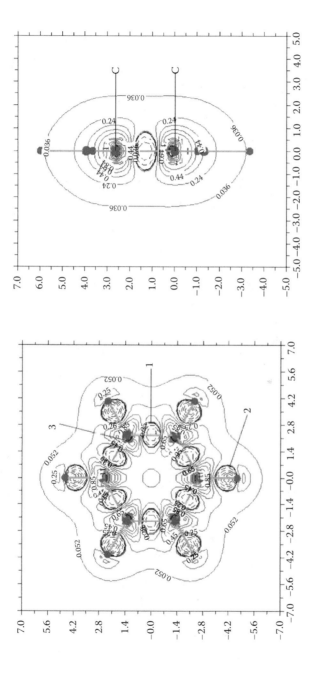

FIGURE 8.10 Contour map of the CG density $f^{\mathrm{madd}}(\boldsymbol{r})$ for benzene: in the molecular plane (left) and in the perpendicular plane passing through the C—C bond (right).

Finally, the bonding patterns in a series of four small propellanes shown in Figure 8.3 are examined in contour maps of Figure 8.11, with each row being devoted to a different propellane in a series: [1.1.1], [2.1.1], [2.2.1], and [2.2.2]. The left panels of each row correspond to the plane-of-section perpendicular to the central bond between the bridgehead carbons (at bond midpoint), while the axial cuts of the right panels involve the carbon bridges. The main results of the density difference and entropy deficiency analyses in Figure 8.4, of the apparent lack of a direct (through-space) bond between the carbon bridgeheads in the [1.1.1] and [2.1.1] systems, and of the presence of a full single bond in the [2.2.1] and [2.2.2] propellanes remain generally confirmed by the CG probe in Figure 8.9, but this transition is now seen to be less sharp, with very small bonding basins between bridgeheads also being observed in the two small molecules. The small bonding basin of the [1.1.1] system is steadily evolving into that attributed to the full bond in the [2.2.2] propellane (Figure 8.11). The bridge C–C and C–H chemical bonds are again seen to be perfectly localized by the closed valence surfaces of the vanishing CG density. One thus concludes that the CG probe of chemical bonds indeed provides an efficient tool for locating direct bonding regions in typical diatomic and polyatomic molecules.

One of the primary goals of theoretical chemistry is to identify the physical sources of chemical bonds. Most existing theoretical interpretations of its origins emphasize, almost exclusively, the potential (interaction) aspect of this phenomenon, focusing on the mutual attraction between the accumulation of electrons between two atoms (negative "bond charge") and the partially screened (positively charged) nuclei. Indeed, the familiar virial theorem decomposition of the diatomic Born–Oppenheimer potential indicates that, for equilibrium bond length, the change in its potential component relative to the dissociation limit, due to an effective contraction of AIM, is ultimately responsible for the net bonding effect.

The ELF and CG criteria, which reflect nonadditive kinetic energy/Fisher information terms, adopt the complementary view by stressing the importance of kinetic energy bond component, which was shown to efficiently identify the bonding regions in molecules. The associated displacement in the total kinetic energy of electrons is bonding only at an earlier stage of the mutual approach by both atoms, when it is dominated by the longitudinal contribution associated with the gradient component along the bond axis; it ultimately assumes a destabilizing character at the equilibrium internuclear separation, mainly due to its transverse contribution associated with the gradient components in the directions perpendicular to the bond axis [71–73]. The overall change in kinetic energy (Fisher information) due to chemical bond formation emphasizes a contraction of the electron density in the presence of the remaining nuclear attractors in the molecule. It blurs the subtle information origins of chemical bonds, since the total kinetic energy component combines the minute bonding (nonadditive) interatomic effects originating from the stabilizing combinations of AO and occupied MO and the dominating nonbonding contributions due to intra-atomic polarization and lone pairs of inner and outer electrons.

Therefore, the displacement in the overall kinetic energy contribution effectively hides the minute changes in the system valence shell, which chemists associate with the chemical bond concept. Some partitioning of this overall energy component is called for to separate the subtle bonding phenomenon from the associated

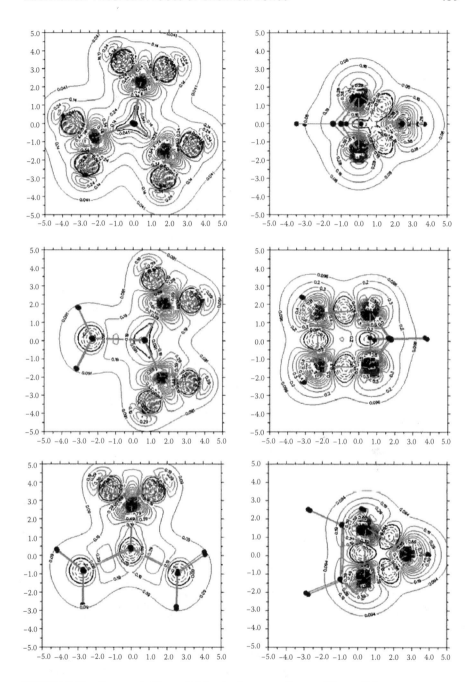

FIGURE 8.11 Same as in Figure 8.6 for the four propellanes of Figure 8.3.

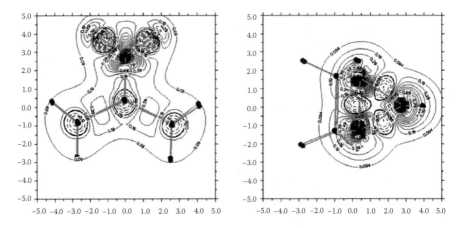

FIGURE 8.11 (*continued*) Same as in Figure 8.6 for the four propellanes of Figure 8.3.

nonbonding promotion of constituent atoms to their effective valence states in molecules. Only by focusing on the nonadditive part of electronic kinetic energy can one uncover in the CG criterion the real information origins of direct chemical bonds, define useful local probes for their localization, and formulate adequate quantitative information descriptors [30].

This virial theorem perspective also indicates that kinetic energy constitutes the driving force of the bond formation process. It follows from the classical analysis by Feinberg et al. [71] and Feinberg and Ruedenberg [72,73] that the ultimate contraction of atomic electron distributions is possible in molecules because of this relative lowering of kinetic energy at large internuclear separations, mainly in the bonding region between the two nuclei. The process of the final redistribution of electrons during chemical bond formation can thus be regarded as being "catalyzed" by the gradient effect of kinetic energy. A similar conclusion follows from the theoretical analysis by Goddard and Wilson [74].

8.4 ENTROPIC BOND MULTIPLICITIES FROM ORBITAL COMMUNICATIONS

This survey of IT probes of chemical bonds continues with some rudiments on the entropic characteristics of the dependent probability distributions and information descriptors of a transmission of signals in communication systems [3,4,7,8]. For two mutually dependent (discrete) probability vectors of two separate sets of events a and b, $P(a) = \{P(a_i) = p_i\} \equiv p$ and $P(b) = \{P(b_j) = q_j\} \equiv q$, one decomposes the joint probabilities of the simultaneous events $a \wedge b = \{a_i \wedge b_j\}$ into these two schemes, $P(a \wedge b) = \{P(a_i \wedge b_j) = \pi_{i,j}\} \equiv \pi$, as products of the "marginal" probabilities of events in one set, say a, and the corresponding conditional probabilities $P(b|a) = \{P(j|i) = \pi_{i,j}/p_i\}$ of outcomes in set b, given that events a have already occurred: $\{\pi_{i,j} = p_i P(j|i)\}$. The relevant normalization conditions for the joint and conditional probabilities then read:

$$\sum_j \pi_{i,j} = p_i, \quad \sum_i \pi_{i,j} = q_j, \quad \sum_i \sum_j \pi_{i,j} = 1, \quad \sum_j P(j|i) = 1 \ (i = 1, 2, \ldots). \tag{8.13}$$

The Shannon entropy of the product distribution

$$S(\boldsymbol{\pi}) = -\sum_i \sum_j \pi_{i,j} \log \pi_{i,j} = -\sum_i \sum_j p_i P(j|i)[\log p_i + \log P(j|i)]$$

$$= -\left[\sum_j P(j|i)\right] \sum_i p_i \log p_i - \sum_i p_i \left[\sum_j P(j|i) \log P(j|i)\right] \tag{8.14}$$

$$\equiv S(\boldsymbol{p}) + \sum_i p_i S(\boldsymbol{q}|i) \equiv S(\boldsymbol{p}) + S(\boldsymbol{q}|\boldsymbol{p})$$

is then expressed as the sum of the average entropy in the marginal probability distribution $S(\boldsymbol{p})$ and the average conditional entropy in $\boldsymbol{q}$ given $\boldsymbol{p}$

$$S(\boldsymbol{q}|\boldsymbol{p}) = -\sum_i \sum_j \pi_{i,j} \log P(j|i)]. \tag{8.15}$$

The latter represents an extra amount of uncertainty about the occurrence of events $\boldsymbol{b}$, given that events $\boldsymbol{a}$ are known to have occurred. In other words, the amount of information obtained by simultaneously observing the events $\boldsymbol{a}$ and $\boldsymbol{b}$ of the two discrete probability distributions equals the amount of information in the marginal set $\boldsymbol{a}$ supplemented by the extra information provided by the occurrence of events in the other set $\boldsymbol{b}$, when events $\boldsymbol{a}$ are known to have occurred already. This is qualitatively illustrated in Figure 8.12.

The common amount of information in two events a_i and b_j, $I(i{:}j)$-measuring the information about a_i provided by the occurrence of b_j, or the information about b_j provided by the occurrence of a_i, is called the mutual information in two events:

$$I(i{:}j) = \log[P(a_i \wedge b_j)/P(a_i)P(b_j)] = \log[\pi_{i,j}/(p_i q_j)] \equiv \log[P(i|j)/p_i] \equiv \log[P(j|i)/q_j] = I(j{:}i). \tag{8.16}$$

It may take on any real value (positive, negative, or zero): it vanishes when both events are independent (i.e., when the occurrence of one event does not influence [or condition] the probability of the occurrence of the other event), and it is negative when the occurrence of one event makes the nonoccurrence of the other event more likely. It also follows from Equation 8.16 that

$$I(i{:}j) = I(i) - I(i|j) = I(j) - I(j|i) = I(i) + I(j) - I(i \wedge j) \text{ or } I(i \wedge j) = I(i) + I(j) - I(i{:}j), \tag{8.17}$$

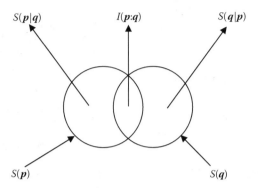

FIGURE 8.12 Entropy for two dependent probability distributions p and q. Two circles enclose areas representing the entropies $S(p)$ and $S(q)$ of two separate probability vectors, while their common (overlap) area corresponds to the mutual information $I(p{:}q)$ in these two distributions. The remaining part of each circle represents the corresponding conditional entropy $S(p|q)$ or $S(q|p)$, measuring the residual uncertainty about events in one set, when one has full knowledge of the occurrence of events in the other set of outcomes. The area enclosed by the envelope of two circles then represents the entropy of the "product" (joint) distribution: $S(\pi) = S(P(a \wedge b)) = S(p) + S(q) - I(p{:}q) = S(p) + S(q|p) = S(q) + S(p|q)$.

where the self-information of the joint event $I(i \wedge j) = -\log \pi_{i,j}$. Thus, the information on the joint occurrence of two events a_i and b_j is the information on the occurrence of a_i plus that on the occurrence of b_j minus the mutual information; for independent events, when $\pi_{i,j} = p_i q_j$, $I(i{:}j) = 0$ and, hence, $I(i \wedge j) = I(i) + I(j)$.

The mutual information of an event with itself defines its self-information: $I(i{:}i) \equiv I(i) = \log[P(i|i)/p_i] = -\log p_i$, since $P(i|i) = 1$. It vanishes when $p_i = 1$ (i.e., when there is no uncertainty about the occurrence of a_i) so that the occurrence of this event removes no uncertainty, hence conveying no information. This quantity provides a measure of the uncertainty about the occurrence of the event (i.e., the information received when the event occurs). Shannon entropy can thus be interpreted as the mean value of self-information in all individual events $S(p) = \sum_i p_i I(i)$. One similarly defines the average mutual information in two probability distributions as the π-weighted mean value of the mutual information quantities for the individual joint events:

$$I(p{:}q) = \sum_i \sum_j \pi_{i,j} I(i{:}j) = \sum_i \sum_j \pi_{i,j} \log\left(\pi_{i,j}/\pi_{i,j}^0\right)$$
$$= S(p) + S(q) - S(\pi) = S(p) - S(p|q) = S(q) - S(q|p) \geq 0,$$
(8.18)

where the equality holds only for the independent distributions when $\pi_{i,j} = \pi_{i,j}^0 = p_i q_j$. Indeed, the amount of uncertainty in q can only decrease when p has been known beforehand, $S(q) \geq S(q|p) = S(q) - I(p{:}q)$, with equality being observed only when the two sets of events are independent (nonoverlapping). These average entropy/ information relations are also illustrated in Figure 8.12.

It should be also observed that the average mutual information is an example of KL entropy deficiency measuring the missing information between the joint probabilities $P(a \wedge b) = \pi$ of the dependent events a and b, and the joint probabilities $P^{ind}(a \wedge b) = \pi^0 = p^T q$ of the independent joint events $I(p:q) = \Delta S(\pi|\pi^0)$. The average mutual information thus reflects a degree of dependence between events defining the two probability schemes. A similar information distance interpretation can be attributed to the conditional entropy $S(p|q) = S(p) - \Delta S(\pi|\pi^0)$.

The conditional entropy and mutual information quantities provide the key descriptors of information channels [3,4,7,8]. The basic elements of such communication devices are shown in Figure 8.13. The signal emitted from n "inputs" $a = (a_1, a_2, \ldots, a_n)$ of the channel source A is characterized by the probability distribution $P(a) = p = (p_1, p_2, \ldots, p_n)$. It can be received at m "outputs" $b = (b_1, b_2, \ldots, b_m)$ of the system receiver B. A transmission of signals in this channel is randomly disturbed, thus exhibiting typical communication noise. Indeed, the signal sent at the given input can, in general, be received with a nonvanishing probability at several outputs. The communication network is thus fully described by the conditional probabilities of the outputs given inputs $P(b|a) = \{P(b_j|a_i) = P(a_i \wedge b_j)/P(a_i) \equiv P(j|i)\}$, where $P(a_i \wedge b_j) \equiv \pi_{i,j}$ stands for the probability of the joint occurrence of the specified pair of the input–output events. The distribution of the output signal among the detection events b is thus given by the output probability distribution $P(b) = q = (q_1, q_2, \ldots, q_m) = pP(b|a)$.

The Shannon entropy $S(p)$ of the input (source) probabilities p determines the channel *a priori* entropy. The average conditional entropy of the outputs given inputs $S(q|p)$ is determined by the scattering probabilities $P(b|a)$. It measures the average noise in the $a \to b$ transmission. The so-called *a posteriori* entropy of the input given output, $H(A|B) \equiv S(p|q)$, is similarly defined by the "reverse" conditional probabilities of the $b \to a$ signals: $P(a|b) = \{P(a_i|b_j) = P(i|j)\}$. It reflects residual indeterminacy about the input signal when the output signal has already been received. The average conditional entropy $S(p|q)$ thus measures the indeterminacy of the source with respect to the receiver, while the conditional entropy $S(q|p)$ reflects the uncertainty of the receiver relative to the source. Hence, an observation of the output signal

Input (Source): **A**	Communication network: $P(b|a)$	*Output* (Receiver): **B**
a_1		b_1
a_2		b_2
$p_i \to a_i$	$P(b_j|a_i) \equiv P(j|i) \longrightarrow$	$b_j \to q_j$
a_n		b_m

FIGURE 8.13 Schematic diagram of the communication system characterized by two probability vectors: $P(a) = \{P(a_i)\} = p = (p_1, \ldots, p_n)$, of the channel "input" events $a = (a_1, \ldots, a_n)$ in the system source A, and $P(b) = \{P(b_j)\} = q = (q_1, \ldots, q_m)$, of the "output" events $b = (b_1, \ldots, b_m)$ in the system receiver B. The transmission of signals in this communication channel is described by the $(n \times m)$ matrix of the conditional probabilities $P(b|a) = \{P(b_j|a_i) \equiv P(j|i)\}$ of observing different "outputs" (columns; $j = 1, 2, \ldots, m$) given the specified "inputs" (rows; $i = 1, 2, \ldots, n$). For clarity, only a single scattering $a_i \to b_j$ is shown in the diagram.

provides, on average, the amount of information given by the difference between the *a priori* and the *a posteriori* uncertainties, $S(p) - S(p|q) = I(p:q)$, which defines the mutual information in the source and the receiver. In other words, the mutual information measures the net amount of information transmitted through the communication channel, while the conditional entropy $S(p|q)$ reflects a fraction of $S(p)$ transformed into noise as a result of the input signal scattering in the information channel. Accordingly, $S(q|p)$ reflects the noise part of $S(q) = S(q|p) + I(p:q)$.

In OCT, the AO basis functions of SCF MO calculations determine a natural resolution level for discussing information contributions to the IT multiplicity (order) of system chemical bonds: $a = \{\chi_i\}$ and $b = \{\chi_j\}$. This AO network describes the probability/information propagation in the molecule, which can be described by the standard quantities developed in IT for real communication devices. Because of electron delocalization throughout the network of chemical bonds, the transmission of "signals" about the electron assignments to these AO events becomes randomly disturbed, thus exhibiting typical communication "noise." Indeed, an electron initially attributed to the given AO in the molecular channel AO "input" a can be later found with a non-vanishing probability at several AO in the molecular AO "output" b. This feature of electron delocalization is embodied in the conditional probabilities of the "outputs given inputs" $\mathbf{P}(b|a) = \{P(\chi_j|\chi_i) \equiv P(j|i)\}$, which define the orbital information channel.

One constructs these direct communication probabilities [46,54–58,75] using the bond-projected superposition principle of quantum mechanics [76] supplemented by "physical" projection onto the subspace of the system-occupied MO, which ultimately determine the molecular network of chemical bonds. Both the molecule as a whole and its constituent subsystems can be adequately described using IT bond indices. The internal and external indices of molecular fragments (groups of AO) can be efficiently generated with the appropriate reduction of the molecular channel by combining selected outputs into larger fragments [10,41,75]. Off-diagonal orbital communications are then related to the familiar Wiberg [77] contributions to molecular bond orders or the related "quadratic" bond multiplicities [78–87] formulated in MO theory. The IT descriptors of such molecular communication systems have been shown to account for the chemical intuition quite well, at the same time providing the resolution of diatomic bond multiplicities into complementary IT-covalent and IT-ionic components [10–12,29,55].

In MO theory, the network of chemical bonds is determined by the occupied MO in the system ground state. For reasons of simplicity, assume the closed-shell (cs) configuration of $N = 2n$ electrons in the standard spin-restricted HF (RHF) description, which involves the n lowest (doubly occupied, orthonormal) MO. In the familiar LCAO MO approach, they are expanded as linear combinations (LC) of the (Löwdin) orthogonalized AO $\chi = (\chi_1, \chi_2, ..., \chi_m) = \{\chi_i\}$ contributed by the system constituent atoms $\langle \chi | \chi \rangle = \{\delta_{i,j}\} \equiv \mathbf{I}, \varphi = (\varphi_1, \varphi_2, ... \varphi_n) = \{\varphi_s\} = \chi\mathbf{C}$, where the rectangular matrix $\mathbf{C} = \{C_{i,s}\} = \langle \chi | \varphi \rangle$ groups the relevant expansion coefficients of MO (i.e., LC of AO [LCAO]) to be determined using the iterative SCF procedure. The molecular electron density

$$\rho(r) = 2\varphi(r)\varphi^{\dagger}(r) = \chi(r)[2\mathbf{C}\mathbf{C}^{\dagger}]\chi^{\dagger}(r) \equiv \chi(r)\gamma\chi^{\dagger}(r) = Np(r) \qquad (8.19)$$

and, hence, the one-electron probability distribution $p(r) = \rho(r)/N$ (the shape factor of ρ) are both determined by the CBO matrix of Equation 8.12:

$$\boldsymbol{\gamma} = 2\langle \boldsymbol{\chi}|\boldsymbol{\varphi}\rangle\langle\boldsymbol{\varphi}|\boldsymbol{\chi}\rangle = 2\mathbf{CC}^\dagger \equiv 2\langle\boldsymbol{\chi}|\hat{P}_\varphi|\boldsymbol{\chi}\rangle = \{\gamma_{i,j} = 2\langle\chi_i|\hat{P}_\varphi|\chi_j\rangle \equiv 2\langle i|\hat{P}_\varphi|j\rangle\}. \qquad (8.20)$$

The latter constitutes the AO representation of the projection operator onto the subspace of all (doubly) occupied MO, $\hat{P}_\varphi = |\boldsymbol{\varphi}\rangle\langle\boldsymbol{\varphi}| = \Sigma_s |\phi_s\rangle\langle\phi_s| \equiv \Sigma_s \hat{P}_s$, and satisfies the idempotency relation

$$(\boldsymbol{\gamma})^2 = 4\langle\boldsymbol{\chi}|\hat{P}_\varphi|\boldsymbol{\chi}\rangle\langle\boldsymbol{\chi}|\hat{P}_\varphi|\boldsymbol{\chi}\rangle = 4\langle\boldsymbol{\chi}|\hat{P}_\varphi^2|\boldsymbol{\chi}\rangle = 4\langle\boldsymbol{\chi}|\hat{P}_\varphi|\boldsymbol{\chi}\rangle = 2\boldsymbol{\gamma}. \qquad (8.21)$$

The CBO matrix reflects the promoted valence state of AO in the molecule, with the diagonal elements measuring the effective electron occupations of the basis functions $\{N_i = \gamma_{i,i} = Np_i\}$ with the normalized probabilities $p = \{p_i = \gamma_{i,i}/N\}$ of the basis functions occupancy in the molecule: $\Sigma_i p_i = 1$.

The orbital information system involves the complete sets of AO events in the channel input $a = \{\chi_i\}$ and the channel output $b = \{\chi_j\}$. In this description, the AO→AO communication network is determined by the conditional probabilities

$$\mathbf{P}(b|a) = \{P(j|i) = P(i \wedge j)/p_i\}, \quad \sum_j P(j|i) = 1, \qquad (8.22)$$

where the joint probabilities of simultaneously observing two AO in the system chemical bonds $\mathbf{P}(a \wedge b) = \{P(i \wedge j)\}$ satisfy the usual normalization relations

$$\sum_i P(i \wedge j) = p_j, \quad \sum_j P(i \wedge j) = p_i, \quad \sum_i \sum_j P(i \wedge j) = 1. \qquad (8.23)$$

These conditional probabilities involve squares of corresponding CBO matrix elements [46]

$$\mathbf{P}(b|a) = \left\{ P(j|i) = \mathcal{N}_i \left|\langle i|\hat{P}_\varphi|j\rangle\right|^2 = (2\gamma_{i,i})^{-1}\gamma_{i,j}\gamma_{j,i} \right\}, \qquad (8.24)$$

where the (closed-shell) constant $\mathcal{N}_i = (2\gamma_{i,i})^{-1}$ follows directly from the normalization condition of Equation 8.22. These probabilities explore the dependencies between AO, which result from their participation in the framework of the occupied MO (i.e., their involvement in the entire network of chemical bonds in the molecule).

The input signal determines the way the channel is used or probed. The orbital channel can thus be exploited using the promolecular $\left(p^0 = \{p_i^0\}\right)$, molecular $(p =$

$\{p_i\}$), or general (ensemble) input probabilities, tailored to extract the desired aspects of the chemical bonding pattern in molecules [10–12,46,55].

The off-diagonal conditional probability of jth AO output given ith AO input is thus proportional to the squared element of the CBO matrix linking the two AO, $\gamma_{j,i} = \gamma_{i,j}$, thus being also proportional to the corresponding AO contribution $M_{i,j} = \gamma_{i,j}^2$ to the Wiberg index [77] of the chemical bond order between two atoms A and B in the molecule

$$M_{A,B} = \sum_{i \in A} \sum_{j \in B} M_{i,j} \qquad (8.25)$$

or to related quadratic descriptors of molecular bond multiplicities [78–87].

In OCT, the entropy/information indices of the covalent/ionic components of chemical bonds represent the complementary descriptors of conditional entropy (average communication noise) and mutual information (amount of information flow, capacity) in the molecular information channel [10–14,35,36,47,54,55,62]. One observes that the molecular input $P(a) \equiv p$ generates the same distribution in the output of the molecular channel

$$q = pP(b|a) = \left\{ \sum_i p_i P(j|i) \equiv \sum_i P(i \wedge j) = p_j \right\} = p, \qquad (8.26)$$

thus identifying p as the stationary probability vector of the molecular ground state. This purely molecular communication channel is devoid of any reference (history) of the chemical bond formation process and generates the average noise index of the molecular overall IT covalency measured by the conditional entropy of the system outputs given inputs:

$$S(P(a)|P(b)) = S(b|a) = -\sum_i \sum_j P(i \wedge j) \log[P(i \wedge j)/p_i] \equiv S. \qquad (8.27)$$

The AO channel with the promolecular input signal $P(a^0) = p^0$ refers to the initial state in the bond formation process. It corresponds to the ground-state (fractional) occupations of the AO contributed by the system constituent (free) atoms before they are mixed into MO. These input probabilities give rise to the average information flow descriptor of the system IT bond ionicity, given by the mutual information in the channel inputs and outputs [62]:

$$I(P(a^0):P(b)) = I(a^0:b) = \sum_i \sum_j P(i \wedge j) \log[P(i \wedge j)/(p_j p_i^0)]$$

$$= S(p) + \Delta S(p|p^0) - S \equiv I^0. \qquad (8.28)$$

This amount of information reflects the fraction of the initial (promolecular) information content $S(p^0)$, which has not been dissipated as noise in the molecular communication system. In particular, for the molecular input, when $p^0 = p$ and, hence, the vanishing information distance $\Delta S(p|p^0) = 0$, $I(p:p) = S(p) - S \equiv I$.

The sum of these two bond components

$$\mathcal{N}(P(a^0);P(b)) = \mathcal{N}(a^0;b) = S + I^0 = S(p^0) = S(p) + \Delta S(p|p^0) \equiv \mathcal{N}^0 \qquad (8.29)$$

measures the overall IT bond multiplicity of all bonds in the molecular system under consideration. For the molecular input this quantity preserves the Shannon entropy of the molecular input probabilities

$$\mathcal{N}(a;b) = S(b|a) + I(a:b) = S(p) \equiv \mathcal{N}. \qquad (8.30)$$

As an illustration, consider the familiar problem of combining the two (Löwdin-orthogonalized) AO, $A(r)$ and $B(r)$ (say, two $1s$ orbitals centered on nuclei A and B, respectively), which contribute a single electron each to form the chemical bond A–B in this 2-AO model. The two basis functions $\chi = (A, B)$ then form the bonding φ_b and antibonding φ_a MO combinations $\varphi = (\varphi_b, \varphi_a) = \chi C$

$$\varphi_b = \sqrt{P} A + \sqrt{Q} B, \quad \varphi_a = -\sqrt{Q} A + \sqrt{P} B, \quad P + Q = 1, \qquad (8.31)$$

where the square matrix

$$C = \begin{bmatrix} \sqrt{P} & -\sqrt{Q} \\ \sqrt{Q} & \sqrt{P} \end{bmatrix} = \begin{bmatrix} C_b & C_a \end{bmatrix} \qquad (8.32)$$

groups the LCAO MO expansion coefficients expressed in terms of the complementary probabilities P and $Q = 1 - P$. The former marks the conditional probabilities $P(A|\varphi_b) = |C_{A,b}|^2 = P(B|\varphi_a) = |C_{B,a}|^2 \equiv P$, while the latter measures the remaining matrix elements of the conditional probability matrix of observing AO in MO: $P(B|\varphi_b) = |C_{B,b}|^2 = P(A|\varphi_a) = |C_{A,a}|^2 = Q$.

In OCT the ground-state configuration of the doubly occupied bonding MO $[\varphi_b^2]$ determines the model CBO matrix, double the density matrix $d = \langle \chi|\varphi_b\rangle\langle\varphi_b|\chi\rangle$,

$$\gamma = \{\gamma_{i,j}\} = 2C_b C_b^\dagger = 2 \begin{bmatrix} P & \sqrt{PQ} \\ \sqrt{QP} & Q \end{bmatrix} \equiv 2d. \qquad (8.33)$$

It generates the associated direct conditional probabilities of AO in the molecular bond system (Equation 8.24)

$$\mathbf{P}(b \mid a) = \left\{ P(j \mid i) = \gamma_{i,j}\gamma_{j,i}/(2\gamma_{i,i}) \right\} = \begin{bmatrix} P & Q \\ P & Q \end{bmatrix}, \qquad (8.34)$$

which determine the $\chi \rightarrow \chi$ communication network shown in Figure 8.14. In this nonsymmetrical binary channel one adopts the molecular input signal $p = (P,Q)$, to extract the molecular bond IT covalency measuring the channel average communication noise, and the promolecular input signal $p^0 = (\frac{1}{2},\frac{1}{2})$, in which the two basis functions contribute a single electron each to form the chemical bond, to determine the model IT ionicity index measuring the channel information capacity.

The bond IT covalency $S(P)$ is thus determined by the binary entropy function $H(P)$ (Figures 8.14 and 8.15), reaching the maximum value $H(P = \frac{1}{2}) = 1$ bit for the symmetric bond $P = Q = \frac{1}{2}$ (e.g., in the σ bond of H_2 or in the π bond of ethylene). It vanishes for the lone-pair molecular configurations when $P = (0,1)$, $H(P = 0) = H(P = 1) = 0$, marking the alternative ion-pair configurations A^+B^- or A^-B^+, respectively, relative to the initial AO occupations $N^0 = (1,1)$ in the assumed promolecular reference.

The complementary descriptor $I^0(P) = 1 - H(P)$ of bond IT ionicity (Figure 8.15), which determines the channel mutual information relative to the promolecular input, reaches the highest value for the two limiting electron transfer configurations, $I^0(P = 0) = I^0(P = 1) = H(\frac{1}{2}) = 1$ bit, and identically vanishes for the purely covalent,

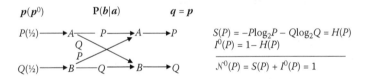

FIGURE 8.14 Communication channel of the 2-AO model of the chemical bond and its entropy/information descriptors (in bits).

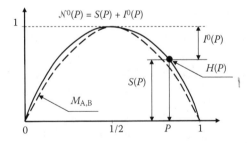

FIGURE 8.15 Conservation of the overall entropic bond multiplicity $N^0(P) = 1$ bit in the 2-AO model of the chemical bond, combining the conditional entropy (average noise, bond covalency) $S(P) = H(P)$ and the mutual information (information capacity, bond ionicity) $I^0(P) = 1 - H(P)$. In MO theory, the direct bond order of Wiberg is represented by the (broken line) parabola $M_{A,B}(P) = 4P(1 - P) \equiv 4PQ$.

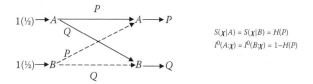

FIGURE 8.16 Elementary (row) subchannels due to atomic inputs A (solid lines) and B (broken lines) in the 2-AO model of the chemical bond and their partial IT covalency/ionicity components.

symmetric bond, $I^0(P = \frac{1}{2}) = 0$. As explicitly shown in Figure 8.15, these two components of chemical bond multiplicity compete with each other, yielding the conserved overall IT bond index $\mathcal{N}^0(P) = S(P) + I^0(P) = 1$ bit, marking a single bond in OCT in the whole range of admissible bond polarizations $P \in [0,1]$. This simple model thus properly accounts for the competition between bond covalency and bond ionicity while preserving the single bond order measure reflected by the conserved overall IT multiplicity of this single chemical bond. Similar effects transpire from the quadratic bond indices formulated in the MO theory [78,79]. For example, the Wiberg bond order plot in Figure 8.15 for the 2-AO model is given by the parabola

$$M_{A,B}(P) = [\gamma_{A,B}(P)]^2 = 4PQ = 4P(1 - P), \qquad (8.35)$$

which closely resembles the IT-covalent plot $S(P) = H(P)$ of the figure.

The bond components of this model can be further decomposed into their corresponding atomic contributions. In communication theory, this partition is accomplished by using the partial (row) subchannels in Figure 8.16, each determining communications originating from the specified AO input. The partial entropy covalency of the given atomic channel is then calculated for the full (unit) probability of its input, thus again recovering the binary entropy estimate in Figure 8.14. The partial mutual information has to reflect the redistribution of electrons in the whole diatomic, since the KL information distance is defined for the same set of events in the channel input and output. Therefore, the promolecular mutual information index reported in Figure 8.14 also characterizes each partial row channel in Figure 8.16.

8.5 DIATOMICS IN MOLECULES

In typical SCF LCAO MO calculations, the lone pairs of the valence and inner electronic shells can strongly affect the overall IT descriptors of the chemical bonds. Elimination of such lone-pair contributions in the resultant IT bond indices requires an ensemble (flexible input) approach [9,10,55]. In this scheme of determining the IT bond descriptors of diatomic fragments in molecules, the joint (bond) probabilities of two AO centered on different atoms constitute the input probabilities of the molecular information channel. Indeed, such probabilities reflect the simultaneous participation of the given pair of basis functions in interatomic chemical bonds so that this ensemble approach effectively projects out the spurious contributions attributable to inner-shell and outer-shell AO, which are excluded from mixing into the delocalized bonding MO combinations. This probability weighting procedure is capable of reproducing the Wiberg bond order in diatomics, at the same time providing the IT-covalent/IT-ionic resolution of this overall bond index.

This weighting procedure can be illustrated in the 2-AO model of Section 8.4. In the bond-weighted approach one uses the elementary subchannels of Figure 8.16 and their partial entropy/information descriptors. The conditional entropy and mutual information quantities for these partial communication systems $\{S(\chi|i)\}$, $I^0(i{:}\chi)$; $i = A,B\}$ are also listed in the diagram. Since these row descriptors represent the IT indices per electron in the diatomic fragment, these contributions have to be multiplied by $N_{AB} = N = 2$ in the corresponding resultant components and in the overall multiplicity of an effective diatomic (localized) bond.

Using the off-diagonal joint probability

$$P(A{\wedge}B) = p_A P(B|A) = P(B{\wedge}A) = p_B P(A|B) = PQ = \gamma_{A,B}\gamma_{B,A}/4 \qquad (8.36)$$

as the ensemble probability for both AO inputs gives the following ensemble average quantities for the 2-AO model of the diatomic system (Figure 8.17)

$$S_{AB} = N[P(A \wedge B)S(\chi|A) + P(B \wedge A)S(\chi|B)] = 4PQH(P) = M_{A,B}H(P)$$

$$I^0_{AB} = N[P(A \wedge B)I^0(A{:}\chi) + P(B \wedge A)I^0(B{:}\chi) = 4PQ[1 - H(P)] = M_{A,B}[1 - H(P)]$$

$$N^0_{AB} = S_{AB} + I^0_{AB} = 4PQ = (\gamma_{A,B})^2 = M_{A,B} \qquad (8.37)$$

The Wiberg index has thus been recovered as the overall IT descriptor of the chemical bond in the 2-AO model, with its covalent (S_{AB}) and ionic (I^0_{AB}) contributions being established at the same time. It follows from Figure 8.17 that these IT covalency/ionicity components compete with one another while conserving the Wiberg bond order as the overall information measure of chemical bond multiplicity

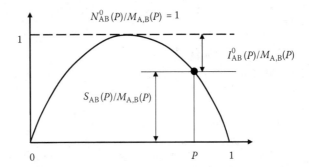

FIGURE 8.17 Variations of the IT-covalent $[S_{AB}(P)]$ and IT-ionic $[I^0_{AB}(P)]$ bond components [in $M_{A,B}(P)$ units] of the 2-AO model with changing MO polarization P, and the conservation of the relative total bond order $N^0_{AB}(P)/M_{A,B}(P) = 1$.

(in bits). This procedure can be generalized to cover general SCF MO calculations in arbitrary basis set [11,12,55]. Illustrative RHF bond orders in diatomic fragments of representative molecules for their equilibrium geometries in the minimum (STO-3G) basis set are compared in Table 8.1. It follows from these predictions that, in poly-atomic systems, the applied weighting procedure gives rise to an excellent agreement with both the Wiberg bond order and chemical intuition.

It can be demonstrated analytically [55] that this ensemble averaging of the local-ized bond descriptors reproduces exactly the Wiberg bond order in diatomic mole-cules. In a series of related compounds (e.g., in hydrides or halides) the trends exhibited by the entropic covalent and ionic components of a roughly conserved overall bond order also agree with intuitive expectations. For example, the single chemical bond between two "hard" atoms in HF appears predominantly covalent, while substantial ionicity is detected for LiF, for which both Wiberg and information-theoretic results predict roughly a 3/2 bond consisting of approximately 1 covalent contribution and ½ ionic contribution. One also observes that the direct carbon–carbon interactions in the benzene ring are properly differentiated. The chemical orders of the multiple bonds in ethane, ethylene, and acetylene are adequately reproduced, and the triple

TABLE 8.1

Comparison of Diatomic Wiberg and Entropy/Information Bond Multiplicity Descriptors in Representative Molecules: RHF Results in Minimum (STO-3G) Basis Set

Molecule	A–B	$M_{A,B}$	N_{AB}^0	S_{AB}	I_{AB}^0
H_2	H–H	1.000	1.000	1.000	0.000
F_2	F–F	1.000	1.000	0.947	0.053
HF	H–F	0.980	0.980	0.887	0.093
LiH	Li–H	1.000	1.000	0.997	0.003
LiF	Li–F	1.592	1.592	0.973	0.619
CO	C–O	2.605	2.605	2.094	0.511
H_2O	O–H	0.986	1.009	0.859	0.151
AlF_3	Al–F	1.071	1.093	0.781	0.311
CH_4	C–H	0.998	1.025	0.934	0.091
C_2H_6	C–C	1.023	1.069	0.998	0.071
	C–H	0.991	1.018	0.939	0.079
C_2H_4	C–C	2.028	2.086	1.999	0.087
	C–H	0.984	1.013	0.947	0.066
C_2H_2	C–C	3.003	3.063	2.980	0.062
	C–H	0.991	1.021	0.976	0.045
C_6H_6[a]	C_1–C_2	1.444	1.526	1.412	0.144
	C_1–C_3	0.000	0.000	0.000	0.000
	C_1–C_4	0.116	0.119	0.084	0.035

Source: Nalewajski, R. F. et al., *Adv. Quantum Chem.*, 2011, *61*, 1.

[a] For the sequential numbering of carbon atoms in the benzene ring.

bond in CO is correctly accounted for. As intuitively expected, the C–H bonds are seen to slightly increase their information ionicity when the number of these terminal bonds increases in a series: acetylene, ethylene, and ethane.

8.6 DIRECT AND INDIRECT BONDS

In OCT the direct communication $\chi_i \to \chi_j$ between the given pair of AO, reflected by the nonvanishing conditional probability $P(j|i) > 0$, generates a "dialogue" between these basis functions. It indicates the existence of direct (through-space) chemical bonding between these orbitals due to their overlap and constructive interference into the bonding MO combination. The Wiberg bond order contribution (Equations 8.25 and 8.35)

$$M_{i,j} = \gamma_{i,j}\gamma_{j,i} = 4d_{i,j}d_{j,i}$$

$$= 4\langle j|\hat{P}_\varphi|i\rangle\langle i|\hat{P}_\varphi|j\rangle = 4\left(\langle j|\hat{P}_\varphi\right)\left[\left(\hat{P}_\varphi|i\rangle\right)\left(\langle i|\hat{P}_\varphi\right)\right]\left(\hat{P}_\varphi|j\rangle\right)$$

$$\equiv 4\langle j^b|\hat{P}_i^b|j^b\rangle = 4\left|\langle i^b|j^b\rangle\right|^2 \equiv 4\left|d_{i,j}^b\right|^2,$$
(8.38)

provides a useful measure of the chemical multiplicity of this interaction. In Equation 8.38, the idempotency property $\hat{P}_\varphi^2 = \hat{P}_\varphi$ has been used, of the projector $\hat{P}_\varphi$ onto the bonding (occupied) subspace of MO, which determines the (nonorthogonal) bond-projections of AO

$$\left|\chi^b\right\rangle = \hat{P}_\varphi\left|\chi\right\rangle = \left\{\left|\chi_i^b\right\rangle\right\}.$$
(8.39)

This contribution to the overall direct bond order (Equation 8.35) is thus seen to measure the magnitude of the overlap $d_{i,j}^b$ between the bond projections of the two interacting basis functions.

However, the communications between χ_i and χ_j can be also realized indirectly, as "gossip" spreads via the remaining AO, $\chi' = \{\chi_{k\neq(i,j)}\}$ (e.g., in the single-step "cascade" propagation: $\chi_i \to \chi' \to \chi_j$). Moreover, since each AO in the molecular channel both emits and receives the signal to and from remaining basis functions, this process can be further extended into any admissible multistep cascade $\chi_i \to \{\chi' \to \chi' \to \ldots \to \chi'\} \to \chi_j$, in which the consecutive sets of AO intermediaries $\{\chi' \to \chi' \to \ldots \to \chi'\}$ form an effective "bridge" for information scattering. The OCT formalism has been extended to tackle such indirect (through-bridge) chemical bonding, using both the generalized Wiberg bond multiplicities and the IT bond descriptors of the cascade information channels [58–61]. In this section the main results of this analysis are summarized.

In MO theory the chemical interaction between, say, two (valence) AO originating from different atoms is strongly influenced by their direct overlap/interaction, which conditions the bonding effect experienced by electrons occupying their bonding combination in the molecule, compared to the nonbonding reference of electrons on separated AO. This through-space bonding mechanism is then associated with a typical accumulation of valence electrons in the region between the two nuclei due to the constructive interference between the two AO functions. Such "shared" bond charge is indeed synonymous with the presence of bond covalency in the direct interaction between the two AO, reflected by the covalent valence bond (VB) structure. In Sections 8.2 and 8.3, a similar effect of the bonding accumulation of the information densities relative to the promolecular distribution has been detected in maps of alternative information densities. The complementary ionicity aspect of the direct chemical bonding is manifested by MO polarization and charge transfer due to the participation of the ionic VB structures in the ground-state wave function. In OCT the bond ionicity descriptor reflects a degree¹ of "localization" (determinicity) in communications between bonded atoms, while bond covalency measures the "delocalization" (noise) aspect of the direct molecular channel.

Thus, the direct ("through-space") bonding interaction between neighboring atoms is associated with the presence of bond charge between the two nuclei. However, for more distant atomic partners such an accumulation of valence electrons can be absent (e.g., in cross-ring π interactions in benzene or between bridgehead carbon atoms in small propellanes). The bonding interaction lacking such an accumulation of the charge (information) can then be realized indirectly through the neighboring AO intermediaries forming a chemical "bridge" for an effective interaction between more distant (terminal) AO (e.g., between two *meta*-carbons or *para*-carbons in benzene, the bridgehead carbons in [1.1.1] propellane [58], or higher neighbors in the polymer chain [60,61]).

In such a generalized outlook on the bond order concept, which emerges from the Wiberg and quadratic difference measures formulated in the MO theory and from the IT bond multiplicities in OCT, one identifies chemical bond multiplicity as a measure of statistical "dependence" (nonadditivity) between orbitals in different atomic centers. On one hand, this dependence between basis functions of different atoms can be realized directly (through-space) by the constructive interference of orbitals (probability amplitudes) on two atoms, which generally increases the electron density between them. On the other hand, it can also have an indirect origin through the dependence on orbitals of the remaining AIM used to construct the system occupied MO. In the nonorthogonal basis of the bond-projected AO, these implicit ("geometrical") dependencies are embodied in the (idempotent) density (overlap) matrix

$$\mathbf{d} = \left\langle \boldsymbol{\chi}^b \middle| \boldsymbol{\chi}^b \right\rangle = \boldsymbol{\gamma}/2 = \mathbf{C}^\circ \mathbf{C}^{\circ\dagger} = \left\{ d_{i,j} = \left\langle i^b \middle| j^b \right\rangle \right\}, \quad \mathbf{d}^n = \mathbf{d}. \tag{8.40}$$

This indirect mechanism was shown to reflect the implicit dependencies between the bond-projected AO of Equation 8.39, reflecting the orbital resultant participation in all chemical bonds in the molecular system under consideration [59]. The orthonormality constraints imposed on the occupied MO indeed imply the implicit dependencies between these AO bond projections on different atoms, which generate indirect contributions to the multiplicities of their chemical interactions, when these orbitals are more widely separated in a molecule, provided that they directly couple to the chemically interacting orbitals of real bridges connecting these terminal orbitals (Figure 8.18).

Each pair of AO or AIM thus exhibits partial through-space and through-bridge components. The bond order of the former quickly vanishes with an increase in interatomic separation or when the interacting AO are heavily engaged in forming chemical bonds with other atoms, while the latter can still assume appreciable values when the remaining atoms form an effective bridge with neighboring chemically interacting atoms, which links the specified AO/AIM. The bridging atoms must be mutually bonded to generate the appreciable through-bridge overlap of the interacting AO so that significant bridges are in fact limited to real chemical bridges of atoms in the molecular structural formula.

The representative indirect overlap $S_{i,j}(k,l,\ldots,m,n)$ through the t-bridge in Figure 8.18 constitutes a natural generalization of its direct (through-space) analog by additionally including the product of bond projections onto the intermediate AO:

$$S_{i,j}(k,l,\ldots,m,n) = \left\langle i^b \left| \prod_{r=1}^{t} \hat{P}_r^b \right| j^b \right\rangle \equiv \left\langle i^b \left| \hat{P}_{t\text{-bridge}}^b \right| j^b \right\rangle. \qquad (8.41)$$

For example, for the single-step and multistep bridges realized by the indicated AO intermediates, this indirect bond overlap is given by the relevant products of direct overlaps in the bridge: $S_{i,j}(k) = d_{i,k}d_{k,j}$, $S_{i,j}(k,l) = d_{i,k}d_{k,l}d_{l,j}$, etc. Hence, for a general t-bridge in Figure 8.18 one finds

$$S_{i,j}(k,l,\ldots,m,n) = d_{i,k}d_{k,l}\ldots d_{m,n}d_{n,j}. \qquad (8.42)$$

The square of this bond overlap defines the associated Wiberg-type bond order for such an implicit interaction between orbitals χ_i and χ_j via the specified t-bridge $(k,l,\ldots,m,n)$

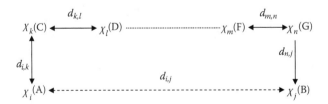

FIGURE 8.18 Direct (through-space) chemical interaction (broken line) between orbitals χ_i and χ_j contributed by atoms A and B, respectively, and indirect (through-bridge) interaction (solid lines), through t AO intermediaries $(\chi_k,\chi_l,\ldots,\chi_m,\chi_n) = \{\chi_r, r = 1,2,\ldots,t\}$, contributed by the neighboring (bonded) atoms (C,D,…,F,G), respectively. The strength of each direct interaction is reflected by the magnitude of the corresponding elements of the density matrix $\mathbf{d} = \langle \chi^b | \chi^b \rangle = \{d_{i,j}\}$, which measure overlaps between bond projections χ^b of AO.

$$M_{i,j}(k,l,\ldots,m,n) = 2^{2t}\left|S_{i,j}(k,l,\ldots,m,n)\right|^2 = \gamma_{i,k}\{\gamma_{k,l}\cdots[\gamma_{m,n}(\gamma_{n,j}\gamma_{j,n})\gamma_{n,m}]\cdots\gamma_{l,k}\}\gamma_{k,i}$$

$$= M_{i,k}M_{k,l}M_{l,m}\ldots M_{m,n}M_{n,j}.$$

$$(8.43)$$

The sum of such contributions due to the most important chemical bridges $\{\alpha\}$

$$M_{i,j}(\text{bridge}) = \sum_{\alpha} M_{i,j}(\alpha) \qquad (8.44)$$

then determines the overall indirect Wiberg-type bond order, which supplements the direct component $M_{i,j}$ in the full quadratic bond multiplicity between orbitals χ_i and χ_j in the presence of all remaining basis functions

$$M(i,j) = M_{i,j} + M_{i,j}(\text{bridge}). \qquad (8.45)$$

As an illustration let us summarize the indirect π bonds between carbon atoms in benzene and butadiene [58] in the Hückel approximation. For the consecutive numbering of carbons in the ring and chain, the relevant CBO matrix elements in benzene read $\gamma_{i,i} = 1$, $\gamma_{i,i+1} = 2/3$, $\gamma_{i,i+2} = 0$, $\gamma_{i,i+3} = -1/3$, while the relevant off-diagonal part of the CBO matrix in butadiene is fully characterized by the elements $\gamma_{1,2} = \gamma_{3,4} = 2/\sqrt{5}$, $\gamma_{1,3} = \gamma_{2,4} = 0$, $\gamma_{1,4} = -1/\sqrt{5}$, and $\gamma_{2,3} = 1/\sqrt{5}$. They generate the associated direct (through-space) bond multiplicities:

benzene: $M_{i,i+1} = 0.44$ (*ortho*), $M_{i,i+2} = 0$ (*meta*), $M_{i,i+3} = 0.11$ (*para*)
butadiene: $M_{1,2} = M_{3,4} = 0.80$, $M_{1,3} = M_{2,4} = 0$, $M_{1,4} = M_{2,3} = 0.20$

These direct bond orders are complemented by following estimates of the resultant multiplicities of the indirect π interactions due to chemical bridges:

benzene: $M_{i,i+1}(\text{bridge}) = 0.06$ (*ortho*), $M_{i,i+2}(\text{bridge}) = 0.30$ (*meta*)
$\qquad M_{i,i+3}(\text{bridge}) = 0.18$ (*para*)
butadiene: $M_{1,2}(\text{bridge}) = M_{3,4}(\text{bridge}) = 0.03$
$\qquad M_{1,3}(\text{bridge}) = M_{2,4}(\text{bridge}) = 0.32$
$\qquad M_{1,4}(\text{bridge}) = M_{2,3}(\text{bridge}) = 0.13$

These two mechanisms give rise to the following resultant bond orders of Equation 8.45:

benzene: $M(para) \cong M(meta) = 0.3 < M(ortho) = 0.5$
butadiene: $M(1-2) = 0.83 > M(1-4) = 0.33 \cong M(1-3) = 0.32$

The artificial distinction in Wiberg's multiplicity scheme of π interactions with the vanishing direct CBO element as being nonbonding, is thus effectively removed when the through-bride contributions are taken into account. One observes the difference in the composition of the resultant indices in benzene: *para* bonds exhibit comparable through-space and through-bridge components, *meta* multiplicities are realized exclusively through bridges, while the strongest *ortho* bonds are of practically direct through-space origin. A similar pattern is observed for butadiene. Of interest also is a comparison of the bond order contributions realized through ring bridges of increasing length:

$$M_{i,i+2}(i + 1) = (M_{i,i+1})^2 = 0.20$$
$$M_{i,i+3}(i + 1, i + 2) = (M_{i,i+1})^3 = 0.09$$
$$M_{i,i+4}(i + 1, i + 2, i + 3) = (M_{i,i+1})^4 = 0.04$$
$$M_{i,i+5}(i + 1, i + 2, i + 3, i + 4) = (M_{i,i+1})^5 = 0.02$$

Thus, the longer the bridge, the smaller the indirect bond order it contributes. The model and HF estimates for representative polymers [60,61] indicate that the range of such bridge interactions is effectively extended up to the third-neighbors in the chain, where the direct interactions practically vanish.

The conditional probabilities of Equation 8.24 determine the molecular information channel for the direct communications between AO, which generate the associated covalency (noise) and ionicity (information flow) descriptors of chemical bonds. One similarly derives the corresponding entropy/information multiplicities for the indirect interactions between the specified (terminal) orbitals χ_i and χ_j from descriptors of the associated AO information cascades, including the most important (chemical) bridges [10,58,60]. These "directed" indices of bridge interactions have been shown to compare favorably with the generalized Wiberg measures of Equation 8.43.

8.7 CONCLUSIONS

IT provides a novel perspective on the origins of chemical bonds. This short overview has introduced IT's key concepts and techniques, which have been subsequently used to explore the electronic structure of prototype molecules in terms of information distribution, bond multiplicities and their ionic/covalent components, as well as through the corresponding entropy/information descriptors of localized (diatomic) chemical interactions. Illustrative numerical results have been presented to validate the claim that alternative information densities increase the understanding of the chemical bond phenomenon and that the communication noise (covalency) and information flow (ionic) measures (in bits) of the IT bond orders indeed reflect chemical intuition quite well. The use of information densities as local probes of equilibrium electronic distributions in molecules has been advocated and the importance of nonadditive entropy/information measures in extracting subtle changes due to bond formation has been stressed. The CG density of nonadditive Fisher information (or electronic kinetic energy) in AO resolution has been advocated as an efficient probe for localizing direct chemical bonds. This development extends the understanding of chemical bonds from the complementary IT viewpoint.

Recent developments in OCT have introduced a novel communication (entropy/information) perspective on several classical issues in the theory of bond multiplicities and their covalent/ionic composition. The off-diagonal conditional probabilities it generates are proportional to the quadratic bond indices formulated in MO description; hence, the strong interorbital communications correspond to strong Wiberg bond orders. The OCT treatment of localized (diatomic) chemical interactions has also been outlined. This approach reproduces Wiberg bond multiplicity in diatomic molecules and allows one to resolve this overall bond order into the associated IT covalent/ionic components. The localized bond-multiplicities in typical polyatomic molecules were shown to approximate the quadratic Wiberg index of quantum chemistry quite well, at the same time providing its IT-covalent/ionic resolution.

A novel "through-bridge" mechanism of bonding interactions in molecules, which has been first conjectured to explain bonding patterns in small propellanes, has also been examined. In the smallest [1.1.1] system, no extra accumulation of electron or information density between bridgehead carbons was observed. This prompted the alternative propositions of the VB-inspired charge shift [88] and through-bridge mechanisms, with the former attributing this bonding effect to instantaneous charge fluctuations between the bridgehead carbons. The bridge mechanism also has important implications for the bonding patterns of π interactions in hydrocarbons: in the π system of benzene, the *ortho*-carbons exhibited a strong Wiberg bond-multiplicity measure of almost exclusively through-space origin, and the cross-ring interactions between the *meta*- and *para*-carbons where shown to be described by much smaller but practically equalized overall bond-orders. The latter are distinguished solely by the direct/indirect composition of these resultant chemical interactions: *meta* bonds are realized exclusively through bridges, while *para* bonds exhibit comparable direct and indirect components. The most efficient bridges for such an implicit bonding mechanism via atomic intermediaries are the real chemical bridges, via the chemically bonded atoms connecting such "terminal" atoms of the AIM chain in question. Therefore, the bonded status of the given pair of orbitals/atoms can be felt even at larger separations, provided that there exists a real bridge of direct chemical bonds connecting them.

In OCT the direct bond component due to the specified pair of interacting AO originates from the mutual probability scattering between these two basis functions, which constructively mix into the bonding MO. Its covalency originates from the finite conditional probability of communications in the molecule, related to the square of the corresponding element of the system density matrix (coupling the two basis functions) and hence also to the associated Wiberg bond order contribution. The direct AO communications are in accordance with the electron delocalization pattern implied by the system occupied (bonding) subspace of MO. The "implicit" (through-bridge) bond component can be similarly viewed as resulting from indirect (cascade) information propagation via the bridging AO. Therefore, while through-space bonding reflects the direct "conversation" between AO, the through-bridge channel can be compared to a chatty talk reporting "hearsay," with the "rumor" spreading between the two AO in question via the connecting chain of the AO intermediaries involved in the effective chemical bridge under consideration.

This novel perspective on the information origins of chemical bonding is very much in line with Wigner's observation, often quoted by Kohn, that understanding in science requires insights from several different points of view. The kinetic energy probe of bonding regions in molecules provides such an additional Fisher information perspective on the genesis of chemical bonds. It complements the familiar potential energy interpretation of textbooks on quantum chemistry. Together, the complementary MO and IT tools give rise to a deeper understanding of the complex phenomenon called chemical bonding.

REFERENCES

1. Fisher, R. A. *Proc. Cambridge Philos. Soc.* 1925, *22*, 700.
2. Frieden, B. R. *Physics From the Fisher Information—A Unification*, 2nd ed., Cambridge University Press, Cambridge, 2004.
3. Shannon, C. E. *Bell Syst. Tech. J.* 1948, *27*, 379, 623.
4. Shannon, C. E. and Weaver, W. *The Mathematical Theory of Communication*, University of Illinois Press, Urbana, 1949.
5. Kullback, S. and Leibler, R. A. *Ann. Math. Stat.* 1951, *22*, 79.
6. Kullback, S. *Information Theory and Statistics*, Wiley, New York, 1959.
7. Abramson, N. *Information Theory and Coding*, McGraw-Hill, New York, 1963.
8. Pfeifer, P. E. *Concepts of Probability Theory*, 2nd ed., Dover, New York, 1978.
9. Hohenberg, P. and Kohn, W. *Phys. Rev.* 1964, *136B*, 864.
10. Nalewajski, R. F. *Information Theory of Molecular Systems*, Elsevier, Amsterdam, 2006.
11. Nalewajski, R. F. *Information Origins of the Chemical Bond*, Nova Science Publishers, New York, 2010.
12. Nalewajski, R. F. *Perspectives in Electronic Structure Theory*, Springer-Verlag, Heidelberg, 2012.
13. Nalewajski, R. F. In *Mathematical Chemistry*, Hong, W. I. (Ed.), Nova Science Publishers, New York, 2011, p. 247.
14. Nalewajski, R. F. In *Chemical Information and Computation Challenges in 21st Century*, Putz, M. V. (Ed.), Nova Science Publishers, New York, 2012, forthcoming.
15. Nalewajski, R. F., de Silva, P., and Mrozek, J. In *Theoretical and Computational Developments in Modern Density Functional Theory*, Roy, A. K. (Ed.), Nova Science Publishers, New York, 2012, forthcoming.
16. Nalewajski, R. F., Świtka, E., and Michalak, A. *Int. J. Quantum Chem.* 2002, *87*, 198.
17. Nalewajski, R. F. and Świtka, E. *Phys. Chem. Chem. Phys.* 2002, *4*, 4952.
18. Nalewajski, R. F. and Broniatowska, E. *J. Phys. Chem. A* 2003, *107*, 6270.
19. Nalewajski, R. F. and Parr, R. G. *Proc. Natl. Acad. Sci. U. S. A.* 2000, *97*, 8879.
20. Nalewajski, R. F. and Parr, R. G. *J. Phys. Chem. A* 2001, *105*, 7391.
21. Nalewajski, R. F. and Loska, R. *Theor. Chem. Acc.* 2001, *105*, 374.
22. Nalewajski, R. F. *Phys. Chem. Chem. Phys.* 2002, *4*, 1710.
23. Nalewajski, R. F. *Chem. Phys. Lett.* 2003, *372*, 28.
24. Parr, R. G., Ayers, P. W., and Nalewajski, R. F. *J. Phys. Chem. A* 2005, *109*, 3957.
25. Nalewajski, R. F. *Adv. Quantum Chem.* 2003, *43*, 119.
26. Nalewajski, R. F. and Broniatowska, E. *Theor. Chem. Acc.* 2007, *117*, 7.
27. Hirshfeld, F. L. *Theor. Chim. Acta* 1977, *44*, 129.
28. Nalewajski, R. F. *Int. J. Quantum Chem.* 2008, *108*, 2230.
29. Nalewajski, R. F. *J. Math. Chem.* 2010, *47*, 667.
30. Nalewajski, R. F., de Silva, P., and Mrozek, J. *J. Mol. Struct. THEOCHEM* 2010, *954*, 57.
31. Nalewajski, R. F., Köster, A. M., and Escalante, S. *J. Phys. Chem. A* 2005, *109*, 10038.
32. Becke, A. D. and Edgecombe, K. E. *J. Chem. Phys.* 1990, *92*, 5397.
33. Silvi, B. and Savin, A. *Nature* 1994, *371*, 683.
34. Savin, A., Nesper, R., Wengert, S., and Fässler, T. F. *Angew. Chem. Int. Ed. Engl.* 1997, *36*, 1808.

35. Nalewajski, R. F. *J. Phys. Chem. A* 2000, *104*, 11940.
36. Nalewajski, R. F. and Jug, K. In *Reviews of Modern Quantum Chemistry: A Celebration of the Contributions of Robert G. Parr*, vol. I, Sen, K. D. (Ed.), World Scientific, Singapore, 2002, p. 148.
37. Nalewajski, R. F. *Struct. Chem.* 2004, *15*, 391.
38. Nalewajski, R. F. *Mol. Phys.* 2004, *102*, 531, 547.
39. Nalewajski, R. F. *Mol. Phys.* 2005, *103*, 451.
40. Nalewajski, R. F. *Mol. Phys.* 2006, *104*, 365, 493, 1977, 2533.
41. Nalewajski, R. F. *Theor. Chem. Acc.* 2005, *114*, 4.
42. Nalewajski, R. F. *J. Math. Chem.* 2005, *38*, 43.
43. Nalewajski, R. F. *J. Math. Chem.* 2008, *43*, 265.
44. Nalewajski, R. F. *J. Math. Chem.* 2008, *44*, 414.
45. Nalewajski, R. F. *J. Math. Chem.* 2009, *45*, 607, 709, 776, 1041.
46. Nalewajski, R. F. *Int. J. Quantum Chem.* 2009, *109*, 425.
47. Nalewajski, R. F. *J. Math. Chem.* 2010, *47*, 709.
48. Nalewajski, R. F. *J. Math. Chem.* 2008, *43*, 780.
49. Nalewajski, R. F. *J. Phys. Chem. A* 2007, *111*, 4855.
50. Nalewajski, R. F. *Mol. Phys.* 2006, *104*, 3339.
51. Nalewajski, R. F. *J. Phys. Chem A* 2003, *107*, 3792.
52. Nalewajski, R. F. *Mol. Phys.* 2006, *104*, 255.
53. Nalewajski, R. F. *Ann. Phys.* 2004, *13*, 201.
54. Nalewajski, R. F. *Adv. Quantum Chem.* 2009, *56*, 217.
55. Nalewajski, R. F., Szczepanik, D., and Mrozek, J. *Adv. Quantum Chem.* 2011, *61*, 1.
56. Nalewajski, R. F. *J. Math. Chem.* 2010, *47*, 692, 709, 808.
57. Nalewajski, R. F. *J. Math. Chem.* 2011, *49*, 592.
58. Nalewajski, R. F. *J. Math. Chem.* 2011, *49*, 371, 546, 806.
59. Nalewajski, R. F. and Gurdek, P. *J. Math. Chem.* 2011, *49*, 1226.
60. Nalewajski, R. F. *Int. J. Quantum Chem.* 2012, *112*, 2355.
61. Nalewajski, R. F. and Gurdek, P. *Struct. Chem.* 2012, *23*, 1383.
62. Nalewajski, R. F. *J. Math. Chem.* 2011, *49*, 2308.
63. Kohn, W. and Sham, L. J. *Phys. Rev.* 1965, *140A*, 1133.
64. Capitani, J. F., Nalewajski, R. F., and Parr, R. G. *J. Chem. Phys.* 1982, *76*, 568.
65. Gordon, R. G. and Kim, Y. S. *J. Chem. Phys.* 1972, *56*, 3122.
66. Cortona, P. *Phys. Rev. B* 1991, *44*, 8454.
67. Wesołowski, T. and Warshel, A. *J. Phys. Chem.* 1993, *97*, 8050.
68. Wesołowski, T. *J. Am. Chem. Soc.* 2004, *126*, 11444.
69. Wesołowski, T. *Chimia* 2004, *58*, 311.
70. von Weizsäcker, C. F. *Z. Phys.* 1935, *96*, 431.
71. Feinberg, M. J., Ruedenberg, K., and Mehler, E. L. *Adv. Quantum Chem.* 1970, *5*, 28.
72. Feinberg, M. J. and Ruedenberg, K. *J. Chem. Phys.* 1971, *54*, 1495.
73. Feinberg, M. J. and Ruedenberg, K. *J. Chem. Phys.* 1971, *55*, 5804.
74. Goddard, W. A. and Wilson, C. W. *Theor. Chim. Acta* 1972, *26*, 195, 211.
75. Nalewajski, R. F. *Int. J. Quantum Chem.* 2009, *109*, 2495.
76. Dirac, P. A. M. *The Principles of Quantum Mechanics*, 4th ed., Clarendon Press, Oxford, 1958.
77. Wiberg, K. A. *Tetrahedron* 1968, *24*, 1083.
78. Nalewajski, R. F., Köster, A. M., and Jug, K. *Theor. Chim. Acta* 1993, *85*, 463.
79. Nalewajski, R. F. and Mrozek, J. *Int. J. Quantum Chem.* 1994, *51*, 187.
80. Nalewajski, R. F., Formosinho, S. J., Varandas, A. J. C., and Mrozek, J. *Int. J. Quantum Chem.* 1994, *52*, 1153.
81. Nalewajski, R. F., Mrozek, J., and Mazur, G. *Can. J. Chem.* 1996, *100*, 1121.
82. Nalewajski, R. F., Mrozek, J., and Michalak, A. *Int. J. Quantum Chem.* 1997, *61*, 589.

83. Mrozek, J., Nalewajski, R. F., and Michalak, A. *Pol. J. Chem.* 1998, *72*, 1779.
84. Nalewajski, R. F. *Chem. Phys. Lett.* 2004, *386*, 265.
85. Gopinathan, M. S. and Jug, K. *Theor. Chim. Acta* 1983, *63*, 497, 511.
86. Jug, K. and Gopinathan, M. S. In *Theoretical Models of Chemical Bonding*, vol. II, Maksić, Z. B. (Ed.), Springer, Heidelberg, 1990, p. 77.
87. Mayer, I. *Chem. Phys. Lett.* 1983, *97*, 270.
88. Shaik, S., Danovich, D., Wu, W., and Hiberty, P. C. *Nat. Chem.* 2009, *1*, 443.

9 Molecular Electrostatic Potentials
Some Observations

Peter Politzer and Jane S. Murray

CONTENTS

9.1   CHARGE SEPARATION IN MOLECULES

Recognition that there is significant charge separation in many molecules has a long history, as do efforts to make this charge separation more predictable. Thus, the concept of electronegativity can be traced back at least to Berzelius in 1835 [1,2]. Perhaps the best known electronegativity scale is that of Pauling [3,4], who assigned relative values to atoms on the basis of the estimated ionic characters of heteronuclear covalent bonds formed by these atoms. Pauling gave his electronegativities to only two significant figures, which is realistic and appropriate for such an inherently approximate concept, but they have been and continue to be quite useful from a practical standpoint.

Over the years, there have been many different formulations of electronegativity to the extent that, in 1961, Iczkowski and Margrave [5] were led to remark that "... there is some confusion as to what physical picture corresponds to the term electronegativity." They pointed out that there is not even an agreement as to what units it should have.

181

With the advent of density functional theory, a new approach to the concept of electronegativity was proposed [6]. It was equated with the negative of the chemical potential, which was interpreted as a measure of the escaping tendency of an electron in a ground-state atom or molecule [7]. (One might ask: How does this differ from ionization energy?) This view of electronegativity is in marked contrast to Pauling's view ("... the power of an atom in a molecule to attract electrons to itself") [8], and it has not been met with universal approval [2,9–12]; strong arguments have been made that electronegativity and chemical potential are two separate and distinct properties [9–11]. Detailed discussions and critiques of the various approaches to electronegativity can be found elsewhere [2,12–16].

Despite the ambiguity associated with electronegativity, there have been numerous attempts to go even further and to assign numerical partial charges to atoms in molecules. (One of the present authors, in his youth, also succumbed to this urge [17].) Meister and Schwarz [18] cited about 30 different schemes for calculating supposed atomic charges. Although there are good correlations between some of them [18,19], that may simply indicate that they are equally misleading. Serious contradictions are also found; for example, six different methods give charges ranging from −0.478 to +0.564 for the carbon in $H_3C–NO_2$ [19].

The basic problem with trying to quantify either electronegativity or atomic charge is that while these are intuitively appealing concepts, they do not correspond to real physical or electronic properties. They are not physical observables and cannot be determined experimentally. There is no rigorous unique definition of either one, which is why so many arbitrary ones have been suggested. Some of these yield predictions that appear to be more (or less) realistic than others, but none can claim to be accurate or correct because these terms are not meaningful in this context; there is no unique definition of "accurate" or "correct" with respect to electronegativity or atomic charge.

Occasionally, one sees references to "experimental electronegativities" or "experimental atomic charges." These terms are misleading. They simply mean that experimental data of some sort were inserted into an arbitrarily proposed electronegativity or atomic charge formula. Electronegativity and atomic charge cannot be determined experimentally because they have no rigorous physical basis.

The significance of the pioneering work of Bonnaccorsi et al. [20] and Scrocco and Tomasi [21,22] in calculating and analyzing molecular electrostatic potentials can be seen from the preceding discussion. The electrostatic potential directly reflects, in a physically rigorous and detailed manner, the distribution of charge in a system of nuclei and electrons. The electrostatic potential is a physical observable. It can be determined experimentally (by diffraction methods) [23,24] and computationally.

The use of the electrostatic potential in interpreting and predicting chemical and physical properties has increased enormously over the past 40 years. A review that appeared in 1981, in a volume edited by Professor B. M. Deb, could still try to encompass every published paper dealing with the electrostatic potential [25]. This would certainly not be a realistic objective today.

In the present chapter, we will look at some specific aspects of atomic and molecular electrostatic potentials. This is not, however, intended to be a comprehensive overview; some recent ones can be found elsewhere [26,27].

9.2 ELECTROSTATIC POTENTIAL

A system of nuclei and electrons (e.g., a molecule) creates an electrostatic potential $V(\mathbf{r})$ at any point $\mathbf{r}$ in the surrounding space, where

$$V(\mathbf{r}) = \sum_A \frac{Z_A}{|\mathbf{R}_A - \mathbf{r}|} - \int \frac{\rho(\mathbf{r}')d\mathbf{r}'}{|\mathbf{r}' - \mathbf{r}|}. \tag{9.1}$$

In Equation 9.1, which is expressed in atomic units, Z_A is the charge on nucleus A located at $\mathbf{R}_A$; the nuclei are treated as stationary (Born–Oppenheimer approximation). The quantity $\rho(\mathbf{r})$ is the electronic density, which is the average number of electrons in each volume element $d\mathbf{r}$. Since the average (static) electronic density is being used, $V(\mathbf{r})$ is labeled the "electrostatic" potential.

By definition, $V(\mathbf{r})$ has units of energy/charge. However, Equation 9.1 can also be viewed as the interaction energy of the system with a unit positive point charge placed at $\mathbf{r}$ (e.g., a proton). It has therefore become customary to express $V(\mathbf{r})$ in units of energy. Thus, $V(\mathbf{r})$ obtained directly from Equation 9.1 would be expressed in hartrees, which might then be converted into other units such as kilocalories per mole.

Equation 9.1 indicates that the sign of $V(\mathbf{r})$ in any given region depends on whether the positive contribution of the nuclei or the negative contribution of the electrons is dominant there. However, certain generalizations have been proven rigorously:

1. The electrostatic potential of a spherically averaged ground-state atom or a positive monoatomic ion is positive everywhere, decreasing monotonically in every radial direction from the nucleus to zero at infinity [28].
2. For a spherically averaged negative monatomic ion, $V(\mathbf{r})$ is positive near the nucleus but decreases radially to a negative minimum at some distance r_m, after which it gradually increases to zero at infinity [28]. The electronic charge encompassed within the sphere of radius r_m exactly equals the nuclear charge and therefore balances it (Gauss's law). It follows that $V(r_m)$ reflects only the excess electronic charge on the ion. $V(r_m)$ accordingly governs the strengths of the ion's interactions. The distances r_m do in fact provide good estimates of crystallographic ionic radii, and the magnitudes of $V(r_m)$ correlate well with experimental lattice energies for a given cation [28,29].
3. Molecules normally have regions of both positive and negative electrostatic potentials. The latter are often associated with lone pairs of the more electronegative elements (nitrogen, oxygen, halogens, etc.) and with π electrons of unsaturated molecules. Such negative regions typically have one or more local minima [i.e., points at which $V(\mathbf{r})$ reaches its most negative values in that region]. However, perhaps somewhat surprisingly, positive regions have local maxima only by the positions of the nuclei. This was proven by Pathak and Gadre [30].

The electrostatic potential of a molecule can be displayed in various ways. Originally, it was often presented as isopotential contours on planes through the molecule [20–22,24]. A more recent practice is to show three-dimensional plots of just a single positive value and a single negative value of $V(\mathbf{r})$. However, this necessarily gives an incomplete picture since other values may also be important.

A particularly effective approach—especially in the context of reactive behavior—is to compute $V(\mathbf{r})$ on the surface of the molecule since this is what is initially encountered by its environment. Unfortunately, there is no rigorous definition of molecular surface. However, a very useful suggestion by Bader et al. [31] is to use an outer contour of the molecule's electronic density $\rho(\mathbf{r})$. This has a very significant advantage in that it reflects the specific features of a particular molecule, such as lone pairs, asymmetries in atoms' charge distributions, π electrons, and strained bonds. The $\rho(\mathbf{r}) = 0.001$ a.u. (electrons/bohr³) contour is frequently used to represent a molecular surface, although other outer contours (e.g., 0.0015 or 0.002 a.u.) could be equally effective, showing the same trends [32]. All these encompass at least 96% of the molecule's electronic charge [31]. $V(\mathbf{r})$ on the 0.001 a.u. surface is commonly labeled $V_S(\mathbf{r})$.

Examples of molecular electrostatic potentials are presented in Figures 9.1 and 9.2. These were computed on the $\rho(\mathbf{r}) = 0.001$ a.u. surfaces of 4-fluoroaniline (1) (Figure 9.1) and 2-bromo-5-hydroxypyrimidine (2) (Figure 9.2). The density functional B3PW91/6-31G(d,p) procedure and Gaussian 09 [33] were used to obtain wave functions, and the WFA-SAS code [34] was used to generate the surface potential $V_S(\mathbf{r})$.

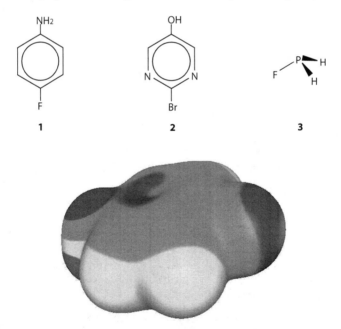

FIGURE 9.1 (See color insert.) Computed electrostatic potential on the 0.001 a.u. molecular surface of (1). The amino group is shown to the left, and fluorine is shown to the right. Color ranges are as follows: red, greater than 18 kcal/mol; yellow, between 18 and 0 kcal/mol; green, between 0 and −18 kcal/mol; blue, more negative than −18 kcal/mol.

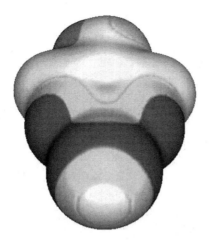

FIGURE 9.2 **(See color insert.)** Computed electrostatic potential on the 0.001 a.u. molecular surface of (**2**). Bromine and azine nitrogens are shown toward the front, and the hydroxyl group is shown toward the back. Color ranges are as follows: red, greater than 30 kcal/mol; yellow, between 30 and 0 kcal/mol; green, between 0 and −13 kcal/mol; blue, more negative than −13 kcal/mol. Note the positive σ-hole (yellow) on bromine.

Figures 9.1 and 9.2 show that each molecular surface has regions of both negative and positive electrostatic potentials. The most negative and most positive potentials in these regions, the local surface minima and maxima, are designated $V_{S,min}$ and $V_{S,max}$, respectively. (Note that these are not global minima and maxima but pertain only to the 0.001 a.u. surfaces. As mentioned earlier, global maxima can be found only by the positions of the nuclei [30].) There are negative regions associated with the lone pairs of amino nitrogen in (**1**) ($V_{S,min}$ = −28 kcal/mol), azine nitrogens in (**2**) ($V_{S,min}$ = −32 and −33 kcal/mol), and hydroxyl oxygen ($V_{S,min}$ = −12 kcal/mol). The surface of fluorine is completely negative as is that of bromine, except for a small but very important positive region ($V_{S,max}$ = 9 kcal/mol) on its outer side; this will be discussed later. Finally, there are negative potentials above and below the ring of (**1**) due to its π electrons, with $V_{S,min}$ of −19 and −15 kcal/mol. However, the regions above and below the ring of (**2**) are positive ($V_{S,max}$ = 11 kcal/mol) primarily because of the electron-withdrawing effect of azine nitrogens. The other positive potentials in (**1**) and (**2**) correspond to the various hydrogens; the most positive are those of the hydroxyl and amino groups, with $V_{S,max}$ of 69 and 38 kcal/mol, respectively.

9.3 ELECTROSTATIC POTENTIAL AND MOLECULAR REACTIVITY

9.3.1 SOME BACKGROUND

A point to keep in mind when considering the electrostatic potential as a guide to reactive behavior is that $V(\mathbf{r})$ for a molecule A is normally determined, via Equation 9.1, from the average equilibrium electronic density of A. Close approach of a reactant B perturbs the electronic density of A to some extent and accordingly changes its electrostatic potential. The $V(\mathbf{r})$ of equilibrium ground state A is then no longer relevant [35]; it is not the $V(\mathbf{r})$ that is seen by nearby B.

It follows that ground-state molecular electrostatic potentials are most meaning-ful, with respect to reactivity, for relatively long-range interactions (i.e., when the reactant B is sufficiently far from A that its perturbing effect on the electronic den-sity of A is minimal). Thus, $V(\mathbf{r})$ has been found to be very useful for interpreting and predicting noncovalent interactions such as hydrogen bonding, σ-hole bonding, and biological recognition processes. For these purposes, computing $V(\mathbf{r})$ upon the 0.001 a.u. contour of $\rho(\mathbf{r})$ is very appropriate because this contour is generally beyond the van der Waals radii of the atoms comprising the molecule (except for hydrogen) [36]. $V(\mathbf{r})$ on this surface (i.e., $V_S(\mathbf{r})$) is therefore reasonably indicative of how the molecule is perceived at long range by an approaching reactant.

9.3.2 Hydrogen Bonding

In 1975, Kollman et al. [37] used $V(\mathbf{r})$ to demonstrate the important role, in hydro-gen bonding, of the electrostatic interaction between positive hydrogens (e.g., OH, NH_2) and negative sites, such as lone pairs (:OH_2, :NH_2R, etc.) or π electrons. More recently, hydrogen $V_{S,max}$ and basic site $V_{S,min}$ have been shown to correlate well with empirical measures of hydrogen-bond–donating and hydrogen-bond–accepting ten-dencies, respectively [38]. A few years ago, attention was focused on "dihydrogen" or "unconventional" hydrogen bonding, A–H—H–B [39–41]. However, this is simply hydrogen bonding in which the negative site also happens to be a hydrogen. While the latter is relatively unusual, well-known examples are alkali and alkaline earth metal hydrides, as well as boron hydrogens in the system H_3B–NH_3, which have a $V_{S,min}$ of −39 kcal/mol at the B3PW91/6-31G(d,p) level [27].

9.3.3 Triumph of Electrostatic Potential: Halogen Bonding and, More Generally, σ-Hole Bonding

It has long been recognized that covalently bonded halogens can often interact favor-ably and highly directionally with negative sites such as lone pairs. (See Politzer et al. [42] for some historical overview.) In the resulting complex R–X—Y (X = halo-gen, Y = negative site), the angle R–X–Y is typically in the neighborhood of 180°. The formation of such complexes was sometimes labeled an enigma since the halo-gens themselves are typically viewed as being negative. Furthermore, it was found that the same halogen in the same molecule could interact with both nucleophiles and electrophiles [43,44].

The "enigma" was resolved in 1992 when Brinck et al. [45] found computation-ally that covalently bonded halogens can have regions of positive electrostatic poten-tial on their outer sides, on the extensions of covalent bonds to them. Through these positive regions (denoted positive σ-holes [46]), the halogens can interact with nega-tive sites, forming what are now commonly called *halogen bonds*. On the other hand, the lateral sides of the halogens are usually negative, accounting for their ability to interact with electrophiles. These features—the positive σ-hole and the negative lat-eral sides—are clearly visible in the bromine atom in Figure 9.2.

A σ-hole is the result of the asymmetry of the electronic charge distribution of a covalently bonded halogen, which reflects a depletion of charge on its outer side [47–49]. In orbital terms, this can be viewed as corresponding to the essentially empty outer (noninvolved) lobe of the p-type orbital forming the covalent bond [50,51]. If the depletion of charge is sufficient, it gives rise to a very directional positive electrostatic potential, focused along the extension of the covalent bond to the halogen. The magnitude of this potential on the 0.001 a.u. surface $V_{S,max}$ has been demonstrated to correlate with the strength of the interaction with a given negative site [52,53].

It has been known for more than 50 years that covalently bonded atoms of groups IV–VI can also form noncovalent complexes with negative sites. Politzer and Murray [54] provided an extensive (but incomplete) list of references. In 2007, it was pointed out that these are very often σ-hole interactions [55–57]. Group IV–VI atoms can have positive σ-holes, analogous to those of the halogens, on the extensions of the bonds to these atoms [51,55–57]. As with the halogens, the σ-holes become more positive (and their directional interactions with negative sites become stronger) in going from the lighter atoms to the heavier (more polarizable) atoms within a given group and as the remainder of the molecule becomes more electron-withdrawing. For example, the fluorine in (**1**) does have a σ-hole (Figure 9.1); there is a region along the extension of the C–F bond (in green) that is less negative than the surrounding surface. However, the outer charge depletion is not enough to create a positive electrostatic potential; the fluorine surface is completely negative, and this fluorine is not expected to halogen bond. When fluorine is in a more electron-withdrawing environment, as is the one bonded to the oxygen in F_3C-OF or to the one in F_2O or F_2, then it does have a positive $V_{S,max}$ and does halogen bond [58–60].

To illustrate a σ-hole on a group V atom, Figure 9.3 shows $V_S(\mathbf{r})$ for fluorophosphine (**3**). There is a positive σ-hole on phosphorus on the extension of the F–P bond,

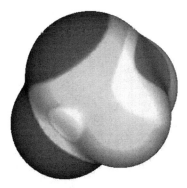

FIGURE 9.3 (**See color insert.**) Computed electrostatic potential on the 0.001 a.u. molecular surface of PH_2F (**3**). The negative region of phosphorus is shown toward the top (blue), while the most positive phosphorus σ-hole (on the extension of the F–P bond) is shown toward the right (red). In the center is a second weaker positive phosphorus σ-hole (yellow) on the extension of an H–P bond. Fluorine is shown toward the left. Color ranges are as follows: red, greater than 21 kcal/mol; yellow, between 21 and 12 kcal/mol; green, between 12 and 0 kcal/mol; blue, negative.

with its $V_{S,max}$ = 34 kcal/mol. The phosphorus also has positive σ-holes on the extensions of the H–P bonds, with $V_{S,max}$ = 16 kcal/mol. (A group IV–VII atom can have a σ-hole on the extension of each covalent bond to it [51,55–57].) In addition, the phosphorus in PH_2F has a negative region, with a $V_{S,min}$ of −14 kcal/mol.

9.3.4 Hydrogen Bonding and σ-Hole Bonding

σ-Hole bonding is competitive with hydrogen bonding [52,61–63]. The importance of the latter is well known; that of the former is increasingly being recognized, particularly in biological chemistry [64–66] and materials design [58,62,65,67].

It has already been suggested that hydrogen bonding can be viewed as a special case of a σ-hole interaction [50,51,66,68]. The σ-hole in this case is the diminished electronic charge in the outer portion of the hydrogen's $1s$ orbital, which is not involved in the covalent bond to the hydrogen. Since the $1s$ orbital is spherical, the σ-hole is not as focused as those corresponding to the outer lobes of p-type bonding orbitals, and hydrogen bonding is consequently not as directional as other σ-hole interactions [50,69].

Both hydrogen bonding and σ-hole bonding can normally be explained quite well as electrostatically driven interactions with some dispersion contributions [70,71]. It should be understood that the electrostatic component includes some degree of polarization of each reactant by the electric field of the other [68]. (The strength of the electric field of the negative site helps determine whether a red shift or a blue shift is observed in the stretching frequency of the covalent bond to the hydrogen or σ-hole atom [72,73].) There is also, of course, a repulsive component of the interaction, which increases as the reactants approach each other.

Such simple and straightforward explanations are, however, often regarded with suspicion. Accordingly, more complicated interpretations of these interactions continue to be presented by ignoring Occam's razor (*lex parsimoniae*: select the hypothesis that makes the fewest new assumptions). It is particularly popular to try to decompose them into supposed contributions from some subset of a whole array of proposed factors: electrostatics, charge transfer, polarization, exchange repulsion, induction, dispersion, orbital interaction, and others. The problem is that these factors are not independent of each other, and it is not physically realistic or meaningful to try to separate them. For instance, polarization is an intrinsic part of the electrostatic interaction, and, as Chen and Martínez [74] pointed out, "… charge transfer is an extreme manifestation of polarization." Since any interaction decomposition scheme is arbitrary, a number of different ones have been proposed, sometimes giving very contradictory results. Thus, for the interactions of H_3C-X and F_3C-X (X = Cl, Br, I) with $O=CH_2$ to form $H_3C-X-O=CH_2$ and $F_3C-X-O=CH_2$, one decomposition procedure led to the conclusion that the primary stabilizing effects are electrostatics and dispersion [70], while another assures us, for exactly the same interactions, that they are dominated by charge transfer and polarization, with electrostatics contributing only "slightly" [75].

Charge transfer is sometimes invoked in connection with hydrogen and halogen bonding and their group IV–VI analogues. There can certainly be gradations in the interactions; as the extent of polarization increases, a noncovalent interaction can be

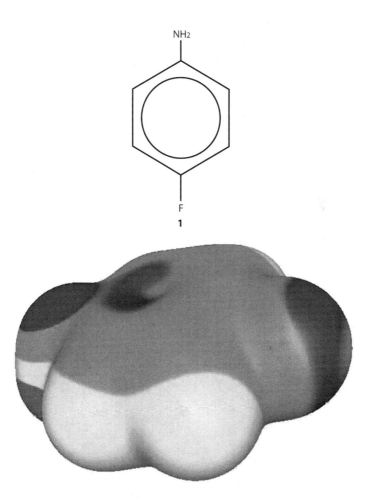

FIGURE 9.1 Computed electrostatic potential on the 0.001 a.u. molecular surface of (**1**). The amino group is shown to the left, and fluorine is shown to the right. Color ranges are as follows: red, greater than 18 kcal/mol; yellow, between 18 and 0 kcal/mol; green, between 0 and −18 kcal/mol; blue, more negative than −18 kcal/mol.

FIGURE 9.2 Computed electrostatic potential on the 0.001 a.u. molecular surface of (**2**). Bromine and azine nitrogens are shown toward the front, and the hydroxyl group is shown toward the back. Color ranges are as follows: red, greater than 30 kcal/mol; yellow, between 30 and 0 kcal/mol; green, between 0 and −13 kcal/mol; blue, more negative than −13 kcal/mol. Note the positive σ-hole (yellow) on bromine.

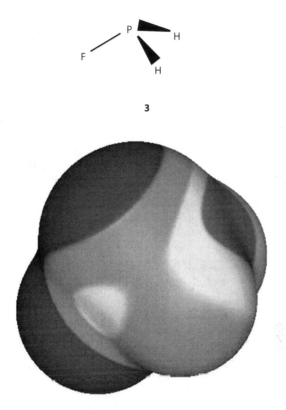

3

FIGURE 9.3 Computed electrostatic potential on the 0.001 a.u. molecular surface of PH$_2$F (**3**). The negative region of phosphorus is shown toward the top (blue), while the most positive phosphorus σ-hole (on the extension of the F–P bond) is shown toward the right (red). In the center is a second weaker positive phosphorus σ-hole (yellow) on the extension of an H–P bond. Fluorine is shown toward the left. Color ranges are as follows: red, greater than 21 kcal/mol; yellow, between 21 and 12 kcal/mol; green, between 12 and 0 kcal/mol; blue, negative.

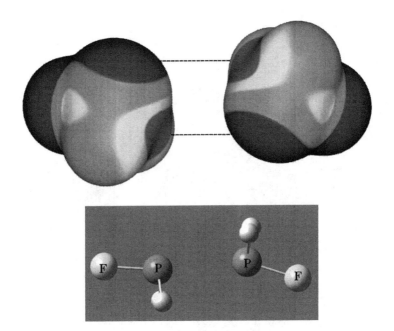

FIGURE 9.4 Double interactions of the FH_2P—PH_2F dimer. On top are surface electrostatic potential plots of the two PH_2F molecules showing the electrostatic forces driving the double interactions. The strong (red) positive σ-hole of each phosphorus interacts with the negative region of the other phosphorus, as indicated by the lines. Below is a structural drawing of the FH_2P—PH_2F dimer.

considered to acquire a degree of coordinate covalent character, with some sharing of electrons. This appears to explain some unusually strong "noncovalent" interactions [76,77], such as those noted by Del Bene et al. [78]. However, analyses of the electronic density rearrangements accompanying examples of normal hydrogen and halogen bonding support polarization [79,80], not charge transfer into a σ^* antibonding orbital, as is sometimes proposed. It should be kept in mind that the charge transfer theory of Mulliken [81] was intended to describe an electronic transition to an excited state, not the nature of bonding in the ground state.

9.3.5 Fallacy of Atomic Charges

None of the usual formulations of atomic charge, which assign a single positive or negative value (determined in some arbitrary manner) to each atom in a molecule, can explain the existence of halogen bonding or σ-hole interactions involving covalently bonded atoms of groups V and VI. They cannot explain why these atoms, which are typically assigned negative charges, can interact favorably with nucleophiles in some directions and with electrophiles in others, as was observed already years ago in extensive surveys of close contacts in crystals [43,44,82]. This can include even "like–like" interactions between an atom in one molecule and the same atom in another identical molecule [83–85]. Figure 9.4 presents an example involving the positive phosphorus σ-hole on the extension of the F–P bond in each PH_2F molecule and the negative portion of the phosphorus in the other molecule. The computed stabilization energy of this complex at the M06-2X/6-311 + G(d,p) level is −5.9 kcal/mol. All these interactions can readily be understood by looking at the electrostatic potentials on the molecular surfaces.

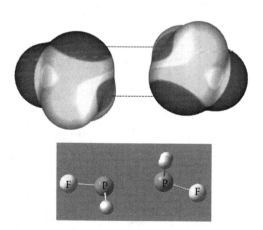

FIGURE 9.4 **(See color insert.)** Double interactions of the FH_2P—PH_2F dimer. On top are surface electrostatic potential plots of the two PH_2F molecules showing the electrostatic forces driving the double interactions. The strong (red) positive σ-hole of each phosphorus interacts with the negative region of the other phosphorus, as indicated by the lines. Below is a structural drawing of the FH_2P—PH_2F dimer.

The fact that covalently bonded atoms may have regions of both positive and negative electrostatic potentials needs to be taken into account in designing force fields [64,84], which have typically used traditional atomic charges. This is currently underway [86–88]. Thus, Jorgensen and Schyman [88] showed several halogen-bonded complexes that their new procedure finds to be stable, contrary to the predictions of older force fields.

9.3.6 Electrostatic Potential on the Entire Molecular Surface

For some purposes, it is necessary to consider not only the local maxima and minima of $V_S(\mathbf{r})$ but also its pattern over an extended portion of the molecular surface. For instance, in designing possible pharmaceutical agents, an effective approach has been to seek characteristic patterns of positive and negative potentials that may enhance or inhibit a particular biological recognition process [24,89–92], as between a drug and a receptor or between an enzyme and a substrate.

As a cautionary note, in considering electrostatically driven interactions, whether they primarily involve potentials at discrete points or over extended areas, it should be kept in mind that the electrostatic potential of the resulting complex will not show the separate positive and negative features that gave rise to the interaction [93,94]. They will have been at least partially neutralized by it.

A very practical application of the electrostatic potential over an entire molecular surface is in predicting the physical properties of condensed phases (liquids, solids, and solutions) that depend on noncovalent interactions. When the molecular $V_S(\mathbf{r})$ is characterized in terms of certain statistical quantities [34] (positive and negative average values and variances, average deviation, maxima, and minima), then good correlations exist between different combinations of these quantities and a variety of properties: heats of phase transitions, boiling points and critical constants, solubilities and solvation energies, partition coefficients, surface tensions, viscosities, diffusion constants, etc. [95,96]. Improved crystal density formulas have also been developed [97,98]. All these condensed phase properties can accordingly be predicted purely computationally from the $V_S(\mathbf{r})$ plus the surface areas or volumes of the individual (gas-phase) molecules. The fact that intermolecular interactions need not be included explicitly is evidently because they reflect the molecular $V_S(\mathbf{r})$.

9.3.7 Covalent Bond Formation

It is tempting to suggest that the $V_{S,max}$ and $V_{S,min}$ on a molecular surface can be viewed as indicative of sites favored for nucleophilic and electrophilic attacks leading to covalent bond formation. Sometimes this is true, but often it is not, and it should not be assumed to be the case. For instance, in substituted benzenes such as aniline, phenol, and phenetole, the most negative $V_{S,min}$ is generally associated with the substituent (Figure 9.1) [99], yet it is well established that electrophilic substitution occurs preferentially on the ring. Predicting the site for such a process requires an indicator that is a measure of the local reactivities of the electrons (e.g., the average local ionization energy) [99,100]. $V_S(\mathbf{r})$ is an effective determinant of noncovalent interactions but is much less reliable for covalent ones. However, it does sometimes serve to steer an approaching reactant, at long range, toward a region favorable for covalent bond formation [99].

9.4 ELECTROSTATIC POTENTIALS AND INTRAMOLECULAR INTERACTIONS

Intramolecular hydrogen bonding is well known to affect molecular conformations, tautomeric equilibria, etc. [101]. However, less attention appears to have been directed toward the consequences of other types of electrostatically driven noncovalent intramolecular interactions, although a recent overview has tried to remedy this [102]. We will briefly mention some examples.

Molecules of the type $Z-C(NO_2)_3$ are of considerable interest because of their potential energetic performance, and they have accordingly been prepared and studied extensively [103]. They offer a wealth of possibilities for intramolecular interactions because of the positive characteristics of the NO_2 nitrogens, the negative oxygens, and the electron-withdrawing power of the NO_2 group, which makes the Z portion of the molecule more positive [104]. Furthermore, the NO_2 groups can rotate to maximize favorable interactions and to minimize repulsive ones. In $Z-C(NO_2)_3$ molecules, the three NO_2 groups are in a propeller-like arrangement (Figure 9.5) [103–105]. The six NO_2 oxygens are in two planes; three (one from each NO_2) are in an upper plane near Z, while the other three lie in a lower plane below the nitrogens. Each of the lower oxygens is in close contact with the positive nitrogen of a neighboring NO_2 group, and these three O—N interactions lock the molecule into the propeller-like structure.

Meanwhile, the three upper oxygens can interact electrostatically with Z if it is sufficiently positive. An especially noteworthy case is when Z is a chlorine atom. $Cl-C(NO_2)_3$ has the distinction of possessing the shortest C–Cl single bond (1.694 Å) found in the Cambridge Structural Database [105]. (For comparison, the experimental C–Cl distance in $H_3C–Cl$ is 1.785 Å [106].) The striking shortening of the C–Cl

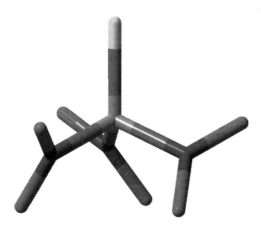

FIGURE 9.5 Framework of chlorotrinitromethane showing chlorine at the top, the two planes of oxygens, and the central plane of nitrogens.

bond in $Cl-C(NO_2)_3$ can be attributed to the attractive interactions between the chlorine (which has a positive electrostatic potential over its entire surface) and the three upper oxygens (Figure 9.5) [105].

The intramolecular nonbonded interactions in $Cl-C(NO_2)_3$ are electrostatic but do not involve σ-holes. Now we will consider some instances of group IV and group VI intramolecular σ-hole interactions.

Mitzel et al. [107,108] studied a series of molecules having the Si–O–N linkage, specifically structures such as (**4**).

They observed that the Si–O–N angles are between 75° and 110°, much smaller than approximately 125° in the isoelectronic (**5**) [107,108]. Furthermore, the magnitude of the Si–O–N angle depends very much on the nature of the substituents X, Y, and Z and on whether they are *anti* (as is X in (**4**)) or *gauche* (as are Y and Z) to the nitrogen.

The contraction of the Si–O–N angle in molecules such as (**4**) has been demonstrated to be the result of the attraction between the lone pair of nitrogen and silicon's positive σ-hole on the extension of the X–Si bond [109]. (This is the σ-hole that is best positioned to interact with the nitrogen lone pair.) Of lesser importance but still significant are secondary (often repulsive) interactions between the lone pair and the substituents Y and Z.

Finally, we will mention the antitumor agent tiazofurin (**6**) and its selenium analogue selenazofurin (**7**). These have S—O and Se—O close contacts that stabilize conformations promoting the binding of the anabolites of (**6**) and (**7**) to cellular targets [110,111]. Many years ago, Goldstein et al. [111] concluded that these S—O and Se—O interactions are electrostatic, and our more recent analysis fully supports this interpretation [112]; the negative oxygens are attracted to positive σ-holes on sulfur and selenium.

9.5 MANY FACES OF ELECTROSTATIC POTENTIAL

The Hohenberg–Kohn theorem focused attention on electronic density as fully determining all the properties of a ground-state system of nuclei and electrons [113].

However, as Ayers [114] pointed out, not only $\rho(\mathbf{r})$ can play this role; among other possibilities, he specifically mentioned the electrostatic potential. This is important because $V(\mathbf{r})$ appears to be more manageable than $\rho(\mathbf{r})$. As Galvez and Porras [115] observed concerning $\rho(\mathbf{r})$, "It is hard to prove mathematical properties of this quantity in a rigorous way."

An immediate example is the radial variation of both $\rho(\mathbf{r})$ and $V(\mathbf{r})$ for a spherically symmetrical ground-state neutral atom. Each decreases monotonically from the nucleus. This is known empirically for $\rho(\mathbf{r})$ [116,117] but, to our knowledge, has never been proven despite considerable effort; for $V(\mathbf{r})$, on the other hand, proof was achieved fairly readily [28].

More significant is the issue of the ground-state energy of a system. According to the Hohenberg–Kohn theorem, this can be expressed exactly as a functional of $\rho(\mathbf{r})$ [113]. While good and extremely useful approximations to this functional have been developed, many of them being incorporated into Gaussian 09 [33], the true functional remains unknown. In contrast, the Hellmann–Feynman theorem [118,119] provides a relatively straightforward route to rigorous formulas for the energies of atoms and molecules in terms of their electrostatic potentials at the positions of their nuclei [120–123].

Another interesting challenge for $\rho(\mathbf{r})$ and $V(\mathbf{r})$ is this: Can one of them identify a physically meaningful boundary between two covalently bonded atoms in a molecule? Since both $\rho(\mathbf{r})$ and $V(\mathbf{r})$ have local maxima by the nuclei, as discussed above, each must pass through an axial minimum at some point along the internuclear axis z. It has been argued [124], and seems intuitively reasonable, that the minimum of $\rho(\mathbf{r})$ along z determines the boundary between the atoms. On the other hand, at the minimum of $V(\mathbf{r})$ along z, we know that $\partial V(z)/\partial z = 0$, which means that an electron located at that minimum feels no electrostatic attraction to either nucleus. This suggests that the axial minimum of $V(\mathbf{r})$ defines a natural boundary between the two atoms [125,126].

Both approaches could be used to obtain atomic covalent radii, and this was performed [125,126]. The results showed that the covalent radii obtained from $V(\mathbf{r})$ manifested, to a much greater extent, the desired transferability from one molecule to another and are also more realistic, in better agreement with empirical values. For instance, the $\rho(\mathbf{r})$ average covalent radii predict that C < N < O, while the predictions from $V(\mathbf{r})$ agree with the observation C > N > O.

9.6 FINAL COMMENT

The fundamental nature of the electrostatic potential, recognized by Feynman [119] in his classic paper "Forces in Molecules," is reflected in its versatility, which we have tried to demonstrate in this chapter. $V(\mathbf{r})$ provides a powerful means of elucidating and predicting the properties and behavior of molecular systems and, more importantly, is a physical observable.

REFERENCES

1. Allen, L. C. 1998. Electronegativity. In *Encyclopedia of Computational Chemistry*, vol. 2, edited by P. v. R. Schleyer, 835–852. New York: Wiley.

2. Hinze, J. 1999. The concept of electronegativity of atoms in molecules. In *Pauling's Legacy: Modern Modelling of the Chemical Bond*, chap. 7, edited by Z. B. Maksic and W. J. Orville-Thomas, 189–212. Amsterdam, The Netherlands: Elsevier.

3. Pauling, L. 1932. The nature of the chemical bond: IV. The energy of single bonds and the relative electronegativity of atoms. *J. Am. Chem. Soc.* 54:3570–3582.

4. Pauling, L.; Yost, D. M. 1932. The additivity of the energies of normal covalent bonds. *Proc. Natl. Acad. Sci. U. S. A.* 18:414–416.

5. Iczkowski, R. P.; Margrave, J. L. 1961. Electronegativity. *J. Am. Chem. Soc.* 83:3547–3551.

6. Parr, R. G.; Donnelly, R. A.; Levy, M.; Palke, W. E. 1978. Electronegativity: the density functional viewpoint. *J. Chem. Phys.* 68:3801–3807.

7. Parr, R. G.; Yang, W. 1989. *Density Functional Theory of Atoms and Molecules*. New York: Oxford University Press.

8. Pauling, L. 1948. *The Nature of the Chemical Bond*, 58. Ithaca, NY: Cornell University Press.

9. Pearson, R. G. 1990. Electronegativity scales. *Acc. Chem. Res.* 23:1–2.

10. Allen, L. C. 1990. Electronegativity scales. *Acc. Chem. Res.* 23:175–176.

11. Allen, L. C. 1990. Chemistry and electronegativity. *Int. J. Quantum Chem.* 49:253–277.

12. Politzer, P.; Shields, Z. P.-I.; Bulat, F. A.; Murray, J. S. 2011. Average local ionization energies as a route to intrinsic atomic electronegativities. *J. Chem. Theory Comput.* 7:377–284.

13. Allen, L. C. 1989. Electronegativity is the average one-electron energy of the valence-shell electrons in ground-state free atoms. *J. Am. Chem. Soc.* 111:9003–9014.

14. Sproul, G. D. 1994. Electronegativity and bond type: 2. Evaluation of electronegativity scales. *J. Phys. Chem.* 98:6699–6703.

15. Bergman, D.; Hinze, J. 1996. Electronegativity and molecular properties. *Angew. Chem. Int. Ed. Eng.* 35:150–163.

16. Politzer, P.; Grice, M. E.; Murray, J. S. 2001. Electronegativities, electrostatic potentials and covalent radii. *J. Mol. Struct. THEOCHEM* 549:69–76.

17. Politzer, P.; Harris, R. R. 1970. Properties of atoms in molecules: I. A proposed definition of the charge on an atom in a molecule. *J. Am. Chem. Soc.* 92:6451–6454.

18. Meister, J.; Schwarz, W. H. E. 1994. Principal components of ionicity. *J. Phys. Chem.* 98:8245–8252.

19. Wiberg, K. B.; Rablen, P. R. 1993. Comparison of atomic charges derived via different procedures. *J. Comput. Chem.* 14:1504–1518.

20. Bonnaccorsi, R.; Scrocco, E.; Tomasi, J. 1970. Molecular SCF calculations for the ground state of some three-membered ring molecules: $(CH_2)_3$, $(CH_2)_2NH$, $(CH_2)_2NH_2^+$, $(CH_2)_2O$, $(CH_2)_2S$, $(CH_2)_2CH_2$, and N_2CH_2. *J. Chem. Phys.* 52:5270.

21. Scrocco, E.; Tomasi, J. 1973. The electrostatic molecular potential as a tool for the interpretation of molecular properties. *Top. Curr. Chem.* 42:95–170.

22. Scrocco, E.; Tomasi, J. 1978. Electronic molecular structure, reactivity and intermolecular forces: an heuristic interpretation by means of electrostatic molecular potentials. *Adv. Quantum Chem.* 11:115–193.

23. Stewart, R. F. 1979. On the mapping of electrostatic properties from Bragg diffraction data. *Chem. Phys. Lett.* 65:335–342.

24. Politzer, P.; Truhlar, D. G., eds. 1981. *Chemical Applications of Atomic and Molecular Electrostatic Potentials*. New York: Plenum Press.

25. Politzer, P.; Daiker, K. C. 1981. Models for chemical reactivity. In *The Force Concept in Chemistry*, edited by B. M. Deb, 294–387. New York: Van Nostrand Reinhold.

26. Politzer, P.; Murray, J. S. 2009. The electrostatic potential as a guide to molecular interactive behavior. In *Chemical Reactivity Theory: A Density Functional View*, edited by P. K. Chattaraj, 243–254. Boca Raton, FL: CRC Press.

27. Murray, J. S.; Politzer, P. 2011. The electrostatic potential: an overview. *Wiley Interdiscip. Rev. Comput. Mol. Sci.* 1:153–163.
28. Sen, K. D.; Politzer, P. 1989. Characteristic features of the electrostatic potentials of singly-negative monoatomic ions. *J. Chem. Phys.* 90:4370–4372.
29. Sen, K. D.; Politzer, P. 1989. Approximate radii for singly-negative ions of 3*d*, 4*d* and 5*d* metal atoms. *J. Chem. Phys.* 91:5123–5124.
30. Pathak, R. K.; Gadre, S. R. 1990. Maximal and minimal characteristics of molecular electrostatic potentials. *J. Chem. Phys.* 93:1770–1773.
31. Bader, R. F. W; Carroll, M. T.; Cheeseman, J. R.; Chang, C. 1987. Properties of atoms in molecules: atomic volumes. *J. Am. Chem. Soc.* 109:7968–7979.
32. Murray, J. S.; Brinck, T.; Grice, M. E.; Politzer, P. 1992. Correlations between molecular electrostatic potentials and some experimentally-based indices of reactivity. *J. Mol. Struct. THEOCHEM* 256:29–45.
33. Frisch, M. J. 2009. *Gaussian 09*. Wallingford, CT; Gaussian, Inc.
34. Bulat, F. A.; Toro-Labbé, A.; Brinck, T.; Murray, J. S.; Politzer, P. 2010. Quantitative analysis of molecular surfaces: areas, volumes, electrostatic potentials and average local ionization energies. *J. Mol. Model.* 16:1679–1691.
35. Alkorta, I.; Perez, J. J.; Villar, H. O. 1994. Molecular polarization maps as a tool for studies of intermolecular interactions and chemical reactivity. *J. Mol. Graphics* 12:3–13.
36. Murray, J. S.; Politzer, P. 2009. Molecular surfaces, van der Waals radii and electrostatic potentials in relation to noncovalent interactions. *Croat. Chem. Acta* 82:267–275.
37. Kollman, P.; McKelvey, J.; Johansson, A.; Rothenberg, S. 1975. Theoretical studies of hydrogen-bonded dimers. complexes involving HF, H_2O, NH_3, HCl, H_2S, PH_3, H_2CS, H_2CO, CH_4, CF_3H, C_2H_2, C_2H_4, C_6H_6, F⁻ and H_3O^+. *J. Am. Chem. Soc.* 97:955–965.
38. Hagelin, H.; Murray, J. S.; Brinck, T.; Berthelot, M.; Politzer, P. 1995. Family-independent relationships between computed molecular surface quantities and solute hydrogen bond acidity/basicity and solute-induced methanol O–H infrared frequency shifts. *Can. J. Chem.* 73:483–488.
39. Liu, Q.; Hoffman, R. 1995. Theoretical aspects of a novel mode of hydrogen–hydrogen bonding. *J. Am. Chem. Soc.* 117:10108–10112.
40. Richardson, T. B.; de Gala, S.; Crabtree, R. H.; Siegbahn, P. E. M. 1995. Unconventional hydrogen bonds: intermolecular B–H—H–N interactions. *J. Am. Chem. Soc.* 117:12875–12876.
41. Grabowski, S. J.; Sokalski, W. A.; Leszczynski, J. 2004. Nature of X–H$^{+\delta}$—$^{-\delta}$H–Y dihydrogen bonds and X–H—σ interactions. *J. Phys. Chem. A* 108:5823–5830.
42. Politzer, P.; Lane, P.; Concha, M. C.; Ma, Y.; Murray, J. S. 2007. An overview of halogen bonding. *J. Mol. Model.* 13:305–311.
43. Murray-Rust, P.; Motherwell, W. D. S. 1979. Computer retrieval and analysis of molecular geometry: 4. Intermolecular interactions. *J. Am. Chem. Soc.* 101:4374–4376.
44. Ramasubbu, N.; Parthasarathy, R.; Murray-Rust, P. 1986. Angular preferences of intermolecular forces around halogen centers: preferred directions of approach of electrophiles and nucleophiles around the carbon–halogen bond. *J. Am. Chem. Soc.* 108:4308–4314.
45. Brinck, T.; Murray, J. S.; Politzer, P. 1992. Surface electrostatic potentials of halogenated methanes as indicators of directional intermolecular interactions. *Int. J. Quantum Chem. Quantum Biol. Symp.* 44(suppl. 19):57–64.
46. Clark, T.; Hennemann, M.; Murray, J. S.; Politzer, P. 2007. Halogen bonding: the σ-hole. *J. Mol. Model.* 13(2):291–296.
47. Stevens, E. D. 1979. Experimental electron density distribution of molecular chlorine. *Mol. Phys.* 37:27–45.
48. Ikuta, S. 1990. Anisotropy of electron-density distribution around atoms in molecules: N, P, O and S atoms. *J. Mol. Struct. THEOCHEM* 205:191–201.

49. Price, S. L.; Stone, A. J.; Lucas, J.; Rowland, R. S.; Thornley, A. E. 1994. The nature of –Cl—Cl– intermolecular interactions. *J. Am. Chem. Soc.* 116:4910–4918.
50. Shields, Z. P.; Murray, J. S.; Politzer, P. 2010. Directional tendencies of halogen and hydrogen bonding. *Int. J. Quantum Chem.* 110:2823–2832.
51. Politzer, P.; Murray, J. S.; Clark, T. 2010. Halogen bonding: an electrostatically-driven highly directional noncovalent interaction. *Phys. Chem. Chem. Phys.* 12:7748–7757.
52. Riley, K. E.; Murray, J. S.; Politzer, P.; Concha, M. C.; Hobza, P. 2009. Br—O complexes as probes of factors affecting halogen bonding: interactions of bromobenzenes and bromopyrimidines with acetone. *J. Chem. Theor. Comput.* 5:155–163.
53. Riley, K. E.; Murray, J. S.; Fanfrlík, J.; Řezáč, J.; Solá, R. J.; Concha, M. C.; Ramos, F. M.; Politzer, P. 2011. Halogen bond tenability: I. The effects of aromatic fluorine substitution on the strengths of halogen-bonding interactions involving chlorine, bromine and iodine. *J. Mol. Model.* 17:3309–3318.
54. Politzer, P.; Murray, J. S. 2012. Halogen bonding and beyond: factors influencing the nature of CN–R and SiN–R complexes with FCl and Cl_2. *Theor. Chem. Acc.* 131:1114(1–10).
55. Murray, J. S.; Lane, P.; Politzer, P. 2007. A predicted new type of directional noncovalent interaction. *Int. J. Quantum Chem.* 107:2286–2292.
56. Murray, J. S.; Lane, P.; Clark, T.; Politzer, P. 2007. σ-Hole bonding: molecules containing group VI atoms. *J. Mol. Model.* 13:1033–1038.
57. Murray, J. S.; Lane, P.; Politzer, P. 2009. Expansion of the σ-hole concept. *J. Mol. Model.* 15:723–729.
58. Politzer, P.; Murray, J. S.; Concha, M. C. 2007. Halogen bonding and the design of new materials: organic bromides, chlorides and perhaps even fluorides as donors. *J. Mol. Model.* 13:643–650.
59. Metrangolo, P.; Murray, J. S.; Pilati, T.; Politzer, P.; Resnati, G.; Terraneo, G. 2011. The fluorine atom as a halogen bond donor, viz. a positive site. *CrystEngComm* 13:6593–6596.
60. Metrangolo, P.; Murray, J. S.; Pilati, T.; Politzer, P.; Resnati, G.; Terraneo, G. 2011. Fluorine-centered halogen bonding: a factor in recognition phenomena and reactivity. *Cryst. Growth Des.* 11:4238–4246.
61. DiPaolo, T.; Sandorfy, C. 1974. On the hydrogen bond breaking ability of fluorocarbons containing higher halogens. *Can. J. Chem.* 52:3612–3622.
62. Corradi, E.; Meille, S. V.; Messina, M. T.; Metrangolo, P.; Resnati, G. 2000. Halogen bonding versus hydrogen bonding in driving self-assembly processes. *Angew. Chem. Int. Ed.* 39:1782–1786.
63. Politzer, P.; Murray, J. S.; Lane, P. 2007. σ-Hole bonding and hydrogen bonding: competitive interactions. *Int. J. Quantum. Chem.* 107:3046–3052.
64. Auffinger, P.; Hays, F. A.; Westhof, E.; Shing Ho, P. 2004. Halogen bonding in biological molecules. *Proc. Natl. Acad. Sci. U. S. A.* 101:16789–16794.
65. Metrangolo, P.; Neukirsch, H.; Pilati, T.; Resnati, G. 2005. Halogen bonding based recognition processes: a world parallel to hydrogen bonding. *Acc. Chem. Res.* 38:386–395.
66. Murray, J. S; Riley, K. E.; Politzer, P.; Clark, T. 2010. Directional weak intermolecular interactions: σ-hole bonding. *Aust. J. Chem.* 63:1598–1607.
67. Metrangolo, P.; Resnati, G. 2008. *Halogen Bonding: Fundamentals and Applications.* Berlin, Germany: Springer.
68. Hennemann, M.; Murray, J. S.; Riley, K. E.; Politzer, P.; Clark, T. 2012. Polarization-induced σ-holes and hydrogen bonding. *J. Mol. Model.* 18:2461–2469.
69. Legon, A. C. 1999. Prereactive complexes of dihalogens XY with Lewis bases B in the gas phase: a systematic case for the halogen analogue B—XY of the hydrogen bond B—HX. *Angew. Chem. Int. Ed.* 38:2686–2714.

70. Riley, K. E.; Hobza, P. 2008. Investigations into the nature of halogen bonding including symmetry adapted perturbation theory analyses. *J. Chem. Theory Comput.* 4:232–242.
71. Riley, K. E.; Murray, J. S.; Fanfrlík, J.; Řezáč, J.; Solá, R. J.; Concha, M. C.; Ramos, F. M.; Politzer, P. 2012. Halogen bond tunability: II. The varying roles of electrostatic and dispersion contributions to attraction in halogen bonds. *J. Mol. Model.* DOI: 10.1007/s00894-012-1428-x.
72. Wang, W.; Wang, N.-B.; Zheng, W.; Tian, A. 2004. Theoretical study on the blueshifting halogen bond. *J. Phys. Chem. A* 108:1799–1805.
73. Murray, J. S.; Concha, M. C.; Lane, P.; Hobza, P.; Politzer, P. 2008. Blue shifts vs. red shifts in σ-hole bonding. *J. Mol. Model.* 14:699–704.
74. Chen, J.; Martínez, T. J. 2007. QTPIE: charge transfer with polarization current equalization. A fluctuating charge model with correct asymptotics. *Chem. Phys. Lett.* 438:315–320.
75. Palusiak, M. 2010. On the nature of the halogen bond—the Kohn–Sham molecular orbital approach. *J. Mol. Struct. THEOCHEM* 945:89–92.
76. Murray, J. S.; Lane, P.; Clark, T.; Riley, K. E.; Politzer, P. 2012. σ-Holes, π-holes and electrostatically-driven interactions. *J. Mol. Model.* 18:541–548.
77. Politzer, P.; Murray, J. S. 2012. Halogen bonding and beyond: factors influencing the nature of CN–R and SiN–R complexes with F–Cl and Cl_2. *Theor. Chem. Acc.* 131:1114.
78. Del Bene, J. E.; Alkorta, I.; Elguero, J. 2010. Do traditional, chlorine-shared, and ion-pair halogen bonds exist? An ab initio investigation of FCl:CNX complexes. *J. Phys. Chem. A* 114:12958–12962.
79. Wang, J.; Gu, J.; Leszczynski, J. 2012. The electronic spectra and the hydrogen-bonding pattern of the sulfur and selenium substituted guanines. *J. Comput. Chem.* 33:1587–1593.
80. Politzer, P.; Bulat, F. A.; Riley, K. E.; Murray, J. S. 2012. Perspectives on halogen bonding: *lex parsimoniae* (Occam's razor). *Comput. Theor. Chem.* 998:2–8.
81. Mulliken, R. S. 1950. Structures of complexes formed by halogen molecules with aromatic and with oxygenated solvents. *J. Am. Chem. Soc.* 72:600–608.
82. Rosenfeld, R. E., Jr.; Parthasarathy, R.; Dunitz, J. D. 1977. Directional preferences of nonbonded atomic contacts with divalent sulfur: 1. Electrophiles and nucleophiles. *J. Am. Chem. Soc.* 99:4860–4862.
83. Guru Row, T. N.; Parthasarathy, R. 1981. Directional preferences of nonbonded atomic contacts with divalent sulfur in terms of its orbital orientations: 2. S—S interactions and nonspherical shape of sulfur in crystals. *J. Am. Chem. Soc.* 103:477–479.
84. Politzer, P.; Murray, J. S.; Concha, M. C. 2008. σ-Hole bonding between like atoms; a fallacy of atomic charges. *J. Mol. Model.* 14:659–665.
85. Bleiholder, C.; Werz, D. B.; Köppel, H.; Gleiter, R. 2006. Theoretical investigations on chalcogen–chalcogen interactions: what makes these nonbonded interactions bonding? *J. Am. Chem. Soc.* 128:2666–2674.
86. Ibrahim, M. A. A. 2011. Molecular mechanical study of halogen bonding in drug discovery. *J. Comput. Chem.* 32:2564–2574.
87. Kolář, M.; Hobza, P. 2012. On extension of the current biomolecular empirical force field for the description of halogen bonds. *J. Chem. Theory Comput.* 8:1325–1333.
88. Jorgensen, W. L.; Schyman, P. 2012. Treatment of halogen bonding in the OPLS-AA force field; application to potent anti-HIV agents. *J. Chem. Theory Comput.* 8:3895–3901.
89. Weiner, P. K.; Langridge, R.; Blaney, J. M.; Schaefer, R.; Kollman, P. A. 1982. Electrostatic potential molecular surfaces. *Proc. Natl. Acad. Sci.* 79:3754–3758.
90. Politzer, P.; Laurence, P. R.; Jayasuriya, K. 1985. Molecular electrostatic potentials: an effective tool for the elucidation of biochemical phenomena. *Environ. Health Perspect.* 61:191–202.

91. Politzer, P.; Murray, J. S. 1991. Molecular electrostatic potentials and chemical reactivity. In *Reviews in Computational Chemistry*, vol. 2, edited by K. B. Lipkowitz and D. B. Boyd, 273–312. New York: VCH Publishers.
92. Naray-Szabó, G.; Ferenczy, G. G. 1995. Molecular electrostatics. *Chem. Rev.* 95: 829–847.
93. Hussein, W.; Walker, C. G.; Peralta-Inga, Z.; Murray, J. S. 2001. Computed electrostatic potentials and average local ionization energies on the molecular surfaces of some tetracyclines. *Int. J. Quantum Chem.* 82:160–169.
94. Politzer, P.; Concha, M. C.; Murray, J. S. 2000. Density functional study of dimers of dimethylnitramine. *Int. J. Quantum Chem.* 80:184–192.
95. Murray, J. S.; Politzer, P. 1998. Statistical analysis of the molecular surface electrostatic potential: an approach to describing noncovalent interactions in condensed phases. *J. Mol. Struct. THEOCHEM* 425:107–114.
96. Politzer, P.; Murray, J. S. 2001. Computational prediction of condensed phase properties from statistical characterization of molecular surface electrostatic potentials. *Fluid Phase Equilib.* 185:129–137.
97. Politzer, P.; Martinez, J.; Murray, J. S.; Concha, M. C.; Toro-Labbé, A. 2009. An electrostatic interaction correction for improved crystal density prediction. *Mol. Phys.* 107:2095–2101.
98. Politzer, P.; Martinez, J.; Murray, J. S.; Concha, M. C. 2010. An electrostatic correction for improved crystal density predictions of energetic ionic compounds. *Mol. Phys.* 108:1391–1396.
99. Politzer, P.; Murray, J. S.; Concha, M. C. 2002. The complementary roles of molecular surface electrostatic potentials and average local ionization energies with respect to electrophilic processes. *Int. J. Quantum Chem.* 88:19–27.
100. Politzer, P.; Murray, J. S.; Bulat, F. A. 2010. The average local ionization energy: a review. *J. Mol. Model.* 16:1731–1742.
101. March, J. 1985. *Advanced Organic Chemistry*, 3rd ed. New York: Wiley-Interscience.
102. Politzer, P.; Murray, J. S. 2012. Non–hydrogen-bonding intramolecular interactions: important but often overlooked. In *Practical Aspects of Computational Chemistry I*, chap. 16, edited by J. Leszczynski and M. Shukla. Amsterdam, The Netherlands: Springer.
103. Göbel, M.; Klapötke, T. M. 2009. Development and testing of energetic materials: the concept of high densities based on the trinitroethyl functionality. *Adv. Funct. Mater.* 19:347–365.
104. Macaveiu, L.; Göbel, M.; Klapötke, T. M.; Murray, J. S.; Politzer, P. 2010. The unique role of the nitro group in intramolecular interactions: chloronitromethanes. *Struct. Chem.* 21:129–136.
105. Göbel, M.; Tchitchanov, B. H.; Murray, J. S.; Politzer, P.; Klapötke, T. M. 2009. Chlorotrinitromethane and its exceptionally short carbon–chlorine bond. *Nat. Chem.* 1:229–235.
106. Lide, D. R. 2006. *Handbook of Chemistry and Physics*, 87th ed. Boca Raton, FL: Taylor & Francis.
107. Mitzel, N. W.; Blake, A. J.; Rankin, D. W. H. 1997. β-Donor bonds in SiON units. An inherent structure-determining property leading to (4 + 4)-coordination in tetrakis-(N,N-dimethylhydroxylamido)silane. *J. Am. Chem. Soc.* 119:4143–4148.
108. Mitzel, N. W.; Vojinovic, K.; Froehlich, R.; Foerster, T.; Robertson, H. E.; Borisenko, K. B.; Rankin, D. W. H. 2005. Three-membered ring of open chain molecule—$(F_3C)F_2SiONMe_2$—a model for the α-effect in silicon chemistry. *J. Am. Chem. Soc.* 127:13705–13713 [and papers cited].
109. Murray, J. S.; Concha, M. C.; Politzer, P. 2011. Molecular surface electrostatic potentials as guides to Si–O–N angle contraction: tunable σ-holes. *J. Mol. Model.* 17:2151–2157.

110. Goldstein, B. M.; Takusagawa, F.; Berman, H. M.; Scrivastava, P. C.; Robins, R. K. 1983. Structural studies of a new antitumor agent: tiazofurin and its inactive analogs. *J. Am. Chem. Soc.* 105:7416–7422.

111. Goldstein, B. M.; Takusagawa, F.; Berman, H. M.; Scrivastava, P. C.; Robins, R. K. 1983. Structural studies of a new antitumor and antiviral agent: selenazofurin and its α anomer. *J. Am. Chem. Soc.* 107:1394–1400.

112. Murray, J. S.; Lane, P.; Politzer, P. 2008. Simultaneous σ-hole and hydrogen bonding by some sulfur- and selenium-containing heterocycles. *Int. J. Quantum Chem.* 108:2770–2781.

113. Hohenberg, P.; Kohn, W. 1964. Inhomogeneous electron gas. *Phys. Rev. B* 136:864–871.

114. Ayers, P. 2007. Using reactivity indicators instead of the electron density to describe Coulomb systems. *Chem. Phys. Lett.* 438:148–152.

115. Galvez, F. J.; Porras, I. 1991. Atomic charge density at the nucleus and inequalities among radial expectation values. *Phys. Rev. A* 44:144–147.

116. Weinstein, H.; Politzer, P.; Srebrenik, S. 1975. A misconception concerning the electronic density distribution of an atom. *Theor. Chim. Acta* 38:159–163.

117. Angulo, J. C.; Yáñez, R. J.; Dehesa, J. S.; Romera, E. 1996. Monotonicity properties of the atomic charge density function. *Int. J. Quantum Chem.* 58:11–21.

118. Hellmann, H. 1937. *Einführung in die Quantumchemie*, 285. Leipzig, Germany: Franz Deuticke.

119. Feynman, R. P. 1939. Forces in molecules. *Phys. Rev.* 56:340–343.

120. Wilson, E. B., Jr. 1962. Four-dimensional electron density function. *J. Chem. Phys.* 36:2232.

121. Politzer, P.; Parr, R. G. 1974. Some new energy formulas for atoms and molecules. *J. Chem. Phys.* 61:4258.

122. Politzer, P.; Murray, J. S. 2002. The fundamental nature and role of the electrostatic potential in atoms and molecules. *Theor. Chem. Acc.* 108:134–142.

123. Politzer, P. 2004. Atomic and molecular energies as functionals of the electrostatic potential. *Theor. Chem. Acc.* 111:395–399.

124. Bader, R. F. W. 1985. Atoms in molecules. *Acc. Chem. Res.* 18:9–15.

125. Wiener, J. M. M.; Grice, M. E.; Murray, J. S.; Politzer, P. 1996. Molecular electrostatic potentials as indictors of covalent radii. *J. Chem. Phys.* 104:5109–5111.

126. Politzer, P.; Murray, J. S.; Lane, P. 2003. Electrostatic potentials and covalent radii. *J. Comput. Chem.* 24:505–511.

10 Extending the Domain of Application of Constrained Density Functional Theory to Large Molecular Systems

Aurélien de la Lande, Dennis R. Salahub, and Andreas M. Köster

CONTENTS

10.1 INTRODUCTION

Over the last decades, density functional theory (DFT) has emerged as the method of choice for the first-principles investigation of molecular systems on increasing spatial and temporal scales. In addition, the variationally fitted Kohn–Sham potential provides a computational framework with advantageous cost–quality ratio, with inclusion of a large part of electron correlation for "normal" situations. Molecules composed of tens to hundreds of atoms are now routinely investigated by DFT. By construction, DFT is restricted to electronic ground states. Yet many interesting chemical problems involve electronic excited states or transitions between electronic states. Among them, electron transfer (ET) reactions, eventually photo-induced, represent one of the most common reactions encountered in chemistry, biology, or materials science. High-precision quantum chemistry methodologies (e.g., the complete active space self-consistent field (CAS SCF) approach with second-order perturbation or multireference-coupled cluster theory can be employed to model electron dynamics but, owing to their computational cost, are often restricted to rather small systems.

Another approach to describing electron dynamics in molecules is based on time-dependent DFT (TDDFT).[1] In the so-called real-time TDDFT approach, the time-dependent Kohn–Sham density is propagated according to the time-dependent Kohn–Sham equations. This introduces a nonadiabatic coupling vector that permits changes in electronic states.[2] A corresponding mean field approach, also known as Ehrenfest dynamics,[3] is well suited for describing electron dynamics on the attosecond time scale. However, for the investigation of chemical processes, a serious drawback of Ehrenfest dynamics is its failure to recover adiabatic states asymptotically after a mixing of electron states. Thus, Ehrenfest dynamics is, in general, incompatible with the Born–Oppenheimer approximation, which is the basis for the definition of molecular structure in chemistry. As a result, potential energy surface (PES) hopping[4] is often preferred for the description of nonadiabatic chemical processes. For ET reactions, constrained DFT (cDFT) can be used to describe PES hopping in its most elementary steps.

cDFT was developed in the 1980s to investigate charge or magnetic fluctuations in solids after perturbation of the electronic density.[5] The pioneering work of Dederichs et al. was itself inspired by the work of Gunnarsson and Lundqvist,[6] who had previously shown that, in addition to the true ground state, DFT can provide information on the electronic ground-state densities of particular molecular symmetries. Dederichs et al.[5] generalized this idea through the introduction of a constraint in the Kohn–Sham energy functional with a Lagrange multiplier technique.[7] In cDFT, the ground-state energy for a system is obtained by minimization of the DFT energy functional with respect to the electronic density, coupled to maximization with respect to the Lagrange multiplier:

$$\mathcal{E}[\rho, \lambda_c] = \min_{\rho} \max_{\lambda_c} \left(E[\rho] + \lambda_c \left[\int \rho(\mathbf{r}) w(\mathbf{r}) d\mathbf{r} - N_c \right] \right) \tag{10.1}$$

with

$$E[\rho] = \sum_i^{occ} \left\langle \psi_i \left| -\frac{1}{2} \nabla^2 \right| \psi_i \right\rangle + \int v_{\text{ext}}(\mathbf{r}) \, \rho(\mathbf{r}) \, d\mathbf{r} + J[\rho] + E_{\text{xc}}[\rho]. \tag{10.2}$$

In these two equations, $\psi_i(\mathbf{r})$ are the molecular orbitals (MOs), $\rho(\mathbf{r})$ is the electronic density, $v_{\text{ext}}(\mathbf{r})$ is the external potential felt by electrons, N_c is the set point for the constraint, and $w(\mathbf{r})$ is a weight function that defines the constraint property. The Coulomb and exchange-correlation energy contributions are denoted by $J[\rho]$ and $E_{\text{xc}}[\rho]$. The cDFT energy equation can be applied, for example, to constrain a given number of electrons N_c to occupy a specific volume Ω. Alternatively, if the weight function is made spin-dependent, Equation 10.1 enables one to constrain the magnetization of atoms or molecular fragments. The variation of Equation 10.1 with respect to MOs—keeping them orthonormalized—leads to a set of modified canonical Kohn–Sham equations:

$$\left(-\frac{1}{2} \nabla^2 + v_{ext}(r) + \int \frac{\rho(r')}{|r-r'|}\, dr' + v_{xc}(r) + \lambda_c w(r) \right) \psi_i(r) = \varepsilon_i \psi_i(r) \; \forall i. \quad (10.3)$$

From these equations, it is apparent that the Lagrange term acts as a supplementary potential on the electronic density. Its role is to guide the SCF procedure to an electronic ground state that satisfies the desired constraint. However, although the weight function is a parameter of the computation, the Lagrange multiplier λ_c is not known beforehand and must be determined within the SCF procedure. In older implementations of cDFT, it was customary to determine the Lagrange multiplier value by trial and error: scans over possible values of λ_c were performed until the desired constraint was fulfilled. This strategy was obviously cumbersome and certainly hampered the spreading of the cDFT technique to the computational chemistry community. This situation changed in 2005 when Wu and Van Voorhis[8] proposed an elegant computational procedure founded on the optimized effective potential theory to optimize the correct value for λ_c (i.e., the one that leads to an electronic density fulfilling the desired constraint). Various algorithms have been described to solve the set of cDFT Kohn–Sham equations, using either plane wave[9] or localized basis sets.[8] Recent reviews on cDFT describe these implementations in detail.[10,11] The method has been made available in various quantum chemistry packages such as QCHEM,[12] NWCHEM,[13] CONQUEST,[14] and deMon2k.[15] In this chapter, we will focus on our implementation of the method in deMon2k in the context of auxiliary DFT (ADFT). The central characteristic of ADFT is its use of variationally fitted electronic densities to calculate the both the Coulomb and the exchange-correlation potentials and energies.[16] The resulting computational efficiency opens the door to geometry optimizations and Born–Oppenheimer molecular dynamics (BOMD) simulations on large molecular systems composed of hundreds to thousands of atoms. In addition, the availability of analytical energy gradients and higher-energy derivatives of ADFT energy renders the computation of many different properties tractable for large molecular systems. It is thus very tempting to extend the ideas of cDFT developed by Dederichs et al.[5] and Wu and Van Voorhis[8] to the ADFT framework.

In the theory part of this chapter, we first introduce briefly the working equations of ADFT (for more detailed reviews, see Janetzko et al.[17] and Calaminici et al.[18]) and then incorporate the density constraint characteristic of cDFT into this framework. Because our main focus is spatial density constraints, we discuss the various population analyses (PA) available in deMon2k for cDFT implementation. In the application part of this chapter, we review some of the most recent works that employed cDFT for modeling ET reactions and dioxygen activation by metallic surfaces or inorganic complexes.

10.2   THEORETICAL BACKGROUND

10.2.1   Auxiliary Density Functional Theory

In deMon2k, the linear combination of Gaussian-type orbital (LCGTO) approximation is used to expand the molecular Kohn–Sham orbitals:

$$\psi_i(r) = \sum_\mu c_{\mu i} \mu(r). \tag{10.4}$$

Here, $c_{\mu i}$ denotes an MO coefficient, and $\mu(r)$ is an atomic orbital represented by a set of contracted Gaussian functions. With this LCGTO expansion for MOs, we find for the electronic Kohn–Sham density

$$\rho(r) = \sum_{\mu,\nu} P_{\mu\nu} \, \mu(r)\nu(r). \tag{10.5}$$

Here, $P_{\mu\nu}$ denotes an element of the (closed-shell) density matrix given by

$$P_{\mu\nu} = 2 \sum_i^{occ} c_{\mu i} c_{\nu i}. \tag{10.6}$$

With the LCGTO expansion of the Kohn–Sham MOs and the density, the unconstrained DFT energy takes the form

$$E_{\text{DFT}} = \sum_{\mu,\nu} P_{\mu\nu} H_{\mu\nu} + \frac{1}{2} \sum_{\mu,\nu} \sum_{\sigma,\pi} P_{\mu\nu} \, P_{\sigma\tau} \, \langle \mu\nu \| \sigma\tau \rangle + E_{xc}[\rho]. \tag{10.7}$$

Throughout this chapter, the Greek letters μ, ν, σ, and τ denote atomic orbitals. In Equation 10.7, the first term is the core energy, with $H_{\mu\nu}$ being an element of the core Hamiltonian, including kinetic energy and nuclear attraction integrals. The second term represents the Coulomb repulsion energy between electrons. It includes four-center electron repulsion integrals (ERIs). The symbol $\|$ denotes the two-electron Coulomb repulsion operator $1/|r - r'|$, which separates the basis functions of electron 1 (lhs) from those of electron 2 (rhs). The calculation of this term introduces a formal N^4 scaling into the DFT energy, where N refers to the number of (contracted) basis functions. The last term of Equation 10.7 is the exchange-correlation energy functional. Its calculation usually involves a numerical integration on an atom-centered grid.[19]

To circumvent the N^4 scaling bottleneck of the ERI calculation, we employed variational fitting of the Coulomb potential in deMon2k. For this purpose, a linear scaling auxiliary density, $\tilde{\rho}(r)$, is introduced. In deMon2k, this auxiliary density is expanded in primitive Hermite Gaussian auxiliary functions (denoted by a bar) that are grouped together in auxiliary function sets with common exponents:

$$\tilde{\rho}(r) = \sum_{\bar{k}} x_{\bar{k}} \bar{k}(r). \tag{10.8}$$

This structure is computationally advantageous for the ERI calculation and the numerical integration of the auxiliary density. The Coulomb fitting coefficients $x_{\bar{k}}$ are obtained by minimizing the following Coulomb energy error:

$$\varepsilon_2 = \iint \frac{[\rho(r) - \tilde{\rho}(r)] \, [\rho(r') - \tilde{\rho}(r')]}{|r - r'|} \, dr \, dr'. \tag{10.9}$$

The minimization procedure leads to a set of inhomogeneous equations for the determination of the fitting coefficients $x_{\bar{k}}$ collected in the vector x:

$$Gx = J. \tag{10.10}$$

To ease the notation, we have introduced the Coulomb matrix with elements

$$G_{\bar{k}\bar{l}} = \langle \bar{k} \| \bar{l} \rangle \tag{10.11}$$

and the Coulomb vector with elements

$$J_{\bar{k}} = \sum_{\mu,\nu} P_{\mu\nu} \langle \mu\nu \| \bar{k} \rangle. \tag{10.12}$$

Because ε_2 is positive definite, the following inequality holds:

$$\frac{1}{2} \sum_{\mu,\nu} \sum_{\sigma,\tau} P_{\mu\nu} P_{\sigma\tau} \langle \mu\nu \| \sigma\tau \rangle \geq \sum_{\mu,\nu} \sum_{\bar{k}} P_{\mu\nu} \langle \mu\nu \| \bar{k} \rangle x_{\bar{k}} - \frac{1}{2} \sum_{\bar{k},\bar{l}} x_{\bar{k}} x_{\bar{l}} \langle \bar{k} \| \bar{l} \rangle. \tag{10.13}$$

Here, the ERI notation for the four-center integrals is extended to three-center and two-center ERIs. Inserting this inequality into the DFT energy expression (Equation 10.7) then yields the DFT energy $\tilde{E}_{DFT}$, with the variational fitting of the Coulomb potential

$$\tilde{E}_{DFT} = \sum_{\mu,\nu} P_{\mu\nu} H_{\mu\nu} + \sum_{\mu,\nu} \sum_{\bar{k}} P_{\mu\nu} \langle \mu\nu \| \bar{k} \rangle x_{\bar{k}} - \frac{1}{2} \sum_{\bar{k},\bar{l}} x_{\bar{k}} x_{\bar{l}} \langle \bar{k} \| \bar{l} \rangle + E_{xc}[\rho]. \qquad (10.14)$$

As this equation shows, the variational fitting of the Coulomb potential eliminates the N^4 bottleneck of the ERI calculation and moves the computationally most demanding task to the numerical integration of the exchange-correlation energy and potential.

Several approximations for the calculation of the exchange-correlation energy and potential have been proposed in the literature to avoid the bottleneck of the numerical integration. For an authoritative review on this research field, we refer the readers to Dunlap et al.[20] One approach is the so-called ADFT methodology in which the auxiliary density from the variational fitting of the Coulomb potential is used also for the calculation of the exchange-correlation energy and potential.[16] This does not avoid the numerical integration for these contributions but considerably simplifies grid work due to the linear scaling of the auxiliary density and the special structure of the employed primitive Hermite Gaussian auxiliary functions. The ADFT energy expression thus takes the form

$$E_{ADFT} = \sum_{\mu,\nu} P_{\mu\nu} H_{\mu\nu} + \sum_{\mu,\nu} \sum_{\bar{k}} P_{\mu\nu} \langle \mu\nu \| \bar{k} \rangle x_{\bar{k}} - \frac{1}{2} \sum_{\bar{k},\bar{l}} x_{\bar{k}} x_{\bar{l}} \langle \bar{k} \| \bar{l} \rangle + E_{xc}[\tilde{\rho}]. \qquad (10.15)$$

The approximation error introduced by ADFT (below 1 pm for optimized bond lengths and around 1 kcal/mol for corresponding binding energies) is well below the intrinsic accuracy of Kohn–Sham LCGTO DFT calculations, and the improvement in performance is substantial. Moreover, the ADFT energy expression is variational (although it will not show strict convergence from above, which gives rise to a MinMax SCF[21]) and therefore permits the formulation of analytical energy derivatives. As an example, we will now derive the Kohn–Sham matrix elements as derivatives of the energy with respect to the density matrix elements:

$$K_{\mu\nu} \equiv \left(\frac{\partial E_{ADFT}}{\partial P_{\mu\nu}} \right)_x = H_{\mu\nu} + \sum_{\bar{k}} \langle \mu\nu \| \bar{k} \rangle x_{\bar{k}} + \frac{\partial E_{xc}[\tilde{\rho}]}{\partial P_{\mu\nu}}. \qquad (10.16)$$

The partial derivative of the (local) exchange-correlation energy with respect to the density matrix elements yields

$$\frac{\partial E_{xc}[\tilde{\rho}]}{\partial P_{\mu\nu}} = \int \frac{\delta E_{xc}[\tilde{\rho}]}{\delta \tilde{\rho}(r)} \frac{\partial \tilde{\rho}(r)}{\partial P_{\mu\nu}} dr. \qquad (10.17)$$

As usual, the functional derivative defines the exchange-correlation potential; however, it is now calculated with the approximated density

$$v_{xc}[\tilde{\rho}] \equiv \frac{\delta E_{xc}[\tilde{\rho}]}{\delta \tilde{\rho}(r)}. \tag{10.18}$$

For the partial derivative of the auxiliary density, it follows that[16]

$$\frac{\partial \tilde{\rho}(r)}{\partial P_{\mu\nu}} = \sum_{\bar{k}} \frac{\partial x_{\bar{k}}}{\partial P_{\mu\nu}} \bar{k}(r) = \sum_{\bar{k},\bar{l}} \langle \mu\nu \| \bar{l} \rangle G_{\bar{l}\bar{k}}^{-1} \bar{k}(r). \tag{10.19}$$

Thus, we find for the partial derivative of the (local) exchange-correlation energy that

$$\frac{\partial E_{xc}[\tilde{\rho}]}{\partial P_{\mu\nu}} = \sum_{\bar{k},\bar{l}} \langle \mu\nu \| \tilde{k} \rangle G_{\bar{k}\bar{l}}^{-1} \langle \bar{l} \mid v_{xc}[\tilde{\rho}] \rangle. \tag{10.20}$$

For convenience, we now introduce exchange-correlation fitting coefficients defined as

$$z_{\bar{k}} \equiv \sum_{\bar{l}} G_{\bar{k}\bar{l}}^{-1} \langle \bar{l} \mid v_{xc}[\tilde{\rho}]. \tag{10.21}$$

Note that these coefficients are spin dependent because of the exchange-correlation potential. The ADFT Kohn–Sham matrix elements are given by

$$K_{\mu\nu} = H_{\mu\nu} + \sum_{\bar{k}} \langle \mu\nu \| \bar{k} \rangle \left(x_{\bar{k}} + z_{\bar{k}} \right). \tag{10.22}$$

Thus, the ADFT Kohn–Sham matrix elements are independent of the density matrix. As a result, only the approximated density and the corresponding density derivatives, in the case of gradient-corrected functionals, are numerically calculated on a grid.

10.2.2 CONSTRAINING ADFT ENERGY

After the introduction of the ADFT formalism, we can now incorporate constraints according to the cDFT formalism to fix, for example, the charge or magnetization of the electronic density in selected regions of a molecule. To do so, we follow the original methodology proposed by Wu and Van Voorhis[8] but substitute the energy expression in $E[\rho,\lambda_c]$ with the E_{ADFT} energy functional. The resulting cDFT functional, now denoted by $\mathcal{E}_{ADFT}[\rho,\lambda_c]$, takes the form[22]

$$
\begin{aligned}
\mathcal{E}_{ADFT}[\rho,\lambda_c] &= E_{ADFT}[\rho] + \lambda_c \left[\int \rho(r)w(r)\,dr - N_c \right] \\
&= \sum_{\mu,\nu} P_{\mu\nu} H_{\mu\nu} + \sum_{\mu,\nu} \sum_{\bar{k}} P_{\mu\nu} \langle \mu\nu \| \bar{k} \rangle x_{\bar{k}} - \frac{1}{2} \sum_{\bar{k},\bar{l}} x_{\bar{k}} x_{\bar{l}} \langle \bar{k} \| \bar{l} \rangle + E_{xc}[\tilde{\rho}] \quad (10.23) \\
&+ \lambda_c \left[\int \rho(r)w(r)\,dr - N_c \right].
\end{aligned}
$$

The corresponding modified Kohn–Sham matrix elements are given by

$$
K_{\mu\nu} \equiv \left(\frac{\partial \mathcal{E}_{ADFT}}{\partial P_{\mu\nu}} \right)_x = H_{\mu\nu} + \sum_{\bar{k}} \langle \mu\nu \| \bar{k} \rangle \left(x_{\bar{k}} + z_{\bar{k}} \right) + \lambda_c W_{\mu\nu}, \qquad (10.24)
$$

with

$$
W_{\mu\nu} \equiv \langle \mu|w|\nu \rangle. \qquad (10.25)
$$

The Kohn–Sham equations now contain two types of Lagrange multipliers, namely, the MO eigenvalues ε_i and λ_c. Instead of solving the modified Kohn–Sham equations for both types of Lagrange multipliers in one step, we use a sequential approach. First, $c_{\mu i}$ and ε_i are determined for a given λ_c. This step is identical to the corresponding step in an unconstrained SCF (i.e., it requires the transformation and diagonalization of the modified Kohn–Sham matrix). Once the MO coefficients are obtained, the density matrix and the corresponding $x_{\bar{k}}$ and $z_{\bar{k}}$ fitting coefficients are calculated and frozen. In the second step, λ_c is optimized such that the density constraint in $\mathcal{E}_{ADFT}$ is fulfilled. Note that for this optimization, $x_{\bar{k}}$ and $z_{\bar{k}}$ are frozen; therefore, the density matrices for the constraint and fitting coefficient calculations are, in general, not the same. However, they converge to the same matrix with the convergence of the SCF. For λ_c optimization, a Newton–Raphson method is most appropriate.[23] Thus, the first and second derivatives of $\mathcal{E}_{ADFT}$ with respect to λ_c are needed. They can be calculated readily as

$$
\left(\frac{d\mathcal{E}_{ADFT}}{d\lambda_c} \right)_{x,z} = \frac{\partial \mathcal{E}_{ADFT}}{\partial \lambda_c} = \sum_{\mu,\nu} P_{\mu\nu} W_{\mu\nu} - N_c \qquad (10.26)
$$

and

$$\left(\frac{d^2\mathcal{E}_{\text{ADFT}}}{d\lambda_c^2}\right)_{x,z} = \sum_{\mu,\nu} P_{\mu\nu}^{(\lambda)} W_{\mu\nu}. \tag{10.27}$$

Whereas the calculation of the first derivative is straightforward, the calculation of the second derivative involves the perturbed density matrix for the constraint term. For perturbation-independent basis and auxiliary function sets, it follows from McWeeny's self-consistent perturbation (SCP) theory[24] for the elements of the (closed-shell) perturbed density matrix:

$$P_{\mu\nu}^{(\lambda)} \equiv \left(\frac{\partial P_{\mu\nu}}{\partial\lambda_c}\right)_{x,z} = 2\sum_{i}^{\text{occ}}\sum_{a}^{\text{uno}} \frac{K_{ia}^{(\lambda)}}{\varepsilon_i - \varepsilon_a}(c_{\mu i}c_{\nu a} + c_{\mu a}c_{\nu i}). \tag{10.28}$$

Here, the i and a indices refer to occupied (occ) and unoccupied (uno) MOs. Thus, the perturbed MO Kohn–Sham matrix can be expanded as

$$K_{ia}^{(\lambda)} = \sum_{\sigma,\tau} c_{\sigma i}c_{\tau a} K_{\sigma\tau}^{(\lambda)}. \tag{10.29}$$

The perturbed Kohn–Sham matrix elements can be calculated straightforwardly from Equation 10.24:

$$K_{\sigma\tau}^{(\lambda)} \equiv \left(\frac{\partial K_{\sigma\tau}}{\partial\lambda_c}\right)_{x,z} = W_{\sigma\tau}. \tag{10.30}$$

Backsubstitution into the expression for the perturbed density matrix elements then yields

$$P_{\mu\nu}^{(\lambda)} = 2\sum_{i}^{\text{occ}}\sum_{a}^{\text{uno}} \frac{W_{ia}}{\varepsilon_i - \varepsilon_a}(c_{\mu i}c_{\nu a} + c_{\mu a}c_{\nu i}) \tag{10.31}$$

with

$$W_{ia} = \sum_{\sigma,\tau} c_{\sigma i}c_{\tau a} W_{\sigma\tau}. \tag{10.32}$$

By inserting this expression for the perturbed density matrix elements into Equation 10.27, we find

$$\left(\frac{d^2 \mathcal{E}_{ADFT}}{d\lambda_c^2}\right)_{x,z} = 2\sum_i^{occ}\sum_a^{uno}\frac{W_{ia}}{\varepsilon_i - \varepsilon_a}\sum_{\mu\nu}(c_{\mu i}c_{\nu a} + c_{\mu a}c_{\nu i})W_{\mu\nu}$$

$$= 4\sum_i^{occ}\sum_a^{uno}\frac{W_{ia}^2}{\varepsilon_i - \varepsilon_a}. \tag{10.33}$$

Since the Kohn–Sham orbitals are populated according to the Aufbau principle, the second $\mathcal{E}_{ADFT}$ derivative is always negative for nondegenerate states, thereby showing that the correct λ_c corresponds to a maximum of the energy.[8] A flowchart of this ADFT/cDFT SCF is presented in Figure 10.1.[25] The ADFT/cDFT SCF algorithm now includes a supplementary loop during which the correct λ_c is determined in each SCF step. A tolerance criterion on the first-order derivative of $\mathcal{E}_{ADFT}$ is considered for declaring the convergence of the inner loop. Its value is itself adapted at every SCF step, depending on the degree of convergence of the MO coefficients, to ensure a smooth convergence of the overall ADFT/cDFT energy.

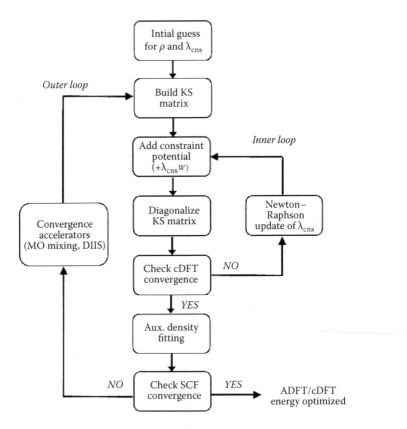

FIGURE 10.1 Double-loop algorithm employed in ADFT/cDFT calculations for the optimization of MO coefficients and cDFT Lagrange multipliers.

Note that each cycle of the inner loop does not require computing a new Kohn–Sham matrix, with the expensive Coulomb and exchange-correlation contributions, so that the time spent for optimizing λ_c is usually not problematic in terms of computational time, especially if an initial guess can be provided. So far, we have not specified the form of the weight function $w(r)$ and its corresponding matrix elements that are used in the cDFT constraint. We now turn to this point.

10.2.3   POPULATION ANALYSIS FOR cDFT

It is convenient to partition the full molecular density into atomic basins to constrain spatial densities. It is, however, well known that atomic charges are not uniquely defined in quantum mechanics and that numerous types of spatial partitioning exist. The richness of PA approaches pops up again for cDFT computations when expressing the **W** matrix. We have gathered in Table 10.1 the most common PA used in cDFT computations. Some partitions are based on an MO analysis such as Mulliken or Löwdin PA, while others are based on the electronic density itself. The latter are probably best suited for cDFT computations, and computational prescriptions for computing the **W** matrix have been reported for the Becke,[26] Hirshfeld,[27] and Voronoi deformation density (VDD)[28] partitioning schemes. In these cases, the **W** matrix elements are computed numerically on a DFT grid at the beginning of the SCF procedure:

$$W_{\mu\nu} = \sum_i a_i w_i \mu(r_i) \nu(r_i),   (10.34)$$

where a_i is the integration weight for grid point i, and w_i is a second weight with a definition depending on the partitioning scheme. Within the Becke approach, the real space is divided into Voronoi cells, each of them comprising all the grid points that are closer to the nucleus under consideration than to any other nucleus. The integrated number of electrons in Voronoi cells is attributed to the atom, thereby defining its charge (Table 10.1). The weights w_i are equal to 1 inside the cell and to 0 outside the cell and a smoothing function is applied at the border between adjacent Voronoi cells to avoid discontinuities in the function w.[26] A drawback of this partitioning scheme is its purely geometrical definition. This sometimes leads to atomic charges that are not satisfactory from a chemical point of view (e.g., the charges of hydrogens are frequently close to -0.5). In that regard, the Hirshfeld approach provides a more satisfactory approach.[27] The definition of Hirshfeld charges and integration weights w_i involves the electronic densities of the isolated atoms ρ^a (Table 10.1). The sum of the atomic densities gives rise to the so-called promolecular density. The Hirshfeld approach is sometimes referred to as the Stockholder approach: within the molecule, each atom gets a fraction of the relaxed electronic density of the molecule in proportion to its contribution to the promolecular density. In deMon2k, we commonly consider the neutral, spherically averaged forms of the atoms. Finally, within the VDD approach of Bickelhaupt et al.,[28] the charge of an atom is defined as the difference of the integrated electronic density from the Voronoi cells of the atom when passing from the promolecular density to the relaxed density.

TABLE 10.1

Five Population Approaches Commonly Used in cDFT Computations to Define Weight Matrix Elements

PA	Definition of Charge in Volume	Weight Matrix Elements	cDFT Weight w_i
Mulliken	$Q_A^{\text{Mulliken}} = Z_A - \sum_{\mu \in A} (PS)_{\mu\mu}$	$W_{\mu\nu} = S_{\mu\nu}$ if μ and $\nu \in A$ $W_{\mu\nu} = \frac{1}{2} S_{\mu\nu}$ if μ or $\nu \in A$ $W_{\mu\nu} = 0$ if μ and $\nu \notin A$	—
Löwdin	$Q_A^{\text{Löwdin}} = Z_\Omega - \sum_{\mu \in A} (S^{1/2}PS^{1/2})_{\mu\mu}$	$W_{\mu\nu} = \sum_\lambda S_{\mu\lambda}^{1/2} S_{\lambda\nu}^{1/2}$	—
Becke	$Q_A^{\text{Becke}} = Z_A - \int_{\text{Becke cell of A}} \rho(r)\,dr$	$W_{\mu\nu} = \sum_i a_i w_i \mu(r_i) v(r_i)$	$w_i^{\text{Becke}} = 1.0$ inside the cell $w_i^{\text{Becke}} = 0.0$ outside the cell + smoothing function
Hirshfeld	$Q_A^{\text{Hirshfeld}} = Z_A - \int \rho(r) \dfrac{\rho_A^a(r)}{\sum_{\text{all atoms}} \rho^a(r)}\,dr$	$W_{\mu\nu} = \sum_i a_i w_i \mu(r_i) v(r_i)$	$w_i = \dfrac{\sum_{\text{atoms in A}} \rho^a(r_i)}{\sum_{\text{all atoms}} \rho^a(r_i)}$
VDD	$Q_A^{\text{VDD}} = \int_{\text{Voronoi cell of A}} \left[\rho(r) - \sum_{\text{all atoms}} \rho^a(r) \right] dr$	$W_{\mu\nu} = \sum_i a_i w_i \mu(r_i) v(r_i)$	$w_i^{\text{VDD}} = 1.0$ inside the cell $w_i^{\text{VDD}} = 0.0$ outside the cell

The working equations of cDFT are, in principle, compatible with other population schemes, such as those based on the topological analysis of well-behaved functions (e.g., the electron density and the electron localization function [ELF][29,30]) or those that make use of iterative procedures (e.g., iterative Hirshfeld[31] or Stockholder schemes[32]). Some of these have been recently implemented in deMon2k for PA.[33] For cDFT computations, more complex algorithms should be devised to calculate the weight matrix elements. In fact, contrary to the cases mentioned in Table 10.1, for which the weight matrix elements are defined according to strictly defined MO criteria or according to the promolecular density, the weight matrix adapted for topological or iterative partitioning schemes would also depend on the relaxed density itself. In other words, the $\mathbf{W}$ matrix would vary at each SCF step and should be recomputed. Despite these difficulties, the adaptation of the cDFT methodology to more advanced PA schemes is doubtlessly an objective worth pursuing.

10.2.4   APPLICATIONS OF cDFT

The strength of the cDFT formulation is its compatibility with various types of constraints: the charge of a molecular fragment, its magnetization, or its electronic shell populations. The scientific topics that can be investigated with cDFT are thus numerous and varied, and cannot be reviewed extensively here. In this chapter, we have chosen to focus on two specific applications of cDFT: ET reactions and dioxygen activation by metal surfaces or inorganic complexes.

10.2.4.1   cDFT Framework and Marcus Theory of ET

ET is one of the most common chemical reactions encountered in chemistry and biochemistry. The Marcus theory of ET, whose earlier developments date back to the 1950s, has emerged over the years as a powerful theoretical framework for rationalizing ET rates between electron donors and acceptors.[34] The Marcus theory relies on the definition of two phenomenological electronic states that correspond to the localization of the transferred electron to a reductant (initial state) or an oxidant (final state). Depending on the electronic coupling strength between diabatic states, adiabatic (strong coupling) and nonadiabatic (weak coupling) rate constant expressions have been developed (Figure 10.2). The former is well suited, for example, for ET between free-moving inorganic complexes in solution, while the latter is best suited for long-range ET (occurring over more than 1 nm) or for symmetry-forbidden ET.

cDFT provides a means to approximate these phenomenological electronic states at the DFT level by imposing a charge difference between the donor and the acceptor fragments. The energies of the two ET states are thus accessible and—thanks to the availability of analytical energy gradients—geometry optimizations or BOMD simulations within the two cDFT potential wells can be carried out. Wu and Van Voorhis[35] showed that both reorganization energy and the driving force of the Marcus theory could be estimated with cDFT geometry optimizations. They also reported mathematical expressions to estimate the electronic coupling matrix elements (H_{12}) between cDFT states.[36] Note that such estimations of H_{12} are possible under the approximation by using Kohn–Sham determinants in place of the true diabatic wave functions. The cDFT method has been applied to the determination of equilibrium structures and ET

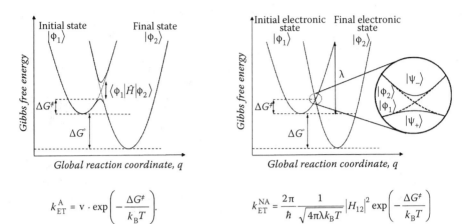

FIGURE 10.2 Adiabatic (left) and nonadiabatic (right) energy profiles and rate constant expressions associated with the limit regimes of the Marcus theory.

Marcus parameters within organic[37] and inorganic[38] mixed valence compounds, as well as to ET between inorganic complexes in solution.[9]

Using an interface between deMon2k and the molecular mechanics (MM) software CHARMM,[39] we have applied the method to evaluate the average electronic coupling and the time-dependent Franck–Condon factors for ET between two lithium ions separated by a polyglycine polymer of increasing length.[22,40] As shown in Figure 10.3, when the bridge length increases, the electronic coupling decays according to

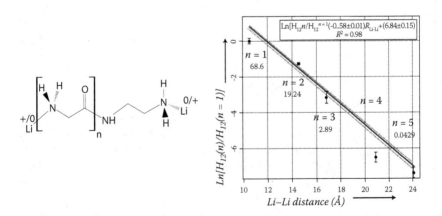

FIGURE 10.3 Evaluation of average electronic coupling for the ET between two lithium atoms separated by a polyglycine bridge of increasing length. Each point corresponds to 1 ps of hybrid cDFT/MM BOMD simulations in explicit water. The electronic coupling values for each n are given in cm^{-1}.

TABLE 10.2

Decay Times of High-Temperature Franck–Condon Factors (in Femtoseconds) for Different Donor–Acceptor Distances (in Angstroms)

		Solv. τ_{dec}	
n	R_{Li-Li}	a_n^{HT}	a_n^{MP2}
1	10.5	1.55 ± 0.13	1.36 ± 0.29
2	14.5	1.93 ± 0.07	1.64 ± 0.05
3	16.8	1.82 ± 0.06	1.56 ± 0.07
4	20.9	1.73 ± 0.02	1.50 ± 0.02
5	24.1	1.78 ± 0.14	1.53 ± 0.12

Note: Uncertainties correspond to a 95% confidence interval. Columns a_n^{HT} and a_n^{MP2} correspond to two choices on the widths of the Gaussian function mimicking nuclear wave packets.

exponential law, as expected for an ET proceeding in the superexchange regime. In the same study, we also calculated the characteristic decay times of the time-dependent Franck–Condon factors (Table 10.2). These numbers reflect the time during which nuclear wave packets remain in the Franck–Condon region before falling into one of the two cDFT potential wells. Our computational procedure was inspired by the work of Prezdho and Rossky,[41] albeit with specific adaptations for hybrid cDFT/MM BOMD simulations.

10.2.4.2 Dioxygen Activation by Cuprous Complexes

Understanding the processes by which dioxygen molecules are activated by coordination to metal surfaces or transition metal complexes is of primary importance for exploiting their oxidative power toward organic molecules. Because of spin selection rules, free dioxygen (a triplet molecule) reacts only very slowly with organic molecules that are usually singlet species. Catalysts are therefore required to flip the dioxygen local spin and speed up the kinetics of the reactions with O_2 under mild conditions. The computational modeling of these reactions is not trivial, and cDFT has been used by various groups in this context.

Behler et al.[42] addressed the question of the dissociative sticking coefficients of triplet O_2 molecules on an aluminum(111) surface as a function of the O_2 incident translational energy. Experimental data indicate sticking coefficients close to 0 for low incidence energies (<100 meV) and a rapid increase in sticking coefficients for higher incidence energies.[43] *Ab initio* molecular dynamics simulations realized on DFT-parameterized PES or "on-the-fly" BOMD simulations fail to reproduce this trend: sticking coefficients equal to 100% are always found, whatever the incidence energies. They attributed this failure to the inability of the adiabatic description provided by DFT to describe the process: "Even at the largest distances the electron chemical potentials of the O_2 molecule and the Al(111) surface align, which is achieved by some electron transfer towards the O_2 molecule."[42] In reality, charge transfer between O_2 and the metallic surface would only occur at shorter distances. Behler et al. thus used a cDFT approach similar to that of Dederichs et al. to constrain the magnetization of the O_2 fragment and to build nonadiabatic PES for both triplet and singlet O_2. On the triplet PES, they highlighted energy barriers of 6.2 and 19.6 kcal/mol when approaching an O_2 molecule to the metal surface (depending on the orientation of the dioxygen molecule with respect to the surface). Such barriers are absent in the adiabatic PES provided by nonconstrained DFT. A sufficient amount of translational incidence energy is therefore required for the O_2 molecules to pass over the barrier before binding to the aluminum surface. *Ab initio* simulations performed on the cDFT-parameterized PES nicely reproduce the experimental trends.[42]

Activation of O_2 molecules can also be achieved through coordination to transition metal complexes. In that regard, metalloenzymes provide a formidable source of inspiration for O_2-activating species. Recently, we reported a systematic investigation of Pd, Fe, Ni, and Cu metal–O_2 adducts with DFT computations and topological analysis using the ELF.[44] We investigated the shape of the dioxygen lone pairs (LP) revealed by the ELF picture as a function of the spin state of the adducts. Figure 10.4 shows, in the case of a mononuclear cuprous complex, a 90° rotation of the orientation of O_2 LP with respect to the CuOO plane when going from triplet to singlet. This rotation is associated with significant variations of LP volumes, making them more accessible to exogenous substrates in the singlet state.

This trend could be qualitatively related to the symmetry of frontier orbitals, whose populations differ between the two spin states. To go deeper into the analysis of this phenomenon, we then employed cDFT to impose spin density at the CuO_2 core and to investigate the ELF topology of the constrained electronic densities. As shown in Figure 10.5, when the spin density at the CuO_2 core increases, LP rotate toward the CuOO plane (+ symbols)—a process associated with the expansion of

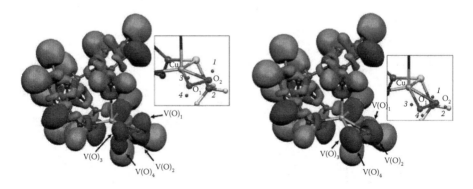

FIGURE 10.4 Orientation of dioxygen LP revealed by ELF topological analysis in the singlet (left) and triplet (right) spin states within a minimal model of noncoupled monooxygenases (isosurface for ELF = 0.8). Insets: CuO_2 core showing the localization of O_2 valence basin attractors. Grey scale: darkest, nonbonding (LP); intermediate, bonding; lightest, protonated bond.

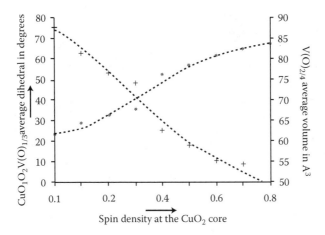

FIGURE 10.5   Evolution of O_2 LP orientations and volumes as a function of spin density at the CuO_2 core.

oxygen valence basins 2 and 4 (* symbols) and the volume reduction of basins 1 and 3. In summary, ELF analysis—thanks to its coupling to the cDFT method—revealed the tight connection between the spin state of adducts and the spatial organization of oxygen LP. We suggested that enzymes could resort to spin-state control to fine-tune the regioselectivity of substrate oxidations.

10.3   CONCLUSIONS AND PERSPECTIVES

This chapter has been devoted to the presentation of the cDFT approach within the ADFT framework. The implementation takes advantage of the methodology of Wu and Van Voorhis for optimization of cDFT Lagrange multipliers and of the advantageous features of the ADFT framework. In the equations, the constraint is incorporated at the level of the orbital electronic density through the definition of a weight matrix $\mathbf{W}$ that depends on the chosen PA. deMon2k supports a large variety of PA compatible with cDFT. The implementation of energy gradients makes it possible to optimize geometries and to perform BOMD simulations on cDFT PES.

At the forefront of future developments, the adaptation of the cDFT formalism to new-generation PA schemes (such as iterative Hirshfeld or iterative Stockholder) should be pursued. These iterative schemes often show improved chemical behavior (e.g., propagation of substituent inductive effects, reproduction of electrostatic potentials) compared to noniterative schemes. Another worthwhile objective would be to derive a fully constrained ADFT formalism, where the constraints would be applied on the auxiliary electronic density itself. Such developments, which are currently underway, would lead to better scaling behavior with system size. There is also much room for future methodological developments aiming at computing interesting properties on cDFT electronic densities such as polarizabilities and hyperpolarizabilities, Nuclear Magnetic Resonance (NMR) shieldings, electronic excitation spectra, and vibrational frequencies. The availability of efficient algorithms for evaluating cDFT energy derivatives is a prerequisite in this perspective.

We have illustrated the relevance of cDFT to the modeling of ET reactions and to the investigation of dioxygen activation processes. Recent extensions of cDFT formalism to TDDFT have been reported to investigate photo-induced ET[45] or exciton charge transfer.[46] Yet, as already mentioned, cDFT has found many applications in diverse scientific areas. The parameterization of model Hamiltonians for modeling singlet fission processes[47] or superconductivity,[48] the computation of charge transfer energies in energy decomposition analysis (EDA) schemes,[49] the control of spin contamination,[50] the calculation of magnetic coupling constants in polynuclear metal complexes,[51] and the investigation of competition between hydride and hydrogen transfers[52] are all examples of recent applications of this powerful tool.

REFERENCES

1. Casida, M. E.; Huix-Rotllant, M. *Annu. Rev. Phys. Chem.* **2012**, 63, 287–323.
2. Niehaus, T. A. *J. Mol. Struct. THOECHEM* **2009**, *914*, 38–49.
3. Doltsinis, N. L.; Marx, D. *J. Theor. Comput. Chem.* **2002**, *1*, 319–349.
4. Tully, J. C.; Preston, R. K. *J. Chem. Phys.* 1971, *55*, 562–572.
5. Dederichs, P. H.; Blügle, S.; Zeller, R.; Akai, H. *Phys. Rev. Lett.* **1984**, *53*, 2512–2515.
6. Gunnarsson, O.; Lundqvist, B. I. *Phys. Rev. B* **1976**, *13*, 4274–4298.
7. Akai, H.; Blügle, S.; Zeller, R.; Dederichs, P. H. *Phys. Rev. Lett.* **1986**, *56*, 2407–2410.
8. Wu, Q.; Van Voorhis, T. *Phys. Rev. A* **2005**, *72*, 024502–024504.
9. Oberhofer, H.; Blumberger, J. *J. Chem. Phys.* **2009**, *131*, 064101–064111.
10. Van Voorhis, T.; Kowalczyk, T.; Kaduk, B.; Wang, L.-P.; Cheng, C.-L.; Wu, Q. *Annu. Rev. Phys. Chem.* **2010**, *61*, 149–170.
11. Kaduk, B.; Kowalczyk, T.; Van Voorhis, T. *Chem. Rev.* **2012**, *112*, 321–370.
12. Shao, Y.; Fusti-Molnar, L.; Jung, Y.; et al. *Phys. Chem. Chem. Phys.* **2006**, *8*, 3172–3191.
13. M. Valieva, M.; Bylaskaa, E. J.; Govind, N.; et al. *Comput. Phys. Commun.* **2010**, *181*, 1477–1489.
14. Bowler, D. R.; Choudhury, R.; Gillan, M. J.; Miyazaki, T. *Phys. Status Solidi B* **2006**, *243*, 989–1000.
15. Koster, A. M.; Geudtner, G.; Calaminici, P.; Casida, M. E.; Dominguez, V. D.; Flores-Moreno, R.; Gamboa, G. U.; et al. *deMon2k, Version 3*. The deMon Developers, CINVESTAV, Mexico City, **2011**.
16. Köster, A. M.; Reveles, J. U.; del Campo, J. M. *J. Chem. Phys.* **2004**, *121*, 3417–3424.
17. Janetzko, F.; Goursot A.; Mineva, T.; Calaminici, P.; Flores-Moreno, R.; Köster, A. M.; Salahub, D. R. *Structure Determination of Clusters: Bridging Experiment and Theory in Nanoclusters: A Bridge Across Disciplines*, edited by P. Jena, A. Castleman Jr. Elsevier, Amsterdam, **2010**.
18. Calaminici, P.; Dominguez-Soria, V. D.; Flores-Moreno, R.; Gamboa, G. U.; Geudtner, G.; Goursot, A.; Salahub, D. R.; Köster, A. M. *Auxiliary Density Functional Theory: From Molecules to Nanostructures in Handbook of Computational Chemistry*, edited by J. Leszczynski. Springer-Verlag, Heidelberg **2011**.
19. Köster, A. M.; Flores-Moreno, R.; Reveles, J. U. *J. Chem. Phys.* **2004**, *121*, 681–690.
20. Dunlap, B. I.; Rösch, N.; Trickey, S. B. *Mol. Phys.* **2010**, *108*(21–23), 3167–3180.
21. Köster, A. M.; del Campo, J. M.; Janetzko, F.; Zúñiga-Gutiérrez, B. *J. Chem. Phys.* **2009**, *130*, 114106–114108.
22. Řezáč, J.; Lévy, B.; Demachy, I.; de la Lande, A. *J. Chem. Theor. Comput.* **2012**, *8*, 418–427.

23. Press, W. H.; Teukolsky, S. A.; Vetterling, W. T.; Flannery, B. P. *Numerical Recipes in Fortran.* Cambridge University Press, Cambridge, **1986**.
24. Dodds, J. L.; McWeeny, J.; Sadlej, A. *J. Mol. Phys.* **1976**, *34*, 1779–1791.
25. de la Lande, A.; Salahub, D. R. *J. Mol. Struct. THEOCHEM* **2010**, *943*, 115–120.
26. Becke, A. D. *J. Chem. Phys.* **1988**, *88*, 2547–2553.
27. Hirshfeld, F. L. *Theor. Chim. Acta* **1977**, *44*, 129–138.
28. Bickelhaupt, F. M.; van Eikema Hommes, N. J. R.; Fonseca Guerra, C.; Baerend, E. J. *Organometallics* **1996**, *15*, 2923–2931.
29. Becke, A. D.; Edgecombe, K. E. *J. Chem. Phys.* **1990**, *92*, 5397–5404.
30. Silvi, B.; Savin, A. *Nature* **1994**, *371*, 683–686.
31. Bultinck, P.; Van Alsenoy, C.; Ayers, P. W. *J. Chem. Phys.* **2007**, *126*, 144111–144119.
32. Lillestolen, T. C.; Wheatley, R. J. *J. Chem. Phys.* **2009**, *131*, 144101–144106.
33. de la Lande, A.; Köster, A. M.; Vela, A.; Lévy, B.; Demachy. I. Submitted.
34. Marcus, R. A.; Sutin, N. *Biochim. Biophys. Acta* **1985**, *811*, 265–322.
35. Wu, Q.; Van Voorhis, T. *J. Phys. Chem. A* **2006**, *110*, 9212–9218.
36. Wu, Q.; Van Voorhis, T. *J. Chem. Phys.* **2006**, *125*, 164105–164109.
37. Hoekstra, R. M.; Telo, J. P.; Wu, Q.; Stephenson, R. M.; Nelsen, S. F.; Zink, J. I. *J. Am. Chem. Soc.* **2010**, *132*, 8825–8827.
38. Lu, Y.; Quardokus, R.; Lent, C. S.; Justaud, F.; Lapinte, C.; Kandel, S. A. *J. Am. Chem. Soc.* **2010**, *132*, 13519–13524.
39. Brooks, B. R.; et al. *J. Comput. Chem.* **1983**, *4*, 187.
40. de la Lande, A.; Babcock, N. B.; Řezáč, J.; Lévy, B.; Sanders, B. C.; Salahub, D. R. *Phys. Chem. Chem. Phys.* **2012**, *14*, 5902–5918.
41. Prezdho, O.; Rossky, P *J. Chem. Phys.* **1997**, *107*, 5863–5878.
42. Behler, J.; Delley, B.; Lorenz, S.; Reuter, K.; Scheffler, M. *Phys. Rev. Lett.* **2005**, *94*, 036104-1–036104-4.
43. Österlund, L.; Zorić, I.; Kasemo, B. *Phys. Rev. B* **1997**, *55*, 15452–15455.
44. de la Lande, A.; Salahub, D. R.; Maddaluno, J.; Scemama, A.; Pilmé, J.; Parisel, O.; Gérard, H.; Caffarel, M.; Piquemal, J.-P. *J. Comput. Chem.* **2011**, *32*, 1178–1182.
45. Chen, H.; Ratner, M. A.; Schatz, G. C. *J. Phys. Chem. C* **2011**, *115*, 18810–18821.
46. Difley, S.; Van Voorhis, T. *J. Chem. Theor. Comput.* **2011**, *7*, 594–601.
47. Greyson, E. C.; Vura-Weis, J.; Michl, J.; Ratner, M. A. *J. Phys. Chem. B* **2010**, *114*, 14168–14177.
48. Anisimov, V. I.; Kurmaev, E. Z.; Moewes, A.; Izyumov, I. A. *Physica C* **2009**, *469*, 442–447.
49. Wu, Q.; Ayers, P. W.; Zhang, Y. *J. Chem. Phys.* **2009**, *131*, 164112–164118.
50. Schmidt, J. R.; Shenvi, N.; Tully, J. C. *J. Chem. Phys.* **2008**, *129*, 114110–114119.
51. Rudra, I.; Wu, Q.; Van Voorhis, T. *J. Chem. Phys.* **2006**, 124, 024103–024109.
52. Gillet, N.; Piquemal, J.-P.; de la Lande, A.; et al. In preparation.

11 Spin and Orbital Physics of Alkali Superoxides
p-Band Orbital Ordering

Ashis Kumar Nandy, Priya Mahadevan, and D. D. Sarma

CONTENTS

11.1 INTRODUCTION

Transition metal (TM) compounds and various rare earth–based compounds have been the workhorse of the strongly correlated community over the past five to six decades, with very interesting properties being found as a consequence of closely coupled spin, charge, orbital, and lattice degrees of freedom [1–7]. P-band materials, with their completely filled bands, were never considered candidates until recent studies brought back alkali (A) metal oxides into the spotlight. Alkali metal oxides have been explored in the 1960s and 1970s by various groups. A detailed review of these materials in 1989 summarized the motivating factors that drove research on these materials. In the 1960s, Hesse et al. [9] concluded that the presence of catenated oxygen (O) species, such as ozonide (O_3^-), hyperoxide (O_2^-), and peroxide (O_2^{2-}), in these ionic compounds led to investigations devoted to the chemical nature of these compounds that had such unusual compositions in their structure. However, in the second period from the 1960s to the late 1980s, the availability of large single crystals led to investigations on crystal chemistry, phase transitions, and physical properties.

The interest in this class of compounds over the past few years has been driven by the search for phenomena similar to $3d$ TM compounds. TM-based materials have been intensively studied because of their rich phase diagrams. For example, manganites show numerous kinds of magnetic ordering, as well as ordering of the charge and orbital degrees of freedom of the electrons associated with structural changes controlled by various external parameters such as temperature, pressure, and doping.

221

Another category of materials that received much attention are cuprates. After the discovery of high-temperature superconductivity in cuprates, many theoretical and experimental attempts have focused on the various aspects of these materials. To find rich phase diagrams in p-band–based materials, we first looked at examples that had an active spin degree of freedom (i.e., those that could be magnetic). Magnetic p-band materials are rare since, in most compounds, p-bands are either broad or completely filled, which will not allow a local moment to be sustained. Thus, we must first find systems with a narrow, partially filled p-band. Recently, examples of p-band materials that exhibit magnetism have been intensively studied. The route to magnetism in these cases has been via defects such as vacancies or doped impurity atoms; in some cases, even high-temperature ferromagnetism is observed. Since one cannot experimentally control defects in materials, the reproducibility of such efforts is difficult. To avoid such a route, we have looked for stoichiometric materials with partially filled p-bands. In that direction, one finds that molecular oxygen contains two electrons in its degenerate π-antibonding molecular orbital levels, which are found to order antiferromagnetically in various phases associated with solid oxygen [8]. Several compounds have oxygen dimers [11,15,16,20]. This route took us to an interesting class of oxides formed by alkali metal atoms. In addition to dioxide (A_2O), we also found superoxide (AO_2), peroxide (A_2O_2), sesquioxide (A_4O_6), and ozonide (AO_3). These compounds surprisingly allow for the realization of variable valencies at the oxygen site, similar to those found at the TM site in TM compounds. Among those oxides, dioxides and peroxides are nonmagnetic owing to filled molecular orbital levels, consistent with experiments. Superoxides such as KO_2, RbO_2, and CsO_2 have been found to exhibit antiferromagnetic (AFM) ordering below their Neel temperatures of 7.1, 15, and 9.6 K, respectively. Sesquioxides such as Rb_4O_6 and Cs_4O_6 both show AFM ordering below 4 K. Ozonides such as KO_3 and RbO_3 show maxima in their temperature-dependent susceptibility at 20 and 17 K, respectively, suggesting AFM ordering, while CsO_3 showed very weak AFM ordering.

Recent studies [10–17] have investigated the implications of closely coupled spin, orbital, and lattice degrees of freedom for this class of d^0 magnets [i.e., magnetic materials where magnetism comes from partially filled p-bands as in alkali superoxides (AO_2)]. All these alkali superoxides have oxygen dimers as an integral part of their crystal structure. Thus, the basic electronic structure is not altered from the molecular orbital picture expected of an oxygen molecule, except for broadening due to solid-state effects. These superoxides (AO_2) represent the realization of electron-doped oxygen molecules (O_2^-) arranged in a lattice. The alkali metal atom A (Na, K, Rb, and Cs) acts as an electron donor. Therefore, each O_2^- anion has nine electrons in $2p$ molecular orbital levels with an electronic configuration of σ_b^2, π_b^4, π_a^3 (Figure 11.1). The extra electron from alkali metals occupies degenerate π-antibonding molecular orbital levels and leaves them partially filled. In calculations that use generalized gradient approximation (GGA) for exchange-correlation functionals, alkali superoxides are found to exhibit a half-metallic ground state due to the partially filled π-antibonding level. In RbO_2, Kovacik and Ederer [10] demonstrated the importance of an onsite Coulomb interaction U in oxygen p-bands, which led to the formation of an orbitally polarized insulating state in contrast to a half-metallic state. However, an earlier work [4] has shown that one cannot have

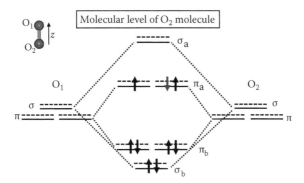

FIGURE 11.1 Schematic diagram of the energy levels of an oxygen molecule and various features in the DOS (Figures 11.3 and 11.4) labeled accordingly. The extra electron (down arrow) from the K atom in KO_2 fills the down-spins π_a level.

degeneracy lifting of orbital ordering(OO) takes place. This is always accompanied by lattice distortions that lower symmetry. Adopting this route for the analogous K compound KO_2, we find that a structural transition determining the nature of orbital ordering (OO) takes place [11]. These results had important implications for the theoretical model—the Kugel–Khomskii model—used to understand OO [18]. In this model, the basic conjecture is that superexchange considerations determine the pattern of OO. We showed that this was not the case and that this class of systems was outside the purview of the Kugel–Khomskii model. It was also pointed out that the OO in the present case destroyed any possibility of long-range magnetic order, which otherwise seemed possible. The Kugel–Khomskii model has been used to study these systems. Ylvisaker et al. [12] examined magnon and orbiton behaviors in localized O_2 antibonding molecular orbitals in RbO_2 using an effective Kugel–Khomskii Hamiltonian derived from a two-band Hubbard model, with hopping parameters taken from ab initio band structure calculations. In that paper, they concluded that the considerable difference between intraband and interband hopping strengths led to a strong coupling between spin-wave dispersions and the orbital ground state, and that proliferation of orbital domains disrupts long-range magnetic order, which results in a significant reduction in the Neel temperature of RbO_2. Kim et al. [13] proposed that the insulating nature of the high-symmetry phase of KO_2 at high temperature arises due to the combined effects of spin–orbital (SO) coupling and electron–electron interactions between the p-electrons in oxygen. A structural transition has been proposed in their GGA level calculations by considering a coherent tilting of O_2^- motifs in the lattice at low temperature. In their calculations, a ferro-orbital order emerged with a band-gap opening due to strong coupling between the SO and the lattice degrees of freedom. The electronic structure of KO_2 at room temperature was investigated by Kang et al. [14] by employing soft X-ray absorption spectroscopy (XAS) and core-level photoemission spectroscopy (PES) at room temperature. Since the main feature in unoccupied states (observed in O $1s$ XAS) was in reasonable agreement with the calculated density of states (DOS) in their GGA + SO + U level calculations, they claimed that the SO effect and the onsite Coulomb interaction ($U = 11.5$ eV) were essential to explaining the observed insulating nature in KO_2. Another puzzling aspect of these alkali metal oxides is their crystal structure at low temperatures. This plays an important role in determining the favored magnetic order. Starting with the high-temperature body-centered tetragonal (bct) structure, we found that ferromagnetism was strongly favored with a large stabilization energy. As the bandwidths associated with the p-bands were ~0.8 eV, electron–electron interaction effects are believed to be important. Calculating the electronic structure as a function of U, one finds a metal-to-insulator transition at $U_c = 2$ eV. In our calculations, we also found orbital polarization as found in RbO_2. This was, however, found to be strongly dependent on the underlying spin arrangement. The partial occupancy of π-antibonding levels makes KO_2 Jahn–Teller active. Halverson [19] pointed out that the O_2^- ion is a Jahn–Teller active ion that is not able to occupy the highly symmetrical D_{4h} site in the high-temperature bct structure, with the anion's (O_2^-) molecular axis coinciding with the crystallographic c-axis of the bct unit cell. He proposed a structural model for KO_2 in the 1960s, with O_2 dimers executing a circular motion about the c-axis in the tetragonal lattice. He speculated that, below some temperature, oxygen dimers may get trapped in certain positions. He also proposed a motion of K atoms in the directions perpendicular to the c-axis. These result in a structural phase transition to a less symmetrical structure. But experimentally, no structural data are available for the low-temperature structure.

In our earlier work, we have shown that the system undergoes a phase transition at U_c. This structural distortion leads to an OO transition for U values above 2 eV. This OO is unique and does not depend on the underlying magnetic ordering. But unfortunately with this OO transition, we have found that all magnetic configurations lie very close in energy. OO usually favors a particular magnetic ordering through superexchange, as proposed by Kugel and Khomskii in the context of manganites, and many systems are found to follow the Kugel–Khomskii model. But the type of OO that we found cannot be explained by the Kugel–Khomskii model. Further various magnetic configurations are found to lie very close in energy. This OO may be the reason why KO_2 has very low magnetic ordering temperatures, although the high-temperature bct structure is likely to be a ferromagnetic metal with a high transition temperature.

11.2 METHODOLOGY

We have carried out first-principles electronic structure calculations for this system within the framework of density functional theory (DFT) using plane-wave pseudopotential implementation [21–23] with projected augmented wave (PAW) potentials [24,25]. GGA [26,27] is used for the exchange. The effects of correlation are treated within GGA + U [28]. We have also verified our results within local spin density approximation (LSDA) + U, and the results have been found to be qualitatively similar. A uniform k-points mesh consisting of $8 \times 8 \times 8$ k-points was used for Brillioun zone (BZ) integrations, and the computed total energies changed by 1 meV when we changed the k-mesh to $12 \times 12 \times 12$. A plane-wave cutoff energy of 875 eV was used for the kinetic energy of the plane waves included in the basis. Spheres of radii 2.0 and 0.8 Å were constructed about the K and O atoms for the computation of atom-projected partial DOS and magnetic moments. Structural optimization was carried out to estimate theoretical lattice parameters. To discuss the magnetism in KO_2, we have constructed a supercell of 24 atoms and considered different magnetic configurations by aligning their spins accordingly. The k-mesh used for the 24-atom supercell is $4 \times 4 \times 4$. To lower the symmetry, we have performed structural relaxations removing all symmetry constraints. While doing the calculations, we fixed the lattice parameters at their optimized/theoretical values, and we allowed the internal coordinates to relax until forces less than 0.01 eV/Å were achieved.

11.3 RESULTS AND DISCUSSION

The magnetic and structural properties of KO_2 had been intensively studied in the 1970s to the 1980s. Early studies on KO_2 [9] revealed a wide variety of structural and magnetic properties that are rather unclear. KO_2 undergoes several structural phase transitions with temperature, and although space groups have been identified, detailed structural information does not exist. Only the structural information for the bct phase is clear. There have been many speculations [13,19] regarding the low-temperature structure, suggesting a tilting of oxygen dimers/molecules in the structure from old experiments. In our calculations, we started with the high-temperature structure shown in Figure 11.2a. The high-temperature KO_2 structure is a bct structure belonging to the space group $I4/mmm$, and the structural information is given in Table 11.1. This bct structure consists of two interpenetrating bct lattices of oxygen dimers/molecules and K atoms. Locally, each oxygen dimer is surrounded by six K atoms, four of which lie in a plane while the remaining two lie above and below the dimer, as shown in the inset of Figure 11.2a. The planar K atoms form a square, and the plane containing them bisects the dimer. The oxygen dimer length is 1.36 Å in KO_2, which is larger than what is found in an oxygen molecule. The oxygen molecular length is 1.22 Å. The four K atoms in the plane that bisects the oxygen

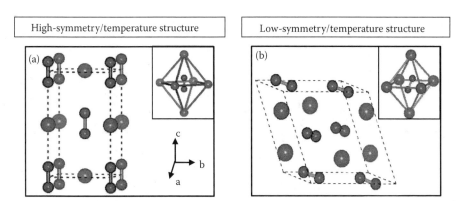

FIGURE 11.2 Crystal structure of KO_2 for (a) high-symmetry/temperature (experimental) and (b) low-symmetry/temperature (theoretically relaxed) structures. The inset of a and b shows the local environment of an oxygen dimer in KO_2 in the high-symmetry/temperature and low-symmetry/temperature phases, respectively.

TABLE 11.1

High-Symmetry/Temperature and Low-Symmetry/Temperature Structural Parameters of KO$_2$

Structural Details	High-Symmetry/ Temperature	Low-Symmetry/ Temperature
Space group	$I4/mmm$	$C2/c$
a (Å)	4.030	8.1769
b (Å)	4.030	7.6320
c (Å)	6.697	4.1396
α (°)	90	90
β (°)	90	90
γ (°)	90	62
K	0.0, 0.0, 0.5	0.0, 0.25, 0.8175
O	0.0, 0.0, 0.0975	0.6857, 0.4748, 0.1889

dimer are at a distance of 2.93 Å from the oxygen atom, while K–O distances of the K atoms above and below the dimer are 2.66 Å. In the high-symmetry phase, KO$_2$ is a ferromagnetic half-metal in our electronic structure calculations. The DOS is given in Figure 11.3. The basic features in the DOS are similar to the molecular orbital levels found in oxygen molecule. These levels are, however, broadened due to solid-state effects. The features are denoted by σ_b, π_b, π_a, and σ_a to facilitate comparison with the molecular orbital levels. For the sake of completeness, we repeat the discussions of molecular orbital formation in an oxygen molecule. In the two oxygen atoms comprising the dimer pointing along the z-direction, the three-fold degenerate p orbitals (p_x, p_y, and p_z) split into singly degenerate σ levels and doubly degenerate

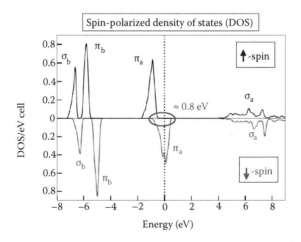

FIGURE 11.3 O p partial DOS calculated for spin-polarized KO$_2$. The zero of energy corresponds to Fermi energy. Various features in the oxygen partial DOS are labeled according to the molecular level diagram depicted in Figure 11.1.

π levels due to strong crystal field effects. The p_z orbitals on each oxygen atom in the dimer are directed toward each other and hence feel a strong electrostatic repulsion compared to the other two p orbitals (p_x and p_y). As a result, the σ levels will lie above the doubly degenerate π levels. This is indicated on the extreme left and right of Figure 11.1 for the two atoms O_1 and O_2 comprising the dimer. The central panel of Figure 11.1 shows the ensuing molecular orbital diagram for the oxygen molecule as one switches on the hopping interactions between the same symmetry levels on O_1 and O_2. Furthermore, the hopping interaction strength is stronger between the σ levels than between the π levels. This results in a reversal in level ordering (with the bonding σ_b levels found to lie below the bonding π_b levels) and in a reverse ordering for antibonding levels. The subscripts b and a have been used to denote bonding and antibonding levels. If one switches on spin polarization, down-spin level ordering is found to follow up-spin levels. An oxygen molecule has the electronic configuration σ_b^2, π_b^4, π_a^2 of its eight electrons. The levels with the labels σ_b and π_b are completely filled in both spin channels, while π_a bands are half-filled. The half-filled π_a band behaves like a half-filled Hubbard band, which results in the AFM ordering found in the different phases of solid oxygen. An analysis of non–spin-polarized DOS (Figure 11.4) suggests that K acts as an electron donor in KO_2 with negligible contributions from K $4s$ and $3p$ states to the DOS in the energy window -7 to 1 eV (Figure 11.4). Thus, KO_2 is an ionic solid in which K remains as a K^+ ion and O_2 becomes charged O_2^- ion. The extra electron fills up the π_a down-spin level partially. Thus, the up-spin channel is insulating, and the down-spin channel is metallic. This explains the half-metallic nature of KO_2 in the high-symmetry phase. This metallic behavior drives the system to be ferromagnetic, which we indeed found in our calculations. The bandwidth of the highest occupied π_a level is ~0.8 eV, which suggests that electron–electron correlation (U) is crucial to understanding its ground-state properties. We

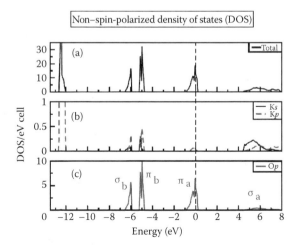

FIGURE 11.4 (a) Total, (b) K s and K p, and (c) O p partial DOS calculated for non–spin-polarized KO_2. The zero of energy corresponds to Fermi energy. The various features in the DOS are labeled according to the molecular level in Figure 11.1.

have used U as a parameter in our GGA + U level calculations and varied it in the range from 0 to 6 eV. Photoemission and Auger experiments placed U in the range of 4–6 eV on the oxygens in TM oxides [29–31]. Interestingly, we found that, for low values of U, where the system is metallic, the favored type of magnetic ground state is ferromagnetic. For larger values of U (>2 eV), the ground state turned out to be antiferromagnetic, which increasingly stabilized with U. Below a critical U value ($U_c \sim 2$ eV), the system was found to be a ferromagnetic metal, and it became an AFM insulator above U_c. We have considered several AFM configurations along with the ferromagnetic one, using a supercell of 24 atoms, and calculated their energies as a function of U. At U_c, the gap opens up at Fermi energy (E_F). We find in our calculations that, irrespective of the nature of underlying magnetic configurations, the system shows large magnetic stabilization energies, except around U_c. Thus, except near U_c, our calculations predict high magnetic ordering temperatures due to large stabilization energies. Experimentally, it is known to be an antiferromagnet below 7.1 K, which contradicts our conclusions. Another phenomenon that we find in the high-symmetry bct structure is orbital polarization (i.e., preferential occupancy of a particular orbital over the notationally degenerate one due to correlation effects). This orbital polarization is not unique and depends on underlying magnetic configurations.

Early experiments predicted symmetry lowering involving a tilting/canting of the oxygen dimers by an angle θ formed with respect to the c-axis of the crystal. The angle of rotation was predicted to be about 20°–30°. But the exact details are not known. Also, from the electronic structure calculations using the high-temperature/symmetrical structure, we conclude that the highest occupied π_a down-spin level is partially filled and, therefore, Jahn–Teller active. The system will distort to break the symmetry of the π_a levels to minimize its energy further. We therefore attempted to lower the symmetry and allowed for a relaxation of the structure without any symmetry constraints. After the relaxation, we found a rotation of the O_2 dimers and a displacement of the K atoms. The rotation angle of the oxygen dimer was found to be 23.5 Å, a value that compares well with experimental prediction. The relaxed structure is given in Figure 11.2b. In the relaxed structure, the local surroundings of the O_2 dimer have changed with a rotation of the dimers, as shown in the inset of Figure 11.2b. Earlier, the dimers were at the center of the squares formed by four K atoms (Figure 11.5a). Those squares became rhombi, with one shorter diagonal and one longer diagonal, which are shown in Figure 11.5b. The dimer is rotated/canted along the long diagonal axis. Also, the top and bottom K atoms are displaced and slide from their locations above and below the dimer in the direction opposite to the tilt of the dimers. The relaxed structure is monoclinic, with the structural details presented in Table 11.1. In the remaining part of the paper, we use the nomenclature ordered structure for the high-temperature structure and disordered structure for the low-temperature structure. The K–O distances are now found to be 2.74 (2.76), 2.76 (2.74), 2.82 (3.42), and 3.42 Å (2.82 Å) for the planar K atoms, and that between the top (bottom) K and the top (bottom) O of the dimer is 2.80 Å. In search for a possible reason behind such a rotation of O_2 dimers, we find that the electrostatic repulsion between the electrons on K and O is responsible for the rotation. As K acts like a donor in this system, the extra electron due to K is shared by the two oxygen atoms comprising the dimer. This increases the electrostatic repulsion between

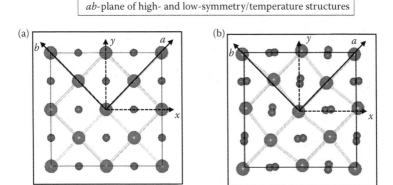

FIGURE 11.5 Top view of the *ab*-plane in KO_2 in the (a) initial high-symmetry/temperature structure and (b) theoretically relaxed low-symmetry/temperature structure. The K atoms (big circles) are found at the corners of the squares/rhombi. The oxygen dimers (small circles) pass through the center of the squares/rhombi. The *x* and *y* axes are the local coordinate axes where DOS (Figure 11.6a and b) have been calculated.

two oxygens compared to that in an oxygen molecule. To reduce the repulsion, the oxygens comprising the dimer move apart, leading to an increased bond length of 1.36 Å. As the oxygens move apart, they feel the presence of electrons on the K atoms lying above and below the dimer. The dimer then rotates, thereby reducing the electrostatic repulsion. This repulsion also results in the movement of the K atoms in a direction opposite to the direction of dimer rotation. Another way by which the system gains energy is through the canting of oxygen dimers along the long diagonal of the rhombi formed by the K atom. To understand the relation between electrostatic repulsion and dimer rotation, we have performed two calculations in which we have increased and decreased the K–O distance along the *c*-axis and examined the change in rotation angle. The increase/decrease is achieved by an appropriate change in the *c* lattice parameter, while increasing (decreasing) the *c* lattice parameter should decrease (increase) the contribution from the electrostatic repulsion between electrons on the K and O atoms of the dimer. Thus, the rotation/canting of oxygen dimers will show changes. In the case where the *c* lattice parameter is increased by 1%, we find that the rotation angle reduces to 22.5°. In the case where the *c* lattice parameter is decreased by 1%, the rotation angle increases to 25°. Thus, the electrostatic repulsion controls the rotation of the oxygen dimers. We examined the electronic structure that we found by relaxations in our calculations. In this low-symmetry structure, we found that the square formed by four planar K atoms displaces to become a rhombus, where one diagonal is longer than the other. The top view of the *ab*-plane for the high-symmetry and low-symmetry structures is shown in Figure 11.5a and b, respectively. This structural distortion leads to an OO transition, which drives the system to an insulating state. This is evident from the O *p* spin-projected partial DOS shown in Figure 11.6a and b, where the doubly degenerate $\pi_a(\pi_b)$ levels split into levels labeled $p_x^a \left(p_x^b \right)$ and $p_y^a \left(p_y^b \right)$. The inset of Figure 11.6b shows the local coordinates with respect to which p_x and p_y have been defined. The real-space charge

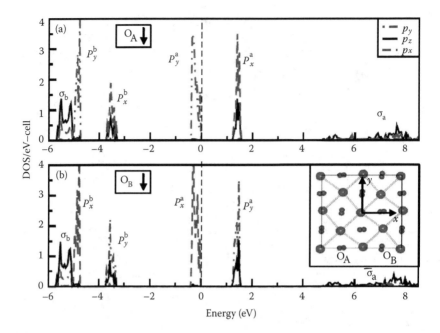

FIGURE 11.6 O p down-spin projected partial DOS calculated for oxygens labeled as (a) O_A and (b) O_B in the inset of (b) for the theoretically relaxed low-symmetry/temperature structure. The zero of energy corresponds to Fermi energy (E_F). Various features in the oxygen partial DOS are labeled according to the molecular level diagram drawn schematically for an oxygen molecule in Figure 11.1, except that doubly degenerate $\pi_a(\pi_b)$ levels are split into $p_x^a\left(p_x^b\right)$ and $p_y^a\left(p_y^b\right)$ levels, and these have been explicitly shown.

density of the feature in the DOS, which is found in the energy window between −0.8 and 0 eV, is shown in Figure 11.7b through e, corresponding to the plane at $c = 0$, $c = 0.25$, $c = 0.5$, and $c = 0.75$ of Figure 11.7a. It is not unusual for structural distortions to lead to such OO.

The consequence of OO on magnetism is dramatic. Earlier, we had large stabilization energies for different types of magnetic order for most values of U, except near U_c. This, however, vanishes; for all values of U considered here, we find that all magnetic states have very close energies and exhibit the same OO. Thus, OO is robust in the low-symmetry phase. We find that all magnetic configurations have collapsed and lie at an interval of ~1.5 meV, implying that long-range spin–spin correlation is absent in the low-symmetry phase. Thus, OO destroys any hope of high ordering temperatures, which is the possible reason why KO_2 has a very low magnetic ordering temperature of 7.1 K.

Kugel and Khomskii proposed a model in which there is a one-to-one correspondence between orbital occupancy and spin alignment (i.e., a particular pattern of magnetic configuration is most favored because of OO). For example, in the parent compound of manganites, $LaMnO_3$, the occupancies of the Mn d state are $t_{2g}\uparrow$

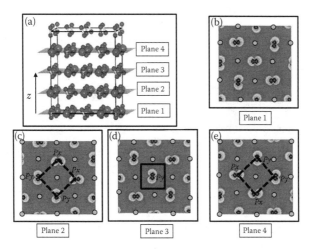

FIGURE 11.7 (a) The theoretically optimised structure as well as the real space charge density for planes (b) 1 (c) 2 (d) 3 and (e) 4 calculated in the energy window – 0.8 eV to 0. The oxygens in planes 2 and 4 which are the nearest neighbors of the oxygen enclosed by the box in plane 3 have been indicated by dashed solid lines.

and $e_g\uparrow$ due to octahedral symmetry. The electron at the level with e_g symmetry, for example, is in a level split by Jahn–Teller distortions. Considering the nearest-neighbor Mn–Mn hopping interactions in the Kugel–Khomskii model, one finds that a ferro arrangement of spins and an anti–ferro-orbital arrangement are favored from superexchange considerations. We set up a tight-binding Hamiltonian with a coupling between the nearest-neighbor oxygen dimers. With an appropriate choice of basis, we find a coupling between like orbitals only in neighboring sites and no cross-couplings. Thus, if the occupied orbital on one oxygen dimer is p_y (indicated by the small square in Figure 11.7d), then the p_x orbitals would be most favored on the neighboring oxygen dimers to maximize the energy gain from superexchange interactions. Here, we find that of the eight neighbors (indicated by the dashed square in Figure 11.7c and e), four have p_x orbitals occupied, while four others have p_y orbitals occupied. This OO is insensitive to the underlying magnetic state considered. Hence, electrostatic considerations that determine lattice distortions and, hence, OO lead to frustration with spin alignment.

11.4 CONCLUSION

Alkali metal superoxides have been examined as candidate systems for spin and orbital physics, similar to what one encounters among $3d$ TM oxides. Considering the case of KO_2, we find an interesting possibility of an OO transition, which precludes the possibility of high magnetic ordering temperatures in this system. The mechanism that we found for OO is different from the Kugel–Khomskii model used to explain OO in manganites and other $3d$ TM compounds.

REFERENCES

1. M. Imada, A. Fujimori, and Y. Tokura, Metal-insulator transitions, *Rev. Mod. Phys.* 70 (1998), 1039–1263.

2. S. Larochelle, A. Mehta, N. Kaneko, P. K. Mang, A. F. Panchula, L. Zhou, J. Arthur, and M. Greven, Nature of e_g electron order in $La_{1-x}Sr_{1+x}MnO_4$, *Phys. Rev. Lett.* 87 (2001), 095502–095505.

3. S. B. Wilkins, P. D. Spencer, P. D. Hatton, S. P. Collins, M. D. Roper, D. Prabhakaran, and A. T. Boothroyd, Direct observation of orbital ordering in $La_{0.5}Sr_{1.5}MnO_4$ using soft X-ray diffraction, *Phys. Rev. Lett.* 91 (2003), 167205–167208.

4. P. Mahadevan, K. Terakura, and D. D. Sarma, Spin, charge, and orbital ordering in $La_{0.5}Sr_{1.5}MnO_4$, *Phys. Rev. Lett.* 87 (2001), 066404–066407.

5. P. Mahadevan, A. Kumar, D. Choudhury, and D. D. Sarma, Charge ordering induced ferromagnetic insulator: $K_2Cr_8O_{16}$, *Phys. Rev. Lett.* 104 (2010), 256401–256404.

6. S. Middey, P. Mahadevan, and D. D. Sarma, Dependence of magnetism on $GdFeO_3$ distortion in the t_{2g} system $ARuO_3$ (A = Sr, Ca), *Phys. Rev. B* 83 (2011), 014416–014420.

7. P. Mahadevan, F. Aryasetiawan, A. Janotti, and T. Sasaki, Evolution of the electronic structure of a ferromagnetic metal: case of $SrRuO_3$, *Phys. Rev. B* 80 (2009), 035106–035109.

8. G. C. DeFotis, Magnetism of solid oxygen, *Phys. Rev. B* 23 (1981), 4714–4740.

9. H. Hesse, M. Jansen, and W. Schnick, Recent results in solid state chemistry of ionic ozonides, hyperoxides, and peroxides, *Prog. Solid State Chem.* 19 (1989), 47–110.

10. R. Kovacik and C. Ederer, Correlation effects in *p*-electron magnets: electronic structure of RbO_2 from first principles, *Phys. Rev. B* 80 (2009), 140411–140414.

11. A. K. Nandy, P. Mahadevan, P. Sen, and D. D. Sarma, KO_2: Realization of orbital ordering in a *p*-orbital system, *Phys. Rev. Lett.* 105 (2010), 056403–056406.

12. E. R. Ylvisaker, R. R. P. Singh, and W. E. Pickett, Orbital order, stacking defects, and spin fluctuations in the *p*-electron molecular solid RbO_2, *Phys. Rev. B.* 81 (2010), 180405–180408.

13. M. Kim, B. H. Kim, H. C. Choi, and B. I. Min, Antiferromagnetic and structural transitions in the superoxide KO_2 from first principles: a $2p$-electron system with spin–orbital–lattice coupling, *Phys. Rev. B* 81 (2010), 100409–100412.

14. J.-S. Kang, D. H. Kim, J. H. Hwang, J. Baik, H. J. Shin, M. Kim, Y. H. Jeong, and B. I. Min, Soft X-ray absorption and photoemission spectroscopy study of superoxide KO_2, *Phys. Rev. B* 82 (2010), 193102–193105.

15. J. Winterlik, G. H. Fecher, C. Felser, C. Mühle, and M. Jansen, Challenging the prediction of anionogenic ferromagnetism for Rb_4O_6, *J. Am. Chem. Soc.* 129 (2007), 6990–6991.

16. J. Winterlik, G. H. Fecher, C. A. Jenkins, C. Felser, C. Mühle, K. Doll, M. Jansen, L. M. Sandratskii, and J. Kubler, Challenge of magnetism in strongly correlated open-shell $2p$ systems, *Phys. Rev. Lett.* 102 (2009), 016401–016404.

17. S. Riyadi, S. Giriyapura, R. A. de Groot, A. Caretta, P. H. M. van Loosdrecht, T. T. M. Palstra, and G. R. Blake, Ferromagnetic order from *p*-electrons in rubidium oxide, *Chem. Mater.* 23 (2011), 1578–1586.

18. K. I. Kugel and D. I. Khomskii, The Jahn–Teller effect and magnetism: transition metal compounds, *Sov. Phys. Usp.* 25(4) (1982), 231–254.

19. F. Halverson, Comments on potassium superoxide structure, *J. Phys. Chem. Solids* 23 (1962), 207–214.

20. A. K. Nandy, P. Mahadevan, and D. D. Sarma, K_2O_2: the most stable oxide of K, *Phys. Rev. B* 84 (2011), 035116–035120.

21. G. Kresse and J. Hafner, Ab initio molecular dynamics for open-shell transition metals, *Phys. Rev. B* 48 (1993), 13115–13118.

22. G. Kresse and J. Furthmuller, Efficient iterative schemes for ab initio total-energy calculations using a plane-wave basis set, *Phys. Rev. B* 54 (1996), 11169–11186.
23. G. Kresse and J. Furthmuller, Efficiency of ab-initio total energy calculations for metals and semiconductors using a plane-wave basis set [original research article], *Comput. Mater. Sci.* 6 (1996), 15–50.
24. P. E. Blochl, Projector augmented-wave method, *Phys. Rev. B* 50 (1994), 17953–17979.
25. G. Kresse and D. Joubert, From ultrasoft pseudopotentials to the projector augmented-wave method, *Phys. Rev. B* 59 (1999), 1758–1775.
26. J. P. Perdew, In *Electronic Structure of Solids '91*, edited by P. Ziesche and H. Eschrig, 11 (Akademie Verlag, Berlin, 1991).
27. J. P. Perdew, J. A. Chevary, S. H. Vosko, K. A. Jackson, M. R. Pederson, D. J. Singh, and C. Fiolhais, Atoms, molecules, solids, and surfaces: applications of the generalized gradient approximation for exchange and correlation, *Phys. Rev. B* 46 (1992), 6671–6687.
28. S. L. Dudarev, G. A. Botton, S. Y. Savrasov, C. J. Humphreys, and A. P. Sutton, Electron energy-loss spectra and the structural stability of nickel oxide: an LSDA + U study, *Phys. Rev. B* 57 (1998), 1505–1509.
29. J. Ghijsen, L. H. Tjeng, J. van Elp, H. Eskes, J. Westerink, G. A. Sawatzky, and M. T. Czyzyk, Electronic structure of Cu_2O and CuO, *Phys. Rev. B* 38 (1988), 11322–11330.
30. A. Chainani, M. Mathew, and D. D. Sarma, Electron-spectroscopy study of the semiconductor–metal transition in $La_{1-x}Sr_xCoO_3$, *Phys. Rev. B* 46 (1992), 9976–9983.
31. A. Chainani, M. Mathew, and D. D. Sarma, Electron spectroscopic investigation of the semiconductor–metal transition in $La_{1-x}Sr_xMnO_3$, *Phys. Rev. B* 47 (1993), 15397–15403.
32. NIST-JANAF Thermochemical Tables 1985.

12 Electronic Stress with Spin Vorticity

Akitomo Tachibana

CONTENTS

12.1 INTRODUCTION

The study of the density functional theory by Hohenberg–Kohn has invoked the new idea of energy in terms of electron density.[1] A natural outcome is the concept of energy density, which has been published in a recent paper of stress tensor and the references cited therein.[2] Electronic stress tensor plays a very significant role as has been originally formulated and reviewed by Pauli[3] in his textbook of quantum mechanics.

It has been shown in the preceding paper that the QED electronic stress tensor plays a fundamentally important role in order to understand the electron spin dynamics; the spin torque and zeta force originated from the chiral nature of electron that is intrinsic to the spin-1/2 Fermion.[4] The dynamics of electron spin with the realization of spin–orbit coupling has recently been of keen interest, particularly in the field of spin torque transfer in spintronics; see recent review and the references cited therein.[5]

It shall be formulated here the energy density concept in terms of stress tensor in general relativity for the unified treatment of spin dynamics and chemical reaction dynamics.

235

12.2 SPIN VORTICITY PRINCIPLE AND ENERGY DENSITY BY GENERAL RELATIVITY

12.2.1 VARIATION PRINCIPLE OF QED IN CURVED SPACETIME

In the variation principle of gravitation in this paper, the semiclassic Einstein–Hilbert gravitational action I_G is added to the system action I_S and made stationary[4]

$$\delta I = 0, \quad I = I_G + I_S. \tag{12.1}$$

First, for the variation $\delta g^{\mu\nu}$ of the symmetric metric tensor $g^{\mu\nu} = g^{\nu\mu}$

$$I_G = \frac{c}{2\kappa} \int R\sqrt{-g}\, d^4 x, \ \ \delta I_G = \frac{c}{2\kappa} \int \left(R_{\mu\nu} - \frac{1}{2} g_{\mu\nu} R \right) \delta g^{\mu\nu} \sqrt{-g}\, d^4 x \tag{12.2}$$

$$I_S = \frac{1}{c} \int L\sqrt{-g}\, d^4 x, \ \ \delta I_S = \frac{1}{2c} \int T_{\mu\nu} \delta g^{\mu\nu} \sqrt{-g}\, d^4 x, \tag{12.3}$$

the Einstein equation is obtained as

$$G_{\mu\nu} = Y_{\mu\nu}, \tag{12.4}$$

where

$$G_{\mu\nu} = R_{\mu\nu} - \frac{1}{2} g_{\mu\nu} R \tag{12.5}$$

$$Y_{\mu\nu} = -\frac{8\pi G}{c^4} T_{\mu\nu}. \tag{12.6}$$

Moreover, in the QED system, the variation principle leads to the Dirac equation of electron

$$\left(i\hbar \gamma^a e_a^\mu D_\mu(g) - mc \right) \psi = 0 \tag{12.7}$$

and the Maxwell equation of photon

$$\nabla_\nu F^{\mu\nu} = \frac{4\pi}{c} j^\mu, \tag{12.8}$$

with the continuity equation of current

$$\partial_\mu j^\mu = 0. \tag{12.9}$$

In terms of the vector potential, the field equation of photon is obtained:

$$\nabla^\kappa \nabla_\kappa A^\mu - R_\nu^\mu A^\nu - \nabla^\mu \nabla_\nu A^\nu = \frac{4\pi}{c} j^\mu. \tag{12.10}$$

Let the Coulomb gauge be given as

$$\nabla_k A^k = 0. \tag{12.11}$$

Then, the Laplace equation

$$\nabla^k \nabla_k A^0 - R_v^0 A^v = \frac{4\pi}{c} j^0 \tag{12.12}$$

and the d'Alembert equation

$$\nabla^\kappa \nabla_\kappa A^k - R_v^k A^v - \nabla^k \nabla_0 A^0 = \frac{4\pi}{c} j^k \tag{12.13}$$

are obtained. The longitudinal and transversal currents may be further introduced as

$$j^k = j_T^k + j_L^k, \tag{12.14}$$

in such a way that Equation 12.13 is reduced to a separable form

$$-\nabla^k \nabla_0 A^0 = \frac{4\pi}{c} j_L^k, \tag{12.15}$$

$$\nabla^\kappa \nabla_\kappa A^k - R_v^k A^v = \frac{4\pi}{c} j_T^k. \tag{12.16}$$

To solve for the retardation interactions in the Heisenberg representation, the initial value problem of Cauchy developed in the Minkowski spacetime may be generalized as shown in Appendix 12.A.

Now, the symmetric energy-momentum tensor $T_{\mu\nu}$ in Equation 12.6 is given as

$$T_{\mu\nu} = -\varepsilon_{\mu\nu}^\Pi - \tau_{\mu\nu}^\Pi(g) - \frac{1}{4\pi} g^{\rho\sigma} F_{\mu\rho} F_{\nu\sigma} - g_{\mu\nu}(L_{EM} + L_e) = T_{\nu\mu}, \tag{12.17}$$

$$T_{\mu\nu} = T_{EM\mu\nu} + T_{e\mu\nu}, \tag{12.18}$$

$$T_{EM\mu\nu} = -\frac{1}{4\pi} g^{\rho\sigma} F_{\mu\rho} F_{\nu\sigma} - g_{\mu\nu} L_{EM} = T_{EM\nu\mu}, \tag{12.19}$$

$$T_{e\mu\nu} = \varepsilon_{\mu\nu}^\Pi - \tau_{\mu\nu}^\Pi(g) - g_{\mu\nu} L_e = T_{e\nu\mu}, \tag{12.20}$$

where $\varepsilon_{\mu\nu}^\Pi$ denotes the symmetry-polarized geometrical tensor and $\tau_{\mu\nu}^\Pi(g)$ denotes the symmetry-polarized electronic stress tensor. $T_{\mu\nu}$ satisfies the conservation law

$$\nabla_\lambda T_\mu^\lambda = 0. \tag{12.21}$$

Also, the antisymmetric angular momentum tensor

$$M^{\lambda\mu\nu} = x^{\mu}T^{\lambda\nu} - x^{\nu}T^{\lambda\mu} = -M^{\lambda\nu\mu} \tag{12.22}$$

satisfies the conservation law

$$\partial_{\lambda}M^{\lambda\kappa\ell} = 0. \tag{12.23}$$

12.2.2 QUANTUM ELECTRON SPIN VORTICITY PRINCIPLE

Because $T_{\mu\nu}$ in Equation 12.17 is symmetric, the antisymmetric component of the geometrical tensor $\varepsilon^{A\mu\nu}$ and the antisymmetric component of the electronic stress tensor $\tau^{A\mu\nu}(g)$ should cancel each other out:

$$\varepsilon^{A\mu\nu} + \tau^{A\mu\nu}(g) = 0. \tag{12.24}$$

The physical meaning is revealed if one takes the limit to the Minkowski spacetime

$$e_{\mu}^{a} \rightarrow \delta_{\mu}^{a}, g_{\mu\nu} \rightarrow \eta_{\mu\nu}, \tag{12.25}$$

when the equation of motion of the Dirac spinor field ψ is reduced from Equation 12.7 to

$$(i\hbar\gamma^{\mu}D_{\mu} - mc)\psi = 0, \tag{12.26}$$

and then Equation 12.24 is reduced to a separable form as follows.[4] The first is the equation of motion of spin $\vec{s}$ with torque $\vec{t}$ and zeta force $\vec{\zeta}$:

$$\frac{\partial}{\partial t}\vec{s} = \vec{t} + \vec{\zeta}, \tag{12.27}$$

$$\vec{s} = \frac{1}{2}\hbar\vec{\sigma} = \frac{1}{2}\hbar\left(\vec{\sigma}_R + \vec{\sigma}_L\right) = \frac{1}{2}\hbar\frac{1}{cq}\vec{j}_5, \tag{12.28}$$

$$t^k = -\varepsilon_{\ell nk}\tau^{A\ell n}, \tag{12.29}$$

$$\zeta^k = -\partial_k\phi_5, \tag{12.30}$$

$$\phi_5 = \frac{\hbar}{2q}j_5^0 = \frac{\hbar c}{2}(N_R - N_L). \tag{12.31}$$

The second is the vorticity of spin:

$$\mathrm{rot}\vec{s} = \frac{1}{2}\left(\overline{\psi}\vec{\gamma}(i\hbar D_0)\psi + h.c.\right) - \vec{\Pi}. \tag{12.32}$$

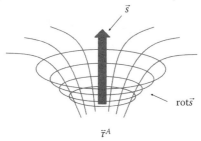

"Antisymmetric electronic stress tensor drives the electron spin through vorticity."

FIGURE 12.1   Quantum electron spin vorticity principle.

Because the vorticity is a solenoidal vector field, the spin $\vec{s}$ itself may be given by integration in the star-like domain using the rotation of torque $\vec{t}$ as the driving force:

$$\frac{\partial}{\partial t}\vec{s}(\vec{r}) = -\vec{r}\times\int_0^1 \text{rot}\frac{\partial}{\partial t}\vec{s}(\lambda\vec{r})\lambda\,d\lambda = -\vec{r}\times\int_0^1 \text{rot}\vec{t}\,(\lambda\vec{r})\lambda\,d\lambda, \qquad (12.33)$$

where the fact that the zeta force $\vec{\zeta}$ is an irrotational conservative vector field is used. Moreover, it should be noted that

$$\text{rot}\vec{t} = -2\text{div}\overleftrightarrow{\tau}^A. \qquad (12.34)$$

Hence, Equation 12.33 is reduced to

$$\vec{s}(t,\vec{r}) - \vec{s}(t_0,\vec{r}) = 2\vec{r}\times\int_{t_0}^t \left(\int_0^1 \text{div}\overleftrightarrow{\tau}^A(t',\lambda\vec{r})\lambda\,d\lambda\right)dt'. \qquad (12.35)$$

This is called the quantum electron spin vorticity principle: the time evolution of the electron spin $\vec{s}$ is driven by the antisymmetric component of the electronic stress tensor $\overleftrightarrow{\tau}^A$ through the vorticity rot$\vec{s}$. The quantum electron spin vorticity principle is schematically shown in Figure 12.1.

The applications to the other particles are also interesting. For example, for chiral spin-1/2 Fermion with the non-Abelian gauge potential, the analogous equation of motion of spin has been found.[2] Another example is the Majorana particle, which is neutral, and the spin properties are shown in Appendix 12.B.

12.2.3   ENERGY DENSITY

In the limit to the Minkowski spacetime, the electronic component of the energy-momentum tensor $T_{e\mu\nu}$ is reduced to

$$T_e^{\mu\nu} \rightarrow \begin{pmatrix} \frac{1}{2}(M + h.c.) & c\left(\vec{\Pi} + \frac{1}{2}\text{rot}\vec{s}\right)_x & c\left(\vec{\Pi} + \frac{1}{2}\text{rot}\vec{s}\right)_y & c\left(\vec{\Pi} + \frac{1}{2}\text{rot}\vec{s}\right)_z \\ c\left(\vec{\Pi} + \frac{1}{2}\text{rot}\vec{s}\right)_x & -\tau_{xx}^S + L_e & -\tau_{xy}^S & -\tau_{xz}^S \\ c\left(\vec{\Pi} + \frac{1}{2}\text{rot}\vec{s}\right)_y & -\tau_{yx}^S & -\tau_{yy}^S + L_e & -\tau_{yz}^S \\ c\left(\vec{\Pi} + \frac{1}{2}\text{rot}\vec{s}\right)_z & -\tau_{zx}^S & -\tau_{zy}^S & -\tau_{zz}^S + L_e \end{pmatrix},$$

$$(12.36)$$

with the mass term M.[6] Also, the electromagnetic component of the energy-momentum tensor $T_{\text{EM}\mu\nu}$ is reduced to

$$T_{\text{EM}}^{\mu\nu} \rightarrow \begin{pmatrix} H_\gamma & cG_x & cG_y & cG_z \\ cG_x & \sigma_{xx} & \sigma_{xy} & \sigma_{xz} \\ cG_y & \sigma_{yx} & \sigma_{yy} & \sigma_{yz} \\ cG_z & \sigma_{zx} & \sigma_{zy} & \sigma_{zz} \end{pmatrix}, \qquad (12.37)$$

with the Poynting vector $\vec{G}$ and the Maxwell stress tensor $\overleftrightarrow{\sigma}$. The conservation law of energy and momentum (Equation 12.21) is then reduced to

$$\nabla_\nu T^{\nu 0} = 0 \rightarrow \frac{\partial}{\partial t} cP^0 + c^2 \text{div}\vec{P} = 0, \qquad (12.38)$$

$$\nabla_\nu T^{\nu k} = 0 \rightarrow \frac{\partial}{\partial t}\vec{P} + \text{div}\left(\overleftrightarrow{\sigma} - \overleftrightarrow{\tau}^S\right) = 0, \qquad (12.39)$$

$$P^\mu = \left(\frac{\frac{1}{2}(M + h.c.) + H_\gamma}{c}, \ \vec{\Pi} + \frac{1}{2}\text{rot}\vec{s} + \vec{G} \right). \qquad (12.40)$$

The conservation law of angular momentum (Equation 12.23) is also reduced to

$$\partial_\lambda M^{\lambda k \ell} = 0 \rightarrow \frac{\partial}{\partial t}\vec{J} + \text{div}(\vec{r} \times (\overleftrightarrow{\sigma} - \overleftrightarrow{\tau}^S)) = 0, \qquad (12.41)$$

$$\frac{1}{c} M^{0 k \ell} \rightarrow \vec{J} = \vec{r} \times \vec{\Pi} + \vec{r} \times \frac{1}{2}\text{rot}\vec{s} + \vec{r} \times \vec{G}. \qquad (12.42)$$

It should be noted that the vorticity $\mathrm{rot}\vec{s}$ appears as a component of the electronic momentum as found in Equation 12.36. This proves the important role of the anti-symmetric component of the electronic stress tensor $\overleftrightarrow{\tau}^A$. It may be further proved that the symmetric component of the electronic stress tensor $\overleftrightarrow{\tau}^S$ plays an important role as tension $\vec{\tau}^S = \mathrm{div}\overleftrightarrow{\tau}^S$ compensating the Lorentz force $\vec{L}$ as

$$\frac{\partial}{\partial t}\left(\vec{\Pi}+\frac{1}{2}\mathrm{rot}\vec{s}\right)=\vec{L}+\vec{\tau}^S. \tag{12.43}$$

Finally, for a finite system, the virial theorem is used to obtain the total energy of the QED system as

$$\int d^3\vec{r}\left\langle T^{00}\right\rangle \to E_{\mathrm{QED}} = mc^2\int d^3\vec{r}\left\langle\overline{\psi}\psi\right\rangle. \tag{12.44}$$

12.3 RIGGED FIELD THEORY

In application to chemical reaction dynamics, the rigged QED theory is used where nuclear degrees of freedom are treated in a unified manner with QED.[2] The approximate treatment of the rigged QED shall be examined here and shall be called the primary rigged QED.

12.3.1 RIGGED QED

Because the nuclear Schrödinger fields have been plugged in, electron may be first distinguished by the obvious suffix "e" in such a way as

$$\left(i\hbar\gamma^a e^{\mu}_a \eta_{\mathrm{e}\mu}(g) - m_{\mathrm{e}}c\right)\psi_{\mathrm{e}} = 0. \tag{12.45}$$

The Schrödinger field equation of the ath nucleus is conventionally put onto the curved spacetime. The procedure is (1) ignore the spin connection[4] in $D_{\mathrm{e}\mu}(g)$ in Equation 12.7, (2) use the Dirac representation with ψ_{e} and approximate the small component as the multiplication of $-\dfrac{1}{2m_{\mathrm{e}}c}i\hbar\sigma^k e^{\mu}_k D_{\mathrm{e}\mu}$ to the large component, and (3) ignore again the spin-dependent terms in the resulting equation, $\sigma^k e^{\mu}_{\kappa} D_{\mathrm{e}\mu}\cdot\sigma^\ell e^{\nu}_\ell D_{\mathrm{e}\nu} \sim \left(e^{\mu}_\kappa D_{\mathrm{e}\mu}\right)^2$, leading to

$$\left(i\hbar e^{\mu}_0 D_{a\mu} - m_a c\right)\psi_a = \frac{(i\hbar)^2}{2m_a c}\left(e^{\mu}_k D_{a\mu}\right)^2\psi_a, \tag{12.46}$$

where the large component for electron is used here as ψ_a for the nuclear Schrödinger field. Note that the mass term is indispensable because it is needed as the source of gravitation. This may be identified as the Schrödinger field equation without a

priori spin and may be used to plug in nuclear degrees of freedom into the formalism, and it may be called the rigged QED in the curved spacetime. In the course to the Minkowski spacetime limit, this equation reduces to the usual Schrödinger field equation plus gravitational potential $m_a\Phi$ as shown in Appendix 12.C.

In the limit to the Minkowski spacetime, the Dirac field equation of electron (Equation 12.26) is combined with the Schrödinger field equation of the ath nucleus as follows:

$$(i\hbar\gamma^\mu D_{e\mu} - m_e c)\psi_e = 0, \tag{12.47}$$

$$i\hbar \frac{\partial}{\partial t}\chi_a = -\frac{\hbar^2}{2m_a}\vec{D}_a^2\chi_a + q_a A_0\chi_a, \tag{12.48}$$

where the gravitational potential $m_a\Phi$ in Equation 12.C3 is neglected (see Appendix 12.C). There is electron spin vorticity here but no spin vorticity for nuclei.

With obvious notation, we obtain the momentum conservation law

$$\frac{\partial}{\partial t}\vec{P}_{\text{Rigged QED}} = -\text{div}\left(\vec{\sigma} - \overleftrightarrow{\tau}_{\text{Rigged QED}}^S\right), \tag{12.49}$$

and the angular momentum conservation law

$$\frac{\partial}{\partial t}\vec{J}_{\text{Rigged QED}} = -\text{div}\left(\vec{r} \times \left(\vec{\sigma} - \overleftrightarrow{\tau}_{\text{Rigged QED}}^S\right)\right). \tag{12.50}$$

The virial theorem is again used to obtain

$$E_{\text{Rigged QED}} = m_e c^2 \int d^3\vec{r}\left\langle \bar{\psi}_e \psi_e\right\rangle - \int d^3\vec{r}\left\langle \sum_a T_a\right\rangle, \tag{12.51}$$

$$T_a = -\frac{\hbar^2}{2m_a} \cdot \frac{1}{2}\left(\chi_a^\dagger \vec{D}_a^2 \chi_a + h.c.\right). \tag{12.52}$$

12.3.2 PRIMARY RIGGED QED

Approximation to electron as with Equation 12.46 is made using symbols $\alpha = e, a$ collectively as

$$\left(i\hbar e_0^\mu D_{\alpha\mu} - m_\alpha c\right)\psi_\alpha = \frac{(i\hbar)^2}{2m_\alpha c}\left(e_k^\mu D_{\alpha\mu}\right)^2 \psi_\alpha. \tag{12.53}$$

In the limit to the Minkowski spacetime, approximation as of Equation 12.48 is further used:

$$ i\hbar \frac{\partial}{\partial t} \chi_\alpha = -\frac{\hbar^2}{2m_\alpha} \vec{D}_\alpha^2 \chi_\alpha + q_\alpha A_0 \chi_\alpha. \tag{12.54} $$

The spin vorticity of electron and the antisymmetric component of the electronic stress tensor have been lost. The momentum conservation law

$$ \frac{\partial}{\partial t} \vec{P}_{\text{Primary Rigged QED}} = -\text{div}\left(\overleftrightarrow{\sigma} - \overleftrightarrow{\tau}_{\text{Primary Rigged QED}}^{S} \right) \tag{12.55} $$

and the angular momentum conservation law

$$ \frac{\partial}{\partial t} \vec{J}_{\text{Primary Rigged QED}} = -\text{div}\left(\vec{r} \times \left(\overleftrightarrow{\sigma} - \overleftrightarrow{\tau}_{\text{Primary Rigged QED}}^{S} \right) \right) \tag{12.56} $$

are then obtained. The virial theorem is then reduced from Equation 12.51 to

$$ E_{\text{Primary Rigged QED}} = -\int d^3\vec{r} \left\langle \sum_\alpha T_\alpha \right\rangle = \frac{1}{2} \int d^3\vec{r} \left\langle \tau_{\text{Primary Rigged QED}}^{Skk} \right\rangle. \tag{12.57} $$

Here, it should be noted that the symmetric stress tensor $\overleftrightarrow{\tau}_{\text{Primary Rigged QED}}^{S}$ gives the tensorial energy density. The eigenvalue of the symmetric stress tensor is the principal stress and the eigenvector is the principal axis as follows:

$$ \overleftrightarrow{\tau}_\alpha^{S} = \begin{pmatrix} \tau_\alpha^{S11} & \tau_\alpha^{S12} & \tau_\alpha^{S13} \\ \tau_\alpha^{S21} & \tau_\alpha^{S22} & \tau_\alpha^{S23} \\ \tau_\alpha^{S31} & \tau_\alpha^{S32} & \tau_\alpha^{S33} \end{pmatrix} \xrightarrow{\text{diag}} \begin{pmatrix} \tau_\alpha^{S11} & 0 & 0 \\ 0 & \tau_\alpha^{S22} & 0 \\ 0 & 0 & \tau_\alpha^{S33} \end{pmatrix}, \quad \tau_\alpha^{S11} \leq \tau_\alpha^{S22} \leq \tau_\alpha^{S33}. \tag{12.58} $$

As shown in Figure 12.2, the electronic tensile stress is visualized as the spindle structure binding a pair of the electronic drop regions R_D's separated from each other with the interface S through the electronic atmosphere region R_A.[2] The spindle structure is mathematically proved to appear at any region where the new Lewis electron pair is formed in association with in-phase overlap of orbitals, such as between a pair of H atoms.[7] The spindle structure is hidden where out-of-phase overlap of orbitals overwhelms the former, such as between a pair of He atoms,[7] where the compressive stress pushes back electron in the remote electronic drop region R_D from the adjacent electronic atmosphere region R_A through the interface S that separates them. The consequence is the no formation of the new Lewis pair of electrons. There appears no spindle structure. The spindle structure is also hidden where a pair of atomic nuclei is so closely combined, such as between a pair of C atoms in C_2H_2.[6] Because R_D, R_A, and S are measures of the kinetic energy density $n_{T_e} = \langle T_e \rangle$, which define physically the intrinsic shape of atoms and molecules, they

"Electronic tensile stress binds a pair of the electronic drop regions R_D's, where the compressive stress is predominant: covalent bond visualization."
"Metallicity emerges as the long-range intrinsic electronic transition state associated with the spindle structure: the long-range Lewis pair formation."

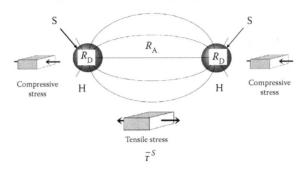

FIGURE 12.2   Spindle structure of the covalent bond and the long-range Lewis pair formation.

are also used to define the intrinsic electronic transition state along the course of the reaction coordinate.[8] For example, in the ground state of hydrogen-like atom, the kinetic energy density is positive around nucleus and negative outside beyond $r > r_S$: if the Schrödinger wave function, the Dirac large component function, and the Dirac wave function are used, then r_S is $2\dfrac{\hbar}{m_e c Z\alpha}$, $\dfrac{\hbar}{m_e c Z\alpha}\left(\sqrt{1-(Z\alpha)^2} + \sqrt[4]{1-(Z\alpha)^2}\right)$ (primary rigged QED), and $\dfrac{\hbar}{m_e c Z\alpha}\left(\sqrt{1-(Z\alpha)^2} + 1\right)$, respectively.

Covalency is the rule of the Lewis electron pair formation. Unlike covalency, metallicity may not be measured by a single bond order. In case of metallic interaction, imagine a pair of metallic atoms with unpaired electron situated far away from each other. The tensile stress pulls up electron in the remote electronic drop region R_D to the adjacent electronic atmosphere region R_A through the interface S that separates them. The consequence is the formation of the new Lewis pair of electron. The spindle structure of covalency is universal even in this sense. However, this fact demonstrates that metal atom itself may not be sufficient to determine metallicity, where the question itself may be even meaningless. This is because metallicity may be the property of the condensed matter. Actually, as the distant pair of metallic atoms comes closer, the metallicity of the condensed matter is the rule of unbinding the Lewis electron pair ever once formed. Finally, in the condensed matter, the spindle structure of covalency may not be observed in its strict sense. However, if an atom is going to be separated from bulk metal, then the spindle structure should emerge. This proves the emergence of covalency prerequisite to condense separated atoms into the bulk metal. The metallicity may be characterized by liquid with isotropic compressive stress in the ultimate case. The electrons contributing to the bulk metallicity behave like gluon that binds quarks in such a way that in metal bulk, as a condensed matter, the bond order may be small and behave as a weak bond, but if an atom is going to be separated from the bulk metal, the spindle structure appears as if the bond should behave to be very strong. In other words, the bulk metallicity

Repulsive electronic tension drives the quantum mechanical electronic diffusion.

Kinetic energy density T defines the intrinsic shape of atoms and molecules

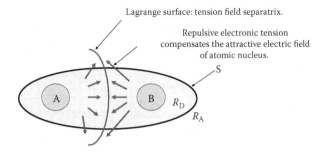

FIGURE 12.3   Local equilibrium in the stationary state with the Lagrange surface.

emerges as the long-range intrinsic electronic transition state associated with the spindle structure: the long-range Lewis pair formation.

In a molecule AB composed of atoms A and B, a universal local equilibrium picture of electronic stationary state is shown in Figure 12.3. The Heisenberg uncertainty principle allows electron to diffuse away from each atomic center it belongs. The diffusive force is the tension $\vec{\tau}^S = \text{div}\overset{\leftrightarrow}{\tau}^S$ compensating the Lorentz force $\vec{L}$ exerting from each atomic center as shown in Equation 12.43. The tension vector fields originated from A and B mutually collide to form separatrix that discriminates each region of atomic center. The separatrix is called the Lagrange surface.[2]

The total angular momentum is conserved as in Equation 12.41; therefore, locally, the particle currents should be deflected in a finite system. This is numerically demonstrated as the complex eigenvalue of the electronic dielectric constant $\overset{\leftrightarrow}{\varepsilon}(\vec{r})$.[9] The complex eigenvalues of $\overset{\leftrightarrow}{\varepsilon}(\vec{r})$, magnetic permeability $\overset{\leftrightarrow}{\mu}(\vec{r})$, effective charge $\overset{\leftrightarrow}{Z}(\vec{r})$, and electric conductance $\overset{\leftrightarrow}{\sigma}(\vec{r})$ mean rotational deflected response of electron toward applied electromagnetic field.[2]

Photon deflection is of course realized by the index of refraction

$$\overset{\leftrightarrow}{n}(\vec{r}) = \sqrt{\overset{\leftrightarrow}{\mu}(\vec{r}) \cdot \overset{\leftrightarrow}{\varepsilon}(\vec{r})}. \tag{12.59}$$

Chiral structure affects the spin density of photon in the wave zone, which is the helicity density. In other words, if atoms and molecules are radiated, then the electrons are affected by torque leading to the imbalance between spin torque and zeta force established in the stationary state. The reaction affects the torque on photon, leading to circular dichroism, the Kerr effect, and the Faraday effect. Forbidden processes may of course be realized due to symmetry of ket vectors.

12.4   CONCLUSION

The energy density concept[2] has been formulated here in terms of stress tensor in general relativity. Electron spin vorticity has been hidden in the energy-momentum tensor and plays a significant role in the dynamics of electron. The time evolution of the electron spin emergence is driven by the antisymmetric component of the electronic stress tensor through the vorticity. This is called the "quantum electron spin vorticity principle." The symmetric component of the electronic stress tensor drives the tensorial energy density of chemical reactivity.

It is the relativity theory of spin from which inherent spin–orbit coupling is realized between spin and orbital-angular momentum. Imagine a bulk magnet. The magnet is phenomenologically composed of many magnetic domains. Each magnetic domain has its unique spin, which is the average of spin density within the domain. For the sake of simplicity, let first electron spin density respond to an applied magnetic field (or even an applied electric field in some cases in recent spintronics) and change through the domain wall, which is called "spin torque transfer" in the experiments of spintronics industry (note the simplification). The prediction is that the spin torque does exist even in the stationary state when the spin torque is exactly cancelled out with the zeta force. In the nonstationary state, however, the external magnetoelectric perturbation disturbs the intrinsic balance between the spin torque and the zeta force established in the stationary state.

Of course, realistically, in addition to the spin of the electrons, the complexity origin of the magnetic spin can be from the motion of either electrons or nuclei, where the spin torque combinations totally can be treated by the equations of motion of angular momentums augmented by the *ad hoc* nuclear spin or more fundamentally the quark spin with the non-Abelian gauge, which were all formulated in the preceding papers.[2]

For future technology of spintronics and photonics, the interaction of chirality of electron spin with another particle, such as nucleus and photon, should play an important role. Furthermore, the general relativity has recently been of vital importance to our daily lives, particularly for ultra-high-precision communication with an artificial satellite (e.g., GPS). The intrinsic formulations of the quantum electron spin vorticity principle and the energy density concept presented in this paper should help us understand the importance of stress tensor in modeling of materials of technological importance and chemical reactions.

ACKNOWLEDGMENTS

This work is supported by Grants-in-Aid for Scientific Research on Priority Areas from the Ministry of Education, Culture, Sports, Science and Technology, Japan.

APPENDICES

Appendix 12.A

In a QED system, all fields are assumed to have definite noninteracting "in" and "out" states asymptotically. In the Cauchy problem of the Maxwell equation in the Minkowski spacetime, integration with respect to $\vec{s}$ at the retarded time $u = t - \dfrac{|\vec{r} - \vec{s}|}{c}$ is performed obeying causality

$$j^{\mu}(\vec{s},u) = 0, \quad u > t \tag{12.A1}$$

under the initial condition

$$j^{\mu}(\vec{s},u) = 0, \quad u < 0. \tag{12.A2}$$

The field in QED is then renormalized step by step, reflecting the time-dependent electron-photon coupling constant.

Any function $F(u)$ satisfying

$$F(u) = 0, \, u < 0, \, u > t \tag{12.A3}$$

may be obtained at $0 < u = t - \dfrac{\left| \vec{r} - \vec{s} \right|}{c} < t$ as

$$F(u)\Big|_{u=t} \frac{\vec{r} - \vec{s}}{c} = \int_{-\infty}^{\infty} du' F(u') \delta \left(u' - \left(t - \frac{\left| \vec{r} - \vec{s} \right|}{c} \right) \right)$$

$$\tag{12.A4}$$

$$= \frac{\left| \vec{r} - \vec{s} \right|}{c\pi} \int_0^t du' \int_{-\infty}^{\infty} d\alpha F(u') e^{i\alpha \left((u'-t)^2 - \frac{(\vec{r}-\vec{s})^2}{c^2} \right)},$$

where the delta function has been used:

$$\delta \left(x^2 - a^2 \right) = \frac{1}{2a} \big(\delta(x-a) + \delta(x+a) \big), \quad a > 0 \tag{12.A5}$$

with

$$\delta \left((u'-t)^2 - \frac{(\vec{r}-\vec{s})^2}{c^2} \right) = \frac{1}{2\pi} \int_{-\infty}^{\infty} d\alpha e^{i\alpha \left((u'-t)^2 - \frac{(\vec{r}-\vec{s})^2}{c^2} \right)}. \tag{12.A6}$$

It follows that

$$\vec{A}(\vec{r},t) = \vec{A}_{\text{radiation}}(\vec{r},t) + \frac{1}{c} \int_{-\infty}^{\infty} du' \int d^3\vec{s} \frac{\vec{j}_T(\vec{s},u')}{\left| \vec{r} - \vec{s} \right|} \delta \left(u' - \left(t - \frac{\left| \vec{r} - \vec{s} \right|}{c} \right) \right)$$

$$= \vec{A}_{\text{radiation}}(\vec{r},t) + \frac{1}{c^2\pi} \int_0^t du' \int_{-\infty}^{\infty} d\alpha \int d^3\vec{s} \vec{j}_T(\vec{s},u') \exp \left(i\alpha \left((u'-t)^2 - \frac{(\vec{r}-\vec{s})^2}{c^2} \right) \right)$$

$$\tag{12.A7}$$

in the rigged QED Cauchy problem.[10]

Appendix 12.B

Currents j^μ or j_5^μ for a neutral particle are $\dfrac{1}{q}$ times those for a charged particle: like the Majorana particle, the chargeless currents are $j^0 = cN = c(N_R + N_L)$ and $\vec{j} = c(\vec{\sigma}_R - \vec{\sigma}_L)$ or $j_5^0 = c(N_R - N_L)$ and $\vec{j}_5 = c\vec{\sigma} = c(\vec{\sigma}_R + \vec{\sigma}_L)$.

The free Majorana particle obeying

$$\left(i\hbar \partial^v \left(\sigma_v\right)^{A\dot{U}} \pm m_L e^{i\delta_L} c e^{AU} K\right)\eta_{\dot{U}} = 0 \tag{12.B1}$$

$$\left(i\hbar \partial_v \left(\sigma^v\right)_{\dot{U}A} \mp m_R e^{i\delta_R} c e_{\dot{U}\dot{A}} K\right)\xi^A = 0 \tag{12.B2}$$

should satisfy the Klein–Gordon equation

$$\left((i\hbar \partial)^2 - (m_L c)^2\right)\eta_{\dot{U}} = 0 \tag{12.B3}$$

$$\left((i\hbar \partial)^2 - (m_R c)^2\right)\xi^A = 0 \tag{12.B4}$$

or the Dirac spinor representation[11]

$$(i\hbar \partial \pm m_L e^{i\delta_L} c)\psi_{M_1} = 0, \quad \psi_{M_1} = \begin{pmatrix} e^{AW} K\eta_{\dot{W}} \\ \eta_{\dot{U}} \end{pmatrix} \tag{12.B5}$$

$$(i\hbar \partial \pm m_R e^{i\delta_R} c)\psi_{M_2} = 0, \quad \psi_{M_2} = \begin{pmatrix} \xi^A \\ -e_{\dot{U}\dot{B}} K\xi^B \end{pmatrix}. \tag{12.B6}$$

Under the charge conjugation transformation, the following are obtained:

$$\psi_{M_1^c} = C\gamma^0 K\psi_{M_1} = -\psi_{M_1}. \tag{12.B7}$$

$$\psi_{M_2^c} = C\gamma^0 K\psi_{M_2} = -\psi_{M_2}. \tag{12.B8}$$

Another representation

$$\Psi_M = \begin{pmatrix} \xi^A \\ \eta_{\dot{U}} \end{pmatrix}, \tag{12.B9}$$

$$\Psi_{M^c} = C\gamma^0 K\Psi_M = (-)\begin{pmatrix} e^{AW} K\eta_{\dot{W}} \\ -e_{\dot{U}\dot{B}} K\xi^B \end{pmatrix}, \tag{12.B10}$$

should obey

$$i\hbar\partial\Psi_M \pm m_M c\left(-\right)\Psi_{M^c} = 0, \qquad (12.\text{B}11)$$

$$m_M c = \begin{pmatrix} m_L e^{i\delta_L} c & 0 \\ 0 & m_R e^{i\delta_R} c \end{pmatrix}. \qquad (12.\text{B}12)$$

Using ψ_{M_1}, the position probability density

$$\left(N_{M_1}\right)_R = \eta_2\eta_2^* + \eta_1\eta_1^*, \qquad (12.\text{B}13)$$

$$\left(N_{M_1}\right)_L = \eta_1^*\eta_1 + \eta_2^*\eta_2, \qquad (12.\text{B}14)$$

and the spin density

$$(\vec{\sigma}_{M_1})_R = \left(-\eta_1\eta_2^* - \eta_2\eta_1^*, -i\eta_1\eta_2^* + i\eta_2\eta_1^*, \eta_2\eta_2^* - \eta_1\eta_1^*\right), \qquad (12.\text{B}15)$$

$$(\vec{\sigma}_{M_1})_L = \left(\eta_2^*\eta_1 + \eta_1^*\eta_2, i\eta_2^*\eta_1 - i\eta_1^*\eta_2, \eta_1^*\eta_1 - \eta_2^*\eta_2\right), \qquad (12.\text{B}16)$$

are obtained. Also, for ψ_{M_2}, the position probability density

$$\left(N_{M_2}\right)_R = \xi^{1*}\xi^1 + \xi^{2*}\xi^2, \qquad (12.\text{B}17)$$

$$\left(N_{M_2}\right)_L = \xi^2\xi^{2*} + \xi^1\xi^{1*}, \qquad (12.\text{B}18)$$

and the spin density

$$\left(\vec{\sigma}_{M_2}\right)_R = \left(\xi^{2*}\xi^1 + \xi^{1*}\xi^2, i\xi^{2*}\xi^1 - i\xi^{1*}\xi^2, \xi^{1*}\xi^1 - \xi^{2*}\xi^2\right), \qquad (12.\text{B}19)$$

$$\left(\vec{\sigma}_{M_2}\right)_L = \left(-\xi^1\xi^{2*} - \xi^2\xi^{1*}, -i\xi^1\xi^{2*} + i\xi^2\xi^{1*}, \xi^2\xi^{2*} - \xi^1\xi^{1*}\right), \qquad (12.\text{B}20)$$

are obtained.

Now, first, assume that $\psi_{M_{1,2}}$ are the usual complex numbers, not the Grassmann numbers; then

$$\left(N_{M_{1,2}}\right)_R = \left(N_{M_{1,2}}\right)_L, \qquad (12.\text{B}21)$$

$$\left(\vec{\sigma}_{M_{1,2}}\right)_R = -\left(\vec{\sigma}_{M_{1,2}}\right)_L, \qquad (12.\text{B}22)$$

and hence the null zeta potential and null spin density

$$N_{M_{1,2}} = 2\left(N_{M_{1,2}}\right)_L \tag{12.B23}$$

$$\phi_{5_{M_{1,2}}} = \frac{\hbar c}{2}\left(\left(N_{M_{1,2}}\right)_R - \left(N_{M_{1,2}}\right)_L\right) = 0, \tag{12.B24}$$

$$\vec{\sigma}_{M_{1,2}} = \left(\vec{\sigma}_{M_{1,2}}\right)_R + \left(\vec{\sigma}_{M_{1,2}}\right)_L = \vec{0}, \tag{12.B25}$$

are obtained.

On the other hand, if $\psi_{M_{1,2}}$ are the Grassmann numbers, then the following are obtained:

$$\left(N_{M_{1,2}}\right)_R = 2\delta_{M_{1,2}} - \left(N_{M_{1,2}}\right)_L, \tag{12.B26}$$

$$\left(\vec{\sigma}_{M_{1,2}}\right)_R = \left(\vec{\sigma}_{M_{1,2}}\right)_L + \vec{\delta}_{M_{1,2}}, \tag{12.B27}$$

and hence the non-null zeta potential and non-null spin density

$$N_{M_{1,2}} = 2\delta_{M_{1,2}} \tag{12.B28}$$

$$\phi_{5_{M_{1,2}}} = \hbar c\left(\delta_{M_{1,2}} - \left(N_{M_{1,2}}\right)_L\right) = -\hbar c\left(\delta_{M_{1,2}} - \left(N_{M_{1,2}}\right)_R\right) \tag{12.B29}$$

$$\vec{\sigma}_{M_{1,2}} = 2\left(\vec{\sigma}_{M_{1,2}}\right)_L + \vec{\delta}_{M_{1,2}} = 2\left(\vec{\sigma}_{M_{1,2}}\right)_R - \vec{\delta}_{M_{1,2}} \tag{12.B30}$$

with

$$2\delta_{M_1} = \left\{\eta_1, \eta_1^*\right\} + \left\{\eta_2, \eta_2^*\right\}, \quad 2\delta_{M_2} = \left\{\xi^1, \xi^{1*}\right\} + \left\{\xi^2, \xi^{2*}\right\}. \tag{12.B31}$$

In other words, the spatial components of the current $j_{M_{1,2}}^\mu(x)$ should be proportional to $\vec{\delta}_{M_{1,2}}$.

Appendix 12.C

Put

$$\psi_a = \chi_a e^{-im_a c^2 t/\hbar} \tag{12.C1}$$

in Equation 12.46 under the weak gravitation condition with only non-Kronecker delta insertion

$$e_0^0 \sim \frac{1}{\sqrt{1 + 2\dfrac{\Phi}{c^2}}} \sim 1 - \frac{\Phi}{c^2}. \tag{12.C2}$$

Then, the following is obtained:

$$i\hbar \frac{\partial}{\partial t}\chi_a = \left(-\frac{\hbar^2}{2m_a}\vec{D}_a^2 + q_a A_0 + m_a \Phi \right)\chi_a. \tag{12.C3}$$

This is the correct Schrödinger equation of neutron, if Φ is identified as the gravitation potential

$$i\hbar \frac{\partial}{\partial t}\chi_{\text{neutron}} = \left(-\frac{\hbar^2}{2m_{\text{neutron}}}\Delta + m_{\text{neutron}}\Phi \right)\chi_{\text{neutron}}. \tag{12.C4}$$

That Equation 12.C4 is correct has been demonstrated experimentally using a neutron interferometer.[12]

REFERENCES

1. Parr, R. G. and Yang, W., *Density Functional Theory of Atoms and Molecules*, Oxford University, New York, 1989.
2. Tachibana, A., *Journal of Molecular Structure: THEOCHEM*, 943, 138–151, 2010.
3. Pauli, W., *Handbuch der Physik, Band XXIV, Teil 1*, Springer, Berlin, Germany, 1933, 83–272, reprinted in *Handbuch der Physik, Vol. 5, Part 1*, Springer, Berlin, Germany, 1958, translated into English in *General Principles of Quantum Mechanics*, Springer, Berlin, Germany, 1980.
4. Tachibana, A., *Journal of Mathematical Chemistry*, 50, 669–688, 2012.
5. Ralph, D. C. and Stiles M. D., *Journal of Magnetism and Magnetic Materials*, 320, 1190–1216, 2008.
6. Tachibana, A., *Journal of Molecular Modeling*, 11, 301–311, 2005.
7. Tachibana, A., *International Journal of Quantum Chemistry*, 100, 981–993, 2004.
8. Tachibana, A., *Journal of Chemical Physics*, 115, 3497–3518, 2001.
9. Doi, K., Nakamura, K., and Tachibana, A., *2006 International Workshop on Nano CMOS*, IEEE, Piscataway, NJ, 2006, 209–235.
10. Tachibana, A., Field energy density in chemical reaction systems, in *Fundamental Perspectives in Quantum Chemistry: A Tribute to the Memory of Per-Olov Löwdin*, Brändas, E. and Kryachko, E., eds., Kluwer, Dordrecht, The Netherlands, 2003, volume 2, 211–239.
11. Weinberg, S., *The Quantum Theory of Fields*, Cambridge University, Cambridge, UK, 1995.
12. Collela, R., Overhauser, A. W., and Werner, S. A., *Physical Review Letters*, 34, 1472–1474, 1975.

13 Single Determinantal Approximations

Hartree–Fock, Optimized Effective Potential Theory, Density Functional Theory

Andreas K. Theophilou

CONTENTS

13.1 INTRODUCTION

All properties of atoms, molecules, and solids can be determined by solving the many-particle Schroedinger equation with external potential arising from the electric charges of the nuclei. The difficulty is that the state of the physical system under consideration is described by a wave function that has as many position variables as that of the number of electrons. At first sight, the situation looks hopeless because one atom may have a hundred of position variables, an organic molecule can have millions, and, for a piece of metal, the number is of the order of 10^{20}. However, after a better look into the Schroedinger equation, one realizes that this is a highly symmetric equation because, first of all, the electrons have the same mass and charge. As a consequence, the permutation of the position variables leaves the equation invariant. This fact implies that one can classify the solutions according to the transformation properties implied by the permutational symmetry. In more technical language, the solutions can be classified according to the irreducible representation (IrRep) of the permutation group. These categories of transformations are a multiple of $N!$, where N is the number of electrons of the system. Fortunately, only one such category exists in nature, and this is the class of antisymmetric solutions, where the interchange of two variables changes the sign of the wave function.

Now consider that, at certain time t_0, the system of all electrons is in the antisymmetric state. Group theory states that the system will be permanently in the antisymmetric state. Thus, instead of $N!$ classes of solutions, only one has to be considered. Then, one can make a proper formulation that considers only the space of antisymmetric wave functions. As shall be seen in a forthcoming section, this formulation is called second quantization representation and simplifies several solutions of the many-particle Schroedinger equation. Further, when no external magnetic field is present, there is spin symmetry and the eigenstates are classified according to the eigenstates of S^2 and S_z.

In addition to the above symmetries, many geometrical symmetries exist in the physical systems, such as rotational symmetry for atoms, point group symmetries for molecules, and translational and rotational symmetries for solids. These symmetries result to further simplification of the solutions of the many-particle Schroedinger equation.

For most problems, finding the ground state (lowest energy state) or other stationary states is of interest. These states are eigenstates of the Schroedinger equation and obey a variational principle. In particular, the ground state obeys the minimum energy principle. This principle facilitates further the solutions and helps in deriving approximations. One such approximation is that of Hartree–Fock (HF), in which one searches for the minimum energy not in the whole space of square integrable functions but restricts the search in the space of Slater determinants (Sldet). The minimization procedure leads to a set eigenvalue equations for the one-particle wave functions (orbitals) of the Sldet. However, the differential equation for the orbitals involves a nonlocal operator, and the problem becomes difficult to solve numerically. For this reason, a simplification was developed, in which the orbital equation is similar to the one-particle Schroedinger equation in an effective potential, which is chosen so that the expectation value of the exact Hamiltonian is minimized. For this reason, this approximation is called the optimized effective potential (OEP).

In general, the development of approximation methods for the solution of the many-electron Schroedinger equation is a challenge for physicists because no exact numerical solutions can be found apart from very few cases of a small number of electrons, such as the helium atom. The main difficulty arises because of the electron–electron interaction, which is a two-particle operator. Thus, increasing the accuracy of solutions implies increasing the computer time needed for the numerical calculations, and the cost becomes prohibitive even for molecules with a few atoms.

Then, for large molecules, one has to apply less accurate methods. A class of such methods restricts the approximation of the wave function to a single Sldet. Such approximations are the HF approximation and the OEP mentioned in the previous paragraph and density functional theory (DFT).

DFT was developed by Kohn and his collaborators [1, 2] and is based on a theorem that states that there is one-to-one correspondence between the exact ground state of the exact Hamiltonian and its electron density. Thus, one can express all terms of the energy in terms of the density and find its minimum. However, there are no good expressions (functionals) of the kinetic energy in terms of the density, and for this reason, Kohn and Sham (KS) [2] introduced a single Sldet whose density is equal to that of the exact one. The advantage is that they had a good approximation for the kinetic energy expressed in terms of this Sldet. The rest of the ground state energy was expressed in terms of the density. The most difficult part was that of the interaction term that has an electrostatic part derived from the classic electrostatic interaction $E_h(\rho)$ of electrons and an additional part called exchange and correlation energy, $E_{xc}(\rho)$. The difficult problem is to determine $E_{xc}(\rho)$ because $E_h(\rho)$ is known from classic electrostatics. Although there are many approximations for $E_{xc}(\rho)$, which give good accuracy for particular systems, a universal expression still does not exist.

In the next section, certain mathematical concepts shall be developed to facilitate the presentation.

13.1.1 MANY-ELECTRON SCHROEDINGER EQUATION

The time-dependent many-electron Schroedinger equation has the form

$$i \frac{\partial}{\partial t} \Psi(\mathbf{r}_1, \dots \mathbf{r}_N, t) = \left(\hat{T} + \hat{V} + \hat{H}_{int} \right) \Psi(\mathbf{r}_1, \dots \mathbf{r}_N, t), \tag{13.1}$$

where $\hat{T}$ is the kinetic energy operator

$$\hat{T} = -\frac{1}{2} \sum_{i=1}^{N} \nabla_i^2, \tag{13.2}$$

$\hat{V}$ is the external potential

$$\hat{V} = \sum_{i=1}^{N} V(\mathbf{r}_i), \tag{13.3}$$

and $\hat{H}_{int}$ is the electron–electron interaction term

$$\hat{H}_{int} = \frac{1}{2} \sum_{i \neq j=1,}^{N} \frac{e^2}{|\mathbf{r}_i - \mathbf{r}_j|}. \tag{13.4}$$

In the above equation, the spin variable was not included to avoid complications at this stage. Moreover, note that a single value of the wave function $\Psi(\mathbf{r}_1, \ldots \mathbf{r}_N, t)$ does not represent the state of a physical system at time t, but the whole range of such values is needed. For this reason, the state at time t will be denoted by $|\Psi(t)\rangle$. Further, the acceptable $|\Psi(t)\rangle$ are those that belong to the space of square integrable wave functions. In the following, normalized wave functions shall be considered unless otherwise stated. Thus, $\langle \Psi(t)|\Psi(t)\rangle = 1$ or explicitly

$$\int d^3 r_1 \ldots \int d^3 r_N \Psi^*(\mathbf{r}_1, \ldots \mathbf{r}_N, t) \Psi(\mathbf{r}_1, \ldots \mathbf{r}_N, t) = 1. \tag{13.5}$$

If $|\Psi(t_0)\rangle$ is an eigenstate of the Hamiltonian $H = \hat{T} + \hat{V} + \hat{H}_{int}$, that is, if

$$H|\Psi(t_0)\rangle = E|\Psi(t_0)\rangle, \tag{13.6}$$

and $\hat{V}$ does not depend explicitly on time, such as the case of the potential due to the nuclear charge of a physical system, then $|\Psi(t)\rangle = \exp(-iE(t - t_0))|\Psi(t_0)\rangle$; thus, the expectation value of any physical property, such as the density, does not depend on time because

$$\langle \Psi(t)|A|\Psi(t)\rangle = \langle \Psi(t_0)|A|\Psi(t_0)\rangle. \tag{13.7}$$

Thus, these states are of special interest since the measurable physical quantities do not depend on time. Further, they are the stable states of the physical system since if

$$H|\Psi\rangle = E|\Psi\rangle, \tag{13.8}$$

then for first order variations i.e. for $|\Psi'\rangle = |\Psi\rangle + \varepsilon|X\rangle$, with $\langle X|\Psi\rangle = 0$,

$$\lim_{\varepsilon \to 0} \left\{ \frac{1}{\varepsilon} [\langle \Psi + \varepsilon X \mid H \mid \Psi + \varepsilon \mid X\rangle - \langle \Psi \mid H \mid \Psi\rangle] \right\}$$
$$= \lim_{\varepsilon \to 0} \{ E(\langle \Psi \mid X\rangle + \langle X \mid \Psi\rangle) + \varepsilon \langle X \mid X\rangle \} \tag{13.9}$$

and after taking the limits, we find

$$\lim_{\varepsilon \to 0} \left\{ \frac{1}{\varepsilon} [\langle \Psi + \varepsilon X \mid H \mid \Psi + \varepsilon \mid X\rangle - \langle \Psi \mid H \mid \Psi\rangle] \right\} = \langle \Psi \mid X\rangle + \langle X \mid \Psi\rangle = 0. \tag{13.10}$$

i.e. the stability condition

$$\lim_{\varepsilon \to 0} \left\{ \frac{1}{\varepsilon} [\langle \Psi + \varepsilon X|H|\Psi + \varepsilon|X\rangle - \langle \Psi|H|\Psi\rangle] \right\} = 0 \tag{13.11}$$

holds for all energy eigenstates. Conversely if the stability condition holds we find by using a similar procedure

$$\langle \Psi|H|X\rangle + \langle X|H|\Psi\rangle = 0. \tag{13.12}$$

By repeating the same procedure with $|X'\rangle = i|X\rangle$ and adding the resulting equation to the above we find $\langle X|H|\Psi\rangle = 0$. But since $H|\Psi\rangle = |\Psi'\rangle$ is orthogonal to all $|X\rangle$ which are orthogonal to $|\Psi\rangle$ it follows that $|\Psi'\rangle$ can have a projection only on $|\Psi\rangle$. Hence $H|\Psi\rangle = |\Psi'\rangle = E|\Psi\rangle$. Thus every stable state satisfies the eigenvalue equation

$$H|\Psi\rangle = E|\Psi\rangle. \tag{13.13}$$

Obviously, the ground state, being the lowest energy state, is the minimum of the functional $E(\Psi) = \langle \Psi|H|\Psi\rangle$.

13.1.2 REDUCTION OF VARIATIONAL PRINCIPLES TO MINIMUM PRINCIPLES

Many theorems concerning the ground state, such as the Hohenberg and Kohn theorem, are based on the minimum principle of the ground state. Most numerical solutions are also based on this principle. Fortunately, in an indirect way, all eigenstates of the Hamiltonian can be derived by a minimum principle. For this purpose, it is necessary to consider a subspace of the square integrable functions, that is, a subspace S of dimension M of the Hilbert space of square integrable functions and the subspace functional

$$E(S) = \sum_{i=1}^{M} \langle \Phi_i \mid H \mid \Phi_i \rangle, \quad \langle \Phi_i \mid \Phi_j \rangle = \delta_{ij}. \tag{13.14}$$

The value of $E(S)$ does not depend on the choice of the orthonormal basis, as one can easily realize by applying a unitary transformation to go from one basis to the other. The above formula can also be expressed as

$$E(S) = Tr_s\{H\}, \tag{13.15}$$

which means taking the trace of H in the subspace S_M of dimension M as the Hilbert space. Now consider the subspace S of the same dimension for which $E(S)$ assumes its minimum value, that is,

$$E(S_M) = min\{E(S)\}. \tag{13.16}$$

Then, $E(S)$ vanishes for first-order variations of the minimizing subspace S_M. Now consider a variation of S_M to a nearby subspace S' by performing first-order variations on $|\Phi_j\rangle$. Then, for first-order variations with $\left|\Phi_j'\right\rangle + \varepsilon\mid X\rangle$ with $|X\rangle$ normal to the space S, that is, $\langle X|\Phi_i\rangle = 0$ for all i, one easily finds that

$$\lim_{\varepsilon\to 0}\frac{1}{\varepsilon}\{E(S') - E(S)\} = \langle \Phi_j \mid H \mid X \rangle = 0. \tag{13.17}$$

One can also choose $|\Phi_j\rangle$ so that first-order variations in the subspace S_M vanish. Then, $\langle \Phi_j|H|\Phi_j\rangle$ are stable with respect to all variations; therefore, $H|\Phi_j\rangle = E_j|\Phi_j\rangle$.

Thus, $E(S_M) = \sum_{i=1}^{M} E_i$; because $E(S_M)$ is a minimum, $|\Phi_i\rangle$ are the M lowest energy eigenstates of the Hamiltonian H. Then,

$$E_M = E(S_M) - E(S_{M-1}),\tag{13.18}$$

and any physical quantity $\langle \Phi_M|A|\Phi_M\rangle$ of an energy eigenstate $|\Phi_M\rangle$ can be obtained from the subspace functionals

$$G(A,S) = \sum_{i=1}^{M} \langle \Phi_i \mid A \mid \Phi_i \rangle, \mid \Phi_i \rangle \varepsilon S \tag{13.19}$$

from the formula

$$\langle \Phi_M|A|\Phi_M\rangle = G(A,S_M) - G(A, S_{M-1}).\tag{13.20}$$

It is concluded that any properties of eigenstates can be derived from minimum principles, although in an indirect way. The above formulation was used by the present author for developing a DFT for excited states [3]. It was also used for the calculation of excited state energies by developing an HF subspace theory [4–7].

13.1.3 PRELIMINARIES ON MANY-PARTICLE WAVE FUNCTIONS

As stated earlier, the many-particle wave function is antisymmetric with respect to the exchange of the indices i, j of the position variables $\mathbf{r}_i$, $\mathbf{r}_j$. Thus, $\Psi(\mathbf{r}_1,...,\mathbf{r}_j,...,\mathbf{r}_i, ...,\mathbf{r}_N) = -\Psi(\mathbf{r}_1,...,\mathbf{r}_i,...,\mathbf{r}_j,...,\mathbf{r}_N)$.

Now consider an orthonormal basis $\phi_i(\mathbf{r})$ for the single-particle states. Then, any N-particle wave function can be expanded in terms of products of $\phi_i(\mathbf{r})$ as follows: $\Psi(\mathbf{r}_1,...\mathbf{r}_i,...\mathbf{r}_j,...\mathbf{r}_N) = \sum_{i,j,...k} A_{i,j,...k}\phi_i(\mathbf{r}_1)\phi_j(\mathbf{r}_2)...\phi_k(\mathbf{r}_N)$; because of antisymmetry, a relation will result between the expansion coefficients $A_{i,j,...,k}$. Thus, if there are only two particles, one will obtain the relation $A_{ji} = -A_{ij}$; then, the products will sum up as $\phi_i(\mathbf{r}_1)\phi_j(\mathbf{r}_2)$

$$\Psi(\mathbf{r}_1,\mathbf{r}_1) = \frac{1}{\sqrt{2}}\{\phi_i(\mathbf{r}_1)\phi_j(\mathbf{r}_2) - \phi_i(\mathbf{r}_2)\phi_j(\mathbf{r}_1)\}.\tag{13.21}$$

One realizes that the antisymmetric products can be expressed as Sldets because

$$\frac{1}{\sqrt{2}}\det\begin{bmatrix} \phi_i(\mathbf{r}_1) & \phi_i(\mathbf{r}_2) \\ \phi_j(\mathbf{r}_1) & \phi_j(\mathbf{r}_2) \end{bmatrix} = \frac{1}{\sqrt{2}}\{\phi_i(\mathbf{r}_1)\phi_j(\mathbf{r}_2) - \phi_i(\mathbf{r}_2)\phi_j(\mathbf{r}_1)\}.\tag{13.22}$$

Proceeding in this way, it is found that $\Psi(\mathbf{r}_1,...,\mathbf{r}_i,...,\mathbf{r}_j,...,\mathbf{r}_N)$ can be expressed as a sum of N-particle Sldets

$$\Phi(\phi_i, \phi_j, \dots \phi_k; \mathbf{r}_1, \mathbf{r}_2 \dots \mathbf{r}_N) = \frac{1}{\sqrt{N!}} \det[\phi_i(\mathbf{r}_j)], \tag{13.23}$$

normalized to unity, where

$$\det[\phi_i(\mathbf{r}_j)] = \det \begin{bmatrix} \phi_i(\mathbf{r}_1) & \phi_i(\mathbf{r}_2) & . & . & . & \phi_i(\mathbf{r}_N) \\ \phi_j(\mathbf{r}_1) & \phi_j(\mathbf{r}_2) & . & . & . & \phi_j(\mathbf{r}_N) \\ . & . & . & . & . & . \\ . & . & . & . & . & . \\ . & . & . & . & . & . \\ \phi_k(\mathbf{r}_1) & \phi_k(\mathbf{r}_2) & . & . & . & \phi_k(\mathbf{r}_N) \end{bmatrix}. \tag{13.24}$$

In the following, these states will be denoted by their one-particle wave functions as

$$|\phi_i, \dots \phi_j, \dots \phi_k\rangle, \tag{13.25}$$

and their explicit form will be used only when necessary. Because $|\phi_i, \phi_j, \dots, \phi_k\rangle$ is a Sldet, it follows that

$$|\phi_j, \phi_i, \dots \phi_k\rangle = -|\phi_i, \phi_j, \dots \phi_k\rangle. \tag{13.26}$$

Further, these are linear with respect to each of their arguments as one can realize from their explicit form. Thus,

$$|\phi_1, \phi_2, \dots \alpha\phi_i + \beta\chi_i, \dots \phi_N\rangle = \alpha|\phi_1, \phi_2, \phi_i, \dots \phi_N\rangle + \beta|\phi_1, \phi_2, \dots \chi_i, \dots \phi_N\rangle. \tag{13.27}$$

In the following, the strict mathematical language will be used, and any mapping that satisfies Equations 13.26 and 13.27 shall be called a determinant. The above determinant maps a set of single-particle wave function to an N-particle wave function. Consider the scalar product of two Sldets:

$$\langle \phi_1, \phi_1, \dots \phi_i, \dots \phi_N \mid \phi_1', \phi_1', \dots \phi_i', \dots \phi_N' \rangle$$

$$= \frac{1}{N!} \int d^3r_1 \dots \int d^3r_N \det\left[\phi_i^*(\mathbf{r}_j)\right] \det\left[\phi_i'(\mathbf{r}_j)\right]. \tag{13.28}$$

The situation looks hopeless as one has to deal with a huge number of products and has to integrate.

However, this is a mapping to a complex number that is linear and antisymmetric with respect to its left and right arguments. Thus, this is a determinantal mapping, and therefore, the only possible form is

$$\left\langle \phi_1, \phi_2, \ldots \phi_i, \ldots \phi_N \mid \phi_1', \phi_e', \ldots \phi_i', \ldots \phi_N' \right\rangle$$

$$= c \det \left[\left\langle \phi_i \mid \phi_j' \right\rangle \right]. \tag{13.29}$$

Its explicit form is next written as follows:

$$\det \left[\left\langle \phi_i \mid \phi_j' \right\rangle \right] = \det \begin{bmatrix} \left\langle \phi_1 \mid \phi_1' \right\rangle & \left\langle \phi_1 \mid \phi_2' \right\rangle & \cdot & \cdot & \cdot & \left\langle \phi_1 \mid \phi_N' \right\rangle \\ \left\langle \phi_2 \mid \phi_1' \right\rangle & \left\langle \phi_2 \mid \phi_2' \right\rangle & \cdot & \cdot & \cdot & \left\langle \phi_2 \mid \phi_N' \right\rangle \\ \cdot & \cdot & \cdot & \cdot & & \cdot \\ \cdot & \cdot & \cdot & \cdot & \cdot & \cdot \\ \cdot & \cdot & \cdot & \cdot & \cdot & \cdot \\ \left\langle \phi_N \mid \phi_1' \right\rangle & \left\langle \phi_N \mid \phi_2' \right\rangle & \cdot & \cdot & \cdot & \left\langle \phi_N \mid \phi_N' \right\rangle \end{bmatrix}. \tag{13.30}$$

Thus, by using only the definition of a determinant, one manages to avoid all products and integrals. It is left to determine the constant c. If one takes $\left| \phi' \right\rangle = \left| \phi_1 \right\rangle$, then the scalar product will be equal to unity because of the normalization condition. Therefore, $c = 1$. From Equation 13.30, it is easily found that, if $\left| \phi_i' \right\rangle = \left| \phi_i \right\rangle$ for $i > 1$, then $\det \left[\left\langle \phi_i \mid \phi_j' \right\rangle \right] = \left\langle \phi_1 \mid \phi_1' \right\rangle$, and if $\left\langle \phi_1 \mid \phi_1' \right\rangle = 0$, then $\det \left[\left\langle \phi_i \mid \phi_j' \right\rangle \right] = 0$. The proof can be done with every index i. Thus,

$$\left\langle \phi_1, \phi_2, \ldots \phi_i, \ldots \phi_N \mid \phi_1, \phi_2, \ldots \phi_{i-1}, \phi_i', \phi_{i+1} \ldots \phi_N \right\rangle = \left\langle \phi_i \mid \phi_i' \right\rangle. \tag{13.31}$$

This result is useful for the future development of the many-particle theory.

13.1.4 SECOND QUANTIZATION REPRESENTATION FOR FERMIONS

Because the Sldets form a basis for expanding any antisymmetric N-particle wave function, it is useful to derive some more simplifications. These simplifications are necessary because one is lost in dealing with determinants that have $N!$ sums, as it is not possible to derive in a smart way matrix elements for such determinants, since one will need to calculate $N!^2$ integrals of N position variables. For this reason, a bird's-eye view of the second quantization representation shall be given, where one has to deal with single products instead of the $N!$ of a single Sldet. In this way, many pages will be saved, and the presentation will be more elegant and less tiresome. For this purpose, define some new operators having in mind that they do not have a numerical expression. Thus, let

$$\hat{a}_i^\dagger \left| \phi_j \right\rangle = \left| \phi_i, \phi_j \right\rangle. \tag{13.32}$$

The operator $\hat{a}_i^\dagger$ is well defined as it maps a single-particle wave function $\left| \phi_j \right\rangle$ to a two-particle wave function represented by the determinant $\left| \phi_i, \phi_j \right\rangle$. The definition of this operator can be extended to many-particle wave functions as follows:

$$\hat{a}_i^\dagger \left| \phi_j, \phi_k \ldots \phi_l \right\rangle = \left| \phi_i, \phi_j, \phi_k \ldots \phi_l \right\rangle. \tag{13.33}$$

Now use the antisymmetry to derive certain relations for these operators. Because

$$\hat{a}_i^\dagger \hat{a}_j^\dagger \left| \phi_k \ldots \phi_l \right\rangle = \left| \phi_i, \phi_j, \phi_k \ldots \phi_l \right\rangle \tag{13.34}$$

and

$$\hat{a}_j^\dagger \hat{a}_i^\dagger \left| \phi_k \ldots \phi_l \right\rangle = \left| \phi_j, \phi_i, \phi_k \ldots \phi_l \right\rangle = -\left| \phi_i, \phi_j, \phi_k \ldots \phi_l \right\rangle, \tag{13.35}$$

by adding, one finds

$$\left(\hat{a}_i^\dagger \hat{a}_j^\dagger + a_j^\dagger \hat{a}_i^\dagger \right) \left| \phi_k \ldots \phi_l \right\rangle = 0, \tag{13.36}$$

and because this happens for all $|\phi_k \ldots, \phi_l\rangle$, it follows that, to have compatibility with the definitions, one must have the anticommutation relation

$$\left[\hat{a}_i^\dagger, \hat{a}_j^\dagger \right]_+ = \hat{a}_i^\dagger \hat{a}_j^\dagger + \hat{a}_j^\dagger \hat{a}_i^\dagger = 0; \tag{13.37}$$

then, $\hat{a}_i^{\dagger\,2} = \hat{a}_i^\dagger \hat{a}_i^\dagger = 0$. This is not strange because, in this case, a determinant with two equal rows is created. These operators are called "creation operators" because acting on an N-particle state creates a state with $N + 1$ particles. Now define the Hermitian conjugates.

Writing the Hermitian conjugate of $\hat{a}_i^\dagger$ as $\hat{a}_i$, the following is obtained:

$$\left\langle \phi_i, \phi_j \right| = \left\langle \hat{a}_i^\dagger \phi_j \right| = \left\langle \phi_j \right| \hat{a}_i \tag{13.38}$$

and

$$\left\langle \phi_i, \phi_j \middle| \phi_k, \phi_j \right\rangle = \left\langle \phi_j \middle| \hat{a}_i \middle| \phi_k, \phi_j \right\rangle = \delta_{ik}.$$

Thus, formally, $\hat{a}_i$ annihilates a particle from the state $|\phi_i, \phi_j\rangle$ mapping it to $|\phi_j\rangle$; in case $i \neq k$, it is mapped to 0. Thus,

$$\hat{a}_i \left| \phi_k, \phi_j \right\rangle = \hat{a}_i \hat{a}_k^\dagger \left| \phi_j \right\rangle = \delta_{ik} \left| \phi_j \right\rangle. \tag{13.39}$$

The same relation holds when applied to a general state with any particle number.

A state can now be defined from which the single-particle states can be created; thereof, by applying creation operators, one is able to create states with any number of particles. This state will be denoted by $|0\rangle$ and will be called the "vacuum state." Thus, the following definition is used:

$$\hat{a}_i^\dagger |0\rangle = |\phi_i\rangle, \tag{13.40}$$

and therefore,

$$\hat{a}_i^\dagger \hat{a}_j^\dagger \hat{a}_k^\dagger \ldots \hat{a}_l^\dagger |0\rangle = |\phi_i, \phi_j, \phi_k \ldots \phi_l\rangle. \tag{13.41}$$

Then, it is necessary to have the relation

$$\hat{a}_i |0\rangle = 0. \tag{13.42}$$

Working in this way, the anticommutation relation is found:

$$\left[\hat{a}_i, \hat{a}_j^\dagger\right]_+ = \hat{a}_i \hat{a}_j^\dagger + \hat{a}_j^\dagger \hat{a}_i = \delta_{ij}. \tag{13.43}$$

From the above, it is realized that one can have an orthonormal basis for many-particle states that can be expressed in terms of products of operators. For example, in many cases, one needs to multiply determinants and construct a state of more particles with certain symmetry Γ of a group; for example, from N_A Sldets $\Phi_a^A(\phi_1, \phi_2, \phi_3; \mathbf{r}_1, \mathbf{r}_2, \mathbf{r}_2)$ and N_B states $\Phi_\beta^B(\phi_4, \phi_5; \mathbf{r}_4, \mathbf{r}_5)$, five-particle Sldets $|\Phi_\gamma^\Gamma\rangle$ must be constructed. In this notation, the upper indices label the IrReps of a group G and the lower indices label their bases. The standard textbook procedure is to express $\Phi(\phi_1, \phi_2, \phi_3; \mathbf{r}_1, \mathbf{r}_2, \mathbf{r}_2)$ into its sum of six products and $\Phi(\phi_4, \phi_5; \mathbf{r}_4, \mathbf{r}_5)$ into its two products and then multiply; after doing this for all $N_A N_B$ determinants, the sum of products that transform according to IrRep Γ of G must be reconstructed. Moreover, the sums must form linear combinations of Sldets. One understands that the procedure becomes hopeless for more particles. However, in the 2nd quantization representation, one has single products of the form that has definite transformation properties, and the proper Clebsch–Gordan coefficients can be used to construct five-particle wave functions having the appropriate transformation properties. An application will be given later for the case of spin symmetry.

The representation of operators on these states can now be found. One can start from one-particle operators, such as the external potential $V(\mathbf{r})$.

We have $V(\mathbf{r})\phi_i(r) = \sum_j V_{ji}\phi_j(r)$, where $V_{ji} = \langle\phi_j|V|\phi_i\rangle$. The above expression in the present notation can be expressed as

$$\hat{V}|\phi_i\rangle = \hat{V}\hat{a}_i^\dagger |0\rangle = \sum_j \langle\phi_j|V|\phi_i\rangle \hat{a}_j^\dagger |0\rangle = \sum_j \langle\phi_j|V|\phi_i\rangle \hat{a}_j^\dagger \hat{a}_i \hat{a}_i^\dagger |0\rangle. \tag{13.44}$$

The last term can be expressed as $\sum_{jk}\langle\phi_j|V|\phi_k\rangle\hat{a}_j^\dagger\hat{a}_k\hat{a}_i^\dagger|0\rangle$, and thus,

$$\hat{V}|\phi_i\rangle=\sum_{jk}\langle\phi_j|V|\phi_k\rangle\hat{a}_j^\dagger\hat{a}_k|\phi_i\rangle. \qquad (13.45)$$

Then, the second quantization representation of $\hat{V}$ is

$$\hat{V}=\sum_{jk}\langle\phi_j|V|\phi_k\rangle\hat{a}_j^\dagger\hat{a}_k. \qquad (13.46)$$

It is easy to verify that the same representation holds for many-particle states. Now consider a unitary operator, for example, a rotation R. Then,

$$R\hat{a}_1^\dagger\hat{a}_2^\dagger\hat{a}_3^\dagger\ldots\hat{a}_N^\dagger|0\rangle=R\hat{a}_1^\dagger R^{-1}R\hat{a}_2^\dagger R^{-1}R\hat{a}_3^\dagger\ldots R^{-1}R\hat{a}_N^\dagger R^{-1}R|0\rangle \qquad (13.47)$$

and $R|0\rangle=|0\rangle$ because the vacuum state is invariant under any geometrical transformation.
However,

$$R\hat{a}_i^\dagger R^{-1}|0\rangle=R|\phi_i\rangle=\sum_j R_{ij}|\phi_j\rangle,$$

and therefore,

$$R\hat{a}_i^\dagger R^{-1}=\sum_j R_{ij}\hat{a}_j^\dagger. \qquad (13.48)$$

In this way one can get transformation properties without having to operate on the products of determinants. In a later section the representation of two particle operators, like the electron–electron interaction $e^2/|\mathbf{r}_1-\mathbf{r}_2|$, will be presented.

13.1.5 FIELD OPERATORS

So far, the operators depended on a definite complete set of one-particle states. However, it is possible to properly define new operators that are independent of the special basis.
Thus, let

$$\hat{\psi}^\dagger(\mathbf{r})=\sum_i \phi_i^*(r)\hat{a}_i^\dagger \qquad (13.49)$$

and $\hat{\psi}(\mathbf{r})$ its Hermitian conjugate

$$\hat{\psi}(\mathbf{r}) = \sum_i \phi_i(r)\hat{a}_i. \tag{13.50}$$

Then,

$$\hat{\psi}(\mathbf{r}')\hat{\psi}^\dagger(\mathbf{r}) = \sum_{ij} \phi_i^*(r)\phi_j(r')\hat{a}_j\hat{a}_i^\dagger \tag{13.51}$$

and $\hat{\psi}^\dagger(\mathbf{r})\hat{\psi}(\mathbf{r}') = \sum_{ij} \phi_i^*(r)\phi_j(r')\hat{a}_i^\dagger\hat{a}_j.$

By summing, the following is obtained:

$$\hat{\psi}^\dagger(\mathbf{r})\hat{\psi}(\mathbf{r}') + \hat{\psi}(\mathbf{r}')\hat{\psi}^\dagger(\mathbf{r}) = \sum_{ij} \phi_i^*(r)\phi_j(r')\left(\hat{a}_i^\dagger\hat{a}_j + \hat{a}_j\hat{a}_i^\dagger\right), \tag{13.52}$$

and because $\hat{a}_i^\dagger\hat{a}_j + \hat{a}_j\hat{a}_i^\dagger = \delta_{ij}$, the right-hand side of this equation becomes $\sum_i \phi_i^*(r)\phi_i(r') = \delta(\mathbf{r} - r')$. Then,

$$[\hat{\psi}^\dagger(\mathbf{r}), \hat{\psi}(\mathbf{r}')]_+ = \hat{\psi}(\mathbf{r}')\hat{\psi}^\dagger(\mathbf{r}) + \hat{\psi}(\mathbf{r}')\hat{\psi}^\dagger(\mathbf{r}) = \delta(\mathbf{r} - r'). \tag{13.53}$$

Thus, the new operators are independent of the special choice of basis. These operators are called "field operators," with $\hat{\psi}^\dagger(\mathbf{r})$ the creation operator and $\hat{\psi}(\mathbf{r})$ the annihilation operator. The next task is to express the one-particle operators in terms of these.

Because

$$\hat{V} = \sum_{jk} \left\langle \phi_j \left| V \right| \phi_k \right\rangle \hat{a}_j^\dagger\hat{a}_k, \tag{13.54}$$

by inserting the explicit form of $\langle \phi_j|V|\phi_k \rangle$ one finds

$$\hat{V} = \sum_{jk} d^3r\phi_j^*(\mathbf{r})V(\mathbf{r})\phi_k(\mathbf{r})\hat{a}_j^\dagger\hat{a}_k = \int d^3rV(\mathbf{r})\sum_j \phi_j^*(\mathbf{r})\hat{a}_j^\dagger \sum_k \phi_k(\mathbf{r})\hat{a}_k, \tag{13.55}$$

and using the definition of the field operators, one finally gets $\hat{V} = \int d^3rV(\mathbf{r})\hat{\psi}^\dagger(\mathbf{r})\hat{\psi}(\mathbf{r})$. To include the spin variable, the $\uparrow$ (up) and $\downarrow$ (down) labels have to be added for the $+\frac{1}{2}$ and $-\frac{1}{2}$ eigenstates of S_z. Then,

$$\hat{V} = \int d^3 r V(\mathbf{r}) \left\{ \hat{\psi}^{\uparrow\dagger}(\mathbf{r}) \hat{\psi}^{\uparrow s}(\mathbf{r}) + \hat{\psi}^{\downarrow\dagger}(\mathbf{r}) \hat{\psi}^{\downarrow s}(\mathbf{r}) \right\}. \tag{13.56}$$

In the same way for the kinetic energy operator, it is found that

$$\hat{T} = \int d^3 r \left\{ \nabla\hat{\psi}^{\uparrow\dagger}(\mathbf{r}) \nabla\hat{\psi}^{\uparrow s}(\mathbf{r}) + \nabla\hat{\psi}^{\downarrow\dagger}(\mathbf{r}) \nabla\hat{\psi}^{\downarrow s}(\mathbf{r}) \right\}, \tag{13.57}$$

and the density operator for the spin-up and spin-down particles $\hat{\rho}^s(\mathbf{r})$ is

$$\hat{\rho}^s(\mathbf{r}) = \hat{\psi}^{s\dagger}(\mathbf{r}) \hat{\psi}^s(\mathbf{r}), \tag{13.58}$$

where s stands for $\uparrow$ (up) and $\downarrow$ (down). Obviously, the total electron density is

$$\hat{\rho}(\mathbf{r}) = \hat{\rho}^{\uparrow}(\mathbf{r}) + \hat{\rho}^{\downarrow}(\mathbf{r}). \tag{13.59}$$

One can also find the form of the two-particle interaction operator that has the form

$$H_{int} = \frac{1}{2} \int d^3 r \int d^3 r' \frac{e^2}{|\mathbf{r} - \mathbf{r}'|} \sum_s \hat{\psi}^{s\dagger}(\mathbf{r}) \hat{\rho}^s(\mathbf{r}') \hat{\psi}^s(\mathbf{r}). \tag{13.60}$$

Thus, all the terms of the many-particle Hamiltonian are expressed in terms of the Fermion field operators.

13.1.6 Matrix Elements of Hamiltonian

As stated earlier, an N-electron state can be expanded in terms of a complete set of Sldets built by an orthonormal set of one-particle states (orbitals). Therefore, it is necessary to find the matrix elements between such determinants. Start first from the diagonal ones and then consider first the Sldet

$$|\phi_1, \phi_2, \dots \phi_i \dots \phi_N\rangle = \hat{a}_1^\dagger \hat{a}_2^\dagger \dots \hat{a}_i^\dagger \dots \hat{a}_N^\dagger |0\rangle. \tag{13.61}$$

Then,

$$\hat{a}_j |\phi_1, \phi_2, \dots \phi_i \dots \phi_N\rangle = \hat{a}_j \hat{a}_1^\dagger \hat{a}_2^\dagger \dots \hat{a}_i^\dagger \dots \hat{a}_N^\dagger |0\rangle. \tag{13.62}$$

It is recalled here that from the anticommutation relation $\hat{a}_j \hat{a}_n^\dagger = -\hat{a}_n^\dagger \hat{a}_j + \delta_{ij}$.

Thus, if $j \neq i$, for all $i = 1\dots, N$, or as is the usual expression, if the single-particle state is not occupied, then $\hat{a}_j \hat{a}_1^\dagger \hat{a}_2^\dagger \dots \hat{a}_i^\dagger \dots \hat{a}_N^\dagger = (-1)^N \hat{a}_1^\dagger \hat{a}_2^\dagger \dots \hat{a}_i^\dagger \dots \hat{a}_N^\dagger \hat{a}_j$, and because $\hat{a}_j |0\rangle = 0$, it follows that

$$\hat{a}_k^\dagger \hat{a}_j |\phi_1, \phi_2, \dots \phi_i \dots \phi_N\rangle = 0, j \neq i. \tag{13.63}$$

Now consider the matrix element

$$\left\langle \phi_1, \phi_2, \dots \phi_i \dots \phi_N \left| \hat{a}_k^\dagger \hat{a}_j \right| \phi_1, \phi_2, \dots \phi_i \dots \phi_N \right\rangle. \tag{13.64}$$

Taking into account that $\left\langle \Phi \left| \hat{a}_k^\dagger \right. = \left\langle \hat{a}_k \Phi \right|$, the action of $\hat{a}_k$ results to a state of $N-1$ particles, and if $k \ne i$, the two determinants on the left and on the right differ by a single orbital and therefore are orthogonal. On the contrary, if $k = j$, then the matrix element of $\hat{a}_j^\dagger \hat{a}_j$ is equal to unity. Thus,

$$\left\langle \phi_1, \phi_2, \dots \phi_i \dots \phi_N \left| \hat{a}_k^\dagger \hat{a}_i \right| \phi_1, \phi_2, \dots \phi_i \dots \phi_N \right\rangle = \delta_{ki}. \tag{13.65}$$

The density matrix operator $\hat{\rho}^s(\mathbf{r}, \mathbf{r}') = \hat{\psi}^{s\dagger}(\mathbf{r}) \hat{\psi}^s(\mathbf{r}')$ can now be written as follows:

$$\hat{\rho}^s(\mathbf{r}, \mathbf{r}') = \hat{\psi}^{s\dagger}(\mathbf{r}) \hat{\psi}^s(\mathbf{r}') = \sum_{kj} \phi_k^*(\mathbf{r}) \phi_j(\mathbf{r}') \hat{a}_k^{s\dagger} \hat{a}_j^s. \tag{13.66}$$

Then, only the indices k, j of occupied states with the same spin will survive and

$$\left\langle \phi_1^s, \phi_2^s, \dots \phi_i^s \dots \phi_N^s \left| \sum_{ki \ occ} \phi_k^*(\mathbf{r}) \phi_i(\mathbf{r}') \hat{a}_k^{s\dagger} \hat{a}_i^s \right| \phi_1^s, \phi_2^s, \dots \phi_i^s \dots \phi_N^s \right\rangle = \sum_{ki \ occ} \delta_{ki} \phi_k^*(\mathbf{r}) \phi_i(\mathbf{r}').$$
$$\tag{13.67}$$

Thus, finally,

$$\left\langle \phi_1^\uparrow, \dots \phi_i^\uparrow \dots \phi_k^\uparrow, \phi_1^{\prime\downarrow}, \dots \phi_j^{\prime\downarrow} \dots \phi_l^{\prime\downarrow} \left| \hat{\rho}^\uparrow(\mathbf{r}, \mathbf{r}') \right| \phi_1^\uparrow, \dots \phi_i^\uparrow \dots \phi_k^\uparrow, \phi_1^{\prime\downarrow}, \dots \phi_j^{\prime\downarrow} \dots \phi_l^{\prime\downarrow} \right\rangle$$

$$= \sum_{i=1}^{k} \phi_i^*(\mathbf{r}) \phi_i(\mathbf{r}'), \tag{13.68}$$

and the density $\rho^\uparrow(\mathbf{r}) = $ follows by setting $\mathbf{r}' = \mathbf{r}$. Obviously, for the spin-down operator $\hat{\rho}^\downarrow(\mathbf{r}, \mathbf{r}')$, the summation is over all occupied spin-down orbitals. Note here that the special parts of the spin-up orbitals are not orthogonal to those with spin-down orbitals because they are already orthogonal due to their spin.

For the kinetic energy operator, both sides of the above equation are multiplied with $\frac{1}{2}\nabla\cdot\nabla'$; afterward, setting $\mathbf{r}' = \mathbf{r}$ and integrating, the following is obtained:

$$\left\langle \phi_1^\uparrow, \dots \phi_i^\uparrow \dots \phi_k^\uparrow, \phi_1^{\prime\downarrow}, \dots \phi_j^{\prime\downarrow} \dots \phi_l^{\prime\downarrow} \left| T \right| \phi_1^\uparrow, \dots \phi_i^\uparrow \dots \phi_k^\uparrow, \phi_1^{\prime\downarrow}, \dots \phi_j^{\prime\downarrow} \dots \phi_l^{\prime\downarrow} \right\rangle$$

$$= \frac{1}{2} \sum_{i=1}^{k} \int d^3r \nabla\phi_i^*(\mathbf{r}) \cdot \nabla\phi_i(\mathbf{r}) + \sum_{i=1}^{l} \int d^3r \nabla\phi_i^{\prime*}(\mathbf{r}) \cdot \nabla\phi_i'(\mathbf{r}). \tag{13.69}$$

Now consider the nondiagonal elements taking ϕ_m orthogonal to all occupied orbitals of the Sldet on the right. Then, one of the determinants can be written as $\hat{a}_j^\dagger \hat{a}_i |\phi_1, \phi_2, \ldots \phi_i \ldots \phi_N\rangle = \hat{a}_j^\dagger \hat{a}_i |\Phi\rangle$ with ϕ_i occupied.

Then, after substituting $V = \sum_{kl} \langle \phi_k |V| \phi_l \rangle \hat{a}_k^\dagger \hat{a}_l$, it is easily found that

$$\left\langle \Phi \left| \sum_{kl} \langle \phi_k |V| \phi_l \rangle \hat{a}_k^\dagger \hat{a}_l \hat{a}_j^\dagger \hat{a}_i \right| \Phi \right\rangle = \sum_{kl} \langle \phi_k |V| \phi_l \rangle \langle \hat{a}_l^\dagger \hat{a}_k \Phi | \hat{a}_j^\dagger \hat{a}_i \Phi \rangle, \tag{13.70}$$

and after taking the orthogonality conditions, it is found that $\langle \hat{a}_l^\dagger \hat{a}_k \Phi | \hat{a}_j^\dagger \hat{a}_i \Phi \rangle = \delta_{ki}\delta_{lj}$. Thus, finally,

$$\langle \Phi |V| \hat{a}_j^\dagger \hat{a}_i \Phi \rangle = \langle \phi_i |V| \phi_j \rangle. \tag{13.71}$$

Similarly, for the density matrix $\hat{\rho}(\mathbf{r}, \mathbf{r}') = \hat{\psi}^\dagger(\mathbf{r}) \hat{\psi}(\mathbf{r}')$, it is found that

$$\langle \Phi |\hat{\rho}(\mathbf{r}, \mathbf{r}')| \hat{a}_j^\dagger \hat{a}_i \Phi \rangle = \langle \Phi |\hat{\psi}^\dagger(\mathbf{r}) \hat{\psi}(\mathbf{r}')| \hat{a}_j^\dagger \hat{a}_i \Phi \rangle = \phi_i^*(\mathbf{r}) \phi_j(\mathbf{r}), \tag{13.72}$$

and for the kinetic energy operator,

$$\frac{1}{2} \left\langle \Phi \left| \int d^3 r \nabla \psi^\dagger(\mathbf{r}) \nabla \hat{\psi}(\mathbf{r}) \right| \hat{a}_j^\dagger \hat{a}_i \Phi \right\rangle = \frac{1}{2} \int d^3 r \nabla \phi_i^*(\mathbf{r}) \nabla \phi_j(\mathbf{r}). \tag{13.73}$$

Finally, if $|\Phi\rangle$ and $|\Phi'\rangle$ differ by two orbitals, then, for all one-particle operators such as V, T, $\hat{\rho}(\mathbf{r}, \mathbf{r}')$, their matrix elements vanish. Thus,

$$\langle \Phi |V| \hat{a}_i^\dagger \hat{a}_j^\dagger \hat{a}_k \hat{a}_l | \Phi \rangle = 0. \tag{13.74}$$

Now check the diagonal matrix elements of a two-particle operator $V_{ijkl} \hat{a}_i^\dagger \hat{a}_j^\dagger \hat{a}_k \hat{a}_l$. We have

$$\langle \Phi |V_{ijkl} \hat{a}_i^\dagger \hat{a}_j^\dagger \hat{a}_k \hat{a}_l | \Phi \rangle = V_{ijkl} \langle \hat{a}_j \hat{a}_i \Phi | \hat{a}_k \hat{a}_l \Phi \rangle, \tag{13.75}$$

and the only case where the above expression does not vanish is when ϕ_k and ϕ_l are occupied and $j = k$, $i = l$ or $j = l$, $i = k$.
Then,

$$\left\langle \Phi \left| \sum_{ijkl} V_{ijkl} \hat{a}_i^\dagger \hat{a}_j^\dagger \hat{a}_k \hat{a}_l \right| \Phi \right\rangle = \sum_{ijkl \in occ} V_{ijkl} (\delta_{jk}\delta_{il} - \delta_{jl}\delta_{ik}) \tag{13.76}$$

or

$$\left\langle \Phi \left| \sum_{ijkl} V_{ijkl} \hat{a}_i^\dagger \hat{a}_j^\dagger \hat{a}_k \hat{a}_l \right| \Phi \right\rangle = \sum_{ij\in occ} V_{ijji} - \sum_{ij} V_{ijij}, \qquad (13.77)$$

where $ijkl \in occ$ means that ϕ_i, ϕ_j, ϕ_k, ϕ_l are occupied orbitals in $|\Phi\rangle$.

Thus, for

$$V_{ijkl} = \frac{1}{2} \int d^3r \int d^3r' \phi_i^*(\mathbf{r})\phi_j^*(\mathbf{r}') \frac{e^2}{|\mathbf{r}-\mathbf{r}'|} \phi_k(\mathbf{r}')\phi_l(\mathbf{r}), \qquad (13.78)$$

we have

$$\left\langle \Phi \left| \hat{H}_{int} \right| \Phi \right\rangle = E_h + E_x, \qquad (13.79)$$

where

$$E_h(\Phi) = \frac{1}{2} \int d^3r \int d^3r' \frac{e^2}{|\mathbf{r}-\mathbf{r}'|} \left\langle \Phi | \hat{\rho}(\mathbf{r}) | \Phi \right\rangle \left\langle \Phi | \hat{\rho}(\mathbf{r}') | \Phi \right\rangle \qquad (13.80)$$

and

$$E_x(\Phi) = -\frac{1}{2} \int d^3r \int d^3r' \frac{e^2}{|\mathbf{r}-\mathbf{r}'|} \sum_{s=\uparrow,\downarrow} \left\langle \Phi | \hat{\rho}^s(\mathbf{r},\mathbf{r}') | \Phi \right\rangle \left\langle \Phi | \hat{\rho}^s(\mathbf{r}',\mathbf{r}) | \Phi \right\rangle. \qquad (13.81)$$

Note that in the electrostatic or Hartree term $E_h(\Phi)$, the density $\left\langle \Phi | \hat{\rho}(\mathbf{r}) | \Phi \right\rangle = \left\langle \Phi | \hat{\rho}^\uparrow(\mathbf{r},\mathbf{r}) | \Phi \right\rangle + \left\langle \Phi | \hat{\rho}^\downarrow(\mathbf{r},\mathbf{r}) | \Phi \right\rangle$, is the total electron density and $E_h(\Phi) > 0$, whereas the exchange energy $E_x(\Phi) < 0$ and the total interaction energy $E_{int}(\Phi) = E_h(\Phi) + E_x(\Phi)$ is always larger than 0. Note the $-$ sign for $E_x(\Phi)$.

One can derive the nondiagonal matrix elements between the state $|\Phi\rangle = |\phi_k, \phi_l, \phi_3, \dots \phi_i \dots \phi_N\rangle$ and $\left| \Phi' \right\rangle = \hat{a}_i^\dagger \hat{a}_j^\dagger \hat{a}_k \hat{a}_l \left| \Phi \right\rangle = \left| \phi_i, \phi_j, \phi_3, \dots \phi_i \dots \phi_N \right\rangle$. The procedure is more copious as one must deal with the expression $\left\langle \Phi \left| \sum_{i'j'k'l'} V_{i'j'k'l'} \hat{a}_{i'}^\dagger \hat{a}_{j'}^\dagger \hat{a}_{k'} \hat{a}_{l'} \hat{a}_i^\dagger \hat{a}_j^\dagger \hat{a}_k \hat{a}_l \right| \Phi \right\rangle$. The final result is

$$\left\langle \Phi \left| \sum_{i'j'k'l'} V_{i'j'k'l'} \hat{a}_{i'}^\dagger \hat{a}_{j'}^\dagger \hat{a}_{k'} \hat{a}_{l'} \hat{a}_i^\dagger \hat{a}_j^\dagger \hat{a}_k \hat{a}_l \right| \Phi \right\rangle = V_{ijkl} - V_{ijlk}. \qquad (13.82)$$

13.2 HF APPROXIMATION

One can easily show that, because of the interaction term, it is not possible to express an eigenstate of a many-electron system as a single Sldet. This can be proved by *reductio ad absurdum.* Thus, consider that $|\Phi\rangle = |\phi_1, \phi_2, \phi_3, \dots \phi_i \dots \phi_N\rangle$ is an eigenstate of the Hamiltonian. By applying the interaction operator H_{int}

$$H_{int}|\Phi\rangle = \sum_{i'j'k'l'} V_{i'j'k'l'} \hat{a}_{i'}^\dagger \hat{a}_{j'}^\dagger \hat{a}_{k'} \hat{a}_{l'} \hat{a}_1^\dagger \dots \hat{a}_i^\dagger \dots \hat{a}_N^\dagger |\Phi\rangle, \tag{13.83}$$

a linear combination of determinants is obtained differing by two orbitals from the initial one. These terms cannot be cancelled out by the kinetic energy and external potential operators of the Hamiltonian because they change the initial Hamiltonian by a single orbital. Thus, they cannot be summed to a single Sldet, and the hypothesis does not hold.

Nevertheless, an Sldet has a very important property, as it is antisymmetric with respect to the position variables. Then, because the ground state is the minimum of the functional $\langle \Phi|H|\Phi\rangle$, one can find the minimum restricting the minimization space to the states that can be expressed in single Sldets. In this way, an approximation to the ground state can be found, which is called the "Hartree–Fock approximation."

As seen earlier, when $|\Phi\rangle$ is a Sldet,

$$\langle \Phi|H|\Phi\rangle = \langle \Phi|T+V|\Phi\rangle + \frac{1}{2}\int d^3r \int d^3r' \frac{e^2}{|\mathbf{r}-\mathbf{r}'|} \langle \Phi|\rho(\mathbf{r})|\Phi\rangle\langle \Phi|\rho(\mathbf{r}')|\Phi\rangle$$

$$- \frac{1}{2}\int d^3r \int d^3r' \frac{e^2}{|\mathbf{r}-\mathbf{r}'|} \langle \Phi|\rho^\uparrow(\mathbf{r},\mathbf{r}')|\Phi\rangle\langle \Phi|\rho^\uparrow(\mathbf{r}',\mathbf{r})|\Phi\rangle \tag{13.84}$$

$$- \frac{1}{2}\int d^3r \int d^3r' \frac{e^2}{|\mathbf{r}-\mathbf{r}'|} \langle \Phi|\rho^\downarrow(\mathbf{r},\mathbf{r}')|\Phi\rangle\langle \Phi|\rho^\downarrow(\mathbf{r}',\mathbf{r})|\Phi\rangle.$$

Then, after minimization under the normalization condition $\langle \Phi|\Phi\rangle = 1$, the HF equation is obtained:

$$H_{HF}(\Phi)|\Phi\rangle = E|\Phi\rangle, \tag{13.85}$$

where $H_{HF}(\Phi) = \hat{T} + \hat{V} + \hat{V}_h(\Phi) + \hat{V}_x(\Phi)$ and

$$\hat{V}_h(\Phi) = \int d^3r \int d^3r' \frac{e^2}{|\mathbf{r}-\mathbf{r}'|} \langle \Phi|\rho^\uparrow(\mathbf{r}') + \rho^\downarrow(\mathbf{r}')|\Phi\rangle(\rho^\uparrow(\mathbf{r}) + \rho^\downarrow(\mathbf{r}))$$

$$\hat{V}_x(\Phi) = -\int d^3r \int d^3r' \frac{e^2}{|\mathbf{r}-\mathbf{r}'|} \langle \Phi|\rho^\uparrow(\mathbf{r},\mathbf{r}')|\Phi\rangle\rho^\uparrow(\mathbf{r}',\mathbf{r}) \tag{13.86}$$

$$- \int d^3r \int d^3r' \frac{e^2}{|\mathbf{r}-\mathbf{r}'|} \langle \Phi|\rho^\downarrow(\mathbf{r},\mathbf{r}')|\Phi\rangle\rho^\downarrow(\mathbf{r}',\mathbf{r})|\Phi\rangle = E'|\Phi\rangle.$$

In this way, because of the nonlinearity of $\hat{V}_h(\Phi)$ and $\hat{V}_x(\Phi)$, a nonlinear equation was obtained, contrary to the exact eigenstate equation that is linear. However, it is a big advantage that $H_{HF}(\Phi)$ is a one-particle operator. The solution of the above equation can be reduced to finding the single-particle orbitals from the equation

$$-\frac{1}{2}\nabla^2\phi_i^s(\mathbf{r}) + \{V(\mathbf{r}) + V_H(\mathbf{r})\}\phi_i^s(\mathbf{r})$$

$$-\int d^3r' \frac{e^2}{|\mathbf{r}-\mathbf{r}'|}\langle\Phi|\rho^s(\mathbf{r},\mathbf{r}')|\Phi\rangle\phi_i^s(\mathbf{r}) = \varepsilon_i\phi_i^s(\mathbf{r}), \tag{13.87}$$

where the superscript s stands for the $\uparrow$ and $\downarrow$ spin and

$$\langle\Phi|\rho^s(\mathbf{r},\mathbf{r}')|\Phi\rangle = \sum_{i=1}^{N_s}\phi_i^{*s}(\mathbf{r})\phi_i(\mathbf{r}'); \tag{13.88}$$

$N_\uparrow = k$, $N_\downarrow = l$ is the number of particles with spin up and spin down, respectively, and $V_H(\mathbf{r})$ is the Hartree potential

$$V_H(\mathbf{r}) = \int d^3r' \frac{e^2}{|\mathbf{r}-\mathbf{r}'|}\{\langle\Phi|\rho^\uparrow(\mathbf{r}')|\Phi\rangle + \langle\Phi|\rho^\downarrow(\mathbf{r}')|\Phi\rangle\}, \tag{13.89}$$

which is due to the electron charge and $\rho^s(\mathbf{r}) = \rho^s(\mathbf{r},\mathbf{r})$.

In this way, a system of eigenvalue equations with a nonlocal term was obtained. Further, the equations are not linear, but the exact one is. Due to nonlinearity, the ground state $|\Phi\rangle$ constructed by the Hartree-Fock orbitals ϕ_i, although eigenstate of S_z with eigenvalue $\frac{1}{2}(N_\uparrow - N_\downarrow)$, is not an eigenstate of $\mathbf{S}^2$, whereas the eigenstates of the exact Hamiltonian are also eigenstates of this operator. The same happens with other symmetries. It is also to be noticed that the spin-up and spin-down states are coupled only through the Hartree potential $V_H(\mathbf{r})$.

In order to see the spin asymmetry, measure of which is the spin contamination defined as

$$S_{con} = \langle\Phi_M|\mathbf{S}^2|\Phi_M\rangle - M(M+1), \tag{13.90}$$

consider the case that $N_\uparrow - N_\downarrow > 0$. Then, the exchange term is different for the spin-up and spin-down orbitals because, if it is assumed that $\langle\Phi|\rho^\uparrow(\mathbf{r},\mathbf{r}')|\Phi\rangle = \langle\Phi|\rho^\downarrow(\mathbf{r},\mathbf{r}')|\Phi\rangle$, then it follows that

$$\int d^3r'\{\langle\Phi|\rho^\uparrow(\mathbf{r},\mathbf{r})|\Phi\rangle - \langle\Phi|\rho^\downarrow(\mathbf{r},\mathbf{r})|\Phi\rangle\} = N_\uparrow - N_\downarrow = 0,$$

which is not true because $N_\uparrow - N_\downarrow > 0$.

Further, even in the case in which $N_\uparrow = N_\downarrow$, one gets numerical solutions with spin orbitals differing by their spatial parts that have less energy than the solutions $\phi_i^\uparrow(\mathbf{r}) = \phi_i^\downarrow(\mathbf{r})$. This is connected with the nonlinearity of HF, but no mathematical proof has been given.

The nonlinearity of the Hamiltonian causes other asymmetry effects beyond those of the spin. Thus, for example, consider a spherically symmetric effective potential such as that of atoms. Then, the eigenstates of the exact Hamiltonian are also the eigenstates of the angular momentum operator L^2. Assume that the HF solutions have this property. Then, the spin orbitals must be the eigenstates of the HF equation (Equation 13.87). Then, for open shells, one shall have densities and density matrices that are not spherically symmetric because

$$\langle \Phi | \rho(\mathbf{r}) | \Phi \rangle = \sum \left| \phi_n^l(r) \right|^2 \sum_{m=-l}^{l} \left| Y_m^l(\Omega) \right|^2 + \left| \phi_n^l(r) \right|^2 \sum_{m=-l}^{l'<l} \left| Y_m^l(\Omega) \right|^2 \qquad (13.91)$$

is not a spherically symmetric density. To see this, take the case of $l = 1$, $l' = -1$. Then, the unfilled shell term $\left| \phi_n^l(r) \right|^2 \sum_{m=-l}^{l'<l} \left| Y_m^l(\Omega) \right|^2$ is not spherically symmetric as one can easily realize by taking $l = 1$ and $l' = -1$. Then, $\left| Y_{-1}^1(\Omega) \right|^2 = c \sin^2 \theta = c(1 - \cos 2\theta)$. Thus, the density in this case is not spherically symmetric because it varies with the angle θ. Then, the Hartree potential will not be spherically symmetric either as it obeys the Poisson's equation

$$\nabla^2 V_H(\mathbf{r}) = 4\pi \langle \Phi | \rho(\mathbf{r}) | \Phi \rangle, \qquad (13.92)$$

and the right-hand side has an angular dependence. The same holds with $\langle \Phi | \rho(\mathbf{r}, \mathbf{r}') | \Phi \rangle$.

Thus, although the hypothesis has been made that the solutions of Equation 13.87 are those of an operator that has no angle dependence, it is found that this hypothesis does not hold. Thus, although one may have a good approximation with HF, the quality of the many-particle wave functions is not very good. To overcome this, one may search for the minimum in a space of Sldets having the appropriate symmetry. This is called "restricted HF." Unfortunately, practice has shown that this choice leads to higher energies than the exact HF, which is called "unrestricted HF."

In the following it will be seen that it is possible to exploit the asymmetry of the HF Sldet to get lower energy and better quality of wave functions.

13.3   IMPROVEMENTS ON THE UNRESTRICTED HF APPROXIMATION

The unrestricted HF Sldet is of the form

$$\left| \Phi \right\rangle = \left| \phi_1^\uparrow, \phi_2^\uparrow, \ldots, \phi_k^\uparrow; \phi_{k+1}^\downarrow, \phi_{k+2}^\downarrow, \ldots, \phi_{k+l}^\downarrow \right\rangle, \qquad (13.93)$$

where the spatial parts of the $\phi_i^{\uparrow}(\mathbf{r})$, $\phi_{k+j}^{\downarrow}(\mathbf{r})$ orbitals are not necessarily mutually orthogonal. Thus, $\langle\phi_i|\phi_{j+k}\rangle$, in general, is not zero, whereas $\langle\phi_i|\phi_j\rangle = \delta_{ij}$ for i and j smaller than $k+1$ and also when both indices are larger than k. This $|\Phi\rangle$, which is an eigenstate S_z with eigenvalue $M = (k-l)/2$, can be expressed in terms of eigenstates of $\mathbf{S}^2$ as follows:

$$|\Phi\rangle = \sum_{S=M}^{N/2} C_M^S \left|\Psi_M^S\right\rangle, \tag{13.94}$$

where $\left|\Phi_M^S\right\rangle$ are mutually orthogonal, $\mathbf{S} = \sum_{i=1}^{N} \mathbf{S}_i$, and

$$\mathbf{S}^2\left|\Psi_M^S\right\rangle = S(S+1)\left|\Psi_M^S\right\rangle. \tag{13.95}$$

The lower bound of the summation sign is M because it is not possible to have $S < M$ as the range of eigenvalues of S_z is from $-S$ to $+S$, and if you take, for example, $S = M - 1$, the upper eigenvalue of S_z is $M - 1$. Then, taking into account that the exact Hamiltonian commutes with $\mathbf{S}^2$ and S_z, it is found that

$$\mathbf{S}^2 H\left|\Psi_M^S\right\rangle = H\mathbf{S}^2\left|\Psi_M^S\right\rangle = S(S+1)H\left|\Psi_M^S\right\rangle, \tag{13.96}$$

and that states with different S are mutually orthogonal, that is, setting $H\left|\Psi_M^S\right\rangle = \left|X_M^S\right\rangle$, it is found that

$$\left\langle\Psi_M^S\left|H\right|\Psi_M^{S'}\right\rangle = \left\langle\Psi_M^S\left|X_M^S\right\rangle = \left\langle\Psi_M^S\left|H\right|\Psi_M^S\right\rangle\delta_{SS'}. \tag{13.97}$$

Then, the HF energy $E(\Phi)$ will contain only diagonal elements:

$$E(\Phi) = \left\langle\Phi\left|H\right|\Phi\right\rangle = \sum_{S=M}^{N/2}\left|C_M^S\right|^2\left\langle\Psi_M^S\left|H\right|\Psi_M^S\right\rangle. \tag{13.98}$$

Thus, because the average is $E(\Phi)$, there must be at least one $\left|\Psi_M^S\right\rangle$ with energy $E\left(\Psi_M^S\right) = \left\langle\Psi_M^S\left|H\right|\Psi_M^S\right\rangle$ less than the HF energy $E(\Phi)$. This state is not a Sldet because otherwise one would have a Sldet with energy less than the minimizing $|\Phi_{UHF}\rangle$, which is not possible. Thus, $\left|\Psi_M^S\right\rangle$ is a linear combination of Sldets that contains nondiagonal terms of H_{int}, which reduce the energy below that of the minimizing $|\Phi\rangle$. For the expansion, Loewdin [9] introduced a spin projection operator that, however, is a many-particle operator; therefore, its use is limited to small systems. The method that is usually followed is to expand the determinant in its products and reconstruct $\left| \Psi_M^S \right\rangle$ by taking linear combinations of mutually orthogonal orbitals. For this purpose, one has to use various tables to construct antisymmetric states with the proper $\mathbf{S}^2$ eigenvalues. Here, the methodology developed in references [10–12] will be used, where the second quantization is adopted and one has a single product of operators instead of the $N!$ products of $|\Phi_{UHF}\rangle$).

13.3.1 Determination of the Expansion Coefficients C_M^S

The explicit form of $\left| \Psi_M^S \right\rangle$ needs to be calculated to be able to calculate the energy $E\left(\Psi_M^S \right)$. The procedure is considerably simplified if the coefficients C_M^S [10] are determined. In case that $|\Phi\rangle$ consists of doubly and singly occupied orbitals, that is, when for $i = 1,\ldots, m$, $|\phi_i\rangle = |\phi_{i+l}\rangle$ and $\langle\phi_i|\phi_j\rangle = \delta_{ij}$ for all i and $j > m$ (singly occupied), a compact form for finding the expansion coefficients C_M^S exists:

$$\left| X_M \right\rangle = \hat{A}^0 \left| \phi_{m+1}^{\uparrow},\ldots,\phi_{m+k}^{\uparrow}; \phi_{k+m+1}^{\downarrow},\ldots\phi_{k+l}^{\downarrow} \right\rangle = \hat{A}^0 \hat{A}_{k/2}^{k/2} \hat{A}_{-l/2}^{l/2} \left| 0 \right\rangle$$

$$= \sum_{S=\left(\frac{k-l}{2}\right)}^{\left(\frac{k+l}{2}\right)} \left[\frac{(2S+1)k!l!}{\left(S+1+\dfrac{(k+l)}{2} \right)!\left(\dfrac{(k+l)}{2}-S \right)!} \right] \left| \Psi_M^S \right\rangle. \tag{13.99}$$

The above expansion is a consequence of the transformation properties of the products of the creation operators acting on the vacuum state. The operator $\hat{A}^0$ denotes the doubly occupied orbitals that do not play any role in the expansion:

$$\hat{A}^0 = \hat{a}_1^{\uparrow\dagger}\hat{a}_1^{\downarrow\dagger}\ldots\hat{a}_m^{\uparrow\dagger}\hat{a}_m^{\downarrow\dagger}. \tag{13.100}$$

Then,

$$\hat{A}_{k/2}^{k/2} = \hat{a}_{m+1}^{\uparrow\dagger}\ldots\hat{a}_{m+k}^{\uparrow\dagger}, \tag{13.101}$$

and

$$\hat{A}_{-l/2}^{l/2} = \hat{a}_{k+m+1}^{\downarrow\dagger}\ldots\hat{a}_{k+m+l}^{\downarrow\dagger}. \tag{13.102}$$

The transformation properties of a group of operators B_i are defined by operating with the group elements g of a group G, and if one gets a linear combination of the form

$$gB_ig^{-1} = \sum_i D_{ji}^{\Gamma}(g)B_j,$$

the set of the operators B_i is called an irreducible set of operators that transform as the $\Phi_j^{'}$ basis vectors of IrRep Γ of the group G. For continuous groups, one has to find the commutation relations of the group generators with the operators. For convenience, only these properties of the operators will be verified by operating on the vacuum state. Thus, because the group elements consist of rotations in the spin space, one can operate with the group generators which are the total spin operators S^α, $\alpha = +, -, z$,

$$S^\alpha = \sum_i S_i^\alpha.$$

The operator $\hat{A}^0$ transforms as an irreducible tensor operator with $S_z = 0$ and $S = 0$ as one can easily verify by operating on $\hat{A}^0 |0\rangle$ with S^+ and taking into account that $\left[S^+, \hat{a}_i^{\uparrow\dagger} \right] = 0$ and $\left[S^+, \hat{a}_i^{\downarrow\dagger} \right] = \hat{a}_i^{\uparrow\dagger}$. Then, $\left[S^+, \hat{a}_i^{\uparrow\dagger} \hat{a}_i^{\downarrow\dagger} \right] = \hat{a}_i^{\uparrow\dagger} \left[S^+, \hat{a}_i^{\downarrow\dagger} \right] = \hat{a}_i^{\uparrow\dagger} \hat{a}_i^{\uparrow\dagger} = 0$ and hence,

$$S^+ \hat{a}_1^{\uparrow\dagger} \hat{a}_1^{\downarrow\dagger} \ldots \hat{a}_m^{\uparrow\dagger} \hat{a}_m^{\downarrow\dagger} |0\rangle = 0. \tag{13.103}$$

This implies that

$$S^2 \hat{a}_1^{\uparrow\dagger} \hat{a}_1^{\downarrow\dagger} \ldots \hat{a}_m^{\uparrow\dagger} \hat{a}_m^{\downarrow\dagger} |0\rangle = 0. \tag{13.104}$$

In the same way, one can find $S_z \hat{A}_{k/2}^{k/2} |0\rangle = \dfrac{k}{2} \hat{A}_{k/2}^{k/2} |0\rangle$, $S^+ \hat{A}_{k/2}^{k/2} |0\rangle = 0$. However, this is the property of the highest eigenvalue of S_z corresponding to $S = \dfrac{k}{2}$. Then,

$$S^2 \hat{A}_{k/2}^{k/2} |0\rangle = \frac{k}{2} \left(\frac{k}{2} + 1 \right) \hat{A}_{k/2}^{k/2} |0\rangle. \tag{13.105}$$

Similarly, $\hat{S}_z A_{-l/2}^{l/2} |0\rangle = -\dfrac{l}{2} A_{-l/2}^{l/2} |0\rangle$, $S^- A_{-l/2}^{l/2} |0\rangle = 0$, and

$$S^2 \hat{A}_{-l/2}^{l/2} |0\rangle = \frac{l}{2} \left(\frac{l}{2} + 1 \right) \hat{A}_{k/2}^{k/2} |0\rangle. \tag{13.106}$$

After having the transformation properties of these operators, the composition of angular momentum operators can be applied using the rotation group Clebsch–Gordan coefficients

$$\hat{A}_{k/2}^{k/2} A_{-l/2}^{l/2} = \sum_s B_M^s \left\langle S, M \middle| \begin{matrix} k/2 & l/2 \\ k/2 & -l/2 \end{matrix} \right\rangle. \tag{13.107}$$

After introducing the explicit form of the Clebsch–Gordan coefficients $\left\langle S, M \middle| \begin{matrix} k/2 \, l/2 \\ k/2, -l/2 \end{matrix} \right\rangle$, Equation 13.99 is obtained. However, for the calculations, it is necessary to know the explicit form of the states $\left|\Psi_M^S\right\rangle$. The procedure followed is to operate on both the left- and right-hand sides of Equation 13.99. Then, $S^+\left|\Psi_M^M\right\rangle = 0$, whereas, for $S > M$,

$$S^+\left|\Psi_M^S\right\rangle = \sqrt{S(S+1)-(M(M+1)}\left|\Psi_{M+1}^S\right\rangle.$$

Thus, by successive applications of S^+, all spin-down indices become spin-up indices, and one finally gets a determinant that is an eigenstate of $\mathbf{S}^2$ with eigenvalue $\dfrac{N}{2}\left(\dfrac{N}{2}+1\right)$. The next step is to start the reverse procedure by applying S^- to the $S = \dfrac{N}{2}$ state and identify the various eigenstates $\left|\Psi_M^S\right\rangle$ of the expansion. More details can be found in reference [11]. An example is given next introducing the values of the corresponding Clebsch–Gordan coefficients:

$$\left|\phi_1^\uparrow,\phi_1^\downarrow,\phi_2^\uparrow,\phi_3^\uparrow,\phi_5^\downarrow\right\rangle = \sqrt{\frac{2}{3}}\left|\Phi_{1/2}^{1/2}\right\rangle + \sqrt{\frac{1}{3}}\left|\Phi_{1/2}^{3/2}\right\rangle. \tag{13.108}$$

By applying S^+ on both sides, it is found that

$$\left|\Phi_{3/2}^{3/2}\right\rangle = \left|\phi_1^\uparrow,\phi_1^\downarrow,\phi_2^\uparrow,\phi_3^\uparrow,\phi_5^\uparrow\right\rangle.$$

Next, by applying S^- obtained from $\left|\Phi_{3/2}^{3/2}\right\rangle$, the following is found:

$$\left|\Phi_{1/2}^{3/2}\right\rangle = \frac{1}{\sqrt{3}}\left\{\left|\phi_1^\uparrow,\phi_1^\downarrow,\phi_2^\downarrow\phi_3^\uparrow,\phi_5^\uparrow\right\rangle + \left|\phi_1^\uparrow,\phi_1^\downarrow,\phi_2^\uparrow\phi_3^\downarrow,\phi_5^\uparrow\right\rangle + \left|\phi_1^\uparrow,\phi_1^\downarrow,\phi_2^\uparrow\phi_3^\uparrow,\phi_5^\downarrow\right\rangle\right\}. \tag{13.109}$$

Inserting this expression in Equation 13.108 and solving with respect $\left|\Phi_{1/2}^{1/2}\right\rangle$, one finds

$$\left|\Phi_{1/2}^{1/2}\right\rangle = \sqrt{\frac{2}{3}}\left|\phi_1^\uparrow,\phi_1^\downarrow,\phi_2^\uparrow\phi_3^\uparrow,\phi_5^\downarrow\right\rangle - \frac{1}{\sqrt{3}}\left\{\left|\phi_1^\uparrow,\phi_1^\downarrow,\phi_2^\downarrow\phi_3^\uparrow,\phi_5^\uparrow\right\rangle + \left|\phi_1^\uparrow,\phi_1^\downarrow,\phi_2^\uparrow\phi_3^\downarrow,\phi_5^\uparrow\right\rangle\right\}. \tag{13.110}$$

Unfortunately, the unrestricted HF Sldet $|\Phi_{UHF}\rangle$ is not of the form $|X\rangle$ described above, as there is nonzero overlap between the spatial parts of the up and down orbitals, which is smaller than unity. Then, to express $|\Phi_{UHF}\rangle$ in terms of $|X_M\rangle$ states, that is, in terms of Sldets that have only singly and doubly occupied orbitals, any common orthonormal basis can be used for the spatial parts of the $\uparrow$ and $\downarrow$ orbitals. Thus, one can take the up HF orthonormal orbitals and add a complementary basis for the spin-down ones. Thus, a space of $k + l$ orthonormal orbitals is needed. Then, after the expansion of $|\Phi_{UHF}\rangle$, a linear combination of $|X\rangle$-type SIdets is obtained that has to be expanded. The number of these SIdets according to reference [11] is $\dfrac{(k+l)!}{k!l!}$, a very high number of terms even for a small number of electrons. However, this odyssey does not end here, as one has to expand each $|X\rangle$ in its spin eigenstates and find the matrix elements between the various terms of the expansion. For this purpose, Loewdin introduced the corresponding orbitals to decrease the number of the expansion SIdets of type $|X\rangle$.

Thus, neglecting the doubly occupied orbitals, a subspace $S^{\uparrow k}$ of the spatial parts of the $\uparrow$ orbitals and a subspace $S^{\downarrow l}$ of the spin-down ones are considered. In the next subsection, a basis will be developed dealt with, the orbitals of which are called "maximum overlap orbitals."

13.3.2 OPTIMUM EXPANSION SET

In choosing a basis, one can search for an optimum choice that gives successively the highest overlap of wave functions. Thus, let $\chi(\mathbf{r}) \in S^{\uparrow k}$ and $\eta(\mathbf{r}) \in S^{\downarrow l}$ and choose the maximum of $|\langle\chi|\eta\rangle|$ with respect to both subspaces; moreover, let $|\chi_1\rangle, |\eta_1\rangle$ be the minimizing orbitals under the normalization conditions. By a proper choice of phases so that $\langle\chi|\eta\rangle$ is real, the following can be obtained:

$$\langle\chi_1|\eta_1\rangle = \max\{\langle\chi|\eta\rangle: |\chi\rangle\in S^k, \eta \in S^l, \langle\chi|\chi\rangle = \langle\eta|\eta\rangle = 1. \tag{13.111}$$

Consider a variation of $|\chi_1\rangle$ to $|\chi_1 + \varepsilon\chi'\rangle$ such that $\langle\chi_1|\chi'\rangle = 0$. Then, because $\langle\chi_1|\eta_1\rangle$ is the minimum, it follows that

$$\lim_{\varepsilon\to 0}\frac{1}{\varepsilon}\left[\langle\chi_1 + \varepsilon\chi'|\eta_1\rangle - \langle\chi_1|\eta_1\rangle\right] = 0. \tag{13.112}$$

After taking the limits, it is found that

$$\langle\chi'|\eta_1\rangle = 0. \tag{13.113}$$

The above relation implies that $|\eta_1\rangle$ is orthogonal to the subspace $S^{\uparrow k-1}$ of $S^{\uparrow k}$, which is orthogonal to $|\chi_1\rangle$. By repeating the same procedure for $|\eta_1 + \varepsilon\eta'\rangle$, it is found that

$$\langle\chi_1|\eta'\rangle = 0, \tag{13.114}$$

that is, $|\chi_1\rangle$ is orthogonal to the subspace $S^{\downarrow l-1}$ of $S^{\downarrow l}$, which is normal to $|\eta\rangle$.

By the above choice, $|\eta_1\rangle$ can be expressed as

$$|\eta_1\rangle = a_1|\chi_1\rangle + b_1|\theta_1\rangle, \langle\chi|\theta_1\rangle = 0, |\chi\rangle \in S^{\uparrow k}. \tag{13.115}$$

A maximum overlap choice can be made for the spaces $S^{\uparrow k-1}$ and $S^{\downarrow l-1}$ naming the minimizing vectors $|\chi_2\rangle$ and $|\eta_2\rangle$ and their orthogonal spaces $S^{\uparrow k-2}$ and $S^{\downarrow l-2}$. Then one can write

$$|\eta_2\rangle = a_2|\chi_2\rangle + b_2|\theta_2\rangle, \ \langle\chi|\theta_2\rangle = 0, \ |\chi\rangle \in S^{\uparrow k}. \tag{13.116}$$

This procedure can be repeated for the new subspace consecutively. Thus, finally, appropriate bases can be found in the $S^{\uparrow k}$ and $S^{\downarrow l}$ subspaces so that

$$|\eta_i\rangle = a_i|\chi_i\rangle + b_i|\theta_i\rangle, \ \langle\chi|\theta_i\rangle = 0, \ |\chi\rangle \in S^{\uparrow k}. \tag{13.117}$$

In the above procedure, a number of new orbitals are introduced, which are orthogonal to $S^{\uparrow k}$. Further, $|\chi_i\rangle$ are as many as the spin-down orbitals. Thus, a subspace $S^{\uparrow k-l}$ of $S^{\uparrow k}$ is left without a basis. For this subspace, one can construct a basis of $|\chi_i\rangle$, $i > l$, which are orthogonal to $S^{\uparrow l}$ and $S^{\downarrow l}$.

In practice, the construction of these orbitals is done by using linear combinations of the HF orbitals $\phi_i(r)$. For this purpose, name the spatial parts of the spin-down orbitals by $\phi_i'(r) = \phi_{i+k}(r)$. It is also considered that the orbitals are real; as in the case of a complete set of orbitals, appropriate linear combinations can be taken so that real orbitals are obtained.

Then, writing

$$|\chi\rangle = \sum_{j=1}^{k} c_j|\phi_j\rangle, |\eta\rangle = \sum_{n=1}^{l} c_n'|\phi_n'\rangle, \tag{13.118}$$

we have

$$\langle\chi|\eta\rangle = \sum_{j=1}^{k}\sum_{n=1}^{l} c_j c_n' \langle\phi_j|\phi_n'\rangle, \tag{13.119}$$

whereas, from the normalization conditions, we have

$$\sum_{j=1}^{k} c_j c_j = 1, \sum_{n=1}^{l} c_n' c_n' = 1. \tag{13.120}$$

From the first-order variational and the normalization condition, the following equations are obtained:

$$\sum_{n=1}^{l} \langle\phi_j|\phi_n'\rangle c_n' = \lambda c_j \tag{13.121}$$

and

$$\sum_{j=1}^{k} \langle\phi_n'|\phi_j\rangle c_j = \lambda' c_n'. \tag{13.122}$$

After using the normalization condition, one finds that $\lambda = \lambda'$.

After multiplying Equation 13.121 by $\langle\phi_m'|\phi_j\rangle$ and summing over j, the right-hand side has the form of Equation 13.122; finally, it is found that

$$\sum_{n=1}^{l} A_{mn} c'_n = \lambda^2 c'_m,$$

where

$$A_{mn} = \sum_{j=1}^{k} \langle \phi'_n | \phi_j \rangle \langle \phi_j | \phi'_m \rangle. \tag{13.123}$$

Similarly from Equation 13.122, after multiplying by c_n and summing over n, the following is obtained:

$$\sum_{n=1}^{l} B_{ij} c_j = \lambda^2 c_i$$

where

$$B_{ij} = \sum_{n=1}^{l} \langle \phi_i | \phi'_n \rangle \langle \phi'_n | \phi_j \rangle. \tag{13.124}$$

Proceeding with the appropriate constraints, one can derive the corresponding orbitals $|x_2\rangle$, $|\eta_2\rangle$. The minimization is under the additional constraints $\langle \chi_2 | \eta_1 \rangle = 0$, $\langle \eta_2 | \chi_1 \rangle = 0$.

Thus, one can express the corresponding Sldets in terms of these orbitals that will be called "optimum choice corresponding orbitals." Then, after neglecting the doubly occupied orbitals for convenience, $|\Phi\rangle$ can be expressed in terms of the new operators

$$|\Phi\rangle = \hat{a}_1^{\uparrow\dagger} \hat{a}_2^{\uparrow\dagger} \dots \hat{a}_N^{\uparrow\dagger} \left(a_1 \hat{a}_1^{\downarrow\dagger} + b_1 \hat{b}_1^{\downarrow\dagger} \right) \dots \left(a_i \hat{a}_i^{\dagger} + b_i \hat{b}_i^{\downarrow\dagger} \right) \dots \left(a_l \hat{a}_l^{\dagger} + b_l \hat{b}_l^{\downarrow\dagger} \right) |0\rangle, \tag{13.125}$$

or after the expansion

$$
\begin{aligned}
|\Phi\rangle = {}& a_1 \dots a_l \hat{a}_1^{\uparrow\dagger} \hat{a}_2^{\uparrow\dagger} \dots \hat{a}_N^{\uparrow\dagger} \hat{a}_1^{\downarrow\dagger} \dots \hat{a}_l^{\downarrow\dagger} |0\rangle \\
& + b_1 a_2 \dots a_l \hat{a}_1^{\uparrow\dagger} \hat{a}_2^{\uparrow\dagger} \dots \hat{a}_N^{\uparrow\dagger} \hat{b}_1^{\downarrow\dagger} \hat{a}_2^{\downarrow\dagger} \dots \hat{a}_l^{\downarrow\dagger} |0\rangle + \dots \\
& a_1 b_i \dots a_l \hat{a}_1^{\uparrow\dagger} \hat{a}_2^{\uparrow\dagger} \dots \hat{a}_N^{\uparrow\dagger} \hat{a}_1^{\downarrow\dagger} \dots \hat{b}_i^{\downarrow\dagger} \dots \hat{a}_l^{\downarrow\dagger} |0\rangle + \dots \\
& b_1 b_i \dots a_l \hat{a}_1^{\uparrow\dagger} \hat{a}_2^{\uparrow\dagger} \dots \hat{a}_N^{\uparrow\dagger} \hat{b}_1^{\downarrow\dagger} \dots \hat{b}_i^{\downarrow\dagger} \dots \hat{b}_l^{\downarrow\dagger} |0\rangle.
\end{aligned}
\tag{13.126}
$$

A simple example is given from the minimizing Sldet of BeH, which can be expressed as follows:

$$\left| \phi_1^{\uparrow}, \phi_1^{\downarrow}, \phi_2^{\uparrow}, \phi_3^{\uparrow}, \phi_2'^{\downarrow} \right\rangle = c \left| \chi_1^{\uparrow}, \chi_1^{\downarrow}, \chi_2^{\uparrow}, \chi_3^{\uparrow}, \chi_2^{\downarrow} \right\rangle + c' \left| \chi_1^{\uparrow}, \chi_1^{\downarrow}, \chi_2^{\uparrow}, \chi_3^{\uparrow}, \eta_2^{\downarrow} \right\rangle, \tag{13.127}$$

where the phases of the orbitals are chosen so that $c = \langle \phi_2 | \phi_2' \rangle$ is positive. Further, $c' = \sqrt{1 - c^2}$, $|\chi_i\rangle = |\phi_i\rangle$ and $\langle \eta_2^\downarrow | \chi_i \rangle = 0$.

The Sldet $|\chi_1^\uparrow, \chi_1^\downarrow, \chi_2^\uparrow, \chi_3^\uparrow, \chi_2^\downarrow \rangle$ has $S = S_z = \frac{1}{2}$ and $|\chi_1^\uparrow, \chi_1^\downarrow, \chi_2^\uparrow, \chi_3^\uparrow, \eta_2^\downarrow \rangle$ is a linear combination of two eigenfunctions of S^2 with $S = \frac{1}{2}$ and $S = \frac{3}{2}$. The expansion is given by Equation 13.108 with

$$\left| \chi_1^\uparrow, \chi_1^\downarrow, \chi_2^\uparrow, \chi_3^\uparrow, \eta_2^\downarrow \right\rangle = \sqrt{\frac{2}{3}} \left| \Psi_{1/2}^{1/2} \right\rangle + \sqrt{\frac{1}{3}} \left| \Psi_{1/2}^{3/2} \right\rangle, \qquad (13.128)$$

with $\left| \Psi_{1/2}^{3/2} \right\rangle$ and $\left| \Psi_{1/2}^{1/2} \right\rangle$ given by Equations 13.109 and 13.110, respectively, with ϕ_i replaced by χ_i and ϕ_5 replaced by η. Thus,

$$\left| \Psi_{1/2}^{3/2} \right\rangle = \frac{1}{\sqrt{3}} \left\{ \left| \chi_1^\uparrow, \chi_1^\downarrow, \chi_2^\downarrow, \chi_3^\uparrow, \eta_2^\uparrow \right\rangle + \left| \chi_1^\uparrow, \chi_1^\downarrow, \chi_2^\uparrow, \chi_3^\downarrow, \eta_2^\uparrow \right\rangle + \left| \chi_1^\uparrow, \chi_1^\downarrow, \chi_2^\uparrow, \chi_3^\uparrow, \eta_2^\downarrow \right\rangle \right\} \quad (13.129)$$

and

$$\left| \Psi_{1/2}^{1/2} \right\rangle = \sqrt{\frac{2}{3}} \left| \chi_1^\uparrow, \chi_1^\downarrow, \chi_2^\uparrow, \chi_3^\uparrow, \eta_2^\downarrow \right\rangle - \frac{1}{\sqrt{3}} \left\{ \left| \chi_1^\uparrow, \chi_1^\downarrow, \chi_2^\downarrow, \chi_3^\uparrow, \eta_2^\uparrow \right\rangle + \left| \chi_1^\uparrow, \chi_1^\downarrow, \chi_2^\uparrow, \chi_3^\downarrow, \eta_2^\uparrow \right\rangle \right\}. \tag{13.130}$$

Therefore, Equation 13.126 becomes

$$\left| \phi_1^\uparrow, \phi_1^\downarrow, \phi_2^\uparrow, \phi_3^\uparrow, \phi_2'^\downarrow \right\rangle = \left\{ c \left| \chi_1^\uparrow, \chi_1^\downarrow, \chi_2^\uparrow, \chi_3^\uparrow, \chi_2^\downarrow \right\rangle_{1/2}^{1/2} + c' \left| \Psi_{1/2}^{1/2} \right\rangle \right\} + c' \left| \Psi_{1/2}^{3/2} \right\rangle. \quad (13.131)$$

Thus, finally, the $S = \frac{1}{2}$ state consists of four Sldets and its energy will include nondiagonal terms. The energy of $\left\langle \Psi_{1/2}^{3/2} \middle| H \middle| \Psi_{1/2}^{3/2} \right\rangle$ is equal to that of $\left| \Psi_{3/2}^{3/2} \right\rangle$ because $S^+ \left| \Psi_{3/2}^{3/2} \right\rangle = \sqrt{3} \left| \Psi_{1/2}^{3/2} \right\rangle$ and S^+ commutes with the Hamiltonian. However, $\left| \Psi_{3/2}^{3/2} \right\rangle$ is a single Sldet, and therefore, its energy is higher than that of the UHF solution. Details and applications in atoms and molecules can be found in references [10–12]. One can also find in references [13–20] other methodologies for the expansion procedure.

13.4   OEP APPROXIMATION

The nonlocal character of the exchange operator in the HF approximation increases a lot the time needed for a self-consistent calculation relative to that of a local operator. For this purpose, a method was developed where one produces the one-particle orbitals by an Sldet satisfying an equation of the form

$$\left(\hat{T} + \hat{V}' \right) |\Phi\rangle = E |\Phi\rangle, \qquad (13.132)$$

and the OEP $V_{oep}(\mathbf{r})$ is chosen so that the energy $E(\Phi_{V'}) = \langle \Phi_{V'} | H | \Phi_{V'} \rangle$,

$$E(\Phi_{V'}) = \langle \Phi_{V'} | \hat{T} + \hat{V} | \Phi_{V'} \rangle + E_h(\Phi_{V'}) + E_x(\Phi_{V'}),$$

is a minimum for all SIdets satisfying Equation 13.131 with any potential V'. Thus, the definition of

$$E(\Phi_{V_{oep}}) = \langle \Phi_{V_{oep}} | H | \Phi_{V_{oep}} \rangle = \min \left\{ \langle \Phi_{V'} | H | \Phi_{V'} \rangle \right\}.$$

Hence, because from the above definition the choice of SIdets is limited to those that are stable states of a Schroedinger equation without an electron interaction term, it follows that

$$E_V(\Phi_{HF}) \le E_V(\Phi_{oep}). \tag{13.133}$$

Suppose now that the density of $|\Phi_{oef}\rangle = |\Phi_o\rangle$ is $\rho(\mathbf{r})$. Then, according to the Hohenberg and Kohn theorem [1], there is an external potential V'_o that reproduces the above density.

Now derive the stability conditions for $V_o(\mathbf{r}) = V_{oep}(\mathbf{r})$. As usual, the variational principle for $E(\Phi_{Vo})$ must be applied; however, the state $|\Phi_{\varepsilon v}\rangle = \left(|\Phi_o\rangle + \varepsilon |X_v\rangle \right) / \sqrt{1 + \varepsilon^2}$ is not arbitrary, but it must be an eigenstate of a noninteracting Hamiltonian with an external potential $V_o(\mathbf{r}) + \varepsilon v(\mathbf{r})$, with the orthogonality condition $\langle X_v | \Phi_o \rangle = 0$ as usual. Thus, some properties concerning V_o and $|\Phi_o\rangle$ must first be examined.

Then, from the variational principle for $|\Phi_o\rangle$, we have

$$\langle X | \hat{T} + \hat{V}_o | \Phi_o \rangle = 0, \ \langle X | \Phi_o \rangle = 0, \tag{13.134}$$

and for the states $|\Phi_{\varepsilon v}\rangle$ and $|Y\rangle$ orthogonal to $|\Phi_{\varepsilon v}\rangle$, that is, $\langle Y | \Phi_o + \varepsilon X_v \rangle = 0$,

$$\langle Y | \hat{T} + \hat{V}_o + \varepsilon \hat{v} | \Phi_o + \varepsilon X_v \rangle = 0 \tag{13.135}$$

or

$$\langle Y | \hat{T} + \hat{V}_o | \Phi_o \rangle + \varepsilon \langle Y | \hat{T} + \hat{V}_o | X_v \rangle + \varepsilon \langle Y | \hat{v} | \Phi_o \rangle + \varepsilon^2 \langle Y | \hat{v} | X_v \rangle = 0. \tag{13.136}$$

By choosing $\langle Y | \Phi_o \rangle = 0$ and $\langle Y | X_v \rangle = 0$, we have $\langle Y | (T + V_o | \Phi_o \rangle = 0$; because potentials in the neighborhood of V_o are of interest, after dividing by ε and taking the limit for $\varepsilon \to 0$, it is found that

$$\langle Y | (T + V^0) | X_v \rangle + \hat{v} | \Phi_o \rangle = 0. \tag{13.137}$$

Because $|Y\rangle$ are orthogonal to both $|X_v\rangle$ and $|\Phi_o\rangle$, it follows that $(T + V^0) | X_v \rangle + \hat{v} \Phi_o \rangle$ can have only nonzero projections on $|X_v\rangle$ and $v | \Phi_o \rangle$.

Thus,

$$(T + V^0) | X_v \rangle + \hat{v} | \Phi_o \rangle = a | X_v \rangle + b | \Phi_o \rangle,$$

and by taking scalar products with $|\Phi_o\rangle$ and $|X_v\rangle$, respectively, it is found that $b = \langle\Phi_o|v|\Phi_o\rangle$, $a = \langle X_v|T + V^0)|X_v\rangle + \langle X_v|v|\Phi_o\rangle$.

Then,

$$(T + V^0 - aI)\big|X_v\big\rangle = (b - \hat{v})\big|\Phi_o\big\rangle.$$

The relation $\langle Y|\Phi_o + \varepsilon X_v\rangle$ is also satisfied by $|Y'\rangle = |\varepsilon\Phi_o - X_v\rangle$, and after substitution in Equation 13.134 and taking the limits for $\varepsilon \to 0$, the following is obtained:

$$E_0 = \big\langle X_v\big|(T + V^0)\big|X_v\big\rangle + \big\langle X_v\big|\hat{v}\big|\Phi_o\big\rangle = 0$$

and therefore

$$a = E_0$$

and then

$$\big|X_v\big\rangle = (T + V^0 - E_0)^{-1}P_o(b - \hat{v})\big|\Phi_o\big\rangle.$$

From the minimum principle of the OEP energy functional for states $|\Phi + \varepsilon|X_v\rangle$ corresponding to a potential $V_0 + \varepsilon v$, the following is obtained:

$$\langle X_v|T + V|\Phi_0\rangle + \langle X_v|V_h(\Phi_0)|\Phi_0\rangle + \langle X_v|V_x|\Phi_0\rangle = 0. \qquad (13.138)$$

Note here that V is the external potential of the physical system and not V_0.
However, $\langle X_v|T + V|\Phi_0\rangle = \langle X_v|T + V_0|\Phi_0\rangle + \langle X_v|V - V_0|\Phi_0\rangle = 0 + \langle X_v|V - V_0|\Phi_0\rangle$; therefore,

$$\{\langle X_v|V - V_0|\Phi_0\rangle + \langle X_v|V_h|\Phi_0\rangle + \langle X_v|V_x|\Phi_0\rangle = 0. \qquad (13.139)$$

The explicit form of $\langle X_v|$ can now be substituted:

$$\langle X_v| = \langle\Phi_0|(b - v)P_0(T + V^0 - E_0)^{-1}$$

and the following is obtained:

$$\big\langle\Phi_o\big|(b - \hat{v})P_o(T + V^0 - E_0)^{-1}(V - V_0)\big|\Phi_0\big\rangle +$$
$$\big\langle\Phi_o\big|(b - \hat{v})P_o(T + V^0 - E_0)^{-1}(V_h + V_x)\big|\Phi_0\big\rangle = 0. \qquad (13.140)$$

In this way, there is a relation for all $|X\rangle$ of $|\Phi\varepsilon| = |\Phi_0\rangle + \varepsilon|X\rangle$ coming from a potential in the ε-neighborhood of the OEP $V_{oef} = V_0$, $V_\varepsilon + \varepsilon v$, and $b = \langle\Phi_o|v|\Phi_o\rangle$. The above condition must be satisfied for all $v(\mathbf{r})$.

It must be noted that this condition was obtained without using perturbation theory. If one has a good complete set of eigenvalues E_i and eigenstates $|\Phi_i\rangle$ of the operator $\hat{T} + \hat{V}_o$, then one can use the identity $\sum|\Phi_i\rangle\langle\Phi_i| = I$ and the resulting condition is

$$\sum_{i=1}^{\infty}(E_o - E_i)^{-1}\langle\Phi_o|\hat{v}|\Phi_i\rangle\langle\Phi_i|(V - V_0)|\Phi_0\rangle +$$

$$\sum_{i=1}^{\infty}(E_o - E_i)^{-1}\langle\Phi_o|\hat{v}|\Phi_i\rangle\langle\Phi_i|(V_h + V_x)|\Phi_0\rangle = 0. \tag{13.141}$$

Although these conditions may be helpful in deriving the OEP V_{oep}, one does not avoid calculating numerically the solutions for many potentials to obtain the one minimizing E_{oep}. For this reason, approximations shall be tried.

13.5   DIRECT MAPPING THEORY

Because it is not easy to determine the exact OEP of HF, another approach can be attempted starting from the fact that an OEP $V_{oep}(\mathbf{r})$ corresponds to each external potential $V(\mathbf{r})$, that is, there is a direct mapping of $V(\mathbf{r})$ to $V_{oep}(\mathbf{r})$:

$$V_{oep}(\mathbf{r}) = (FV)(\mathbf{r}). \tag{13.142}$$

Obviously, if a constant is added to the external potential $V(\mathbf{r})$, the Sldet $|\Phi_{oep}\rangle$ must not change because $E(\Phi_{hf})$ does not change. Thus, $V_{oep}(\mathbf{r})$ cannot depend directly on $V(\mathbf{r})$. Next, if dependence on $\nabla V(\mathbf{r})\cdot\mathbf{f}(\mathbf{r})$ is considered, it means that a rotation of the potential will give a different $V_{oep}(\mathbf{r})$ and hence a different energy. The next dependence is on $\nabla^2 V(\mathbf{r})$, which is acceptable because, for atoms, molecules, and solids, $\nabla^2 V(\mathbf{r}) = -4\pi\rho^+(\mathbf{r})$, where $\rho^+(\mathbf{r})$ is the positive charge density due to the nuclei:

$$\rho^+(\mathbf{r}) = \sum_i Z_i\delta(\mathbf{r} - \mathbf{R}_i). \tag{13.143}$$

The next possible dependence is on $\nabla^4 V(\mathbf{r})$, but this gives $\sum_i Z_i\nabla^2\delta(\mathbf{r} - \mathbf{R}_i)$, which is not a physical potential. Taking into account that translations and rotations of the external potential must result to translations and rotations of the eigenfunctions and therefore of $V_{oep}(\mathbf{r})$, it is concluded that, in first-order approximation, the following form holds:

$$K(|\mathbf{r}|; N) = 1 - \frac{N-1}{Z} + \frac{N-1}{Z}e^{-\zeta|\mathbf{r}|}. \tag{13.144}$$

After searching for the asymptotic conditions of this mapping, it is concluded that

[21, 22] $K(\hat{u}r\hat{u}; N) = 1 - \dfrac{N-1}{Z} + \dfrac{N-1}{Z}e^{-\zeta|\mathbf{r}|}.$

Then, the potential is

$$V_{eff}(\mathbf{r}) = -\sum_k\frac{Z_k}{|\mathbf{r} - \mathbf{R}_k|} + \frac{N-1}{Z}C\sum_k Z_k\frac{1 - \exp(-\zeta_k|\mathbf{r} - \mathbf{R}_k|)}{|\mathbf{r} - \mathbf{R}_k|}, \tag{13.145}$$

where the parameters c and Z_k must be determined by the optimization procedure. Obviously, one can include nonlinear terms of the form

$$\delta V_{oep}(\mathbf{r}) = \int d^3 r_1' K_1(|\mathbf{r}_1' - \mathbf{r}|) \nabla^2 V(\mathbf{r}_1') \int d^3 r_2' K_2(|\mathbf{r}_1' - \mathbf{r}_2'|) \nabla^2 V(\mathbf{r}_2'), \qquad (13.146)$$

but, in application, it turned out that the linear term gives quite good results.

The above method was developed initially by Theophilou [23], was used and extended by Glushkov and coworkers, and gave quite good results that compared well with the experimental ones [21, 22, 24–26]. Recently, Glushkov and Assfeld [27] extended successfully the method for excited states with quite good results.

13.6   DENSITY FUNCTIONAL THEORY

The fact that the many-electron state is described by a function of as many position variables as the number of particles was considered a serious obstacle for calculating the properties of atoms, molecules, and solids. For this reason, attempts were made for simplifications, and density was considered as a potential candidate for this matter. Thus, Thomas [29] and, later, Fermi [30] tried to formulate the many-electron problem in this respect. Other attempts were also made to express the energy in terms of the density, but it was until Hohenberg and Kohn [1] showed rigorously, by their well-known theorem, that the density can represent the ground state. It was really a surprise to show, by strict mathematics, that there is one-to-one correspondence between the density $\rho(\mathbf{r})$ and the wave function of an N-particle system, which is a function of N such variables. An important implication of this theorem is that every physical property of the ground state could be expressed in terms of the density, and one could minimize the energy density functional and get the density equation for the ground state.

The above procedure involved the definition of the space of densities in which one had to search for the minimum of the energy density functional. Moreover, one had to find the expression of the kinetic and interaction energy density functionals. For this purpose, one could generalize the corresponding expressions of the uniform electron gas, where the density is constant and for which there were sufficiently accurate results. Unfortunately, the kinetic energy approximation $T(\rho)$ was found extremely poor, and other efforts, beyond those based on the free electron gas, failed. Kinetic energy has a strong spin dependence as one can conclude from the case of two particles, which is the easiest example. Thus, for spin eigenstates with $S = 0$, the Sldet is

$$\Phi^0(\mathbf{r}_1, s_1, \mathbf{r}_2, s_2) = \phi(\mathbf{r}_1)\phi(\mathbf{r}_2)\sqrt{\frac{1}{2}} |\uparrow_1 \downarrow_2 - \downarrow_1 \uparrow_2\rangle. \qquad (13.147)$$

Then, $\rho(\mathbf{r}) = 2|\varphi(\mathbf{r})|^2$ and the kinetic energy functional per particle is

$$T^0(\rho) = \frac{1}{8} \int |\nabla \rho(\mathbf{r})|^2 / \rho(\mathbf{r}). \qquad (13.148)$$

Contrary to the above, an explicit expression does not exist for $S = 1$ Sldet $\Phi_1^1(r_1, s_1, r_2, s_2)$,

$$\Phi_1^1(r_1, s_1, r_2, s_2) = \sqrt{\frac{1}{2}} [\phi_1(r_1)\phi_2(r_2) - \phi_1(r_2)\phi_2(r_1)] | \uparrow_1, \uparrow_2 \rangle, \qquad (13.149)$$

where $\rho(r) = |\varphi_1(r)|^2 + |\varphi_2(r)|^2$ and $\langle \varphi_i | \varphi_j \rangle = \delta_{ij}$ and the kinetic energy is

$$t(\varphi_1, \varphi_2) = \frac{1}{2} \int d^3r \{ \nabla \varphi_1^*(r) \cdot \nabla \varphi_1(r) + \nabla \varphi_2^*(r) \cdot \nabla \varphi_2(r). \qquad (13.150)$$

From the above, it is concluded that because the ground states have different spin, searching for density functionals without taking into account the spin is a failure. The same holds true for the exchange and correlation energy functional that has a stronger spin dependence.

Obviously, the kinetic energy density functional becomes more complicated as the number of particles increases. Further, the exact many-electron wave function is a linear combination of Sldets; thus, the problem is even more complicated. Moreover, there are cases where the functional derivatives do not exist. Thus, consider a $\delta\rho_i(r)$ due to $\phi_i(r)$ for a noninteracting system energy eigenstate. Then,

$$\frac{\delta T(\rho)}{\delta \rho_i(r)} = \varepsilon_i - V(r). \qquad (13.151)$$

Thus, the functional derivative depends on the way the density is changed, whereas the theory demands independence from the way the density is changed, that is, the so-called Frechet derivatives are needed.

To bypass this difficulty, KS reintroduced the wave function in the form of a Sldet $|\Phi\rangle$,

$$\Phi(r_1, r_1, \ldots r_N) = \frac{1}{\sqrt{N!}} \det\{\varphi_i(r_j)\} \qquad (13.152)$$

imposing the condition that this determinant should have the same density $\rho(r)$ as that of the exact ground state. In this way, one could have a kinetic energy expression in a simple form, namely,

$$\langle \Phi | T | \Phi \rangle = \frac{1}{2} \sum \int d^3r \nabla \varphi_i^*(r) \cdot \nabla \varphi_i(r), \qquad (13.153)$$

where T is the kinetic energy operator and $\phi_i(r)$ are the spin orbitals of the Sldet $|\Phi\rangle$, which in the following will be called the KS state or KS determinant. Then, one could express the energy functional in terms of $|\Phi\rangle$ and minimize in the space of determinants. The advantage of the introduction of $|\Phi\rangle$ is that the equations resulting after minimization are one-particle Schroedinger-type equations with an effective potential. Thus, it is easy to deal numerically with these equations. The drawback is that one had to find an explicit form for the exchange and correlation energy. In addition, some correction to the kinetic energy is necessary because one can prove that the KS kinetic energy is smaller than the exact one. The way the KS equation was obtained was to minimize the density expression of the total energy, writing the exchange and correlation energy as a function of density. This procedure, however, includes functional derivatives that may not exist; further, one has to define the space of densities.

In the present work, a different formulation shall be used that does not make use of functional derivatives.

Formally, the KS determinant $|\Phi\rangle$ can be considered as the many-particle state, which minimizes the kinetic energy under the constraint that its density $\rho_\Phi(\mathbf{r})$ is equal to that of the exact state $|\Psi\rangle$ [28], that is,

$$\langle \Phi|T|\Phi\rangle = \min\left\{\langle\Phi'|T|\Phi'\rangle : \langle\Phi'|\hat{\rho}(\mathbf{r})|\Phi'\rangle = \rho_\Psi(\mathbf{r})\right\}; \qquad (13.154)$$

therefore,

$$\langle\Phi|T|\Phi\rangle < \langle\Psi|T|\Psi\rangle. \qquad (13.155)$$

Because each $\rho(\mathbf{r})$ corresponds to a single KS state $|\Phi\rangle$, there is one-to-one correspondence between the exact ground state and the KS state; therefore, the energy functional $E(\Psi)$ can be expressed in terms of $|\Phi\rangle$ and the new functional $F(\Phi)$ can be minimized.

To get a good approximation, one needs a reasonable exchange and correlation energy functional $E_{xc}(\rho)$ better than that of the free electron gas. Finding an explicit form of $E_{xc}(\rho)$ is a difficult problem, and there are now hundreds of such functionals that give good results for particular categories of physical systems; however, these functionals are still far from the universal one, which, in theory, exists. Now go into more detail concerning the rigorous formulation of the KS theory.

The many-particle Hamiltonian has the form

$$T + H_{int} + \hat{V}, \qquad (13.156)$$

and the explicit form of the various terms was given in the previous section:

$$H = T + H_{int} + \int d^3r \hat{\rho}(\mathbf{r})V(\mathbf{r}) \qquad (13.157)$$

and the ground state obeys the relations

$$\langle\Psi|H|\Psi\rangle = \min\langle\Psi'|H|\Psi'\rangle \qquad (13.158)$$

under the normalization condition $\langle\Psi'|\Psi'\rangle = 1$.

Now consider $|\Psi\rangle$ as the ground state for external potential $V(\mathbf{r})$ and $|\Psi'\rangle$ as that of potential $V'(\mathbf{r})$ with $V'(\mathbf{r}) - V(\mathbf{r})$ not a constant. Then,

$$\langle\Psi|T + H_{int}|\Psi\rangle + \int d^3r\rho_\Psi(\mathbf{r})V(\mathbf{r}) \prec \langle\Psi'|T + H_{int}|\Psi'\rangle + \int d^3r\rho_{\Psi'}(\mathbf{r})V(\mathbf{r}). \quad (13.159)$$

Writing the corresponding equation for $|\Psi'\rangle$ on the left and $|\Psi\rangle$ on the right, the following is obtained:

$$\langle\Psi'|T + H_{int}|\Psi'\rangle + \int d^3r\rho_{\Psi'}(\mathbf{r})V'(\mathbf{r}) \prec \langle\Psi|T + H_{int}|\Psi + \int d^3r\rho_\Psi(\mathbf{r})V(\mathbf{r}), \quad (13.160)$$

and after adding by parts and transferring everything on the left, the following inequality is obtained:

$$-\int d^3r\Delta\rho(\mathbf{r})\Delta V(\mathbf{r}) > 0, \quad (13.161)$$

where $\Delta\rho(\mathbf{r}) = \rho'_{\Psi'}(\mathbf{r}) - \rho_\Psi(\mathbf{r})$ and $\Delta V(\mathbf{r}) = V'(\mathbf{r}) - V(\mathbf{r})$. This means that if the potential is deepened in some region, the density will increase there, that is, the electron charge concentrates in the region where the potential is deep.

By the above inequality derived initially in reference [3], it is easy to get the Hohenberg and Kohn theorem because, if it is supposed that the two densities are equal for all $\mathbf{r}$, a contradiction is obtained. Thus,

$$\rho(\mathbf{r}) \rightleftarrows V(\mathbf{r}), \quad (13.162)$$

and thereof the relation

$$\rho(\mathbf{r}) \rightleftarrows \Psi(\mathbf{r}_1,\mathbf{r}_2,...\mathbf{r}_N) \quad (13.163)$$

because different ground states correspond to two different potentials. In case of degeneracy to the same external potential corresponding to many densities, however, the one-to-one correspondence between $\rho(\mathbf{r})$ and $\Psi(\mathbf{r}_1, \mathbf{r}_2, ..., \mathbf{r}_N)$ holds.

For the rigorous formulation of the KS theory, Levy [28] introduced the following density function:

$$Q(\rho) = \min\{\langle\Psi|T + H_{int}|\Psi\rangle : \langle\Psi|\hat{\rho}(\mathbf{r})|\Psi\rangle = \rho(\mathbf{r})\}. \quad (13.164)$$

As always, in this paper, all $|\Psi\rangle$ are normalized to unity. Following the proof by Levy, [28] we have the following theorem.

Theorem 13.1

The ground state energy corresponding to an external potential V can be expressed as a density functional minimum, namely,

$$E(V) = \min\left\{Q(\rho) + \int d^3r\rho(\mathbf{r})V(\mathbf{r})\right\}. \qquad (13.165)$$

Proof

Identify $|\Psi_\rho\rangle$ as the minimizing $|\Psi\rangle$ for the density $\rho(\mathbf{r})$ and let

$$Q(\rho) = \langle\Psi_\rho|T + H|\Psi_\rho\rangle \qquad (13.166)$$

be the kinetic, exchange, and correlation functional that, in the following, shall be called the Levy functional. Let $|\Psi_V\rangle$ be the ground state for an external potential V and $\rho_{\Psi V}(\mathbf{r}) = \langle\Psi_V|\hat\rho(\mathbf{r}|\Psi_V\rangle$ its ground state density.

Because the ground state energy for an external potential V is

$$E(\Psi_V) = \left\langle\Psi_V\left|T + H_{int} + \int d^3r\hat\rho(\mathbf{r})V(\mathbf{r})\right|\Psi_V\right\rangle$$

$$= \min\left\{\left\langle\Psi\left|T + H_{int} + \int d^3r\hat\rho(\mathbf{r})V(\mathbf{r})\right|\Psi\right\rangle\right\}, \qquad (13.167)$$

it follows that the ground state inequality holds also for $|\Psi_\rho\rangle$, that is,

$$E(\Psi_V) \leq \left\langle\Psi_\rho\left|T + H_{int}\right|\Psi_\rho\right\rangle + \int d^3r\rho(\mathbf{r})V(\mathbf{r}), \qquad (13.168)$$

and for $\rho(\mathbf{r})$ equal to the ground state density, $\rho_{\Psi_V}(\mathbf{r}) = \langle\Psi_V|\rho(\mathbf{r})|\Psi_V\rangle$,

$$E(\Psi_V) \leq Q(\rho_{\Psi_V}) + \int d^3r\rho_{\Psi_V}(\mathbf{r})V(\mathbf{r}). \qquad (13.169)$$

However, by the definition of $Q(\rho)$,

$$Q(\rho_{\Psi_V}) \leq \left\langle\Psi_V\left|T + H_{int}\right|\Psi_V\right\rangle. \qquad (13.170)$$

By adding $\int d^3r\rho_{\Psi_V}(\mathbf{r})V(\mathbf{r})$, the following is obtained:

$$Q(\rho_{\Psi V}) + \int d^3r\rho_{\Psi V}(r)V(r) \leq \left\langle\Psi_V\left|T + H_{int}\right|\Psi_V\right\rangle$$

$$+ \int d^3r\rho_{\Psi_V}(r)V(r) = E(V). \qquad (13.171)$$

Then,

$$Q(\rho_{\Psi_V}) + \int d^3r \rho_{\Psi_V}(r)V(r) \le E(\Psi_V), \tag{13.172}$$

and from inequalities 13.169 and 13.172, it follows that the equality holds. Thus, for each external potential V,

$$E(V) = E(\Psi_V) = Q(\rho_{\Psi_V}) + \int d^3r \rho_V(\mathbf{r})V(\mathbf{r}) = \min\left\{Q(\rho) + \int d^3r \rho(\mathbf{r})V(\mathbf{r})\right\}. \tag{13.173}$$

Thus, the ground state energy for a certain potential V is the minimum of a density functional. However, as stated earlier, one has to define the space of densities. This is the space defined by all $\rho_\Psi(\mathbf{r}) = \langle \Psi|\hat{\rho}(\mathbf{r})|\Psi\rangle$, but this is an indirect definition. (Note here that the densities are labeled by their $|\Psi\rangle$.) However, as shown in references [4] and [31], functionals can be defined over all space of square integrable wave functions $\mathcal{L}^2$ as follows:

$$\hat{Q}(\Psi) = \min\left\{\langle \Psi'|T + H_{int}|\Psi'\rangle : \rho_{\Psi'}(\mathbf{r}) = \rho_\Psi(\mathbf{r}), |\Psi'\rangle \in \mathcal{L}^2\right\}. \tag{13.174}$$

This functional is well defined for every normalized $|\Psi\rangle$ in $\mathcal{L}^2$, the space of all square integrable N-particle wave functions. In the above definition, $\hat{Q}(\Psi')$ is the same for all $|\Psi'\rangle$ having the same density as $|\Psi\rangle$. Thus, $\hat{Q}(\Psi)$ changes only when the density changes. Further, $\hat{Q}(\Psi)$ is not the kinetic plus interaction energy of $|\Psi\rangle$ because

$$\hat{Q}(\Psi) \le \Psi|T + H_{int}|\Psi >. \tag{13.175}$$

A symbol similar to $Q(\rho)$ is used to denote that these mappings are closely related because $\hat{Q}(\Psi) = Q(\rho_\Psi)$. However, $\hat{Q}(\Psi)$ is a mapping of wave functions, whereas $Q(\rho)$ is a mapping of the density that belongs to a different space. The functional $\hat{Q}(\Psi)$ has an advantage over $Q(\rho)$ because the space of wave functions is well defined, whereas the space of densities might include densities that are not representable by N-particle wave functions. In the same way, a kinetic energy functional $K(\Psi)$ is defined as follows:

$$K(\Psi) = \min\{\Psi'|T|\Psi'\rangle : \rho_{\Psi'}(\mathbf{r}) = \rho_\Psi(\mathbf{r})\} \tag{13.176}$$

The next functional defined is

$$G(\Psi) = \hat{Q}(\Psi) + \int d^3r \rho_\Psi(\mathbf{r})V(\mathbf{r}) + \langle \Psi|T|\Psi\rangle - K(\Psi). \tag{13.177}$$

Then, by using a similar procedure to that of Theorem 1, one can show that the minimum of $G(\Psi)$ is attained when

$$\langle \Psi | T | \Psi \rangle = K(\Psi). \tag{13.178}$$

This means that the minimizing $|\Psi\rangle$ is a Sldet $|\Phi\rangle$ and not the exact ground state wave function as shall be seen later. Then,

$$E(V) = G(\Phi) = \min G(\Psi) = \langle \Psi_0 | H | \Psi_0 \rangle. \tag{13.179}$$

It will be shown that the minimizing $|\Psi\rangle$ obeys a Schroedinger-type equation with an external potential $V_{KS}(\mathbf{r}, \rho)$, which will be called KS potential. Because, by Equation 13.178,

$$K(\Psi) = \min\{\langle \Psi' | T | \Psi' \rangle : \rho\Psi\ (\mathbf{r}) = \rho\Psi\ (\mathbf{r})\}, \tag{13.180}$$

write $\rho_\Psi(\mathbf{r}) = \rho(\mathbf{r})$, and instead of the power of the continuum constraints, reduce it to an equivalent denumerable set by choosing a set of $v_i(\mathbf{r})$ for which

$$\left| \int d^3 r \rho_\Psi(\mathbf{r}) v_i(r) \right| = c_i < \infty, \tag{13.181}$$

and c_i defines uniquely the density, that is, for two different densities $\rho(\mathbf{r})$, $\rho'(\mathbf{r})$, there is at least one $v_i(r)$ for which

$$\left| \int d^3 r [\rho'(\mathbf{r}) - \rho(\mathbf{r})] v_i(r) \right| > 0. \tag{13.182}$$

Now assume that the minimum defined by Equation 13.180 under the constraint

$$\int d^3 r \langle \Psi' | \hat{\rho}(\mathbf{r}) | \Psi \rangle v_i(r) = \int d^3 r \rho(\mathbf{r}) v_i(r) \tag{13.183}$$

is obtained by $|\Phi\rangle$. Then, if $\left| \Phi_\varepsilon \right\rangle = \dfrac{1}{\sqrt{1+\varepsilon^2}} | \Phi + \varepsilon | X \rangle$ with $\langle \Phi | X \rangle = 0$, the first-order variations must vanish, that is,

$$\lim_{\varepsilon \to 0} \frac{1}{\varepsilon} \left[\left\langle \Phi_\varepsilon \left| T \right| \Phi_\varepsilon \right\rangle - \left\langle \Phi \left| T \right| \Phi \right\rangle \right] = 0. \tag{13.184}$$

After inserting the explicit form of $|\Phi_\varepsilon\rangle$, $\langle X | T | \Phi \rangle + \langle \Phi | T | X \rangle = 0$ is found, and in the usual way by taking $|X'\rangle = i|X\rangle$, $\langle X | T | \Phi \rangle - \langle \Phi | T | X \rangle = 0$ is also found. Then,

$$\langle X | T | \Phi \rangle = 0. \tag{13.185}$$

After using the same procedure for the constraints, one finds

$$\langle X| \int d^3 r \hat{\rho}(\mathbf{r}) v_i(r) |\Phi\rangle = 0. \tag{13.186}$$

Then, the space M of $|X\rangle$ is orthogonal not only to $|\Phi\rangle$ but also to the space N of $\left|\Phi_i\right\rangle = \int \hat{\rho}(\mathbf{r}) v_i(r) |\Phi\rangle$, and because $T|\Phi\rangle$ is orthogonal to M, it follows that it has nonzero projections only in the space N. Thus,

$$T|\Phi\rangle = \sum_i \lambda_i \int d^3 r v_i(r) \hat{\rho}(\mathbf{r}) |\Phi\rangle + E |\Phi\rangle, \tag{13.187}$$

and the eigenstates $|\Phi(\lambda_1, ..., \lambda_i, ...)\rangle$ must satisfy the constraint in Equation 13.183. In this way, the arbitrary coefficients λ_i can be determined. Then, there is a potential

$$V_{KS}(\mathbf{r}) = -\sum_i \lambda_i v_i(\mathbf{r}), \tag{13.188}$$

for which the minimum of the kinetic energy is obtained.

Then, the antisymmetric solution of the equation $T|\Phi\rangle + V|\Phi\rangle = E|\Phi\rangle$ as is well known from quantum mechanics is a Sldet

$$|\Phi\rangle = |\phi_1 ... \phi_i ... \phi_N\rangle, \tag{13.189}$$

the orbitals of which obey the equation

$$-\frac{1}{2} \nabla^2 \phi_i(\mathbf{r}) + V_{ks}(\mathbf{r}) \phi_i(\mathbf{r}) = \epsilon_i \phi_i(\mathbf{r}). \tag{13.190}$$

Formally, it can be concluded that because the density is uniquely defined by the external potential V, the potential V_{KS} is also uniquely defined by V. Thus, one can work in the same way as in the case of the HF OEP V_{oep}. This theory is called direct mapping DFT (DMDFT). These potentials are closely related and V_{oep} is a good approximation to V_{KS} as shown by Gidopoulos [32]. In this case, one has to do a direct mapping of the external potential to the KS potential. DMDFT was applied successfully by Glushkov and coworkers in finding the energies of ground and excited states of atoms and molecules [21, 22, 24, 25, 27].

Concerning the functionals $Q(\rho)$ or equivalently $\hat{Q}(\Psi)$, as stated in Section 13.1, there are many numerous approximations, giving sufficient accuracy for certain classes of physical systems, but the determination of the universal one is still an open problem.

13.7   QUALITY OF KS WAVE FUNCTIONS

Thus far, the existence of a Sldet has been shown that can give the same density and energy as the exact ground state. However, in applications, one has to examine whether the KS many-particle state has the same qualitative features as the exact ground state; for example, when the exact Hamiltonian is invariant under a group of transformations G, then the exact eigenstates transform according to an IrRep of G. However, this is not the case with the KS states as the KS potential is not invariant under G, because the ground state density $\langle \Psi | \rho(\mathbf{r}) | \Psi \rangle$ is a bilinear functional of $| \Psi \rangle$ and it belongs to certain IrRep of G, only in the case that the IrRep of G is one-dimensional. As emphasized in the section on HF, spin symmetry is a symmetry that is always present. In this case, the spin-up potential is different from that of spin-down potential and the KS states are not eigenstates of S^2 and S_z. Then, one has to develop KS theories that preserve symmetry and develop new functionals. However, such theories give higher energies because of the symmetry constraint. However, one can analyze the KS state into states transforming according to the symmetry group of the Hamiltonian, as in the case of HF, and use the density of such states for the calculation of the energy from the energy functionals. Unfortunately, in this case, the kinetic energy density functional will give poor results. Therefore, this must be calculated by means of the wave function, but it is not sure whether a lower kinetic energy will be obtained.

After the above discussion, it looks like the application of subspace theory is unavoidable. Take the example of rotational symmetry that is always present in atoms and their ions in the absence of external fields. Then, in analyzing $| \Phi_{KS} \rangle$ in eigenstates of L^2, one does not get states with the same density. However, in applying the subspace theory, considering the whole subspace S^l of the states $\left| \Psi_m^l \right\rangle$ with the same l but different eigenstate m, one obtains the subspace functional

$$E(S^l) = \sum_{M=-l}^{l} \left\langle \Psi_m^l \left| H \right| \Psi_m^l \right\rangle = (2l+1)E^l, \tag{13.191}$$

and the subspace density

$$\rho(\mathbf{r}, S^l) = \sum_{M=-S}^{S} \left\langle \Psi_m^l \left| \hat{\rho}(\mathbf{r}) \right| \Psi_m^l \right\rangle \tag{13.192}$$

has spherical symmetry. In this way, one can develop a DFT for subspaces and derive the KS subspace S_{KS}^l, which has the same subspace density as S^l. Then, the KS states $\left| \Phi_m^l \right\rangle$ that will be obtained after minimization will correspond to noninteracting states, although they may be linear combinations of determinants. After determining the subspace energy of S_{KS}^l, one can derive the energy of a single state because $E^l = G\left(S_{KS}^l\right)/(2l+1)$.

By using subspace minimization of $E(S)$, one can derive in a straightforward way the Hohenberg and Kohn[1] theorem [1] for subspaces, that is, the one-to-one correspondence between subspace density and minimizing subspace of certain dimension M. The next step is to derive the equation for the minimizing subspace of the noninteracting Hamiltonian corresponding to the KS ground state theory [2]. The steps that one has to follow are similar to those of the ground state, with the single-state functionals replaced by subspace functionals [3].

The Achilles heel of DFT in applications is the lack of an accurate-enough expression of the exchange and correlation energy as a functional of the density. The same holds true for the subspace theory. Some work concerning this matter was done by Kohn [38] and Stoddart and Davis [39]. Spin dependence of the exchange potential was also used with quite good results in atomic excited states [4]. Kohn et al. [33] developed a statistical ensemble theory by means of which one can get excited states. The equiensemble limit of this theory coincides with the subspace theory. The applications of the ensemble theory, as far as we know, are limited to the subspace theory. The difficulty with an ensemble theory of unequal weights is that it yields distorted forms of statistical densities that give symmetry-breaking KS potentials. Thus, for example, take the case of degenerate states. The subspace density of these states has the symmetry G of the external potential because

$$\rho(g\mathbf{r}, S) = Tr_S \left\{ g\hat{\rho}(\mathbf{r})g^{-1} \right\}, \tag{13.193}$$

where g is an element of the symmetry group G of the Hamiltonian, for example, a rotation. Because $Tr_S\{AB\} = Tr_S\{BA\}$, it follows that

$$\rho(g\mathbf{r}, S) = Tr_S \left\{ \hat{\rho}(\mathbf{r})g^{-1}g \right\} = \rho(\mathbf{r}, S). \tag{13.194}$$

Nevertheless, a density not having the symmetry of the external potential gives KS potentials that cannot reproduce the correct density unless additional constraints are imposed [35]. This problem was discussed in reference [34], and it was shown that symmetry-breaking KS potentials could not reproduce the correct density. Thus, a single-state density of an exact eigenstate may have an asymmetric density, for example, a density of the form [35]

$$\rho(r) = \rho_0(r) + \rho_1(r)\cos\theta. \tag{13.195}$$

This density depends on the direction, whereas the external potential does not. One could argue that this density corresponds to a degenerate state, but the degeneracy of the ground state can be lifted by adding a weak potential of the form $v_1(r) \cos \theta$ to the external potential. This addition is inconsistent with the Poisson's equation for the Hartree potential as the term $\rho_1(r)$ of the density does not go to zero when $v_1(r)$ tends to 0. In addition, $v_1(r) \cos \theta$ will produce higher spherical harmonic terms to the KS orbitals; therefore, one will not obtain a self-consistent density. Similar arguments hold for the exchange and correlation part of the KS potential. These inconsistencies imply that either the functional derivatives of the energy density functionals do not always exist [36] or there are KS potentials that do not come from functional derivatives. Examples of such potentials have been given recently in reference [37].

The above comments do not imply that there is a serious problem with the KS theory, as one can formulate a KS theory without using functional derivatives.

REFERENCES

1. Hohenberg, P. and Kohn, W., *Phys. Rev. B*, 136, 864, 1964.
2. Kohn, W. and Sham L. J., *Phys. Rev. A*, 140, 1133, 1965.
3. Theophilou, A. K., *J. Phys. C*, 12, 5419, 1979.
4. Theophilou, A. K. and Papaconstantinou, P., *Phys. Rev. A*, 61, 022502, 2000.
5. Theophilou, A. K. and Glushkov, V. N., *Phys. Rev. A*, 64, 064501, 2001.
6. Glushkov, V. N. and Theophilou A. K., *J. Phys. B At. Mol. Opt. Phys.*, 35, 2313, 2002.
7. Nagy, A., *Phys. Rev. A*, 57, 1672, 1998.
8. Gidopoulos, N. I. and Theophilou, A. K., *Philos. Mag. B*, 69, 1067, 1994.
9. Loewdin, P. O., *Phys. Rev.*, 97, 1509, 1955.
10. Thanos, S. and Theophilou, A. K., *J. Chem. Phys.*, 124, 204109, 2006.
11. Theophilou, I., Thanos, S., and Theophilou, A. K., *J. Chem. Phys.*, 127, 234103, 2007.
12. Theophilou, I. and Thanos, S., *Mol. Phys.*, 109, 1495, 2011.
13. Li, X. and Paldus, J., *Int. J. Quant. Chem.*, 109, 1756, 2009.
14. Kotani, M., Amemiya, A., Ishigura, E., and Kimura, T., *Table of Molecular Integrals*, Maruzer, Tokyo, 1955.
15. Goddard, W. A. III, *J. Chem. Phys.*, 48, 450, 1968.
16. Manne, R., *Theor. Chim. Acta*, 116, 6, 1966.
17. Smith, V. H. and Harris, F. E., *J. Math. Phys.*, 10, 771, 1969.
18. Pauncz, R., *J. Mol. Struct.*, 199, 257, 1989.
19. Amos, A. T. and Hall, G. G., *Proc. R. Soc. A*, 263, 483, 1961.
20. Karadakov, P., *Int. J. Quant. Chem.*, 27, 699, 1985.
21. Theophilou, A. K. and Glushkov, V. N., *Int. J. Quant. Chem.*, 104, 538, 2005.
22. Theophilou, A. K. and Glushkov, V. N., *J. Chem. Phys.*, 124, 034105, 2006.
23. Theophilou, A. K., The fundamentals of electron density, density matrix and density functional theory, in *Atoms, Molecules and the Solid State*, Gidopoulos, N. I. and Wilson, S., eds., Kluwer, Dordrecht, The Netherlands, 2003.
24. Glushkov, V. N. and Fesenko, S. I., *J. Chem. Phys.*, 125, 234111, 2006.
25. Glushkov, V. N., Fesenko, S. I., and Polatoglou, H. M. *Theor. Chem. Accounts*, 124, 365, 2009.
26. Glushkov, V. N. *Int. J. Quant. Chem*, 2012, DOI: 10.1002/qua.24019.
27. Glushkov, V. N. and Assfeld, X., *J. Chem. Phys.*, 132, 204106, 2010.
28. Levy, M., *Proc. Natl. Acad. Sci. U.S.A.*, 76, 6062, 1979.
29. Thomas, L. H., *Proc. Cambridge Phil. Soc.*, 23, 542, 1926.
30. Fermi, E., *Z. Phys.*, 48, 73, 1928.
31. Theophilou, A. K., *Int. J. Quant. Chem.*, 69, 461, 1998.
32. Gidopoulos, N. I., *Phys. Rev. A* 83, 040502(R), 2011.
33. Gross, E. K. U., Oliveira, L. N., and Kohn, W., *Phys. Rev. A*, 37, 2805, 1988; ibid. 37, 2809, 1988; ibid. 37, 2821, 1988.
34. Theophilou, A. K., *J. Mol. Struct.*, 943, 42, 2010.
35. Fertig, H. A. and Kohn, W., *Phys. Rev. A*, 62, 052511, 2000.
36. Nesbet, R. K., *Phys. Rev. A*, 65, 010502, 2001.
37. Galduk, A. and Staroverov, V. N., *J. Chem. Phys.*, 131, 044107, 2009.
38. Kohn, W., *Phys. Rev. A*, 140, 1133, 1979.
39. Stoddart, J. C. and Davis, K., *Solid State Commun.*, 42, 147, 1982.

14 Analysis of Generalized Gradient Approximation for Exchange Energy*

José L. Gázquez, Jorge M. del Campo, Samuel B. Trickey, Rodrigo J. Alvarez-Mendez, and Alberto Vela

CONTENTS

14.1 INTRODUCTION

The framework provided by the Kohn–Sham[1] (KS) form of density functional theory[2] for the electronic structure description of atoms, molecules, and solids has become of great importance. Currently available approximations to the exchange-correlation (XC) energy functional allow the study of small, medium, and large systems with a reasonable computational effort and quality of outcomes.[3–7] However, the accuracy in the prediction of structural, thermodynamic, and kinetic properties, among others, with comparatively simple functionals needs to be increased beyond present limits to achieve a reliable method for the description of a wide variety of systems with different characteristics.[8]

A few years ago, Perdew and Schmidt[9] were able to organize what had been the historical development and some of the issues confronting any attempt at systematic increase of the accuracy of the XC functional, through what they called *Jacob's ladder*. On it, the local spin density approximation (LDA), in which the functional

---

* We dedicate this work to Prof. B. M. Deb, on his 70th birthday, for his great contributions to the development of theoretical chemistry.

is determined solely in terms of the local values of the spin-up and spin-down electronic densities via the expression for the homogeneous electron gas, is placed on the first rung. The second rung consists of functionals that depend, additionally, on the magnitude of the gradients of the spin-up and spin-down densities to account for inhomogeneities. Because the form of that dependence is not simply that of the gradient expansion, such functionals are usually referred to as generalized gradient approximations (GGAs). The third rung comprises functionals that incorporate the Laplacian of the density and the KS kinetic energy density. Through the latter, one introduces an explicit dependence on the KS orbitals in addition to the one that comes from the expression of the electronic density in terms of these orbitals. On the fourth rung, a fraction or the full exact exchange is incorporated. Recall that the X functional is defined from the KS orbitals through their implicit dependence on the density. Global and local hybrid functionals correspond to this level. On the fifth rung, explicit dependence on the unoccupied KS orbitals is added. The second and third rung functionals are said to be of semilocal nature (although they are strictly one-point functionals), whereas the fourth and fifth rung functionals have explicitly nonlocal components.

In the framework of this systematization, the usual expectation, or at least hope, is for the accuracy to increase as one climbs the ladder, although the complexity and computational effort will also increase. However, an interesting aspect is that the actual accuracy limits for a given rung are not necessarily exhausted by the current state-of-the-art at that rung. Lacking an analytical framework by which to assess the ultimately achievable accuracy for some quantity (e.g., bond lengths or atomization energies), functional developers are forced to explore the improvement of existing successful approximations.

In this context, the search for better descriptions at the GGA rung continues to be a very active research area. In part, this is because such functionals provide good balance between accuracy and computational effort. Interest in GGAs also persists because of the quest for implementation of orbital-free density functional theory.[10,11] Additionally, the GGA functionals usually are ingredients of the higher-rung approximations, so that it is important to achieve a better understanding of their general behavior and the constraints that they could or should satisfy.

The objectives of the present work are to analyze the behavior of GGA functionals from the perspective of the constraints related to the low and large reduced density gradient regions, taking as a starting point the approximation of Perdew, Burke, and Ernzerhof (PBE)[12]; to analyze the behavior of the reduced density gradient as a function of the density; and to establish some of the implications it may have in the development of improved GGA functionals.

14.2 PBE FORMULATION OF GGA FOR EXCHANGE ENERGY FUNCTIONAL

14.2.1 PBE AND ITS REVISIONS

The exchange energy functional in the GGA usually is expressed, for the spin non-polarized case, in the form

$$E_x[\rho] = \int d\mathbf{r}\rho(\mathbf{r})\varepsilon_{xLDA}[\rho(\mathbf{r})]F_x(s), \qquad (14.1)$$

where the X energy density in the LDA is

$$\varepsilon_{xLDA}[\rho(\mathbf{r})] = -C_X\rho^{1/3}(\mathbf{r}) = -(3/4)(3/\pi)^{1/3}\rho^{1/3}(\mathbf{r}), \qquad (14.2)$$

and the reduced density gradient is defined as

$$s(\mathbf{r}) = |\nabla\rho(\mathbf{r})|/[2(3\pi^2)^{1/3}\rho^{4/3}(\mathbf{r})]. \qquad (14.3)$$

The enhancement factor $F_X(s)$ describes deviations from homogeneous electron gas behavior. For PBE,[12] it is written as

$$F_X^{PBE}(s) = 1 + \kappa - \frac{\kappa}{1 + \dfrac{\mu s^2}{\kappa}}, \qquad (14.4)$$

where κ and μ are parameters.

The PBE enhancement factor form, together with the values assigned to the parameters κ and μ, fulfills the following constraints:

1. Under uniform density scaling, $\rho(\mathbf{r}) \to \lambda^3\rho(\lambda\mathbf{r})$, the exchange energy scales as λ.

2. From Equation 14.4, for $s = 0$, $F_X^{PBE}(0) = 1$, to recover the homogeneous electron gas exchange.

3. Although written for the spin nonpolarized case, the spin-polarized expressions satisfy the spin-scaling relationship

$$E_x[\rho_\uparrow,\rho_\downarrow] = (E_x[2\rho_\uparrow] + [2\rho_\downarrow])/2, \qquad (14.5)$$

   where $\rho_\uparrow$ and $\rho_\downarrow$ are the spin-up and spin-down electron densities, respectively.

4. Being a GGA, the exchange energy and its functional derivative do not diverge for atoms and molecules at the exponential tail of the charge distribution.

5. The Lieb–Oxford bound[13,14] that establishes the most negative value that the exchange energy may attain for a given electron density, namely,

$$E_X[\rho_\uparrow,\rho_\downarrow] \geq -1.679\int d\mathbf{r}\rho^{4/3}(\mathbf{r}), \qquad (14.6)$$

   is satisfied by imposing the condition that the local values of $F_X(s)$ should not grow beyond a value of 1.804. Because, for the form given by Equation 14.4, the maximum value of $F_X(s)$ is $1 + \kappa$ (the limit when $s \to \infty$), one finds that $\kappa = 0.804$.

6. From Equation 14.4, one has that in the small s regime,

$$F_X(s) \to 1 + \mu s^2, \qquad (14.7)$$

which indicates that μ is directly related to the weight given to the density gradient. Now, the LDA provides a very good description of the linear response of the spin-unpolarized uniform electron gas. Thus, in the PBE functional, μ was set to cancel the second-order gradient contribution to the correlation energy in the high-density limit through the relationship

$$\mu = \pi^2 \beta / 3, \qquad (14.8)$$

where, according to Ma and Brueckner,[15] $\beta = 0.066725$; therefore, $\mu_{PBE} = 0.21951$.

The form of the enhancement function given by Equation 14.4 was first proposed by Becke,[16] who determined the values of the parameters empirically, through a least squares fit to the Hartree–Fock X energies of the noble gas atoms He–Xe, using Hartree–Fock densities to evaluate the integral in Equation 14.1. The values that he found were $\kappa = 0.967$ and $\mu = 0.235$.

The PBE XC functional leads to a reasonable description of a wide variety of properties of molecules and solids. It has become one of the most heavily used approximations in electronic structure calculations. However, several modifications have been proposed with the objective of improving the calculated values of various sets of properties.[17–26]

A general finding that emerges from those modifications is that the magnitude of μ is correlated with the prediction of structural, thermodynamic, and kinetic properties. The smallest nonempirical value of μ corresponds to the gradient expansion approximation (GEA) value,[27] $\mu_{GEA} = 10/81 \approx 0.12346$, whereas the largest nonempirical value of μ comes from imposition of the constraint that, for the hydrogen atom, the exchange energy cancels the Coulomb repulsion energy. The latter constraint leads[26] to $\mu_{xH} = 0.27583$. Large values of μ are better for atomization energies and worse for crystalline lattice constants, whereas low values of μ generate the opposite trend. In fact, Zhao and Truhlar[20] also showed that the magnitude of the coefficient μ correlates with the behavior of the cohesive energies of solids, reaction barrier heights, and nonhydrogenic bond distances in small molecules.

It is important to note that, in general, when the value of μ is changed, the value of β is also changed, through Equation 14.8 or through other arguments, so that, in addition to the exchange energy, the correlation energy also has an impact on this behavior.

14.2.2 LARGE REDUCED DENSITY GRADIENT LIMIT

Most of the PBE modifications have essentially the same behavior at large values of the reduced density gradient, that is, they are built in such a way that $F_x(s) \to 1 + \kappa$ when $s \to \infty$. Their main differences occur in the interval $0 \le s \le 3$, although, for larger values of s, the difference among the different approximations may also come from the different values of κ. Nonetheless, one may wonder about the behavior of the enhancement function in this limit, which arises from the exponential tail of the electronic density of finite systems far away from the nuclei. This was the viewpoint adopted in the development of the VMT[28] X energy functional, which can be viewed as a candidate for superseding PBE. A key analysis in the development of the VMT enhancement factor form involves the outer regions of an atom or a molecule. There the electronic density and its gradient vary slowly, and the density is small compared to its peak value. Thus, one should recover the homogeneous electron gas type behavior, and therefore, VMT was built imposing the condition $F_{xVMT}(s) \rightarrow 1$ when $s \rightarrow \infty$. However, Levy and Perdew[14] have shown that the exact asymptotic behavior is given by

$$\lim_{s \to \infty} s^{1/2} F_{XC}(\rho, s) < \infty. \tag{14.9}$$

Because the unpolarized enhancement factor may be expressed in terms of its exchange and correlation components, Equation 14.9 separates into

$$\lim_{s \to \infty} s^{1/2} F_X(s) < \infty, \tag{14.10}$$

and

$$\lim_{s \to \infty} s^{1/2} F_C(\rho, s) < \infty. \tag{14.11}$$

The limit expressed in Equation 14.10 indicates that $F_x(s)$ should decay asymptotically to zero at a rate equal or faster than $s^{-1/2}$. To our knowledge, only the GGA functionals PW91,[29] LG,[30] and the recently proposed VT{8,4},[31] based on VMT, satisfy Equation 14.10. Although the primary ranges of interest for real systems, that is, $0 \le s \le 3$[32-34] and $0 \le r_s \le 10$[32,33], where $r_s = (3/4\pi\rho)^{1/3}$, are generally accepted, it has been shown[31] that the imposition of the large-s constraint induces subtle but important changes in the region $0 \le s \le 3$.

One may go back to the original PBE form (Equation 14.4) and apply the reasoning that led from VMT to VT{8,4} to satisfy the constraint given by Equation 14.10. A simple approach that allows one to achieve this goal is to express the enhancement function in the form

$$F_X^{PBE-LS}(s) = F_X^{PBE}(s) - (\kappa + 1)\left(1 - e^{-\alpha s^2}\right), \tag{14.12}$$

because the product in the second term of the right-hand side of this equation leads to a term that cancels the $s \rightarrow \infty$ limit of $F_X^{PBE}(s)$ and a second term that decays as $(\kappa + 1)e^{-\alpha s^2}$, hence fulfilling Equation 14.10.

Substitution of Equation 14.4 in Equation 14.12 gives

$$F_X^{PBE-LS}(s) = 1 + \kappa\left(1 - \frac{1}{1 + \dfrac{\mu s^2}{\kappa}}\right) - (\kappa + 1)\left(1 - e^{-\alpha s^2}\right), \tag{14.13}$$

and therefore, in the limit $s \to 0$,

$$F_X^{PBE-LS}(s) = 1 + \left(\mu - \alpha(\kappa + 1)\right)s^2 + O(s^4). \tag{14.14}$$

Thus, incorporation of the correct large-s limit gives an additional parameter α, besides μ and κ. To proceed nonempirically, one may make use of two constraints introduced recently in consideration of modifying the original PBE enhancement factor parameters.

One of the constraints was developed by Constantin et al.,[24] who made use of the asymptotic expansion of the semiclassic neutral atom,[35–37] which, for the exchange energy, adopts the form

$$E_X = E_X^{LDA} + d_1 Z + d_2 Z^{2/3} + \dots, \tag{14.15}$$

where Z is the number of electrons (equal to the nuclear charge for a neutral atom), and the coefficients are determined from a numerical analysis that leads to $d_1 = -0.2240$ and $d_2 = 0.2467$.[36,37] Within this approach, one finds that the value of the coefficient for $d_1 = -0.2240$ can be obtained from a modified second-order gradient expansion (MGEA) in which $\mu_{MGEA} = 0.260$. The PBE exchange functional with this value of μ, with $\kappa = 0.804$, and with the PBE correlation functional with a value of $\beta = 3\mu_{MGEA}/\pi^2 = 0.07903$ was named APBE. It provides an improvement in the description of atomization energies compared to the original PBE.

The other constraint[26] addresses the issue of inadequate cancellation of spurious self-interaction. The constraint is that, for one-electron systems with $\rho_1(\mathbf{r})$, the one-electron density, the exact exchange energy must cancel the Coulomb repulsion energy:

$$E_X[\rho_1] = -J[\rho_1]. \tag{14.16}$$

One can use this equation with the hydrogen atom electron density to fix the value of μ, for $\kappa = 0.804$, with the enhancement function of original PBE form (Equation 14.4). The value obtained from that procedure is $\mu_{xH} = 0.27583$. The use of those μ and κ parameters in the PBE X form, together with the PBE correlation with $\beta = 3\mu_{xH}/\pi^2 = 0.08384$, leads to a functional that is called PBEmol, because at the GGA rung, it provides a good description of several molecular properties, especially the atomization energies. In essence, PBEmol is the molecular counterpart to PBEsol in the sense summarized above. Increasing μ favors finite system accuracy at the cost of worsening extended system accuracy and conversely.

We return to the task of setting the three parameters, μ, κ, and α, in Equation 14.13. We assume first that the correct value of the coefficient of the second-order gradient expansion in the limit $s \to 0$ is the one that arises from the asymptotic expansion of the semiclassic neutral atom. Therefore, from Equation 14.14, one has $\mu_{MGEA} = \mu - \alpha(\kappa + 1) = 0.26$. The second constraint we enforce is freedom from one-electron self-interaction error (Equation 14.16) for the hydrogen atom density $\rho_H(r) = \pi^{-1}e^{-2r}$. The third constraint is to enforce the local Lieb–Oxford bound at the maximum s_{max} of the function given by Equation 14.13, which is $F_X^{PBE-LS}(s_{max}) = 1.804$.

These three requirements yield $\mu = 0.26151$, $\kappa = 0.9403$, and $\alpha = 0.00078$. Within this procedure, the value of the coefficient associated with the second-order gradient correction to the correlation energy β could be fixed through Equations 14.8 and 14.14, so that $\beta_{PBE-LS} = 3\mu_{MGEA}/\pi^2 = 0.07903$.

A comparison between several forms of the X enhancement function is given in Figures 14.1 and 14.2. In both, we present a plot for a large interval of values of s and a plot for the interval $0 \leq s \leq 3$ to distinguish differences in this region.

Figure 14.1 compares PBE, PBE-LS, VMT, and VT{8,4}. One sees that, in the interval $0 \leq s \leq 3$, PBE-LS is larger than all of the others and that VMT and VT{8,4}, which are similar to each other, also lie above PBE. However, as s grows, PBE-LS goes to zero more rapidly than VT{8,4}, VMT tends to unity (by design), and PBE tends to 1.804. Figure 14.2 provides comparison among PBE, PBE-LS, B88[38], and OPTX.[39] The enhancement function for the latter functional, proposed by Handy and Cohen, is

$$F_X^{OPTX}(s) = a_1 + a_2 c_2 \left[\frac{\gamma c_1 s^2}{1 + \gamma c_1 s^2} \right]^2, \tag{14.17}$$

where $c_1 = 4(6\pi^2)^{2/3}$, $c_2 = (2^{1/3}C_X)^{-1}$, and the parameters a_1, a_2, and γ are fixed through a fitting of the Hartree–Fock energy for the atoms H–Ar. The results are $a_1 = 1.05151$, $a_2 = 1.43169$, and $\gamma = 0.006$. Although this functional has a semiempirical nature, when combined with the correlation energy functional of Lee, Yang, and Parr[40] (LYP), it leads to a very good description of several molecular properties, so it is interesting to compare it with the PBE-type approximations to analyze the differences among them. Thus, one can see (Figure 14.2) that, at $s = 0$, the OPTX enhancement function differs from unity and that, for values of s between 0 and approximately 0.5, OPTX remains practically constant. Thus, in this approximation

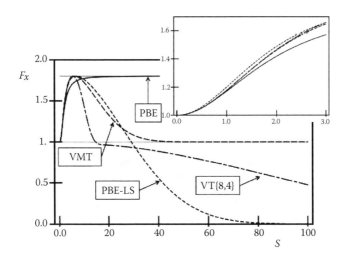

FIGURE 14.1   Plot of the enhancement factor as a function of the reduced density gradient for PBE, VMT, VT{8,4}, and PBE-LS. (Inset) Zoom in the range of interest for real systems.

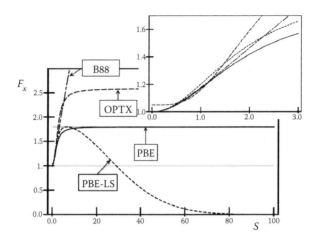

FIGURE 14.2 Plot of the enhancement factor as a function of the reduced density gradient for PBE, OPTX, B88, and PBE-LS. (Inset) Zoom in the range of interest for real systems.

and over this interval, the contributions due to the nonuniformities of the density are essentially constant and larger than those provided by X functionals with quadratic behavior. For values of s beyond 0.5, OPTX grows abruptly toward its $s \to \infty$ limiting value $F_X^{OPTX}(s) \to a_1 + a_2 c_2 = 2.59$. On the other hand, one can see that PBE and PBE-LS, in comparison with OPTX, grow slower, starting from unity, in the interval $0 \le s \le 3$. In contrast, the widely used B88 X enhancement function[38] was formulated to yield the correct asymptotic limit of the exchange energy density per electron, namely, $\lim_{r \to \infty} \varepsilon_x = -1/2r$. However, it has been shown[41] that, for a GGA, this condition is equivalent to $\lim_{s \to \infty} F_{xGGA}(s) \to s/\ln s$, which clearly is irreconcilable with Equation 14.10. In Figure 14.2, one can see that B88 lies between PBE and PBE-LS in the interval from $s = 0$ to $s \approx 2.5$, and after that, it grows abruptly. Separately, it has been shown[42] that this asymptotic growth constraint $\lim_{s \to \infty} F_{xGGA}(s) \to s/\ln s$ is, in fact, irrelevant to the energetic behavior of a GGA, which is why both VT{8,4} X and the PBE-LS proposed here ignore that constraint and use the relevant one (Equation 14.10).

14.2.3 RESULTS FOR SEVERAL ATOMIC AND MOLECULAR PROPERTIES

To analyze the performance of the functionals described in the previous section, we consider several atomic and molecular properties. The calculations were done with a developmental version of NWChem version 6.0.[43]

Table 14.1 shows the exchange energies for the atoms H–Ar obtained from exchange-only calculations with a Def2-QZVP basis set for several functionals in comparison with the Hartree–Fock values obtained for the same basis set. One can see, as expected, that the empirical functionals OPTX and B88, whose parameters were set through least squares fits to the exchange energy of atoms He–Ar, lead to the lowest mean absolute errors (MAEs). In fact, OPTX with three adjustable

TABLE 14.1

Exchange Energies and MAEs with Respect to the Hartree–Fock Values, in Hartree Units, as Determined with Several GGA Functionals and One Meta-GGA through Exchange-Only Calculations

Atom	Exchange Energy							
	HF	OPTX	B88	PBE	VMT	VT{8,4}	PBE–LS	revTPSS
H	−0.313	−0.308	−0.306	−0.301	−0.304	−0.304	−0.310	−0.311
He	−1.026	−1.019	−1.016	−1.002	−1.010	−1.011	−1.029	−1.025
Li	−1.781	−1.775	−1.768	−1.748	−1.761	−1.762	−1.791	−1.783
Be	−2.667	−2.663	−2.652	−2.627	−2.645	−2.647	−2.690	−2.671
B	−3.770	−3.754	−3.748	−3.716	−3.738	−3.740	−3.799	−3.775
C	−5.077	−5.054	−5.048	−5.008	−5.034	−5.036	−5.111	−5.065
N	−6.607	−6.584	−6.569	−6.521	−6.549	−6.551	−6.643	−6.578
O	−8.218	−8.190	−8.188	−8.131	−8.162	−8.165	−8.279	−8.198
F	−10.045	−10.016	−10.021	−9.954	−9.988	−9.990	−10.125	−10.012
Ne	−12.108	−12.088	−12.087	−12.009	−12.044	−12.047	−12.201	−12.060
Na	−14.017	−13.989	−13.977	−13.891	−13.930	−13.933	−14.108	−13.944
Mg	−15.994	−15.968	−15.954	−15.863	−15.905	−15.908	−16.105	−15.907
Al	−18.092	−18.068	−18.055	−17.952	−17.997	−18.000	−18.220	−17.992
Si	−20.304	−20.281	−20.261	−20.147	−20.194	−20.198	−20.440	−20.180
P	−22.642	−22.627	−22.593	−22.467	−22.517	−22.521	−22.784	−22.493
S	−25.034	−25.014	−24.976	−24.837	−24.889	−24.893	−25.180	−24.865
Cl	−27.544	−27.529	−27.481	−27.329	−27.384	−27.388	−27.698	−27.349
Ar	−30.185	−30.185	−30.119	−29.953	−30.011	−30.015	−30.347	−29.964
MAE		0.017	0.034	0.109	0.076	0.073	0.080	0.071

parameters leads to the best description followed by B88 that has only one adjustable parameter. On the other hand, one can also see that the three nonempirical functionals VMT, VT{8,4}, and PBE-LS, whose enhancement function tends to one, zero, and zero, respectively, in the $s \rightarrow \infty$ limit, provide an improvement over the original PBE functional, which is rather similar for the three functionals. We have also included the nonempirical functional revTPSS,[44] which is a meta-GGA and therefore belongs to the next rung in the Perdew–Schmidt Jacob's ladder. One can see that the results provided by revTPSS are close to the ones obtained from the three large-s corrected GGA functionals.

In Table 14.2, we present the results for heats of formation, ionization potentials, electron affinities, proton affinities, binding energies of weakly interacting systems, barrier heights for hydrogen and nonhydrogen transfer reactions, bond distances, and harmonic frequencies for some well-known test sets designed to validate approximate energy density functionals. Those have been described in previous works.[26,31] In these cases, correlation energy functionals have been added to the exchange energy functionals discussed. Thus, OPTX and B88 X have been combined, as usual, with LYP C, whereas PBE, VMT, VT{8,4}, and PBE-LS have been combined with PBE C to maintain their nonempirical nature. As described previously, for PBE-LS,

TABLE 14.2

MAEs for Several Properties as Calculated with Several GGA Functionals and One Meta-GGA

Property	MAE						
	OLYP	BLYP	PBE	VMT	VT{8,4}	PBE-LS	revTPSS
Heats of formation (223)	5.51	9.64	21.21	10.53	9.98	9.39	4.55
Ionization potentials (13)	2.61	4.20	3.47	3.26	3.25	3.65	3.06
Electron affinities (13)	3.63	2.97	2.64	2.48	2.48	2.61	2.45
Proton affinities (8)	1.66	1.78	1.39	1.07	1.09	1.23	1.80
Binding energies of weakly interacting systems (31)	2.27	1.67	1.64	1.52	1.53	1.49	1.41
Reaction Barrier Heights							
Hydrogen transfer forward (19)	6.02	7.81	9.49	8.31	8.23	7.26	6.63
Hydrogen transfer backward (19)	6.06	7.85	9.72	8.66	8.59	7.77	7.72
Nonhydrogen transfer forward (19)	7.74	10.49	10.38	9.82	9.79	9.45	11.18
Nonhydrogen transfer backward (19)	7.21	10.03	9.96	9.49	9.46	9.31	10.08
Bond distances (96)	0.0198	0.0240	0.0179	0.0209	0.0211	0.0216	0.0204
Frequencies (82)	40.42	56.25	43.30	45.41	45.57	45.74	39.58

Note: All energies are in kcal/mol, bond distances in Å, and frequencies in cm^{-1}. The number of cases for each test set appears in parentheses.

$\beta_{PBE-LS} = 3\mu_{MGEA}/\pi^2 = 0.07903$. One can see that, except for the electron affinities, the proton affinities, and the binding energies of weakly interacting systems, OLYP provides the best description. Particularly, the MAE for the heats of formation is about 1 kcal/mol above the value for revTPSS, and the barrier heights have the lowest MAE, including even revTPSS. In the case of the nonempirical functionals, it may be observed that heats of formation and barrier heights show smaller MAEs for VMT, VT{8,4}, and PBE-LS than PBE, with the best results corresponding to PBE-LS. However, PBE provides smaller MAEs for bond distances and frequencies.

14.3 BEHAVIOR OF REDUCED DENSITY GRADIENT AS A FUNCTION OF DENSITY

The foregoing analysis indicates that it is important to get a closer picture of the behavior of the density and its gradient in different regions of a molecule. To do this, one can follow the procedure of Johnson et al.,[45,46] in which the reduced gradient given by Equation 14.3 is plotted against the electronic density. Although their objective was to reveal the noncovalent interactions, the plot provides significant information about all regions of a molecule. Thus, in Figure 14.3, we present these types of plots for the water molecule, its dimer, and its tetramer, and in Figure 14.4, we present the plots corresponding to the reactant, transition state, and product of the isomerization reaction of HCN. The calculations reported in this section were carried out with a developmental version of deMon2k, version 3.x.[47]

For the analysis of these plots, it is important to consider the behavior of the electronic density near the nuclei and far from them, that is,[48,49] for a spherically averaged density, we have that, at a given nucleus a of nuclear charge Z_a,

$$\left.\frac{\partial \rho(r)}{\partial r}\right|_{r=0} = -2Z_a\rho(0), \qquad (14.18)$$

and, asymptotically, at large distances from the nuclei,[50,51]

$$\rho(r)\xrightarrow[r\to\infty]{} e^{-2(2I)^{1/2}r}, \qquad (14.19)$$

where I is the first ionization potential.

The overall shape of the graph follows a $\rho^{-1/3}$ pattern due to the piecewise exponential nature of the electronic density. With respect to the values of ρ, one can identify three approximate regions:

• Region I, where ρ takes values between 0 and approximately 0.1
• Region II, where ρ takes values between approximately 0.1 and 0.8
• Region III, where ρ takes values beyond approximately 0.8

Thus, the regions near the nuclei will appear toward the right edge of the graph (in region III) and will be governed by an exponential fall that satisfies Equation 14.18 at each nucleus, whereas the regions far from the nuclei will appear in the upper part of the left edge (in region I) and will be governed by the exponential form given in Equation 14.19. Note that, in this limit, the reduced density gradient diverges because the density raised to the 4/3 power (denominator) decays faster to zero than the gradient of the density (numerator).

The spikes that appear are a very important aspect in the regions that correspond to small-s values because these correspond to the different chemical interactions present. For all cases, the spikes show that, in these regions, the density gradient approaches zero and becomes equal to zero at the critical point. In Figure 14.3, the spikes located in region II reveal the covalent interactions that correspond to regions with intermediate values of the density and low values of the gradient of the density, whereas the spikes located in region I reveal the noncovalent interactions. They correspond to regions with low values of both the density and its gradient. In Figure 14.4, one can see that the covalent bond between carbon and nitrogen basically remains constant at a value of the density slightly over 0.4, whereas the H–C bond in the reactant and the N–H bond in the product appear slightly below and slightly above a value of the density of 0.3, respectively. However, at the transition state, this

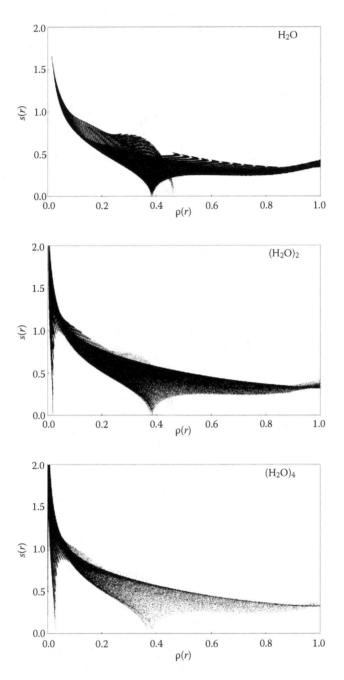

FIGURE 14.3 Plot of the reduced density gradient versus the electronic density for water molecule (top), its dimer (middle), and its tetramer (bottom).

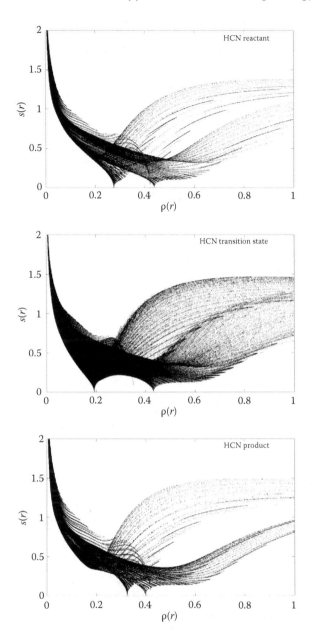

FIGURE 14.4 Plot of the reduced density gradient versus the electronic density for the reactant (top), transition state (middle), and product (bottom) of the isomerization reaction of the HCN molecule.

spike moves to a value slightly below 0.2, that is, it moves toward the noncovalent interaction region, because, at this stage of the interaction, the H–C bond is breaking and the N–H bond is forming, so that, in general, one could infer that there is a weak interaction of the entity formed by carbon and nitrogen with hydrogen.

Let us consider now the density of the reduced density gradient, $g(s)$. It provides, for a given electronic density, the number of electrons with the values of s that lie between s and $s + ds$. Zupan et al.[32,33] have defined this distribution as

$$g(s) = \int d\mathbf{r} \rho(\mathbf{r}) \delta\big(s - s(\mathbf{r})\big),$$ (14.20)

where $\delta(s - s(\mathbf{r}))$ is the Dirac delta function. In Figures 14.5 and 14.6, we present plots of $g(s)$ for the water systems considered here and the isomerization reaction of

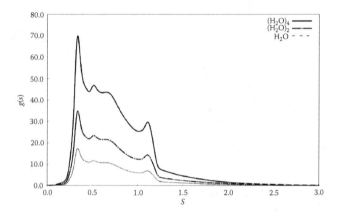

FIGURE 14.5   Plot of the density of the reduced density gradient as a function of s for the water systems considered in this work.

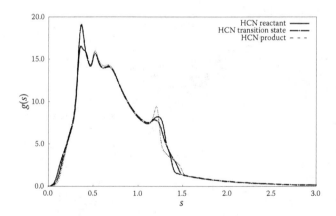

FIGURE 14.6   Plot of the density of the reduced density gradient as a function of s for the isomerization reaction of HCN.

HCN, respectively (details of the numerical procedure are given in reference 34). It is clear from these plots, and from the plots for many others we have done, that the region of physical interest for real systems lies primarily in the interval $0 \leq s \leq 3$.

14.4   CONCLUDING REMARKS

The study of the GGA functionals considered in this work, and the analysis of the behavior of the reduced density gradient as a function of the electronic density, together with its density, allows one to observe two important aspects for the possible improvement of exchange energy functionals that only depend on the density and its gradient.

The first one is that the contribution to the exchange energy seems to be dominated by the behavior of the enhancement function in the interval $0 \leq s \leq 3$ and that its description in this region is crucial to improve the accuracy of the nonempirical GGAs known to date. However, this behavior does not imply that the imposition of constraints related to large values of s, such as the one given by Equation 14.10, is not important. Indeed, we have found in a previous work[31] and in the present one that, by imposing the large-s constraint, one induces changes in the interval $0 \leq s \leq 3$, which may have an important impact on the prediction of diverse properties. In fact, the results presented in this work, together with the ones reported in reference 26, indicate that PBE-LS provides the best description for most of the properties considered. Note also that, in its development, the constraints associated with Equations 14.15 and 14.16 were also incorporated, adding to the ones already present in PBE.

The second important aspect of the present analysis comes from the plots of the reduced density gradient versus the density. It was shown that there are two regions of the electronic density that give rise to low values of s. One of those corresponds to the covalent interactions, and one corresponds to the noncovalent or to the transition-state-type interactions. This situation could indicate that the weight given to the gradients of the density could be different, depending on the values of the density itself, and that the limit for small s could be revised and modified. Recall that the functional OPTX, which leads to a rather good description of the heats of formation, behaves as s^4 for small s, in contrast with PBE that behaves as s^2.

In summary, the analysis presented here points at some of the issues that may be studied to improve the accuracy of the GGAs known to date.

ACKNOWLEDGMENTS

We thank the Laboratorio de Supercómputo y Visualización of Universidad Autónoma Metropolitana-Iztapalapa and DGTIC-UNAM for the use of their facilities. JMC and RJAM were supported in part by Conacyt through a postdoctoral and a doctoral fellowship, respectively. AV, JLG, and JMC were supported in part by the Conacyt project grant 128369. JMC was also supported in part by DGAPA-UNAM under grant no. IA101512. SBT was supported in part by the US Department of Energy grant DE-SC-0002139. Part of this work was done while AV was on a sabbatical leave at UAM-Iztapalapa, occupying the "Raul Cetina Rosado" chair, and while JMC was an invited professor in this same institution. We thank all the members of the Fisicoquímica Teórica group for their warm hospitality and the intense discussions during their stay.

REFERENCES

1. Kohn, W. and Sham, L. J., *Phys. Rev.*, 140, A1133, 1965.
2. Hohenberg, P. and Kohn, W., *Phys. Rev. B*, 136, B864, 1964.
3. Parr, R. G. and Yang, W. T., *Density-Functional Theory of Atoms and Molecules*, Oxford University Press, New York, 1989.
4. Dreizler, R. M. and Gross, E. K. U., *Density Functional Theory*, Springer, Berlin, 1990.
5. Perdew, J. P. and Kurth, S., in *A Primer in Density Functional Theory*, Fiolhais, C., Nogueira, F., and Marques, M. A. L., eds., Springer, Berlin, 2003, 1.
6. Perdew, J. P., Ruzsinszky, A., Tao, J. M., Staroverov, V. N., Scuseria, G. E., and Csonka, G. I., *J. Chem. Phys.*, 123, 062201, 2005.
7. Scuseria, G. E. and Staroverov, V. N., in *Theory and Applications of Computational Chemistry: The First Forty Years*, Dykstra, C., Frenking, G., Kim, K. S., and Scuseria, G. E., eds., Elsevier, Amsterdam, 2005, 669.
8. Cohen, A. J., Mori-Sanchez, P., and Yang, W. T., *Chem. Rev.*, 112, 289, 2012.
9. Perdew, J. P. and Schmidt, K., in *Density Functional Theory and Its Application to Materials*, Van Doren, V. E., Van Alsenoy, C., and Geerlings, P., eds., AIP, Melville, New York, 2001, 1.
10. Dufty, J. W. and Trickey, S. B., *Phys. Rev. B*, 84, 125118, 2011.
11. Trickey, S. B., Karasiev, V. V., and Vela, A., *Phys. Rev. B*, 84, 075146, 2011.
12. Perdew, J. P., Burke, K., and Ernzerhof, M., *Phys. Rev. Lett.*, 77, 3865, 1996, erratum, 78, 1396, 1997.
13. Lieb, E. H. and Oxford, S., *Int. J. Quantum Chem.*, 19, 427, 1981.
14. Levy, M. and Perdew, J. P., *Phys. Rev. B*, 48, 11638, 1993, erratum, 55, 13321, 1997.
15. Ma, S. K. and Brueckner, K. A., *Phys. Rev.*, 165, 18, 1968.
16. Becke, A. D., *J. Chem. Phys.*, 84, 4524, 1986.
17. Zhang, Y. K. and Yang, W. T., *Phys. Rev. Lett.*, 80, 890, 1998.
18. Hammer, B., Hansen, L. B., and Norskov, J. K., *Phys. Rev. B*, 59, 7413, 1999.
19. Perdew, J. P., Ruzsinszky, A., Csonka, G. I., Vydrov, O. A., Scuseria, G. E., Constantin, L. A., Zhou, X. L., and Burke, K., *Phys. Rev. Lett.*, 100, 136406, 2008.
20. Zhao, Y. and Truhlar, D. G., *J. Chem. Phys.*, 128, 184109, 2008.
21. Swart, M., Sola, M., and Bickelhaupt, F. M., *J. Chem. Phys.*, 131, 094103, 2009.
22. Cooper, V. R., *Phys. Rev. B*, 81, 161104, 2010.
23. Fabiano, E., Constantin, L. A., and Della Sala, F., *Phys. Rev. B*, 82, 113104, 2010.
24. Constantin, L. A., Fabiano, E., Laricchia, S., and Della Sala, F., *Phys. Rev. Lett.*, 106, 186406, 2011.
25. Fabiano, E., Constantin, L. A., and Della Sala, F., *J. Chem. Theory Comput.*, 7, 3548, 2011.
26. M. del Campo, J., Gázquez, J. L., Trickey, S. B., and Vela, A., *J. Chem. Phys.*, 136, 104108, 2012.
27. Antoniewicz, P. R. and Kleinman, L., *Phys. Rev. B*, 31, 6779, 1985.
28. Vela, A., Medel, V., and Trickey, S. B., *J. Chem. Phys.*, 130, 244103, 2009.
29. Perdew, J. P., in *Electronic Structure of Solids '91*, Ziesche, P. and Eschrig, H., eds., Akademie, Berlin, 1991, 11.
30. Lacks, D. J. and Gordon, R. G., *Phys. Rev. A*, 47, 4681, 1993.
31. Vela, A., Pacheco-Kato, J. C., Gázquez, J. L., M. del Campo, J., and Trickey, S. B., *J. Chem. Phys.*, 136, 144115, 2012.

32. Zupan, A., Perdew, J. P., Burke, K., and Causà, M., *Int. J. Quantum Chem.*, 61, 835, 1997.
33. Zupan, A., Burke, K., Ernzerhof, M., and Perdew, J. P., *J. Chem. Phys.*, 106, 10184, 1997.
34. M. del Campo, J., Gázquez, J. L., Alvarez-Mendez, R. J., and Vela, A., *Int. J. Quantum Chem.*, 112, 3594, 2012.
35. Perdew, J. P., Constantin, L. A., Sagvolden, E., and Burke, K., *Phys. Rev. Lett.*, 97, 223002, 2006.
36. Elliott, P. and Burke, K., *Can. J. Chem.*, 87, 1485, 2009.
37. Lee, D., Constantin, L. A., Perdew, J. P., and Burke, K., *J. Chem. Phys.*, 130, 034107, 2009.
38. Becke, A. D., *Phys. Rev. A*, 38, 3098, 1988.
39. Handy, N. C. and Cohen, A., *J. Mol. Phys.*, 99, 403, 2001.
40. Lee, C. T., Yang, W. T., and Parr, R. G., *Phys. Rev. B*, 37, 785, 1988.
41. van Leeuwen, R. and Baerends, E., *J. Phys. Rev. A*, 49, 2421, 1994.
42. Engel, E., Chevary, J. A., Macdonald, L. D., and Vosko, S. H., *Z. Phys. D*, 23, 7, 1992.
43. Valiev, M., Bylaska, E. J., Govind, N., Kowalski, K., Straatsma, T. P., Van Dam, H. J. J., Wang, D., Nieplocha, J., Apra, E., Windus, T. L., and de Jong, W., *Comput. Phys. Commun.*, 181, 1477, 2010.
44. Perdew, J. P., Ruzsinszky, A., Csonka, G. I., Constantin, L. A., and Sun, J. W., *Phys. Rev. Lett.*, 103, 026403, 2009, erratum, 106, 179902, 2011.
45. Johnson, E. R., Keinan, S., Mori-Sanchez, P., Contreras-Garcia, J., Cohen, A. J., and Yang, W. T., *J. Am. Chem. Soc.*, 132, 6498, 2010.
46. Contreras-Garcia, J., Johnson, E. R., Keinan, S., Chaudret, R., Piquemal, J. P., Beratan, D. N., and Yang, W. T. *J. Chem. Theory Comput.*, 7, 625, 2011.
47. Koster, A. M., Geudtner, G., Calaminici, P., Casida, M. E., Domínguez, V. D., Flores-Moreno, R., Gamboa, G. U., Goursot, A., Heine, T., Ipatov, A., Janetzko, F., M. del Campo, J., Reveles, J. U., Vela, A., Zuniga-Gutierrez, B., and Salahub, D. R., *deMon2k, Version 3.x*, The deMon Developers, Cinvestav, Mexico City, 2011.
48. Kato, T., *Commun. Pure Appl. Math.*, 10, 151, 1957.
49. Steiner, E., *J. Chem. Phys.*, 39, 2365, 1963.
50. Morrell, M. M., Parr, R. G., and Levy, M., *J. Chem. Phys.*, 62, 549, 1975.
51. Katriel, J. and Davidson, E. R., *Proc. Natl. Acad. Sci. USA*, 77, 4403, 1980.

15 Intermolecular Interactions through Energy Decomposition
A Chemists' Perspective*

R. Mahesh Kumar, Dolly Vijay,
G. Narahari Sastry, and V. Subramanian

CONTENTS

15.1 INTRODUCTION

Chemists have always strived to comprehend how atoms combine to form compounds of both simple and complex architecture. The variety of compounds which surround us is formed by the difference in the chemical bonding between the mere 118 atoms of the periodic table. There is no denial that these strong chemical bonds that hold the atoms of a molecule together are important,[1-7] but equally important is to understand how these molecules are held together to form large clusters and supramolecular assemblies.[8-15] Unraveling the structure of these assemblies and correlating with their functions have been the central theme of many chemical and biological investigations.[16-22] In such a scenario, it becomes crucial not only to know the geometrical disposition of the system but also to have a clear idea about the various forces that hold together these assemblies and govern its stability. The forces holding molecules together in such large assemblies play an important

---

* The authors dedicate this chapter to Prof. B. M. Deb for his outstanding contributions to develop chemistry, especially theoretical chemistry, in India.

role in determining the properties of organic, inorganic, organometallic, biomolecular, and supramolecular systems.[23-27] In addition, the role of intermolecular interactions in specificity and selectivity of chemical reactions and biocatalysis, molecular recognition phenomenon, and drug–receptor interactions has also been identified.[28-38] Although these intermolecular forces are significantly weaker than the chemical bond, a thorough understanding and quantification of these forces are of utmost importance. Such vital information will not only help to modify the structure and hence the function of drugs and materials but also aid in the design of new functional molecules, nanomaterials, and molecular devices with improved efficacy.[39-43] Thus, the study of intermolecular interactions has received intensive attention of researchers and has been the main topic of research in the past 100 years.

Ever since the discovery of "intermolecular interactions," innumerable research articles, reviews, books, and monographs have appeared on this topic.[44-53] It is found from various analytical tools that the strength of the intermolecular interaction varies from weak to van der Waals type to covalent interaction smoothly without any border.[54-56] Analysis of energies of intermolecular systems reveals the presence of a continuum of strength ranging from 1.0 to >100 kcal/mol for weak and van der Waals systems to strong hydrogen-bonded systems (covalent limit).[57] Advancement in quantum chemical methods and tools facilitates the pursuit of finding a new type of weak interactions.[53,58] Quantification of such intermolecular forces will provide insight to understand the stability of clusters or assemblies. Evaluating the interaction energy (INT) or complexation energy is one of the most straightforward approaches to gauge the stability of a molecular system. The most simple and conventional method to calculate INT is the supermolecule approach wherein the INT is expressed as the difference in the total energy of the complex and monomers:

$$INT = (E_{COMPLEX} - (E_{MONOMER1} + E_{MONOMER2})), \tag{15.1}$$

where $E_{COMPLEX}$ and $E_{MONOMER1}$ and $E_{MONOMER2}$ are the total energies of complex and monomer, respectively. Although the supermolecule approach is very simple and has universal applicability, it has some inherent limitations. This approach assumes that the internal coordinates of the monomers are essentially the same in both isolated and bonded states. This leads to the neglect of the deformation in the monomer geometries upon complexation. Basis set superposition, which causes an unphysical lowering of the complex energy, is another stumbling block that is not accounted for in this approach.[59-65] Apart from these, a fundamental lacuna of this approach is that it fails to provide any insight into the nature of various components (intermolecular forces) that contribute to the total INT. There are a number of schemes that have been developed over the years to partition the total INT into various conceptual and physically meaningful forces such as electrostatic, dispersion, charge transfer, polarization, and exchange repulsion. The charge–charge, charge–dipole, dipole–dipole, dipole–quadrupole, and quadrupole–quadrupole interactions and other interactions involving multipoles are categorized as electrostatic. The induced dipole–dipole interaction is dispersion. Understanding the contributions to the total stabilization of the intermolecular interactions is the primary motive of a number of previous investigations. These schemes employ either the variational or the perturbative approach to decompose the INT. Some of the popular schemes based on the variational approach include the Kitaura–Morokuma (KM) scheme, which forms the basis for many of the other energy decomposition analysis (EDA) schemes; reduced variational space (RVS) analysis; the constrained space orbital variation method; natural EDA; Ziegler's extended transition state scheme; and localized molecular orbital-energy decomposition analysis (LMO-EDA) by Su and Li, to name a few.[66–78] Along with these methods, Wu et al.[79] developed density-based EDA approach using a constrained density functional theory (DFT) formalism. Reinhardt et al.[80] proposed an EDA scheme based on fragment-localized Kohn–Sham orbitals. Dapprich and Frenking[81] developed another approach called charge decomposition analysis. The block-localized wavefunction method proposed by Mo et al. is yet another popular EDA tool based on valence bond theory.[82,83] An alternative tool for decomposing the INT is based on the perturbation theory.[84–87] The symmetry-adapted perturbation theory (SAPT) is one of the most popular schemes in this category.[88–90] Although this approach originally was developed with the wave function theory, it has now been modified for DFT also.[91–94] In this chapter, we will highlight the KM scheme and proceed to show how the definition of various energy components of the KM scheme is redefined to give rise to new EDA schemes, specifically RVS and LMO-EDA. Further, a brief overview of SAPT is also provided.

15.2 KM SCHEME

The first and very popular energy decomposition scheme that is used to decompose the total INT into various contributing factors was developed by Kitaura and Morokuma in the late 1970s.[66–68] This method was mainly developed to decompose the INT of hydrogen-bonded systems within the Hartree–Fock approximation. Since then, it has also been successfully applied to donor–acceptor pairs, π and σ interactions in transition metal complex, and decomposition of electron density.[95,96] A broad outline of the key steps involved in the analysis of various components of the INT is provided in the following section. Before we proceed to the KM analysis in detail, it is important to look at the initial decomposition scheme developed by Morokuma.[66] We start the discussion by taking a dimer AB into consideration.

Let $\mathcal{A}\psi_A^0$ and $\mathcal{A}\psi_B^0$ represent the monomer wavefunction in their isolated geometry and E_A^0 and E_B^0 represent the corresponding energies. The two wavefunctions are antisymmetric with $\mathcal{A}$ antisymmetrizer. We define E_0 as the sum of the energies of the isolated system:

$$E_0 = E_A^0 + E_B^0. \tag{15.2}$$

In the case of dimer, four wavefunctions are calculated for a given dimer geometry. Ψ_1 is defined as the non-antisymmetrized Hartree product of isolated monomer wavefunctions, represents pure classic coulombic/electrostatic interaction (ELE) between the two rigid bodies, and allows no exchange of electrons between the monomers. Clearly, the interaction of all permanent charges and multipoles will contribute to electrostatic energy (E_{ELE}).

$$\Psi_1 = \mathcal{A}\psi_A^0 \cdot \mathcal{A}\psi_B^0. \tag{15.3}$$

If E_1 is the energy associated with wavefunction Ψ_1, then the energy difference between E_1 and E_0 will amount to the electrostatic energy. Therefore, E_{ELE} can be defined as

$$E_{ELE} = E_1 - E_0. \tag{15.4}$$

It is to be noted that Ψ_1 violates the Pauli exclusion principle, as it is not an anti-symmetric wavefunction. Hence, the energy E_1 can be lower than the Hartree–Fock energy.

Ψ_2 is defined again as the non-antisymmetrized product of the monomer wave-functions, wherein the wave functions are allowed to distort in the presence of the electric field of the other, although the exchange of electrons is still not permitted:

$$\Psi_2 = \mathcal{A}\psi_A \cdot \mathcal{A}\psi_B. \tag{15.5}$$

Such a description now invokes the polarization concept. Let E_2 be the energy of Ψ_2, and then polarization energy (E_{POL}) is defined as

$$E_{POL} = E_2 - E_1. \tag{15.6}$$

The exchange energy (E_{EX}) is evaluated from wavefunction Ψ_3, which is an antisymmetrized product of the isolated molecular wavefunctions. Ψ_3 allows the exchange of electrons not only within the fragments but also between the separated fragments. It is to be noted that Ψ_3 clearly satisfies the Pauli exclusion principle:

$$\Psi_3 = \mathcal{A}\left(\psi_A^0 \cdot \psi_B^0\right). \tag{15.7}$$

If E_3 is the energy associated with Ψ_3, then E_{EX} can be expressed as

$$E_{EX} = E_3 - E_1. \tag{15.8}$$

E_{EX} is a repulsive term, and the origin of the repulsive character is evident from the combination of two-electron exchange term and intermolecular overlap integrals. The overlap integral terms that generally result in repulsion dominate, imparting the net repulsive character to E_{EX}.

Finally, the self-consistent field wave function for the dimer AB is defined as $\Psi_4(\mathcal{A}\Psi_{AB})$ and E_4 as the energy.

The interaction energy E_{INT} is given as

$$E_{INT} = E_4 - E_0. \tag{15.9}$$

The charge transfer (CT) is not derived explicitly but rather as the difference between the INT and sum of the other three energy terms, that is,

$$E_{CT} = E_{INT} - (E_{ELE} + E_{POL} + E_{EX}). \tag{15.10}$$

The above decomposition scheme proposed by Morokuma suffers from many drawbacks.[97,98] One of the limitations is the definition of change transfer, which when calculated as energy difference term will include all the coupling terms between the components. Again, during the evaluation of polarization term, a non-antisymmetrized wavefunction is used due to which the polarization term will vary drastically as the distance between fragments is decreased or as the basis set becomes more complete.

To overcome these drawbacks, the scheme was modified by Morokuma and Kitaura and is the popular EDA tool currently used.[67] Similar to the Morokuma analysis in KM scheme, the energies of isolated monomers are first evaluated and the energy of the unperturbed complex is expressed as the sum of their energies:

$$E^0 = E_A^0 + E_B^0 = \left\langle \psi_A^0 \left| H_A \right| \psi_A^0 \right\rangle + \left\langle \psi_B^0 \left| H_B \right| \psi_B^0 \right\rangle, \tag{15.11}$$

where ψ_A^0 and ψ_B^0 are the antisymmetric HF wavefunctions of A and B with $\mathcal{A}$ antisymmetrizer and H_A and H_B are their Hamiltonians.

The wavefunction of the complex (ψ_i) is expressed as the linear combination of the molecular orbitals $\left(\phi_k^A \text{ and } \phi_\mu^B \right)$ of the isolated monomers:

$$\psi_i = \sum_k C_{ik} \phi_k^A + \sum_\mu C_{i\mu} \phi_\mu^B. \tag{15.12}$$

Once the wavefunction is defined, the HF Roothan equation is set in terms of the Fock matrix $(\mathcal{F}^0)$ at infinite separation and the molecular interaction matrix (Σ^x) and coefficient matrix C^x as expressed below:

$$(\mathcal{F}^0 - \varepsilon 1 + \Sigma^x) C^x = 0, \tag{15.13}$$

where x is the particular interaction component. The solution to the above equation gives the energy term

$$E^x = \sum_i^{Occ} C_i^{x\dagger} \left(F_x + H^x \right) C_x^i. \tag{15.14}$$

The stabilization due to a particular interaction x is given by

$$E_x = E^x - E^0. \tag{15.15}$$

There are four fundamental building blocks of the INT as defined below:

1. ELE: It arises from the pure classic coulombic interaction between charge densities of occupied orbitals of each monomer. This is generally an attractive term and dominates in polar interactions.
2. POL: The electrons in the occupied orbitals of each monomer can be promoted to its higher vacant orbitals. Mixing of the occupied and vacant orbitals due to intramolecular delocalization of electrons imparts stability to the system that is defined as POL.
3. CT: The intermolecular delocalization of electrons, which is responsible for mixing of occupied orbitals of one monomer with vacant orbitals of another monomer, results in the attractive CT interaction.
4. EX: EX results from the exchange and delocalization of electrons between the molecules. This is generally a repulsive term.

With the various energy terms now defined, let us look at the molecular interaction matrix (Σ^x; Figure 15.1) whose elements are the occupied and unoccupied orbitals of the monomers.

Clearly, the ELE arises from the interaction between occupied (occ) and virtual (vir) molecular orbitals. In the above matrix, all blocks along the diagonal with this type of interaction represent the ES along with a part of EX (ESX). Blocks highlighted in blue represent the polarization term along with some exchange interaction (POLX). The polarization interaction arises due to the mixing of occupied and virtual orbitals within each molecule. Based on the definition of CT and EX stated above, the corresponding blocks are highlighted in pink and orange, respectively. The molecular interaction matrix in terms of the energy components is shown in Figure 15.1b.

The energy terms can be evaluated by retaining only selected components in the interaction matrix and the rest of the matrix elements is set to zero. For example, preserving only the diagonal elements of the interaction matrix will result in the evaluation of ESX term, which includes both electrostatic and part of exchange interaction. The pure electrostatic energy (ELE) can be evaluated by neglecting the differential overlap between the molecular orbitals. The choice of the block elements to

MO	A_{occ}	A_{vir}	B_{occ}	B_{vir}
A_{occ}	$\sum_{oo}^{AA}$	$\sum_{ov}^{AA}$	$\sum_{oo}^{AB}$	$\sum_{ov}^{AB}$
A_{vir}	$\sum_{vo}^{AA}$	$\sum_{vv}^{AA}$	$\sum_{vo}^{AB}$	$\sum_{vv}^{AB}$
B_{occ}	$\sum_{oo}^{BA}$	$\sum_{ov}^{BA}$	$\sum_{oo}^{BB}$	$\sum_{ov}^{BB}$
B_{vir}	$\sum_{vo}^{BA}$	$\sum_{vv}^{BA}$	$\sum_{vo}^{BB}$	$\sum_{vv}^{BB}$

(a)

MO	A_{occ}	A_{vir}	B_{occ}	B_{vir}
A_{occ}	ESX	$POLX_A$	EX	$CT_{A \rightarrow B}$
A_{vir}	$POLX_A$	ESX	$CT_{B \rightarrow A}$	EX
B_{occ}	EX	$CT_{B \rightarrow A}$	ESX	$POLX_B$
B_{vir}	$CT_{A \rightarrow B}$	EX	POLX	ESX

(b)

FIGURE 15.1 Molecular interaction (Σ^x) in terms of the (a) interacting molecular orbitals and (b) corresponding energy terms.

be retained for evaluating CT, POL, and EX is explicitly indicated in Figure 15.2. It should be noted that, as the diagonal elements cannot be set to zero, the ESX terms appear inherently in evaluation of the POL, CT, and EX terms, which need to be subtracted to obtain their exclusive contribution.

Having looked at the mathematical derivation of the KM decomposition method, let us apply it to small hydrogen-bonded molecular clusters. The KM analysis has been extensively used by researchers to decompose the INT of many complexes.[99–110] We have taken some prototypical dimers as examples that are often discussed in the context of hydrogen bonding interaction. All dimers were subjected to geometry optimization at the MP2/aug-cc-pVDZ level of computation followed by frequency calculation to ensure that the obtained stationary points are minima on the respective potential energy surface. All optimized geometries are given in Figure 15.3. The KM analysis was carried out at the HF/6-311++G** level on MP2/aug-cc-pVDZ optimized geometries. The various energy terms obtained from decomposition analysis are presented in Table 15.1.

Water dimer $(W)_2$, which is perhaps the most extensively studied hydrogen-bonded dimer, is an archetypal model of moderately strong hydrogen bond.[111–117] Let us first discuss various energy terms included in the KM scheme by taking water dimer as an example. It is well known that hydrogen-bonded water dimer is linear in shape. The hydrogen bond between water dimer is often designated as O–H$\cdots$O, where O, which possesses the lone pairs of electron, is called the hydrogen bond acceptor and O–H bond is called the hydrogen bond donor. It is of immense interest to understand the nature of O–H$\cdots$O interaction. The coulombic interaction between

MO	A_{occ}	A_{vir}	B_{occ}	B_{vir}
A_{occ}	ESX	0	0	0
A_{vir}	0	ESX	0	0
B_{occ}	0	0	ESX	0
B_{vir}	0	0	0	ESX

(a)

MO	A_{occ}	A_{vir}	B_{occ}	B_{vir}
A_{occ}	ESX	$POLX_A$	0	0
A_{vir}	$POLX_A$	ESX	0	0
B_{occ}	0	0	ESX	$POLX_B$
B_{vir}	0	0	$POLX_B$	ESX

(b)

MO	A_{occ}	A_{vir}	B_{occ}	B_{vir}
A_{occ}	ESX	0	0	$CT_{A\rightarrow B}$
A_{vir}	0	ESX	$CT_{B\rightarrow A}$	0
B_{occ}	0	$CT_{B\rightarrow A}$	ESX	0
B_{vir}	$CT_{A\rightarrow B}$	0	0	ESX

(c)

MO	A_{occ}	A_{vir}	B_{occ}	B_{vir}
A_{occ}	ESX	0	EX	0
A_{vir}	0	ESX	0	EX
B_{occ}	EX	0	ESX	0
B_{vir}	0	EX	0	ESX

(d)

FIGURE 15.2 Schematic representation of the orbitals to be retained for evaluating the (a) ESX, (b) POLX, (c) CT, and (d) EX terms, respectively.

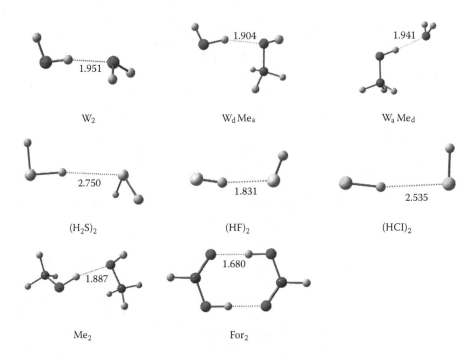

FIGURE 15.3 Representative hydrogen-bonded dimers.

TABLE 15.1

Results of KM Analysis Carried Out on Representative Hydrogen-Bonded Systems at the HF/6-311++G Level**

	$(W)_2$	W_dMe_a	W_aMe_d	$(H_2S)_2$	$(HF)_2$	$(HCl)_2$	Me_2	For_2
ELE	−9.7	−10.61	−9.6	−3.13	−7.57	−2.84	−10.73	−34.23
EX	7.1	8.8	7.4	4.05	5.01	3.56	9.3	34.56
POL	−1.14	−1.99	−1.26	−0.6	−0.95	−0.49	−2.17	−13.17
CT	−1.64	−2.14	−1.7	−1.16	−1.34	−1.15	−2.25	−11.54
MIX	0.57	1.29	0.61	0.58	0.41	0.46	1.39	8.43
IE	−4.82	−4.66	−4.55	−0.25	−4.44	−0.46	−4.45	−15.95
$CT_{d \to a}$	−0.42	−0.57	−0.44	−0.2	−1.08	−0.98	−0.65	−5.77
$CT_{a \to d}$	−1.22	−1.57	−1.26	−0.95	−0.25	−0.16	−1.61	−5.77
$POL_{d \to a}$	−0.38	−0.47	−0.5	−0.18	−0.72	−0.36	−0.72	−5.81
$POL_{a \to d}$	−0.68	−1.38	−0.66	−0.38	−0.2	−0.12	−1.29	−5.81

Note: All values are in kcal/mol.

partially negative charged oxygen atom of one water molecule with partially positive charged hydrogen atom of the other water molecule results in the manifestation of the classic ELE term. The EX is generally repulsive and arises from repulsive forces. The large ELE is attenuated by the repulsive exchange-repulsion energy. However, it can be noted from Table 15.1 that the summation of electrostatic and exchange-repulsion terms gives rise to the attractive INT, which further suggests that the ELE is the driving force for the hydrogen bond formation. As stated before, the orbital interaction between two water molecules results in two energy terms called POL and CT energy. The summation of POL and CT energy is often called as delocalization energy. POL arises because of the orbital stabilization between two water molecules due to the formation of hydrogen bond. The charge transfer between two water molecules results in the formation of CT energy. It should be noted that the principal contribution to CT came from the hydrogen bond acceptor to the hydrogen bond donor. Recall from previous section that the CT energy arises due to the transfer of electron density from HOMO of monomer-1 to LUMO of monomer-2. During the hydrogen bond formation, the oxygen atom in one of the water molecules (hydrogen bond acceptor) transfers the electron density (its lone pair of electrons) to the anti-bonding orbital of the $-OH$ bond in the other water molecule (hydrogen bond donor). It is also worth mentioning that due to CT the $-OH$ bond in the donor water molecule is slightly elongated resulting in the associated red-shift and enhanced intensity in the IR spectrum.

To reinforce the current discussion, the characteristic nature of hydrogen bond can be further understood by examining the two possible scenarios in methanol–water dimer.[118–120] In the first case, water molecule can behave as the hydrogen bond acceptor and methanol as the donor (W_aMe_d), whereas in the second case, water behaves as the donor (W_dMe_a). Clearly, the calculated intermolecular distance and hydrogen bond energy reveal that methanol prefers to act as the hydrogen bond acceptor rather than as the donor evidently from the fact that it is a stronger base but a weaker acid than water in the gas phase. Further, the calculated CT energy from hydrogen bond acceptor to hydrogen bond donor ($CT_{a\rightarrow d}$) is higher in the case where methanol acts as the hydrogen bond acceptor. This can be attributed to the positive inductive effect of the methyl group. The methyl group in the methanol increases the electron density at the oxygen center, and hence, it can actively participate as the hydrogen bond acceptor rather than the donor.

Although water and hydrogen sulfide (H_2S) share similarity in structure, it is well known that H_2O exists in liquid form, whereas H_2S exists in gaseous form. This is mainly because sulfur is not nearly as electronegative as oxygen, so that H_2S is not quite as polar as water. Hence, comparatively weak intermolecular forces exist for H_2S, and the melting and boiling points are much lower than they are in water. Drastic decrease in hydrogen bond INT along with various energy terms between dimers of H_2O and H_2S (Table 15.1) reflects similar findings. Similar to water dimer, hydrogen fluoride is another classic example of moderate strong hydrogen bond. Again, chlorine is not nearly as electronegative as fluorine. Hence, HCl dimer shows weak hydrogen bond INT compared to hydrogen fluoride dimer.

Formic acid is the simplest carboxylic acid and acts as a prototypical motif that can form two strong O–H$\cdots$O hydrogen bonds.[121–124] Although a variety of conformations exist for formic acid dimer, we have taken only the cyclic dimer conformation, which is the most stable conformation in gas phase. The cyclic dimer of formic acid contains two strong O–H···O=C hydrogen bonds. This kind of hydrogen bond is also often called as resonance-assisted hydrogen bond (RAHB).[125–128] In RAHB, hydrogen atom is attached to the two oxygen atoms through two π-conjugate double bonds. In fact, formic acid dimer can be considered as the simplest supramolecular synthon constructed with strong hydrogen bonds.

15.3   RVS SCHEME

As mentioned above, the method of KM is the most widely used variational energy decomposition method and works well for hydrogen-bonded systems, but it is not completely foolproof. Some of the shortcomings of the KM scheme include its limitation to only the HF method; the final INT obtained is not corrected for unphysical lowering of energy caused by the basis set superposition error (BSSE). Also, molecular complexes separated by short distances, for instance, the cation–π complexes of benzene with Li^+ and Mg^{2+}, exhibit numerical instabilities in POL and CT energies. Several alternative schemes have been proposed to avoid problems of the KM EDA, one among which is the RVS analysis developed by Stevens and Fink and has been successfully applied to molecular complexes with various kinds of interactions.

Akin to KM analysis, the RVS analysis decomposes the total INT into chemically intuitive components. This analysis utilizes the McWeeney's group function approach, which offers the advantage of diagonalization and solving smaller Fock matrix.[129,130] The function space is separated into the variational and nonvariational space with the flexibility to add or remove the unoccupied orbitals of the monomers in the functional space. Such a procedure helps to evaluate various conceptual INT terms.

A concise summary of the RVS analysis is schematically represented in Figure 15.4. In the RVS analysis, the energies of the monomers are first calculated from the respective isolated wavefunction, and the dimer wavefunction is then generated by considering only the occupied orbitals of the monomers. Such an approach hence does not allow any variational flexibility to the dimer wavefunction and results in a combined electrostatic and exchange-repulsion term (ESX), which represents the Heitler–London interaction. Hence,

$$E_1 = E_0 + ESX, \tag{15.16}$$

where E_0 represents the sum of the monomer energies ($E_0 = E_X + E_Y$).

The evaluation of the POL and CT terms requires the construction of frozen and variational space. Excluding the occupied orbitals of the monomer for which the POL and CT are to be evaluated, the frozen subspace is constructed by including all other monomer-occupied orbitals, which are orthogonalized by the Gramm–Schmidt procedure; then the subspace is frozen by retaining only the diagonal elements and setting the off-diagonal elements to zero. The reduced subspace (which includes only the occupied and virtual orbitals of only monomers excluded in the frozen subspace,

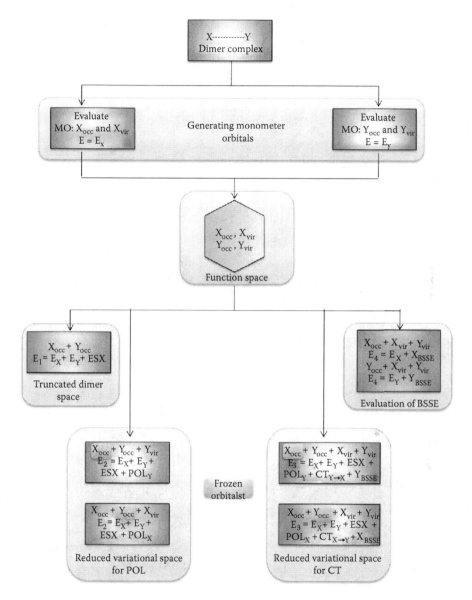

FIGURE 15.4 Overview of the RVS scheme.

which is orthogonal to the frozen occupied subspace) is then variationally solved to yield the POL term:

$$E_2 = E_1 + POL_n \text{ (where } n = \text{X or Y).} \tag{15.17}$$

On similar ground, the CT term can be evaluated by including the virtual orbitals of all monomers in the variational subspace:

$$E_3 = E_2 + CT_n + BSSE_n \text{ (where } n = \text{X or Y).} \tag{15.18}$$

The BSSE term is evaluated from wavefunction generated from the occupied and virtual orbitals of a given monomer along with only the vacant orbital of the other monomers:

$$E_4 = E_n + n_{BSSE} \text{ (where } n = \text{X or Y).} \tag{15.19}$$

Hence, the summation of ESX, POL, CT, and BSSE terms will yield the final INT:

$$INT = ESX + POL_X + POL_Y + CT_X + CT_Y + X_{BSSE} + Y_{BSSE}. \tag{15.20}$$

For the current discussion, RVS analysis has been carried out on cation–π complexes of benzene–metal complexes.[131–140] An overview of the results obtained for the RVS analysis carried out at the HF/6-31G* level on the cation–π complexes of benzene with monovalent and bivalent metal ions such as Li^+, Na^+, K^+, Mg^{2+}, and Ca^{2+} is presented in Figure 15.5. Generally, cation–π complexes are held by pure electrostatic effects due to strong attraction between opposite charges. Hence, bivalent metal ions show stronger interaction than the monovalent metal ions. However, the contribution of ion–quadrupole interactions in complex stabilization is still being

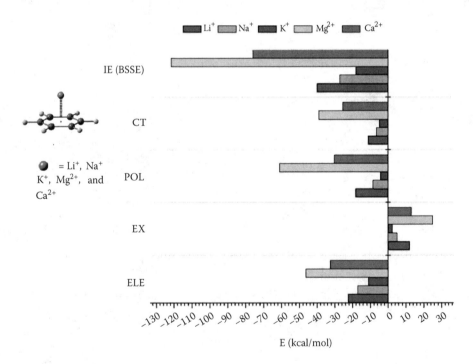

FIGURE 15.5 RVS decomposition carried out on various benzene–metal ion complexes at the HF/6-31G* level.

debated. Although cyclohexane is more polarizable than benzene, it has been shown that the calculated interaction between cyclohexane and metal ion is less compared to that of benzene. It can be seen from Figure 15.5 that, in the cases of Li^+, Na^+, and K^+ complexes, the electrostatic and polarization energies dominate the total INT. Metal ions can polarize the benzene molecule effectively and the POL solely arises because of benzene. In fact, in the case of Mg^{2+}, the POL is more than the electrostatic contribution. The CT between metal ion and aromatic ring is reflected as CT energy. As alkali and alkaline earth metal ions cannot donate electron density to the benzene molecule, again, the entire CT contribution arises only because of transfer of electron density from π-cloud to metal ion. Overall, POL and ELE are predominant factors in the stabilization of the cation–π complexes.

15.4 COMPARISON OF KM AND RVS SCHEMES

The RVS analysis when applied to hydrogen-bonded systems behaves in many ways similar to the KM analysis, especially in the ELE and EX terms. The variation can generally be seen in the POL and CT terms, which are generally overestimated in the KM analysis. The overestimation becomes more apparent with the addition of diffuse function to the basis set. Results presented in Table 15.2 reinforce the above-mentioned fact. A similar observation has also been seen in the case of water dimer and trimers by Chen and Gordon.[141]

Having looked at some of the similarities between the two popular EDA schemes, let us now divert our attention to some of their differences. One of the major flaws of the KM analysis is its definition of POL and CT terms that are not evaluated by employing proper antisymmetrized intermediate wavefunctions. For example, when two subsystems are separated by short intermolecular distances such as in cation–π complexes, the POL and CT terms in KM analysis are unusually blown up (Table 15.3). This problem, however, gets rectified in the RVS scheme as all wavefunctions satisfy the Pauli exclusion rule by using proper antisymmetrized wavefunctions.

15.5 LMO-EDA SCHEME

The energy decomposition scheme developed by Su and Li is popularly known as the LMO-EDA.[78] This again is a variational approach and can be considered as an extension and modification of the KM. Unlike the previous two methods mentioned, the LMO-EDA scheme calculates the electrostatic, exchange, and repulsion terms separately. This method can successfully be applied to both open and closed shell systems and is insensitive to the quality of the basis set employed. An added advantage of this method is its applicability to high-level quantum chemical methods, such as MP2, CCSD, and CCSD(T), making it a robust model. Inclusion of such methods enhances the EDA to the wide range of intermolecular interactions.[142] However, large complexes that contain more number of atoms cannot be handled such high-level methods. Another advantage of LMO-EDA is that EDA can be performed with DFT methods also.[143] As stated before, LMO-EDA can be applied to open shell systems. Hence, covalent bond analysis also can be carried out with the LMO-EDA method.

TABLE 15.2
Results of KM and RVS Analysis Exemplifying the Role of Diffuse Functions

			ELE	EX	POL	CT	IE
Phe-W	KM	6-31G*	-11.63	8.15	-1.18	-2.48	-7.39
		6-311++G**	-12.81	10.15	-2.11	-2.54	-6.12
	RVS	6-31G*	-11.63	8.15	-1.49	-1.91	-7.03
		6-311++G**	-12.81	10.15	-1.55	-1.18	-5.47
Phe-NH$_3$	KM	6-31G*	-17.60	14.25	-1.98	-3.54	-9.00
		6-311++G**	-18.56	17.26	-4.20	-4.97	-7.21
	RVS	6-31G*	-17.60	14.25	-2.33	-2.51	-8.40
		6-311++G**	-18.56	17.26	-2.56	-2.60	-6.64

1.864

Phe-W

1.854

Phe-NH$_3$

TABLE 15.3

Results of KM and RVS Analyses Carried Out on Cation–π Complexes at HF/6-31G* Level

	BEN-Li⁺		BEN-K⁺	
	KM	RVS	KM	RVS
ELE	−22.39	−22.39	−11.14	−11.14
EX	11.90	11.90	2.41	2.41
POL	*	−18.20	−4.46	−4.27
CT	−293.32	−11.24	−7.03	−4.92
IE	−41.85	−39.95	−19.52	−18.02

In this scheme, the INT is partitioned into electrostatic, exchange, repulsion, polarization, and dispersion. Let us now look at the mathematical derivation of the various components. For a molecular system (AB) with wavefunction Φ and total energy Hamiltonian operator H, the expectation value E_{AB}^{HF} is given as

$$E_{AB}^{HF} = \langle \Phi | H | \Phi \rangle. \tag{15.21}$$

This expectation value can also be rewritten in terms of the one- and two-electron integrals and the nuclear repulsion energy (E^{nuc}) as follows:

$$E_{AB}^{HF} = \sum_{i \in AB}^{\alpha,\beta} h_i + \frac{1}{2} \sum_{i \in AB}^{\alpha,\beta} \sum_{j \in AB}^{\alpha,\beta} \langle ii \mid jj \rangle - \frac{1}{2} \sum_{i \in AB}^{\alpha} \sum_{j \in AB}^{\alpha} \langle ij \mid ij \rangle - \frac{1}{2} \sum_{i \in AB}^{\beta} \sum_{j \in AB}^{\beta} \langle ij \mid ij \rangle + E^{nuc} \tag{15.22}$$

where α and β represent the HF spin orbitals of the dimer orthonormal to each other, the one-electron integral for the ith electron is represented as h_i, and the two-electron coulomb and exchange integrals of the ith and jth electron are represented as $\langle ii|jj \rangle$ and $\langle ij|ij \rangle$, respectively.

For a dimer AB, the HF INT in accordance to the supermolecule approach can be written as

$$\Delta E^{HF} = \langle \Phi_{AB} | H_{AB} | \Phi_{AB} \rangle - \left(\langle \Phi_A | H_A | \Phi_A \rangle + \langle \Phi_B | H_B | \Phi_B \rangle \right), \tag{15.23}$$

where wavefunction and corresponding Hamiltonians for the dimer and monomers are given as Φ_{AB}, Φ_A, and Φ_B and H_{AB}, H_A, and H_B, respectively.

To evaluate the electrostatic energy, we replace the energy expression (Equation 15.22) that represents the HF energy of the dimer with an approximate energy expression obtained by replacing the exchange integrals between the monomers with the exchange integrals of each monomer as shown in Equation 15.24:

$$E_{AB}^{1} = \sum_{i}^{\alpha,\beta} h_i + \frac{1}{2} \sum_{i \in AB}^{\alpha,\beta} \sum_{j \in AB}^{\alpha,\beta} \langle ii \mid jj \rangle$$

$$- \frac{1}{2} \left(\sum_{i \in A}^{\alpha} \sum_{j \in A}^{\alpha} \langle ij \mid ij \rangle + \sum_{i \in A}^{\beta} \sum_{j \in A}^{\beta} \langle ij \mid ij \rangle \right. \tag{15.24}$$

$$\left. + \sum_{i \in B}^{\alpha} \sum_{j \in B}^{\alpha} \langle ij \mid ij \rangle + \sum_{i \in B}^{\beta} \sum_{j \in B}^{\beta} \langle ij \mid ij \rangle \right) + E^{nuc}.$$

The electrostatic energy is now calculated as the difference in the energy of the dimer evaluated using the approximate energy expression and the sum of the isolated monomer energies:

$$\Delta E^{ELE} = E_{AB}^{1} - (E_A + E_B) \tag{15.25}$$

$$= \frac{1}{2} \sum_{i \in AB}^{\alpha,\beta} \sum_{j \in AB}^{\alpha,\beta} \langle ii \mid jj \rangle - \frac{1}{2} \left(\sum_{i \in A}^{\alpha,\beta} \sum_{j \in A}^{\alpha,\beta} \langle ii \mid jj \rangle + \sum_{i \in B}^{\alpha,\beta} \sum_{j \in B}^{\alpha,\beta} \langle ii \mid jj \rangle \right) + E_{AB}^{nuc} - E_A^{nuc} - E_B^{nuc}. \tag{15.26}$$

Equation 15.26 can also be approximated by considering the orthonormal HF spin orbitals of the monomers that may or may not be orthonormal to each other. The energy expression now takes the form

$$E_{AB}^{2} = \sum_{i \in AB}^{\alpha,\beta} h_i + \frac{1}{2} \sum_{i \in AB}^{\alpha,\beta} \sum_{j \in AB}^{\alpha,\beta} \langle ii \mid jj \rangle - \frac{1}{2} \sum_{i \in AB}^{\alpha} \sum_{j \in AB}^{\alpha} \langle ij \mid ij \rangle - \frac{1}{2} \sum_{i \in AB}^{\beta} \sum_{j \in AB}^{\beta} \langle ij \mid ij \rangle + E^{nuc}. \tag{15.27}$$

Unlike ΔE^{ELE}, the EX (ΔE^{EX}) is calculated as the difference in the HF energy of the dimer (Equation 15.22) and the approximate energy shown in Equation 15.24:

$$\Delta E^{EX} = E_{AB}^{2} - E_{AB}^{1} \tag{15.28}$$

$$= -\frac{1}{2} \left(\sum_{i \in AB}^{\alpha} \sum_{j \in AB}^{\alpha} \langle ij \mid ij \rangle + \sum_{i \in AB}^{\beta} \sum_{j \in AB}^{\beta} \langle ij \mid ij \rangle \right)$$

$$+ \frac{1}{2} \left(\sum_{i \in A}^{\alpha} \sum_{j \in A}^{\alpha} \langle ij \mid ij \rangle + \sum_{i \in A}^{\beta} \sum_{j \in A}^{\beta} \langle ij \mid ij \rangle + \sum_{i \in B}^{\alpha} \sum_{j \in B}^{\alpha} \langle ij \mid ij \rangle + \sum_{i \in B}^{\beta} \sum_{j \in B}^{\beta} \langle ij \mid ij \rangle \right). \tag{15.29}$$

The evaluation of the repulsion energy (ΔE^{REP}) requires the energy to be expressed in terms of the monomer orbitals that are orthonormal to one another. Orthonormality between the monomer orbitals is enforced by the inverse of the overlap matrix (S^{-1}). An approximate energy expression in terms of the monomer spin orbitals (i, j, k, and l) and the inverse overlap matrix corresponding to the monomer spin orbitals can be expressed as

$$E^3_{AB} = \sum_{i \in AB}^{\alpha, \beta} \sum_{j \in AB}^{\alpha, \beta} h_{ij}(S^{-1})_{ij} + \frac{1}{2} \sum_{i \in AB}^{\alpha, \beta} \sum_{j \in AB}^{\alpha, \beta} \sum_{k \in AB}^{\alpha, \beta} \sum_{l \in AB}^{\alpha, \beta} \langle ij \mid kl \rangle$$

$$\times (S^{-1})_{ij}(S^{-1})_{kl} - \frac{1}{2} \sum_{i \in AB}^{\alpha} \sum_{j \in AB}^{\alpha} \sum_{k \in AB}^{\alpha} \sum_{l \in AB}^{\alpha} \langle ik \mid jl \rangle \tag{15.30}$$

$$\times (S^{-1})_{ij}(S^{-1})_{kl} - \frac{1}{2} \sum_{i \in AB}^{\beta} \sum_{j \in AB}^{\beta} \sum_{k \in AB}^{\beta} \sum_{l \in AB}^{\beta} \langle ik \mid jl \rangle \times (S^{-1})_{ij}(S^{-1})_{kl} + E_{nuc}$$

$$\Delta E^{REP} = E^3_{AB} - E^2_{AB}. \tag{15.31}$$

The polarization term is defined as the difference in the HF energy of the supermolecule and the E^3_{AB} energy term:

$$\Delta E^{POL} = E^{HF}_{AB} - E^3_{AB}. \tag{15.32}$$

Hence, the INT can be expressed as the sum of various components obtained above as

$$\Delta E^{HF}_{INT} = \Delta E^{ELE} + \Delta E^{EX} + \Delta E^{REP} + \Delta E^{POL}. \tag{15.33}$$

The dispersion energy is included by evaluating the difference in the energy of the system calculated using the MP2 or CCSD(T) methods and the HF approach.

$$\Delta E^{DISP} = \Delta E^{MP2}_{AB} - \Delta E^{HF}_{AB} \tag{15.34}$$

or

$$\Delta E^{DISP} = \Delta E^{CCSD(T)}_{AB} - \Delta E^{HF}_{AB}. \tag{15.35}$$

On similar grounds, the energy evaluated using various DFT functionals can also be decomposed into various chemically intuitive components.

In this section, a detailed LMO-EDA on various molecular clusters [S_m where S is water (W), methanol (M), and ethylene glycol (EG) and $m = 2–4$] is discussed (Figure 15.6). All clusters were optimized at the M05-2X/6-311++G** level of theory, and the obtained geometries are in good agreement with previously reported geometrical values. Let us first take the case of water cluster. The calculated INT at the MP2/6-31G** level using the LMO-EDA for water dimer is –5.00 kcal/mol and is in good agreement with previously reported CCSD(T)/CBS result of 5.01 kcal/mol.[144,145] Similarly, the CCSD(T)/CBS and MP2/CBS interaction energies of the trimer are –15.82 and –15.80 kcal/mol,whereas the MP2/CBS interaction energies predicted for water tetramer is –27.63 kcal/mol.[146,147] The interaction energies predicted by LMO-EDA at the MP2/6-31G** level for trimer (–16.01 kcal/mol) and tetramer (–28.91 kcal/mol) agree well with those reported at a higher level of computation. The energy decomposition of the water clusters has also been carried out by Su and Li[78] at the CCSD(T)/ aug-cc-pVQZ (for dimer) and CCSD(T)/aug-cc-pVTZ (for trimer and tetramer). Figure 15.7 illustrates a good correlation between the various energy terms predicted at the CCSD(T) and MP2 level using a basis set of different sizes, exemplifying the insensitivity of the decomposition to the quality of the basis set employed.[78]

Another interesting feature would be to analyze the variation in the various INT terms as a function of cluster size (Figure 15.8). It can be noted from Figure 15.8 that although ELE INT is one of the dominant contributors to the total stabilization of all

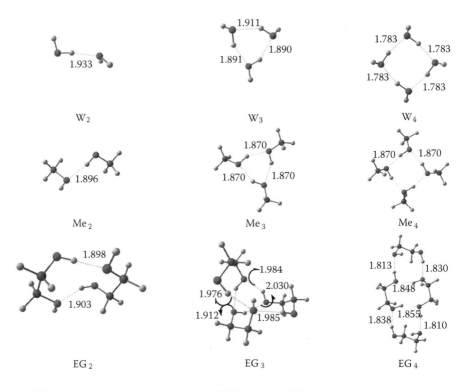

FIGURE 15.6   Optimized geometries (M05-2X/6-311++G**) of the molecular clusters.

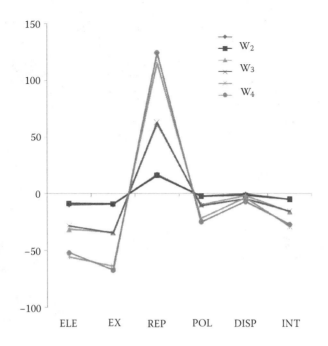

FIGURE 15.7   Correlating the various energy terms at the CCSD(T) and MP2 levels.

clusters, its effect, however, is compensated by the EX-REP terms. This effect gets pronounced as the cluster size increases and the polarization and dispersion terms contribute significantly to the total INT. The influence of the alkyl group to the hydrogen bonding interaction can be analyzed by comparing the energies of water, methanol, and ethylene glycol clusters. It is worth mentioning here that the increase in the alkyl groups enhances the contribution from the electrostatic energy term.

The concept that a particular configuration of single and multiple covalent bonds has a total energy greater than the sum of the energies of the individual bonds is generally recognized in chemistry. This extra energy is referred to as resonance energy and delocalization energy in valence bond and molecular orbital theories, respectively. The same concept applies to certain patterns of H-bonded network, where it is known as nonadditivity or cooperativity.[148–150] The concept of cooperativity in hydrogen-bonded systems has received widespread attention.[118] Cooperativity can be classified as (a) positive (synergistic), when the subsequent binding of another molecule is higher than that for the previous one; (b) negative (interfering), when the binding is lower; and (c) noncooperative (additive), when the binding is identical. These cooperative effects are sensitively dependent on the number, strength, and orientation of the H-bonds. For example, it is well known that the addition of a water molecule to the water dimer leads to a water trimer with a cyclic structure that is more stable than the dimer. The third water molecule enhances the strength of the H-bonding interaction between initial two water molecules through cooperativity. It is of immense interest to analyze the factors that contribute to the cooperativity;

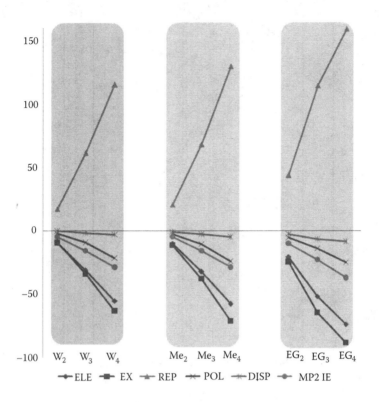

FIGURE 15.8 Variation in INT terms as a function of cluster size.

hence, pairwise additivity of the energy terms has been calculated using the LMO-EDA. The cooperativity has been calculated using the equation

$$E_{COOP} = INT_{COMP} - \sum_{\substack{i=1 \\ j>i}}^{4} E_{i,j}, \tag{15.36}$$

where E_{COOP} is the cooperativity, INT_{COMP} is the INT of the complex, and ΔE_{ij} is the pairwise binding energy (where $i, j = 1$–4).

The discussion here is limited to the tetramers alone. The pairwise interactions of six pairs of dimers that contribute to the cooperativity in every tetramer were calculated. Figure 15.9 depicts the contribution of various energy terms to the cooperativity. Among the three clusters considered, the cooperativity is highest for the water cluster. It can be seen from Figure 15.9 that POL has the predominant contribution to the cooperativity that clearly indicates that the total cooperativity is mainly due to the polarization effect.

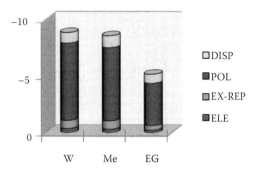

FIGURE 15.9 Contribution of the various energy terms to the cooperativity of tetrameric clusters.

15.6 SAPT

The various EDA techniques discussed so far were based on the variational principle. Another route to partition the INT is by employing the perturbation method.[84–87] Among the various perturbational tools used to analyze the individual energy components contributing to the stabilization of intermolecular complexes, many-body SAPT has received widespread attention.[88–90,151–153] Jeziorski et al.[90] have reviewed various aspects of wavefunction-based SAPT theory.

A brief overview of the SAPT is as follows. For a dimer AB, let H_A and H_B represent the electronic Hamiltonians of the monomers A and B. The unperturbed Hamiltonian (H_0) of the dimer can be expressed as a sum of H_A and H_B.

$$H_0 = H_A + H_B. \tag{15.37}$$

From the perturbation theory, the Hamiltonian of the dimer (H) can be written in terms of the unperturbed Hamiltonian (H_0) and the intermolecular interaction operator V:

$$H = H_0 + V. \tag{15.38}$$

According to Møller–Plesset[154] partitioning of monomer Hamiltonian,

$$H_A = F_A + W_A \text{ and } H_B = F_B + W_B, \tag{15.39}$$

where F_A/F_B and W_A/W_B are the Fock and intramonomer correlation operators of the monomers, respectively.

Hence, in SAPT, the total Hamiltonian of the dimer is expressed as the sum of three operators as given below:

$$H = F + W + V, \tag{15.40}$$

where $F = F_A + F_B$, and $W = W_A + W_B$. Here, F serves as the unperturbed operator, and V and W are the perturbations, thus making SAPT a double perturbation theory.

The Schrodinger equation for dimer AB will be

$$(F + \zeta V + \lambda W)\Phi_{AB}(\zeta,\lambda) = E_{AB}(\zeta,\lambda)\Phi_{AB}(\zeta,\lambda) \tag{15.41}$$

where ζ and λ define the order of perturbation of V and W, respectively.

E_{INT} can be expressed as a perturbative series

$$E_{INT} = E_{INT} = \sum_{n=1}^{\infty}\sum_{j=0}^{\infty} E_{pol}^{(nj)} + E_{exch}^{(nj)}, \tag{15.42}$$

where n and j are the orders of perturbation in V and W, respectively. Depending on the orders of perturbation in V and W, one can define a large number of energy components.[155–157]

The HF INT $\left(E_{INT}^{HF}\right)$ can be approximated to the sum of the polarization and exchange corrections of the zeroth order in W:

$$E_{INT}^{HF} = E_{ELE}^{(10)} + E_{EX}^{(10)} + E_{IND,resp}^{(20)} + E_{EX-IND,resp}^{(20)} + \delta E_{INT,resp}^{HF}. \tag{15.43}$$

The above equation clearly indicates that the HF INT includes contributions from the first-order polarization (electrostatic, ELE) and exchange terms (EX) and also from the second-order induction (IND) and exchange-induction (EX-IND) terms. The third- and higher-order IND and EX-IND terms are collected in $\delta E_{INT,resp}^{HF}$. The subscript "resp" indicates that the coupled perturbed Hartree–Fock–type response is included in the correction terms. To evaluate the correlation part $\left(E_{INT}^{Corr}\right)$ of the INT, one must include the intramolecular correlation at least up to the second order:

$$E_{INT}^{Corr} = E_{ELE,resp}^{(12)} + E_{EX}^{(11)} + E_{EX}^{(12)} + E_{EX-IND,resp}^{(20)} + E_{EX-DISP}^{(20)} + E_{DISP}^{(20)}. \tag{15.44}$$

The total INT can be expressed as

$$E_{INT} = E_{INT}^{HF} + E_{INT}^{Corr}. \tag{15.45}$$

One of the main advantages of the MB-SAPT method is that each term in the perturbation series can be meaningfully interpreted. In addition, the calculated INT is closer to the value as obtained from the CCSD(T) method.[158,159] However, the application of the MB-SAPT is limited by its scaling behavior. For a chosen dimer, both CCSD(T) and MB-SAPT scale as N^7, where N is the measure of the size of the systems. Therefore, the computational demand of the MB-SAPT has been reduced by combining with the computationally less expensive DFT. Hesselmann et al.[94] have provided the detailed account of the theory and applications of the DFT-SAPT method. In this approach, the first-order term provides exact electrostatic Coulomb interaction. The second-order induction energy contribution is obtained from the solution of the coupled perturbed Kohn–Sham equation. The second-order dispersion energy is derived from the solution of frequency-dependent coupled perturbed Kohn–Sham equations (functional theory, TDDFT). As the exact xc potential is not available, the intermolecular exchange corrections to these energies have to be approximated with DFT-SAPT. Systematic comparison of results from MB-SAPT and DFT-SAPT reveals that DFT-SAPT performs well provided that well-balanced asymptotically corrected model of xc potential is employed.[91–93] An analysis of performance of DFT-SAPT shows that it scales as N^6. The application of DFT-SAPT method to larger dimer requires further improvement in the scaling behavior. An attempt to improve the scaling has been made by Hesselmann et al.[94] using the density fitting (DF) approach. It is also called the resolution-of-the-identity approach. In this approach, certain intermediate steps are approximated using expansion in an auxiliary basis set. In DF-DFT-SAPT, the electron repulsion integrals are approximated using the above-mentioned DF technique. Thus, N^5 and N^3 scalings have been achieved with reference to the size of the systems and the basis set. Therefore, large dimer systems can be handled with the help of DF-DFT-SAPT.[160,161] Recently, Hohenstein and Sherrill[162] have applied the DF procedure to the wavefunction-based SAPT approach. In this approach, the two-electron integrals are approximated using the expansion with the help of an auxiliary basis set. This approach has been benchmarked against the S22 test set of Hobza et al. It is found that this method can routinely be applied to large dimer systems.

In the remaining part of the chapter, the usefulness of DF-DFT-SAPT has been illustrated by taking benzene–carbon dioxide as a model system. In this case study, an attempt has been made to understand the nature of interaction between a gas molecule such as CO_2 and a variety of organic ligands including benzene and its isoelectronic analogues. High-level *ab initio* calculation with electron correlation at the MP2/aug-cc-pVDZ level was performed to obtain the stationary points for each complex. The nature of the interaction between gas molecules with various organic linkers was characterized by decomposition of the net INT into electrostatic, exchange (repulsion), induction, and dispersion contributions using the DF-DFT-SAPT method. The optimized geometries of organic linker–CO_2 complexes are given in Figure 15.10. It can be seen that the molecular axis of CO_2 is parallel to the plane of the aromatic ring. The distance between the center of the aromatic ring to the carbon atom of CO_2 for various complexes varies from 2.82 to 3.25 Å. The origin of the dispersive force in these complexes is mainly due to the interaction of quadrupole of the CO_2 with delocalized π-electron density of the planar aromatic ring.

To assess the nature of interactions between organic linkers and gas molecules, total INT was decomposed into first-order E^1, second-order E^2, and higher-order terms (δHF) by using the DF-DFT-SAPT method. E^1 is the sum of the electrostatic INT (E^1_{pol}) and the first-order EX (E^1_{exch}). The net induction energy (E_{ind}) is considered as the summation second-order induction (E^2_{ind}), second-order exchange-induction (E^2_{ex-ind}), and δHF terms, where δ(HF) is assumed as dominated by third- and higher-order induction and exchange-induction contributions. Addition of second-order dispersion (E^2_{disp}) and second-order exchange dispersion gives the net dispersion energy (E_{disp}), and E^2 denotes the sum of second-order induction

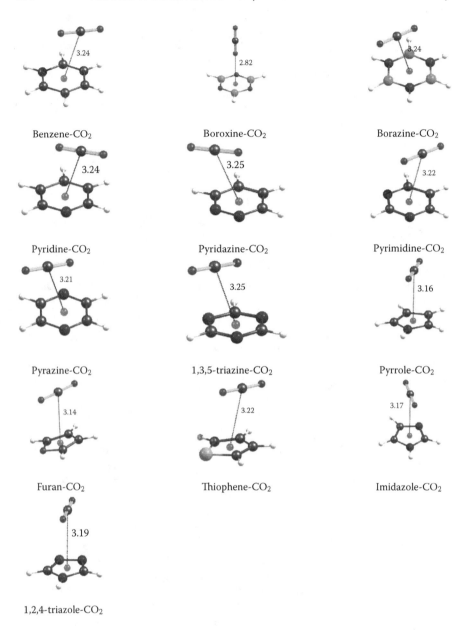

Benzene-CO$_2$	Boroxine-CO$_2$	Borazine-CO$_2$
Pyridine-CO$_2$	Pyridazine-CO$_2$	Pyrimidine-CO$_2$
Pyrazine-CO$_2$	1,3,5-triazine-CO$_2$	Pyrrole-CO$_2$
Furan-CO$_2$	Thiophene-CO$_2$	Imidazole-CO$_2$

1,2,4-triazole-CO$_2$

FIGURE 15.10 Optimized geometries of the various organic linker–CO$_2$ complexes calculated at the MP2/aug-cc-pVDZ level of theory.

energy (E_{ind}) and second-order dispersion energy (E_{disp}). In DF-DFT-SAPT calculations, the LBE0AC that uses the localized Hartree–Fock scheme in the exchange part of the PBE0AC functional was employed. DF approximation employing the aug-cc-pVTZ basis set for orbitals and the aug-cc-pVQZ for JK fitting was used to calculate both first-order and induction contributions, and aug-cc-pVQZ for MP2 fitting was used for evaluating dispersion and exchange-dispersion contributions. The asymptotic behavior of the PBE0 functional was corrected using the gradient-regulated asymptotic correction approach of Grüning et al.[163] For this approach, a shift parameter that approximates the difference between the first vertical ionization potential and the negative HOMO energy was obtained from the PBE0/aug-cc-pVDZ level of calculation. The calculated shift parameters for benzene, pyridine, pyridazine, pyrimidine, pyrazine,1,3,5-triazine, pyrrole, imidazole, 1,2,4-triazole, furan, thiophene, borazine, boroxine,and carbondioxide are 0.0718, 0.0753, 0.0787, 0.0780, 0.0792, 0.0804, 0.0767, 0.0804, 0.0860, 0.0797, 0.0734, 0.0716, 0.0812, and 0.1118 a.u., respectively.

Interaction energies calculated using the MP2/aug-cc-pvtz and DF-DFT-SAPT analysis are given in Table 15.4.

It is possible to note that the dispersion interaction plays the major role in the stabilization of different complexes. The interplay of electrostatic forces is also clearly evident from the DF-DFT-SAPT results. The interaction of carbon dioxide with the organic linkers has significant contribution from electrostatic and induction terms in the stabilization of these complexes. It is also worth noting that the interaction energies obtained from DF-DFT-SAPT results are in very good quantitative agreement with the BSSE-corrected MP2/aug-cc-pVTZ level of theory. This clearly indicates the usefulness of the method to decompose weakly bound complexes.

TABLE 15.4
DF-DFT-SAPT Decomposition of Binding Energy (kcal/mol) for Complexes Formed by CO_2 with Various Ligands in Metal-Organic Frameworks

Complex	E_{pol}^1	E_{exch}^1	E^1	F_{ind}	F_{disp}	E^2	E_{BE}	$E_{BE}^{CP}(MP2)$
Benzene–CO_2	−3.06	6.50	3.44	−0.73	−4.83	−5.20	−2.12	−2.92
Boroxine–CO_2	−2.12	3.66	1.54	−0.38	−2.81	−3.19	−1.65	−1.87
Borazine–CO_2	−2.16	4.84	2.68	−0.42	−4.13	−4.36	−1.88	−2.12
Pyridine–CO_2	−2.73	5.94	3.22	−0.66	−4.65	−4.96	−2.08	−2.82
Pyridazine–CO_2	−2.44	5.67	3.24	−0.66	−4.21	−4.57	−1.95	−2.75
Pyrimidine–CO_2	−1.98	5.74	3.76	−0.57	−4.52	−4.82	−1.33	−2.22
Pyrazine–CO_2	−2.37	6.04	3.67	−0.65	−4.6	−4.97	−1.66	−2.53
1,3,5-Triazine–CO_2	−1.78	5.08	3.30	−0.50	−4.30	−4.58	−1.52	−2.21
Pyrrole–CO_2	−4.50	7.28	2.89	−0.89	−5.00	−5.41	−3.03	−3.65
Furan–CO_2	−3.57	6.73	3.16	−0.69	−4.70	−5.08	−2.22	−2.88
Thiophene–CO_2	−3.38	6.78	3.42	−0.76	−4.89	−5.28	−2.24	−3.05
Imidazole–CO_2	−4.05	6.66	2.61	−0.83	−4.7	−5.17	−2.93	−3.52
1,2,4-Triazole–CO_2	−4.24	6.52	2.28	−0.94	−4.52	−5.15	−3.17	−3.72

Note: Counterpoise-corrected binding energies at the MP2/aug-cc-pVTZ level are also given for comparison.

15.7 CONCLUSION

Although several publications have appeared on EDA schemes, there is no comprehensive report on these schemes. Hence, in this chapter, an attempt has been made to provide a brief account of various EDA schemes that are routinely used to unravel the energetics of different types of noncovalent interactions. It is clearly evident from the previous reports and the case study presented in this chapter that the energetics of the classic H-bonded systems could be easily understood with the help of KM decomposition analysis within the Hartree–Fock framework. However, for van der Waals and weak H-bonded interactions wherein dispersion interaction (intermolecular electron correlation) plays a major role in the stabilization of the complexes, it is necessary to look for other alternative approaches. In this context, both LMO-EDA method and SAPT approach are extremely useful. Between these two methods, the success of SAPT method has been highlighted in a number of studies as well as in the case study considered in this chapter. Results obtained from the DF-DFT-SAPT method for various model systems are very close to INT obtained from the CCSD(T) level of calculation with a reasonably large basis set. In addition, various contributions to the total energy give meaningful information on the nature of interactions that are involved in the stabilization of the systems. One of the important features of the DF-DFT-SAPT (or wavefunction theory-based DF-SAPT) is that it provides valuable information on the complete spectrum of noncovalent interactions from weak van der Waals to covalent limit. As mentioned in the previous section, this method does not suffer from the drawback of BSSE correction. As mentioned in previous reviews and research articles, each scheme has its own merits and demerits. Chemists have to use their chemical intuitions and the physics behind the problem to analyze the outcome of the EDA.

ACKNOWLEDGMENTS

V.S. and G.N.S. thank the CSIR Network Program of the Centre for Excellence Project (NWP-53) for financial support. G.N.S. acknowledges financial support from Department of Biotechnology and the Swarnajayanthi fellowship from the Department of Science and Technology. R.M.K. and D.V. thank the CSIR New Delhi for financial support in the form of senior research fellowship and CSIR-Nehru Science postdoctoral fellowship, respectively.

REFERENCES

1. Lewis, G. N., *J. Am. Chem. Soc.*, 38, 762–785, 1916.
2. Lewis, G. N., *Valence and the Structure of Atoms and Molecules*, American Chemical Society Monograph Series, New York, 1923.
3. Heitler, W. and London, F., *Z. Phys.*, 44, 455–472, 1927.
4. Pauling, L., *The Nature of the Chemical Bond*, Cornell University Press, Ithaca, NY, 1939.
5. Bader, R. F. W., Hernández-Trujillo, J., and Cortés-Guzmán, F., *J. Comput. Chem.*, 28, 4–14, 2007.
6. Frenking, G. and Krapp, A., *J. Comput. Chem.*, 28, 15–24, 2007.

7. Shaik, S., *J. Comput. Chem.*, 28, 51–61, 2007.
8. Lehn, J. M., *Supramolecular Chemistry—Concepts and Perspectives*, VCH, Weinheim, 1995.
9. Atwood J. L., Davies J. E. D., MacNicol D.M., Vogtle F., and Lehn J. M., *Comprehensive Supramolecular Chemistry*, Pergamon, Oxford, 1996.
10. Steed, J. W. and Atwood, J. L., *Supramolecular Chemistry*, Wiley, Chichester, 2000.
11. Ariga, K. and Kunitake, T., *Supramolecular Chemistry—Fundamentals and Applications*, Springer, New York, 2006.
12. Lehn, J. M., *Angew. Chem. Int. Ed.*, 29, 1304–1319, 1990.
13. Lehn, J. M., *Polym. Int.*, 51, 825–839, 2002.
14. van der Waals, J. D., Doctoral dissertation, Leiden, 1873.
15. Muller-Dethlefs, K. and Hobza, P., *Chem. Rev.*, 100, 143–167, 2000.
16. Barthélemy, P., *Compt. Rend. Chim.*, 12, 171–179, 2009.
17. Jeffrey, G. A. and Saenger, W., *Hydrogen Bonding in Biology and Chemistry*, Springer-Verlag, Berlin, 1991.
18. Meyer, E. A., Castellano, R. K., and Diederich, F., *Angew. Chem. Int. Ed.*, 42, 1210–1250, 2003.
19. Dougherty, D. A., *Science*, 271, 163–168, 1996.
20. Kim, K. S., Tarakeshwar, P., and Lee, J. Y., *Chem. Rev.*, 100, 4145–4185, 2000.
21. Lee, E. C., Kim, D., Jurečka, P., Tarakeshwar, P., Hobza, P., and Kim, K. S., *J. Phys. Chem. A*, 111, 3446–3457, 2007.
22. Černý, J. and Hobza, P., *Chem. Phys.*, 9, 5291–5303, 2007.
23. Desiraju, G. R. and Steiner, T., *The Weak Hydrogen Bond in Structural Chemistry and Biology*, Oxford University Press, Oxford, 1999.
24. Desiraju, G. R., *Angew. Chem. Int. Ed.*, 46, 8342–8356, 2007.
25. Hoeben, F. J. M., Jonkheijm, P., Meijer, E. W., and Schenning, A. P. H. J., *Chem. Rev.*, 105, 1491–1546, 2005.
26. Davis, A. V., Yeh, R. M., and Raymond, K. N., *Proc. Natl. Acad. Sci. USA*, 99, 4793–4796, 2002.
27. Schneider, H. and Strongin, R. M., *Acc. Chem. Res.*, 42, 1489–1500, 2009.
28. Burley, S. K. and Petsko, G. A., *J. Am. Chem. Soc.*, 108, 7995–8001, 1986.
29. Burley, S. K. and Petsko, G. A., *Science*, 229, 23–28, 1985.
30. Burley, S. K. and Petsko, G. A., *FEBS Lett.*, 203, 139–143, 1986.
31. Ma, C. J. and Dougherty, D. A., *Chem. Rev.*, 97, 1303–1324, 1997.
32. Sunner, J., Nishizawa, K., and Kebarle, P., *J. Phys. Chem.*, 85, 1814–1820, 1981.
33. Castonguay, L. A., Rappé, A. K., and Casewit, C. J., *J. Am. Chem. Soc.*, 113, 7177–7183, 1991.
34. Pietsch, M. A. and Rappé, A. K., *J. Am. Chem. Soc.*, 118, 10908–10909, 1996.
35. Kolb, H. C., Andersson, P. G., and Sharpless, K. B., *J. Am. Chem. Soc.*, 116, 1278–1291, 1994.
36. Gilman, A. G., Rall, T. W., Mies, A. S., and Taylor, P., *The Pharmaceutical Basis of Therapeutics*, McGraw Hill, Inc., New York, 1993.
37. Buckingham, A. D., Legon, A. C., and Roberts, S. M., *Principles of Molecular Recognition*, Blackie Academic & Professional, London, 1993.
38. Desiraju, G. R., *Angew. Chem. Int. Ed.*, 34, 2311–2327, 1995.
39. Desiraju, G. R., *Crystal Engineering: The Design of Organic Solids*, Elsevier, Amsterdam 1989.
40. Nishio, M., Hirota, M., and Umezawa, Y., *The CH/π Interaction*, Wiley-VCH, New York, 1998.
41. Chen, R. J., Zhang, Y., Wang, D., and Dai, H. J., *J. Am. Chem. Soc.*, 123, 3838–3839, 2001.
42. Elango, M., Subramanian, V., Rahalkar, A. P., Gadre, S. R., and Sathyamurthy, N., *J. Phys. Chem. A*, 112, 7699–7704, 2008.

43. Britz, D. A. and Khlobystov, A. N., *Chem. Soc. Rev.*, 35, 637–659, 2006.
44. Hobza, P. and Zahradnik, R., *Chem. Rev.*, 88, 871–897, 1988.
45. Chalasinski, G. and Szczgsniak, M. M., *Chem. Rev.*, 94, 1723–1765, 1994.
46. Hobza, P. and Sponer, J., *Chem. Rev.*, 99, 3247–3276, 1999.
47. Hobza, P. and Havlas, Z., *Chem. Rev.*, 100, 4253–4264, 2000.
48. Rappe, A. K. and Bernstein, E. R., *J. Phys. Chem. A*, 104, 6117–6128, 2000.
49. Sherrill, C. D., in *Reviews in Computational Chemistry*, Lipkowitz, K. B. and Cundari, T. R., eds., Wiley, New York, 2009, 1–38.
50. Sinnokrot, M. O. and Sherrill, C. D., *J. Phys. Chem. A*, 110, 10656–10668, 2006.
51. Zhao, Y. and Truhlar, D. G., *J. Chem. Theory Comput.*, 2, 1009–1018, 2006.
52. Zhou, H.-X. and Gilson, M. K., *Chem. Rev.*, 109, 4092–4107, 2009.
53. Riley, K. E., Pitonak, M., Jurecka, P., and Hobza, P., *Chem. Rev.*, 110, 5023–5063, 2010.
54. Desiraju, G. R., *Acc. Chem. Res.*, 35, 565–573, 2002.
55. Munshi, P. and Guru Row, T. N., *Cryst. Eng. Comm.*, 7, 608–611, 2005.
56. Parthasarathi, R., Subramanian, V., and Sathyamurthy, N., *J. Phys. Chem. A*, 110, 3349–3351, 2006.
57. Scheiner, S., *Hydrogen Bonding: A Theoretical Perspective*, Oxford University Press: Oxford, 1997.
58. Sherrill, C. D., *J. Chem. Phys.*, 132, 110902:1–7, 2010.
59. Boys, S. F. and Bernardi, F., *Mol. Phys.*, 19, 553–566, 1970.
60. Riley, K. E. and Hobza, P., *J. Phys. Chem. A*, 111, 8257–8263, 2007.
61. Sinnokrot, M. O. and Sherrill, C. D., *J. Phys. Chem. A*, 110, 10656–10668, 2006.
62. Janowski, T. and Pulay, P., *Chem. Phys. Lett.*, 447, 27–32, 2007.
63. Sinnokrot, M. O. and Sherrill, C. D., *J. Am. Chem. Soc.*, 126, 7690–7697, 2004.
64. van Duijneveldt, F. B., van Duijneveldt-van de Rijdt, J. G. C. M., and van Lenthe, J. H., *Chem. Rev.*, 94, 1873–1885, 1994.
65. Vijay, D., Sakurai, H., and Sastry, G. N., *Int. J. Q. Chem.*, 111, 1893–1901, 2011.
66. Morokuma, K., *J. Chem. Phys.*, 55, 1236–1244, 1971.
67. Kitaura, K. and Morokuma, K., *Int. J. Q. Chem.*, 10, 325–340, 1976.
68. Morokuma, K., *Acc. Chem. Res.*, 10, 294–300, 1977.
69. Stevens, W. J. and Fink, W. H., *Chem. Phys. Lett.*, 139, 15–22, 1987.
70. Bagus, P. S. and Illas, F., *J. Chem. Phys.*, 96, 8962–8970, 1992.
71. Glendening, E. D. and Streitwieser, A., *J. Chem. Phys.*, 100, 2900–2909, 1994.
72. Reed, A. E., Curtiss, L. A., and Weinhold, F., *Chem. Rev.*, 88, 899–926, 1988.
73. Reed, A. E. and Weinhold, F., *Isr. J. Chem.*, 31, 277–285, 1991.
74. Glendening, E. D., *J. Phys. Chem. A*, 109, 11936–11940, 2005.
75. Ziegler, T. and Rauk, A., *Theor. Chem. Acc.*, 46, 1–10, 1977.
76. Bickelhaupt, F. M. and Baerends, E. J., in *Rev. Comp. Chem.*, Lipkowitz, K. B. and Boyd, D. B., eds., Wiley-VCH, New York, 1999, 1–86.
77. te Velde, G. T., Bickelhaupt, F. M., Baerends, E. J., Guerra, C. F. S., Van Gisbergen, J. A., Snijders, J. G., and Ziegler, T., *J. Comput. Chem.*, 22, 931–967, 2001.
78. Su, P. and Li, H., *J. Chem. Phys.*, 131, 014102:1–15, 2009.
79. Wu, Q., Ayers, P. W., and Zhang, Y. K., *J. Chem. Phys.*, 131, 164112:1–8, 2009.
80. Reinhardt, P., Piquemal, J.-P., and Savin, A., *J. Chem. Theory Comput.*, 4, 2020–2029, 2008.
81. Dapprich, S. and Frenking, G., *J. Phys. Chem.*, 99, 9352–9362, 1995.
82. Mo, Y. and Peyerimhoff, S. D., *J. Chem. Phys.*, 109, 1687–1697, 1998.
83. Mo, Y., Baob, P., and Gao, J., *Phys. Chem. Chem. Phys.*, 13, 6760–6775, 2011.
84. Diner, S., Malrieu, J. P., and Claverie, P., *Theor. Chem. Acc.*, 13, 1–10, 1969.
85. Diner, S., Malrieu, J. P., Jordan, F., and Gilbert, M., *Theor. Chem. Acc.*, 15, 100–110, 1969.
86. Hayes, I. C. and Stone, A. J., *Mol. Phys.*, 53, 83–105, 1984.

87. Stone, A. J., *The Theory of Intermolecular Forces*, Oxford University Press, New York, 1996.
88. Szalewicz, K. and Jeziorski, B., *Mol. Phys.*, 38, 191–208, 1979.
89. Chalasinski, G. and Szczesniak, M. M., *Mol. Phys.*, 63, 205–224, 1988.
90. Jeziorski, B., Moszynski, R., and Szalewicz, K., *Chem. Rev.*, 94, 1887–1930, 1994.
91. Hesselmann, A. and Jansen, G., *Chem. Phys. Lett.*, 357, 464–470, 2002.
92. Hesselmann, A. and Jansen, G., *Chem. Phys. Lett.*, 362, 319–325, 2002.
93. Hesselmann, A. and Jansen, G., *Chem. Phys. Lett.*, 367, 778–784, 2003.
94. Hesselmann, A., Jansen, G., and Schütz, M., *J. Chem. Phys.*, 122, 014103, 2005.
95. Morokuma, K. and Kitaura, K., *Chemical Applications of Atomic and Molecular Electrostatic Potentials*, Politzer, P. and Truhlar, D. G., eds., Plenum, New York, 1981, 215–242.
96. Sakaki, S., Sato, H., Imai, Y., Morokuma, K., and Ohkubo, K., *Inorg. Chem.*, 24, 4538–4544, 1985.
97. Frey, R. F. and Davidson. E. R., *J. Chem. Phys.*, 10, 5555–5562, 1989.
98. Cybulski, S. M. and Scheiner, S., *Chem. Phys. Lett.*, 166, 57–64, 1990.
99. Kitaura, K. and Morokuma, K., in *Molecular Interactions*, Rateiczak, H. and Orville-Thomas, W. J., eds., Wiley, Chichester, 1980.
100. Gonzales, J. M., Pak, C., Cox, R. S., Allen, W. D., Schaefer, H. F., Csaszar, A. G., and Tarczay, G., *Chem. Eur. J.*, 9, 2173–2192, 2003.
101. Gordon, M. S. and Jensen, J. H., in *Encyclopedia of Computational Chemistry*, Schleyer, P. V. R., ed., Wiley, Chichester, 1998, 3198–3214.
102. Takahashi, H., Yuki, K., and Nitta, T., *Fluid Phase Equilib.*, 194, 153–160, 2002.
103. Pejov, L. and Hermansson, K., *J. Chem. Phys.*, 119, 313–324, 2003.
104. Wang, J., Gut, J., and Leszczynski, J., *J. Phys. Chem. B*, 109, 13761–13769, 2005.
105. Johansson, P., Abrahamsson, E., and Jacobsson, P., *J. Mol. Struct.*, 717, 215–221, 2005.
106. Gupta, M., Maity, D. K., Singh, M. K., Nayak, S. K., and Ray. A. K., *J. Phys. Chem. B*, 116, 5551–5558, 2012.
107. Chen, W. and Gordon M. S., *J. Phys. Chem.*, 100, 14316–14328, 1996.
108. Danilova, V. I., Anisimovb, V. M., Kuritac, N., and Hovoruna, D., *Chem. Phys. Lett.*, 412, 285–293, 2005.
109. Philip, V., Harris, J., Adams, R., Nguyen, D., Spiers, J., Baudry, J., Howell, E. E., and Hinde, R. J., *Biochemistry*, 50, 2939–2950, 2011.
110. Gadre, S. R. and Pingale, S. S., *J. Am. Chem. Soc.*, 120, 7056–7062, 1998.
111. Kollman, P. A. and Buckingham, A.D., *Mol. Phys.*, 21, 561–570, 1972.
112. Parthasarathi, R., Subramanian, V., and Sathyamurthy, N., *J. Phys. Chem. A*, 199, 843–351, 2005.
113. Ludwig, R., *Angew. Chem. Int. Ed.*, 40, 1808–1827, 2001.
114. Feyereisen, M. W., Feller, D., and Dixon, D. A., *J. Phys. Chem.*, 100, 2993–2997, 1996.
115. Scheiner, S., *Annu. Rev. Phys. Chem.*, 45, 23–56, 1994.
116. Joseph, J. and Jemmis, E. D., *J. Am. Chem. Soc.*, 129, 4620–4632, 2007.
117. Neela, Y. I., Mahadevi, A. S., and Sastry, G. N., *J. Phys. Chem. B*, 114, 17162–17171, 2010.
118. Mandal, A., Prakash, M., Kumar, R. M., Parthasarathi, R., and Subramanian, V., *J. Phys. Chem. A*, 114, 2250–2258, 2010.
119. Gonzalez, L., Mo, O., and Yanez, M., *J. Chem. Phys.*, 109, 139–150, 1998.
120. Fileti, E. E. and Canuto, S., *Int. J. Q. Chem.*, 102, 554–564, 2005.
121. Hayashi, S., Umemura, J., Kato, S., and Morokuma, K., *J. Phys. Chem.*, 88, 1330–1334, 1984.
122. Miura, S., Tuckerman, M. E., and Klein, M. L., *J. Chem. Phys.*, 109, 5290–5299, 1998.
123. Chojnacki, H., Andzelm, J. A., Nguyen, D. T., and Sokalski, W. A., *Comput. Chem.*, 19, 181–187, 1995.

124. Balabin, R. M., *J. Phys. Chem. A*, 113, 4910–4918, 2009.
125. Gilli, P., Bertolasi, V., Ferretti, V., and Gilli, G., *J. Am. Chem. Soc.*, 116, 909–915, 1994.
126. Shetty, S., Pal, S., Kanhere, D. G., and Goursot, A., *Ind. J. Chem.*, 45A, 202–212, 2006.
127. Gilli, G., Bellucci, F., Ferretti, V., and Bertolasi, V., *J. Am. Chem. Soc.*, 111, 1023–1028, 1989.
128. Bertolasi, V., Gilli, P., Ferretti, V., and Gilli, G., *J. Am. Chem. Soc.*, 113, 4917–4925, 1991.
129. McWeeney, R., *Proc. R. Soc.*, A253, 242–259, 1959.
130. McWeeney, R. and Ohno, K.A., *Proc. R. Soc.*, A255, 367–381, 1960.
131. Reddy A. S. and Sastry, G. N., *J. Phys. Chem. A*, 109, 8893–8903, 2005.
132. Reddy, A. S., Vijay, D., Sastry, G. M., and Sastry, G. N., *J. Phys. Chem. B*, 110, 2479–2481, 2006.
133. Rodgers, M. T. and Armentrout, P. B., *Int. J. Mass Spectrom.*, 185/186/187, 359–380, 1999.
134. Vijay, D. and Sastry, G. N., *Chem. Phys.*, 10, 582–590, 2008.
135. Vijay, D. and Sastry, G. N., *J. Phys. Chem. A*, 110, 10148–10154, 2006.
136. Vijay, D., Sakurai, H., Subramanian, V., and Sastry, G. N., *Phys. Chem. Chem. Phys.* 14, 3057–3065, 2012.
137. Amicangelo, J. C. and Armentrout, P. B., *J. Phys. Chem. A*, 104, 11420–11432, 2000.
138. Zhu, W., Luo, X., Puah, C. M., Tan, X., Shen, J., Gu, J., Chen, K., and Jiang, H., *J. Phys. Chem. A*, 108, 4008–4018, 2004.
139. Ryzhov, V. and Dunbar, R. C., *J. Am. Soc. Mass Spectrom.*, 11, 1037–1046, 2000.
140. Dunbar, R. C., *J. Phys. Chem. A*, 104, 8067–8074, 2000.
141. Chen, W. and Gordon, M. S., *J. Phys. Chem.*, 100, 14316–14328, 1996.
142. Romero, J., Reyes, A., Davidb, J., and Restrepo, A., *Phys. Chem. Chem. Phys.*, 13, 15264–15271, 2011.
143. Wang, F.-F., Jenness, G., Al-Saidi, W., and Jordan, K. D., *J. Chem. Phys.*, 132, 134303:1–8, 2010.
144. Klopper, W., van Duijneveldt-van de Rijdt, J., and van Duijneveldt, F. B., *Chem. Phys.*, 2, 2227–2234, 2000.
145. Min, S. K., Lee, E. C., Lee, H. M., Kim, D. Y., Kim, D., and Kim, K. S., *J. Comput. Chem.*, 29, 1208–1221, 2008.
146. Xantheas, S. S., Burnham, C. J., and Harrison, R. J., *J. Chem. Phys.*, 116, 1493–1499, 2002.
147. Anderson, J. A., Crager, K., Fedoroff, L., and Tschumper, G. S., *J. Chem. Phys.*, 121, 11023–11029, 2004.
148. Jeffrey, G. A., *An Introduction to Hydrogen Bonding*, Oxford University Press, Inc., Oxford, New York, 1997.
149. Vijay, D. and Sastry, G. N., *Chem. Phys. Lett.*, 485, 235–242, 2010.
150. Vijay, D., Zipse, H., and Sastry, G. N., *J. Phys. Chem. B*, 112, 8863–8867, 2008.
151. Cybulski, H. and Sadlej, J., *J. Chem. Theory Comput.*, 4, 892–897, 2008.
152. Sinnokrot, O. M. and Sherrill, D. C., *J. Am. Chem. Soc.*, 126, 7690–7697, 2004.
153. Szalewicz, K., *WIREs Comput. Mol. Sci.*, 2, 254–272, 2012.
154. Møller, C. and Plesset, M. S., *Phys. Rev.*, 46, 618–622, 1934.
155. Stone, A. J., *The Theory of Intermolecular Forces*, 1st ed., Clarendon Press, Oxford, 1996.
156. Szalewicz, K. and Jeziorski, B., *Mol. Phys.*, 38, 191–208, 1979.
157. Rybak, S., Jeziorski, B., and Szalewicz, K., *J. Chem. Phys.*, 95, 6576–6601, 1991.
158. McBane, G. C. and Cybulski, S. M., *J. Chem. Phys.*, 110, 11734–11741, 1999.
159. Misquitta, A. J., Bukowski, R., and Szalewicz, K., *J. Chem. Phys.*, 112, 5308–5319, 2000.

160. Korona, T., Hesselmann, A., and Dodziuk, H., *J. Chem. Theory Comput.*, 5, 1585–1596, 2009.
161. Hesselmann, A., Jansen, G., and Schütz, M., *J. Am. Chem. Soc.*, 128, 11730–11731, 2006.
162. Hohenstein, E. G. and Sherrill, C. D., *J. Chem. Phys.*, 132, 184111:1–10, 2010.
163. Grüning, M., Gritsenko, O. V., van Gisbergen, S. J. A., and Baerends, E. J., *J. Chem. Phys.*, 114, 652–660, 2001.

16 Perfectly Periodic Table of Elements in Nonrelativistic Limit of Large Atomic Number*

John P. Perdew

CONTENTS

16.1 INTRODUCTION: PERIODIC TABLE

The regular variation of chemical properties along a row or down a column of the periodic table is a cornerstone of chemistry. The modern periodic table is a completion of the table sketched by Mendeleev in 1869. Mendeleev knew that certain groups (columns of the table) have chemically similar properties, such as the alkali metals or the rare gases. He organized the elements within a group or column in order of increasing atomic weight, and the columns were then assembled into a rectangular table in which the atomic weight increased roughly along each row. The modern periodic table achieves a more regular organization by using the atomic number Z (the number of electrons or protons in the neutral atom) instead of the atomic weight. Mendeleev left gaps in his table, and the gaps predicted unknown elements that were later discovered. There are now about 118 known elements ($1 \le Z \le 118$), with the first 98 occurring naturally.

Elements in the same group or column tend to have the same valence. Thus, it is the quantum mechanical shell structure of the atom that creates the periodic variation of chemical properties. For $Z \le 36$, relativistic corrections and nonzero nuclear radii have only small effects, but these effects increase with increasing Z. The periodic table can further be broken down into an *s* block, with valence *s* electrons,

---

* This article is dedicated to Professor B. M. Deb on the occasion of his 70th birthday. Professor Deb has done much interesting related work on the Z dependence of the energies of atoms, Thomas–Fermi–like theories, electron density–based dynamics, atoms in strong magnetic or laser fields, etc. Strong static fields might change the periodic table in interesting ways.

345

a p block, and various d, f, etc., transition blocks. The two columns of the s block and the six columns of the p block (the chemical "octaves") shall be discussed here. The discussion assumes that the atomic subshells fill according to the empirical Madelung rule [1], in which orbitals with principal quantum number n and orbital angular momentum quantum number l fill in order of increasing $n + l$.

Moving across a row in order of increasing Z, from the ns^1 alkali metal to the np^6 rare-gas configuration, the first ionization energy I tends to increase and the valence-shell radius tends to shrink as valence orbitals of fixed n experience a growing attraction to the nucleus. The electron affinity A also tends to increase through the np^5 halogen configuration. These effects occur in the Hartree, Hartree–Fock, correlated wave function, and Kohn–Sham density functional descriptions but not in the orbital-free Thomas–Fermi density functional approximation. All levels of description, however, agree with experiment in the trend down a fixed column, where I and A decrease and the valence-shell radius increases as n increases.

What happens to the nonrelativistic point-nucleus periodic table in the limit of large Z? Experiment cannot answer this question because one cannot turn off the relativistic effects or the effects of nuclear size and instability. Correlated wave functions cannot answer this question either because they cannot be performed for more than about 100 electrons. The orbital-free Thomas–Fermi approximation might seem appealing because it can treat tens of thousands of electrons and because it yields the correct $-cZ^{7/3}$ leading behavior [2] of the total energy in the limit $Z \to \infty$. The Thomas–Fermi theory also predicts that the valence-shell radius tends to a constant limit (somewhat bigger than that for realistic Z) and a limiting constant ionization energy of 1.3 eV (somewhat smaller than that for realistic Z) [3]. However, the Thomas–Fermi theory has no shell structure and does not bind one atom to another [4]. The shell-structure effects are known to enter the total energy to a lower order than $Z^{7/3}$. Do they really disappear in the large-Z limit?

16.2   KOHN–SHAM CALCULATIONS IN THE LIMIT OF LARGE Z

To answer the question posed at the end of the previous section, Constantin et al. [3] solved the Kohn–Sham equations self-consistently for neutral sp block atoms with up to 3000 electrons. The Kohn–Sham equations [5]

$$\left[\frac{1}{2}\nabla^2 + v(\vec{r}) + \int d^3r'n(\vec{r}')/|\vec{r}' - \vec{r}| + \delta E_{xc}/\delta n_\sigma(\vec{r})\right]\psi_{\alpha\sigma}(\vec{r}) = \varepsilon_{\alpha\sigma}\psi_{\alpha\sigma}(\vec{r}), \quad (16.1)$$

$$n_\sigma(\vec{r}) = \sum_\alpha \theta(\mu - \varepsilon_{\alpha\sigma})\left|\psi_{\alpha\sigma}(\vec{r})\right|^2, \quad n(\vec{r}) = n_\uparrow(\vec{r}) + n_\downarrow(\vec{r}), \quad (16.2)$$

$$E = T_s[n_\uparrow, n_\downarrow] + \int d^3rn(\vec{r})v(\vec{r}) + \frac{1}{2}\int d^3rn(\vec{r})\int d^3r'n(\vec{r}')/|\vec{r}' - \vec{r}| + E_{xc}[n_\uparrow, n_\downarrow], \quad (16.3)$$

$$T_s[n_\uparrow, n_\downarrow] = \int d^3r\sum_{\alpha\sigma}\theta(\mu - \varepsilon_{\alpha\sigma})\frac{1}{2}\left|\nabla\psi_{\alpha\sigma}(\vec{r})\right|^2 \quad (16.4)$$

yield in principle the exact ground-state energy E and electron spin densities $n_\uparrow(\vec{r})$ and $n_\downarrow(\vec{r})$ for N electrons in the presence of external potential $v(\vec{r})$ (in this case, the electron-nuclear attraction $-Z/r$). In practice, the spin-density functional $E_{xc}[n_\uparrow, n_\downarrow]$ for the exchange-correlation energy must be approximated. The considered approximations were the local spin density approximation (LSDA) [5], a generalized gradient approximation (GGA) [6], and exact exchange without correlation. Interestingly, at the exchange-only level in the large Z limit, the computed valence electron properties were nearly the same in LSDA, GGA, or exact. Probably the most accurate results are LSDA or GGA, including correlation.

Within each column of the periodic table, the ionization energy was computed as

$$I(Z) = E(N = Z - 1, Z) - E(N = Z, Z), \qquad (16.5)$$

and an ionization density was computed similarly as

$$\Delta n(Z, \vec{r}) = n(N = Z, Z, \vec{r}) - n(N = Z - 1, Z, \vec{r}). \qquad (16.6)$$

From the latter, an ionization radius (valence electron radius)

$$R(Z) = \int d^3 r \Delta n(Z, \vec{r}) r \qquad (16.7)$$

was also computed. The results up to $Z = 3000$ were then carefully extrapolated to infinite Z.

The extrapolation showed clearly that each column of the periodic table has its own limiting ionization energy and valence electron radius. Thus, the periodic table becomes perfectly periodic in the limit of large Z. As more electrons and neutralizing protons are added, the valence region for each column reaches a saturation limit with a large finite radius R and a small nonzero ionization energy I. For the first column of the s block, the limiting I is 1.8 eV and the limiting R is 14.1 bohr from GGA. For the last column of the p block, the limiting I is 5.1 eV and the limiting R is 8.8 bohr also from GGA. The chemistry of the heavy nonrelativistic atoms is thus presumably a universal chemistry of long, weak bonds. The real periodic table is, in a sense, a reflection of the perfect periodicity that would be achieved in the non relativistic limit of a large atomic number.

It remains to check if the Madelung rule for subshell filling always leads to the Aufbau principle expressed by the step function $\theta(\mu - \varepsilon_{\alpha\sigma})$ in Equations 16.2 and 16.4. Also a limiting electron affinity still remains to be extracted for each column of the periodic table, in the same careful way used to extract a limiting ionization energy.

ACKNOWLEDGMENTS

This work was supported in part by the National Science Foundation under grant DMR-0854769 and under cooperative agreement EPS-1003897, with additional support from the Louisiana Board of Regents.

REFERENCES

1. Madelung, E., in *Die Matematischen Hilfsmittel des Physikers*, 3rd ed., Springer, Berlin, 1936, 359.
2. Lieb, E.H., *Rev. Mod. Phys.*, 48, 553, 1976.
3. Constantin, L.A., Snyder, J.C., Perdew, J.P., and Burke, K., *J. Chem. Phys.*, 133, 241103, 2010.
4. Teller, E., *Rev. Mod. Phys.*, 34, 627, 1962.
5. Kohn, W. and Sham, L.J., *Phys. Rev.*, 140, A1133, 1965.
6. Perdew, J.P., Burke, K., and Ernzerhof, M., *Phys. Rev. Lett.*, 77, 3865, 1996.

17 Quantum Similarity

Ramon Carbó-Dorca

CONTENTS

17.1 QUANTUM SIMILARITY INDICES AND MEASURES

Since the publication of what can be considered as the first reference on quantum similarity (QS) in 1980,[1] there has been a steady increasing amount of devoted papers,[2–4] where an up-to-date bibliographic account on relevant QS contributions is given. In this 1980 original work, the initial idea was planned to emphasize the dissimilarity or, what is the same, the Euclidean distance between two quantum mechanical density functions (DFs) instead of their similarity or cosine of the subtended angle, although both possibilities were indeed open for practical use in reference 1 as a first instance. It was decided afterward to put more emphasis on similarity than on dissimilarity parameters, however; nevertheless, the importance of the Euclidean distance has not been forgotten (see reference 5 for an application example). The reason for this accent put on similarity can be understood by, first, a referee suggestion at the time of publication of reference 1 and, second, a quotation made by Hodgkin and Richards[6] (HR) later on, which proposed a new similarity index and named simultaneously for the first time as Carbó similarity index (CSI), the already defined cosine of reference 1. Then again, a recent study has analyzed the HR index, arriving at the conclusion that it is equivalent to a scaled Euclidean distance,[7] certainly the dissimilarity index already defined in reference 1 too.

All these peculiar characteristics have been commented here to introduce as soon as possible the reader to the evidence that similarity and dissimilarity indices are connected to each other in an inverse manner and are somehow interchangeable. Such a connection has been studied several years ago.[8,9]

Moreover, it might be also interesting to note now that a DF of any kind that is properly normalized can be transformed into probability distributions.[10] Then again, it has been long time ago since the work of Burbea and Rao[11] has discussed in deep the question of comparing two statistical distributions, encompassing the earlier work of Good,[12] Jeffreys,[13] and Kullback and Leibler.[14] Burbea and Rao proposed a general convex function, which by an appropriate choice of the structural parameters produced a distance, among other possible issues. In this way, such a result connects QS indices with statistical comparison of probability DF.

17.2 SIMILARITY MEASURES

Although QS has started within such similarity–dissimilarity index premises, essentially the fact is that the elementary QS computational element building block reduces to the well-known scalar product of two DFs, a so-called similarity measure. Indeed, given two quantum systems, say $\{A,B\}$, the familiar quantum mechanical theoretical basis permits to obtain their attached wavefunctions via solving the respective Schrödinger equations. From the system wavefunctions, a pair of associated DF $\{\rho_A(\mathbf{r}), \rho_B(\mathbf{r})\}$ can be simply set up, with the vector $\mathbf{r}$ representing some number of particle coordinates. In molecular QS studies, the usual DF chosen is the first-order one; thus, vector $\mathbf{r} = (x,y,z)$ corresponds to one-electron position coordinate only. Then, the similarity measure between the system pair of DF is simply defined as the overlap similarity integral*:

$$z_{AB} = \left\langle \rho_A \rho_B \right\rangle = \int_D \rho_A(\mathbf{r}) \rho_B(\mathbf{r}) d\mathbf{r} \in \mathbf{R}^+, \qquad (17.1)$$

where $A = B$ transforms into a so-called self-similarity measure. The integrals of this type are always positive definite because over any integration domain D, both DFs are positive definite, and as a consequence, the integral (Equation 17.1) will always yield a positive real number. Also, these integrals become symmetrical: $z_{AB} = z_{BA}$.

---

* In this chapter, the symbol $\langle f \rangle$, where f is an appropriate function, is used to indicate an integral over a given well-defined domain D: $\langle f \rangle = \int_D f(\mathbf{x}) d\mathbf{x}$.

17.2.1 GENERALIZED SIMILARITY MEASURES

A similarity measure can be easily generalized, choosing any positive-definite operator written as $\Omega(\mathbf{r},\mathbf{r}') > 0$, so the previous overlap similarity measure (Equation 17.1) can transform into the operator weighted scalar product:

$$z_{AB}[\Omega] = \left\langle \rho_A \Omega \rho_B \right\rangle = \int_D \int_D \rho_A(\mathbf{r})\Omega(\mathbf{r},\mathbf{r}')\rho_B(\mathbf{r}')\,d\mathbf{r}\,d\mathbf{r}' \in \mathbf{R}^+. \qquad (17.2)$$

From the integral Equation 17.2 above, choosing the weighting operator as a Dirac's delta function $\Omega(\mathbf{r},\mathbf{r}') = \delta(\mathbf{r} - \mathbf{r}')$ one obtains the former integral (Equation 17.1). Such a relation between the measures 17.1 and 17.2 permits one to simplify the notation of this concept; it allows to use the abridged one appearing in the definition of the similarity integral (Equation 17.1), unless it is necessary to explicitly choose the weight operator in the context where the integrals are employed. It is only necessary to accept Equation 17.1, which also means the general measure Equation 17.2, unless a chosen specific operator is explicitly written. Among the positive-definite operators one can choose, such as the coulomb operator $|\mathbf{r} - \mathbf{r}'|^{-1}$, any DF can also be selected in the role of a weight operator. Any DF attached to a third quantum system C $\rho_c(\mathbf{r})$ can be chosen for this purpose, producing in this way a triple-density QS measure:

$$z_{ACB} = \left\langle \rho_A \rho_C \rho_B \right\rangle = \int_D \rho_A(\mathbf{r})\rho_C(\mathbf{r})\rho_B(\mathbf{r})\,d\mathbf{r} \in \mathbf{R}^+, \qquad (17.3)$$

which, due to the nature of the involved DF, has a hypersymmetric property: $z_{ACB} = z_{ABC} = z_{CAB} = z_{CBA} = z_{BCA} = z_{BAC}$.

Such QS measures permit to construct tensor (or hypermatrix)[15] discrete descriptions of quantum systems, representing them in terms of other known quantum systems. Obviously enough, a triple-density self-similarity measure corresponds to an integral of type Equation 17.3, where the three DFs are the same, for example,

$$z_{AAA} = \left\langle |\rho_A|^3 \right\rangle = \int_D |\rho_A(\mathbf{r})|^3 \, d\mathbf{r}.$$

17.2.2 QS GEOMETRIC POINT OF VIEW

One must be aware that nowadays the mathematical concept of distance has evolved into an intricate labyrinth of alternative definitions and variants[16]; however, one can safely rely on the classic Euclidean concepts for practical QS purposes. From the QS point of view, any DF can be studied as a function belonging to a vector semispace.*[4,17-18] Furthermore, DF can be seen as vectors belonging to infinite-dimensional Hilbert semispaces and thus can be also subject to comparative measures of distances and angles between the two of them. A pair of DF may, in this way, be considered as vectors subtending an angle α and situated in a plane

---

* A vector semispace is a vector space without reciprocal elements and thus defined over the positive real field.

defined by both DF vectors, with their endpoints being a Euclidean distance apart:

$$D_{AB} = \left| \rho_A - \rho_B \right| = \left\langle (\rho_A - \rho_B)^2 \right\rangle^{\frac{1}{2}}.$$

17.3   QS SIMILARITY–DISSIMILARITY INDICES

As has been pointed out previously, both similarity–dissimilarity indices can be computed just knowing the corresponding QS measures. The CSI can be written using a generalized cosine expression of the angle subtended by two DFs:

$$R_{AB} = \cos(\alpha) = \left(\left\langle \rho_A \rho_A \right\rangle \left\langle \rho_B \rho_B \right\rangle \right)^{-\frac{1}{2}} \left\langle \rho_A \rho_B \right\rangle = \left(z_{AA} z_{BB} \right)^{-\frac{1}{2}} z_{AB} \in [0,1], \quad (17.4)$$

so the DF subtended angle can be easily evaluated, whereas in the same way, the Euclidean distance between two DFs can be now written as

$$D_{AB} = \left(\left\langle \rho_A \rho_A \right\rangle + \left\langle \rho_B \rho_B \right\rangle - 2 \left\langle \rho_A \rho_B \right\rangle \right)^{\frac{1}{2}} = \left(z_{AA} + z_{BB} - 2 z_{AB} \right)^{\frac{1}{2}} \in [0, +\infty]. \quad (17.5)$$

It is trivial to see that a scheme about behavior and interconnection between both similarity–dissimilarity indices can be easily obtained. For example, (1) when $R_{AB} = \pm 1 \leftrightarrow D_{AB} = \min$, this corresponds to a collinear DF situation and results into maximal similarity and minimal dissimilarity, and (2) another extremal situation can also be observed: $R_{AB} = 0 \leftrightarrow D_{AB} = \max$, when the DF becomes orthogonal (a not plausible situation unless the DFs are manipulated in some way) and thus possesses minimal similarity and maximal dissimilarity. Thus, CSI and Euclidean distances are somehow inversely related, as has also been commented earlier. Moreover, whereas Euclidean distances are difficult to be defined involving more than two DFs, CSI is easily extended to any number of DFs.[19]

17.3.1   ORIGIN SHIFT OF TWO DFS

Nevertheless, until recently,[20] the role of the origin location in this kind of DF relationship from the QS point of view has been overlooked. Indeed, taking the DF set $\{\rho_A(\mathbf{r}), \rho_B(\mathbf{r})\}$ again as described in a vector semispace context, unless explicitly stated, the DF set has to be considered possessing an arbitrary common origin.* It must be also taken into account that an origin translation shift related to the nature of both DFs will make the Euclidean distance between them invariant, but undoubtedly such an operation will modify the subtended angle. To observe such a geometric characteristic, it is only necessary to construct a centroid DF by the simple formulation:

$\rho_C = \dfrac{1}{2}(\rho_A + \rho_B)$, transforming afterward both DFs according to a straightforward function difference: $\forall I = A, B : p_I = \rho_I - \rho_C \rightarrow \theta_A = \dfrac{1}{2}(\rho_A - \rho_B) \wedge \theta_B = \dfrac{1}{2}(\rho_B - \rho_A) = -\theta_A.$

---

* In fact, one can consider the null DF as origin, that is, $\Theta(\mathbf{r})$, defined as $\forall \mathbf{r} : \Theta(\mathbf{r}) = 0$.

Such manipulation provides a new pair of origin-shifted DFs (OSDFs): $\{\theta_A,$ $\theta_B\}$, which cannot behave anymore as DF does, and, obviously enough, they become linearly dependent pointing to opposite directions. Then, irrespective of the nature of the initial DF pair, the OSDF CSI will in any circumstance

yield the value $r_{AB}\left(\langle\theta_A\theta_A\rangle\langle\theta_B\theta_B\rangle\right)^{-\frac{1}{2}}\langle\theta_A\theta_B\rangle = -1$. Moreover, expressing now by v_I the number of particles involved in any Ith DF, the following can be obtained:

$$\forall I = A, B : v_I = \langle\rho_I\rangle = \int_D \rho_I(\mathbf{r})\,d\mathbf{r} \rightarrow \langle\theta_A\rangle = \frac{1}{2}(v_A - v_B) = -\langle\theta_B\rangle.$$

Showing that over the OSDF, the Minkowski norm, within the chosen integral volume domain, becomes a pseudonorm yielding some real value, the field where the OSDF is now defined. From a geometrical viewpoint, though, Euclidean distances will become

invariant, that is, $D_{AB}^2 = \langle(\rho_A - \rho_B)^2\rangle = \langle\left((\rho_A - \rho_C) - (\rho_B - \rho_C)\right)^2\rangle = \langle(\theta_A - \theta_A)^2\rangle.$

The previously discussed behavior, shown by two DFs after an origin shift, precludes that it is not wise to use CSI for studying the similarity of a given pair of quantum DF only, whereas the Euclidean distance seems adequate to observe their dissimilarity relationship in any circumstance. However, when the set to be compared consists of three or more DFs, then there is no other problem than to be aware of the DF angle variance when using CSI and OSDF. Such angle variance affecting CSI can be assumed as the fact that, by changing the origin, one looks at the set of DFs from another point of view. Another characteristic of the origin shift corresponds to the loss of one degree of freedom of the studied DF set as a whole because the number of linearly independent DFs is reduced in origin translations by one unit.[20] This has been already seen in the case of a pair of DFs. Such a property will be studied in general below.

17.3.2 QUANTUM OBJECT SETS AND QUANTUM POINT CLOUDS

By a quantum object (QO), it can be understood that a two-tuple is formed by some specific submicroscopic object identification and an attached quantum DF. For instance, within the two quantum systems $\{A,B\}$ and their respective DF used along the previous discussion, a set possessing two QOs: $\{\omega_A = (A,\rho_A); \omega_B = (B,\rho_B)\}$, can be formed. Thus, the resultant set $\mathbb{O} = \{\omega_A; \omega_B\}$ can be named a QO set (QOS) with a cardinality of two. When constructing such composite sets, one can imagine a broader mathematical structure encompassing the QOS definition. Thus, as defined above, any QOS can be also considered as a tagged set,[21] made by the Cartesian product of some well-defined submicroscopic object set and a set of tags attached to them. See also, for instance, reference 22, where a particular kind of tagged sets is described, and reference 18 for more details on QOS. Within this definition, a QOS is nothing else but a tagged set whose tags are the DFs attached in quantum theoretical fashion to a submicroscopic object moiety. When the number of QOs grows up to a cardinality N, a QOS of this generic size can be defined as a whole by means of the symbols $\mathbb{O}_N = \{\omega_I = (O_I; \rho_I)|I = 1, N\}$. At that point, with such a general feature established, any QOS DF tag set can be also seen as a set of vectors defined in an infinite-dimensional semispace, forming an N vertex polyhedron. This generalized polyhedron might be also referred to as an infinite-dimensional complex. This is the same as saying that the QOS can be considered a triangle when $N = 3$, a tetrahedral complex when $N = 4$, and so on. Thus, a QOS from a geometrical point of view might be associated to some polyhedral body within the Hilbert semispace, where the DF tags belong. From this geometrical viewpoint when considering a QOS, one may also say that the set of DF tags forms a quantum point cloud,[4] which corresponds to an N-vertex complex defined in infinite-dimensional Hilbert semispaces.

17.3.3 DISCRETE REPRESENTATIONS OF QUANTUM POINT CLOUDS

Returning again to the QOS made by two DFs only, it is easy to consider that, with the DF tags, one can compute the set of all QS measures. Knowing these QS measures, a (2×2) matrix can be finally constructed as

$$\mathbf{Z} = \begin{pmatrix} z_{AA} & z_{AB} \\ z_{BA} & z_{BB} \end{pmatrix},$$ (17.6)

which, irrespective of the nature of the QOS matrix, becomes symmetrical. In general, when considering QOS of type $\mathbb{O}_N$, the matrix made of the whole set of similarity measures involving the QOS element pairs becomes a symmetrical $(N \times N)$ matrix and can be called a QS matrix (QSM). The diagonal elements of the QSM, which belong to the quantum self-similarity measure set $\{z_{II}|I = 1,N\}$, act as Euclidean norms of the associated DF. Such a norm role can be easily inferred from the CSI definition as shown in Equation 17.4 when considered as a cosine of the subtended angle between two DFs: $z_{AB} = \left(z_{AA}z_{BB}\right)^{\frac{1}{2}} \cos(\alpha) \equiv |\rho_A|_2 |\rho_B|_2 \cos(\alpha)$. In the same way, the off-diagonal elements of the QSM can be associated to scalar products of the involved DF. Therefore, the QSM (Equation 17.6) and their generalized counterparts correspond to metric matrices, generated by the linearly independent DF tag set. By discussing again the QSM (Equation 17.6) features, it can be easily seen that the columns (or rows), which can be reordered as two vectors:

$$\left|\mathbf{z}_A\right) = \begin{pmatrix} z_{AA} \\ z_{BA} \end{pmatrix} \leftrightarrow \rho_A \wedge \left|\mathbf{z}_B\right) = \begin{pmatrix} z_{AB} \\ z_{BB} \end{pmatrix} \leftrightarrow \rho_B,$$ (17.7)

might be also associated to discrete representations of the respective DF. In general, it is evident that a QOS of N-elements generates an $(N \times N)$ QSM: $\mathbf{Z} = \{z_{IJ}\}$ whose columns (or rows) $\{|\mathbf{z}_I\rangle\} = \{z_{JI}|J = 1,N\}| I = 1,N\}$ can be considered in a one-to-one correspondence with the DF tag elements of the attached QOS, that is, it can be written as $\forall I = 1,N : |\mathbf{z}_I\rangle \leftrightarrow \rho_I$. The QSM column (or row) set can be alternatively considered as forming discrete tags, associated in turn to the submicroscopic objects of the initial QOS. In this way, transforming any QOS into a discrete QOS (DQOS), defined formally as $\mathbb{D}_N = \{w_I = (O_I;|\mathbf{z}_I\rangle) \,|\, I = 1,N\}$. In the same manner as before, the geometric perspective indicates that, when considering the set of N-dimensional discrete vectors $\{|\mathbf{z}_I\rangle | I = 1,N\}$, then some discretely defined N-dimensional point cloud is formed in this way, constituting in fact an N-dimensional complex. Such an N-dimensional point cloud does not necessarily preserve the form of the original infinite-dimensional one, being a projection of this infinite-dimensional set of points into an N-dimensional subspace. To see this, one can consider the two-dimensional column vector example given in Equation 17.7; computing their metric matrix, one can admit it as coincident

with the square of the original QSM: $\mathbf{Z}^2 = \begin{pmatrix} \langle \mathbf{z}_A | \mathbf{z}_A \rangle & \langle \mathbf{z}_A | \mathbf{z}_B \rangle \\ \langle \mathbf{z}_B | \mathbf{z}_A \rangle & \langle \mathbf{z}_B | \mathbf{z}_B \rangle \end{pmatrix}$.

Accordingly, the angle subtended between the DQOS tag elements now becomes $R_{AB}^{(Z)} = \left(\langle \mathbf{z}_A | \mathbf{z}_A \rangle \langle \mathbf{z}_B | \mathbf{z}_B \rangle \right)^{-\frac{1}{2}} \langle \mathbf{z}_A | \mathbf{z}_B \rangle$ and the Euclidean distance between them can be associated to the expression $D_{AB}^{(Z)} = \left(\langle \mathbf{z}_A | \mathbf{z}_A \rangle + \langle \mathbf{z}_B | \mathbf{z}_B \rangle - 2 \langle \mathbf{z}_A | \mathbf{z}_B \rangle \right)^{\frac{1}{2}}$.

This new quantum similarity–dissimilarity index formulation does not necessarily have to coincide with the original similarity–dissimilarity indices (Equations 17.4 and 17.5). Therefore, both matrices (the original metric of the QOS DF $\mathbf{Z}$ and the metric of the DQOS tag set elements $\mathbf{Z}^2$) might provide complementary geometrical and topological information about the associated QO point cloud. A discussion on the nature of the QS metric matrices has been recently published.[23] More information on this QS feature can be obtained.

17.4   ILLUSTRATIVE EXAMPLE: QS ANALYSIS OF THREE p-TYPE FUNCTIONS

The set made by the three p-type functions in a Gaussian type orbital (GTO) framework is sufficiently simple to be used as an example but at the same time provides a yet unexplored context, which can show how QS techniques can handle quantum system states, even degenerate ones. The set of p-type Gaussian functions can be collected into a vector, which can be associated in turn to three degenerate model wavefunctions $|p\rangle = \eta \mathbf{r} \exp(-a|\mathbf{r}|^2)$ where η is a normalization factor, $\mathbf{r} = (x,y,z)$, and a is an arbitrary positive-definite parameter. The set of attached DF can be also written as a vector $|d\rangle = \eta^2 (\mathbf{r} * \mathbf{r}) \exp(-2a|\mathbf{r}|^2)$, where the position vector product symbol is defined employing the inward product[24]: $(\mathbf{r} * \mathbf{r}) = (x^2, y^2, z^2)$. Thus, the components of the DF vector can be written as $|d\rangle = (d_x, d_y, d_z)$, considering $d_x = \eta^2 x^2 \exp(-2a|\mathbf{r}|^2)$, with the same convention for the other components but changing accordingly the variable x by the variables y and z. Under these considerations, the QS metric matrix for the p-type DF vector can be written as

$$\mathbf{Z} = \begin{pmatrix} \alpha & \beta & \beta \\ \beta & \alpha & \beta \\ \beta & \beta & \alpha \end{pmatrix} \leftarrow \alpha = \langle d_x | d_x \rangle = \langle d_y | d_y \rangle \ldots \wedge \beta = \langle d_x | d_y \rangle = \langle d_x | d_y \rangle \ldots \quad (17.8)$$

The two different needed integrals can be easily calculated,* although in the following reasoning their analytical or numerical values are not essential to understand the QS relationship structure of this triple degenerate set of functions. In the first place, it is now easy to obtain the Euclidean distance between the three p-type DFs, which will take the form $D^{(p)} = (2\alpha - 2\beta)^{\frac{1}{2}} = \sqrt{2}(\alpha - \beta)^{\frac{1}{2}}$ irrespective of which pair of functions is considered. The CSI can be also computed at this level as $R^{(p)} = \alpha^{-1}\beta$, an expression that is independent of the function pair considered. Such a behavior corresponds to what must be expected for three points forming an equilateral triangle structure irrespective of the dimension of the space where the three points are considered to belong. Moreover, the metric Equation 17.8 can be associated to a topological matrix bearing exactly the same structure as a Hückel molecular orbital (HMO) Hamiltonian matrix picture, associated to an equilateral triangular structure made of three equal atoms. The three metric matrix rows or columns, as it has been earlier discussed, can be considered discrete representations of the three p-type DFs in question: they correspond to three linearly independent vectors referred to the zero function as origin. Hence, the topological structure of the discrete QS representation of the three degenerate p-type DFs is coincident with an equilateral triangle. Considering the three column vectors of the matrix representation Equation 17.8: $\mathbf{Z} = (|\mathbf{z}_x\rangle, |\mathbf{z}_y\rangle, |\mathbf{z}_z\rangle)$ the discrete representation of the system centroid can be easily computed as the vector sum: $|\mathbf{z}_c\rangle = \frac{1}{3}\left(|\mathbf{z}_x\rangle + |\mathbf{z}_y\rangle + |\mathbf{z}_z\rangle\right) = \frac{1}{3}(\alpha + 2\beta)\begin{pmatrix} 1 \\ 1 \\ 1 \end{pmatrix} = \frac{1}{3}(\alpha + 2\beta)|\mathbf{1}\rangle$, so the origin shifted columns of the matrix become, using the unity matrix symbol,

$$\mathbf{1} = \{1_{IJ} = 1\} : \mathbf{A} = \frac{\gamma}{3}\begin{pmatrix} 2 & -1 & -1 \\ -1 & 2 & -1 \\ -1 & -1 & 2 \end{pmatrix} = \frac{\gamma}{3}(3\mathbf{I} - \mathbf{1}) \leftarrow \gamma = \alpha - \beta.$$

As commented earlier, the three columns of the origin shifted matrix A are linearly dependent now, and it is easy to verify that $Det\,|\mathbf{A}| = 0$. The Euclidean distances between the columns of the original metric Z and those of the origin shifted column matrix A, irrespective of the column pairs considered, can be written respectively as $D^Z = D^A = \gamma\sqrt{2}$. Both distances are coincident and scaled by $\sqrt{\gamma}$ with respect to the original one, directly obtained from the DF in the Hilbert semispace. In this way, one can see how the QS discrete numerical picture of the DF can provide an accurate geometrical picture of the DF relationships. The equilateral triangle shape associated to the DF trio is homothetically preserved from continuous infinite-dimensional vectors to discrete three-dimensional space. Moreover, although this p-type GTO example constitutes a very simple scheme, it is plausible to conjecture that such a QS representation can be obtained in all the cases, where a set of triple degenerate wavefunctions is analyzed; that is, the topology of threefold degenerate state wavefunctions from the QS viewpoint will be topologically invariant irrespective of the nature

---

* In fact, one can use $\alpha = \eta^4 \int_D x^4 \exp\left(-4a|\mathbf{r}|^2\right)d\mathbf{r} \wedge \beta = \eta^4 \int_D x^2 y^2 \exp\left(-4a|\mathbf{r}|^2\right)d\mathbf{r}$, with η being some normalization factor.

of the three involved DFs. The CSI in this shifted situation can be also described as $R^{(A)} = -2^{-1} \equiv \cos\left(\dfrac{2\pi}{3}\right)$, showing that the angles of an equilateral triangle are correctly measured from the centroid.

17.4.1 DISCRETE QS REPRESENTATION OF N-FOLD DEGENERATE WAVEFUNCTIONS

A generalization of the threefold degenerate case can be achieved without effort. Suppose that an n-fold degenerate orthonormalized set of wavefunctions is known: $\{|I\rangle\}$; their DF can be written as the symbols $\{\rho_I = |I\rangle^*|I\rangle\}$; therefore, the DF metric matrix will have as before only two kinds of possible elements:

$\alpha = \left\langle |\rho_I|^2 \right\rangle = \langle \rho_I \rho_I \rangle \wedge \beta = \langle \rho_I \rho_J \rangle$. The metric constructed in this way, $\mathbf{Z} = \alpha\mathbf{I} + \beta(1 - \mathbf{I})$,

provides a generalized discrete picture of the initial wavefunctions once transformed into DF. The distances between any pair of degenerate DFs can be written again as

in the threefold degenerate case: $D^{(\rho)} = \sqrt{2}(\alpha - \beta)^{\frac{1}{2}}$. This result necessarily precludes that, in any circumstance, $\alpha > \beta$ will hold; otherwise, one might encounter purely imaginary Euclidean distances. Also, the matrices Z and A can be constructed in the same way as in the threefold degenerate case, providing distances given by the same

values as in the triangular case: $D^{(Z)} = D^{(A)} = \sqrt{2}\left(\alpha - \beta\right)$. The emerging picture, from the threefold degenerate case up to the general n-fold degenerate situation, consists of considering an n-vertex polyhedron with equidistant vertices. Such a geometric object can always be represented in a space that is one dimension less than the original order of the degeneracy. Thus, it can be said that the geometric and topologic QS representation of any n-fold degenerate set of wavefunctions corresponds to an $(n - 1)$ dimensional simplex. To finish with this application example, the CSI can be also studied. In the original metric matrix, it is easy to see that all indices will be the same: $R^{(\rho)} = \alpha^{-1}\beta$, whereas, in the QS projected metric, taking into account that one can write $\langle \mathbf{z}_I | \mathbf{z}_I \rangle = \alpha^2 + (n-1)\beta^2 \wedge \langle \mathbf{z}_I | \mathbf{z}_J \rangle = 2\alpha\beta + (n-2)\beta^2$, the CSI will become, in general,

$$R^{(Z)} = \frac{2\alpha\beta + (n-2)\beta^2}{\alpha^2 + (n-1)\beta^2} = \frac{2\alpha\beta^{-1} + n - 2}{\alpha^2\beta^{-2} + n - 1} = \frac{2\left(R^{(\rho)}\right)^{-1} + n - 2}{2\left(R^{(\rho)}\right)^{-2} + n - 1}.$$

17.4.2 QS OF BINARY STATES

When considering some QO pairs, such as in the case of two conformers of the same molecule, the attached DF might possess the same number of electrons; therefore, one will obtain $\langle \rho_A \rangle = \langle \rho_B \rangle = v$. Even in this restricted case, the pair of DFs can be considered essentially diverse, as they possess a different set of nuclear coordinates: $\{\mathbf{R}_A; \mathbf{R}_B\}$. In case of need, this condition can be symbolically expressed explicitly and written within each DF as $\{\rho_A(\mathbf{r}|\mathbf{R}_A);\rho_B(\mathbf{r}|\mathbf{R}_B)\}$. This kind of situation can be common for cases such as R–S stereoisomers and also for staggered-eclipsed, boat-tub, or *cis-trans* conformations. Alternatively, such a QS framework can be used to compare two different electronic states or, say, ions with the neutral molecule. In any situation whatsoever, constructing the DF mean centroid corresponds to defining a function that possesses a Minkowski norm yielding the mean number of electrons of both structures. In the more restricted case, when both DFs bear the same number of electrons, the centroid norm possesses the same number of electrons as the initial DF couple, as it can be written as $\langle\rho_c\rangle = \frac{1}{2}\left(\langle\rho_A\rangle + \langle\rho_B\rangle\right) = \frac{1}{2}(v_A + v_B)$; then, one will also have $v_A = v_B = v \rightarrow \langle\rho_c\rangle = v$. Therefore, the centroid DF might be used to shift the original densities in the manner already described. Then, any binary QO system tags might be referred to a common origin by a centroid translation procedure. As it has been discussed before, the OSDF $\{\theta_A;\theta_B\}$ are no longer linearly independent and cannot bear the positive-definite nature of the original DF tag set. Quantum self-similarity of the centroid shifted pair, being defined as a Euclidean norm, clearly becomes the same measure for both OSDF,

$$\left\langle|\theta_A|^2\right\rangle = \left\langle|\theta_B|^2\right\rangle = \frac{1}{4}\left\langle|\rho_A - \rho_B|^2\right\rangle = \frac{1}{4}\left(\left\langle|\rho_A|^2\right\rangle + \left\langle|\rho_B|^2\right\rangle - 2\langle\rho_A\rho_B\rangle\right) = \frac{1}{4}D_{AB}^2,$$ and coin-

cides with the scaled squared Euclidean distance between the original DF pair, D_{AB}^2. Therefore, most of the properties of these binary systems, such as rotation barriers and specific optical rotation angles, might be somehow correlated with the squared distance or the distance between their original DF tags.

17.5 STEREOISOMER PARTICULAR CASE OF QUANTUM MECHANICAL EXPECTATION VALUES AND QUANTUM QSPR

In binary stereoisomers, one has the original DF self-similarity property, $\langle|\rho_R|^2\rangle = \langle|\rho_S|^2\rangle$, with the wavefunctions and DF per se indistinguishable; consequently, it can be also written as $\frac{1}{4}D_{RS}^2 = \frac{1}{2}\left(\left\langle|\rho_R|^2\right\rangle - \langle\rho_R\rho_S\rangle\right) = \frac{1}{2}\left(\left\langle|\rho_S|^2\right\rangle - \langle\rho_R\rho_S\rangle\right)$, because the scalar product of the two stereoisomer DFs has to be essentially different from the equal corresponding self-similarities. Thus, the specific optical rotation angle $[\alpha]_D$, which is a crucial property attached to stereoisomers, might be correlated with the integral expression difference $\langle|\rho_R|^2\rangle - \langle\rho_R\rho_S\rangle$ or its square root. Then again, one can see that the dissimilarity index D_{RS} measures the existing difference between the elements of the attached DF pair. Knowing this possibility to measure the differences between R and S molecular specular images, one can suppose that the property $[\alpha]_D$ can be computed as an expectation value implying some unknown operator.

17.5.1 SETTING UP QQSPR OPERATOR

In general, this is similar to considering at some first-order level the construction of a QQSPR operator. In this approximate framework, the unknown operator, giving rise to the property to be obtained as an expectation value in a binary QO case, can be expressed as a linear combination of the OSDF. It can be written in a simplified QQSPR manner as[25,26]

$$\Omega \approx w_A \theta_A + w_B \theta_B = (w_A - w_B)\theta_A = \lambda \theta_A. \tag{17.9}$$

Reflecting that as the shifted DF pair is no longer linearly independent, there is only one OSDF needed as a basis set. The expectation value obtained by applying the quantum mechanical expression using the approximate operator Equation 17.9 can be written as

$$\pi_A \approx \langle \Omega \theta_A \rangle = \lambda \langle |\theta_A|^2 \rangle = \lambda \langle |\rho_A|^2 - 2\rho_A \rho_B + |\rho_B|^2 \rangle$$
$$= \lambda \left(\langle |\rho_A|^2 \rangle - 2\langle \rho_A \rho_B \rangle + \langle |\rho_B|^2 \rangle \right) = \lambda \left(z_{AA} - 2z_{AB} + z_{BB} \right) = \lambda D_{AB}^2. \tag{17.10}$$

When considering the stereoisomer case, it can be also written as $\pi_R = [\alpha]_D \approx 2\lambda \left(Z_{RR} - Z_{RS} \right) = \lambda D_{RS}^2$, a resultant situation that is coincident with the previous discussion about the self-similarity of the OSDF.

17.5.2 Assorted Molecular Structure-Specific Polarization Angle QQSPR

To test the possibility of finding a relation between specific optical rotation angles $[\alpha]_D$ and QS measures for amino acids, some initial attempts have been already published by Mezey et al.[27] and more recently by other authors.[28] However, in both studies, the following points were not taken into account: (1) the previous quantum expectation value reasoning might be used to obtain a possible QQSPR; (2) for different molecular stereoisomer pairs to be comparable and compared, the associated DF pairs might possess the same centroid origin; (3) the fact that stereoisomer values of $[\alpha]_D$ have two opposite signs; and (4) optically inert molecules possess null $[\alpha]_D$, which means the zero point needs to be present in any model trying to connect $[\alpha]_D$ with some parameter bearing the DF R–S difference. Therefore, taking into consideration all these points, it might seem natural to use at first instance not D_{RS}^2 but $\pm D_{RS}$, allowing the dissimilarity parameter to possess the two signs associated to the square root to be related with $\pm[\alpha]_D$ values. Because both stereoisomers are present in the QS distance expression, both signs in the experimental specific optical rotation angle expectation values $\pi_{R,S} = \pm[\alpha]_D$ must be also taken into account. To deal with this two-sign framework between the two variables to be related, a straightforward least squares procedure provides a linear function of the type $[\alpha]_D = aD_{RS} = \pm a\sqrt{D_{RS}^2}$, where the constant a is evaluated with the experimental specific rotation angle values for the molecular set studied and the corresponding computed QS Euclidean distance index.

17.5.3 QQSPR for Amino Acids

At first instance, such a relationship has been tested for 10 amino acids, the set of the previous studies plus some add-ons. See Table 17.1 to grasp the considered molecules

TABLE 17.1
Studied Molecules with Specific Rotation and Quantum Euclidean Distance Values

	Molecule	[α]	D_{RS}	Reference
1	Alanine	2.7	14.8772	A
2	Asparagine	4.7	19.2614	A
3	Glutamine	11.5	20.2	A
4	Lysine	14.6	19.4	A
5	Valine	6.42	17.7691	A
6	Leucine	10.8	16.3077	A
7	Serine	6.83	16.7786	A
8	Methionine	8.11	19.5254	A
9	Arginine	17	21.8826	B
10	Cysteine	9.4	17.9232	C
11	Canavanine	7.9	23.0826	D
12	(+)Pulagone	21	17.442	D
13	Versimide	22.3	23.9536	D
14	(+)Carvone	62.3	18.0235	D
15	Methamphetamine	18.9	16.6801	E
16	MDMA	15.2	20.8278	F
17	LSD	217	28.6062	G
18	Testosterone	110	26.0204	H
19	Estradiol	81	24.4703	I
20	Ibuprofen	58	22.7788	J
21	4-Terpenol	48.3	12.3158	D
22	Limonene	56	13.7752	K
23	β-Pinene	21.8	16.2480	K
24	cis-Pinane	88.4	15.4670	K
25	Propylene oxide	26.4	9.0259	K
26	2-Butanol	13.52	8.6060	L
27	Tartaric acid	12.5	21.6742	M

Note: A: Weast, R. C., *Handbook of Chemistry and Physics*, 63rd ed., CRC Press, 1982–1983. B: Miller, H. K. and Andrews, J. C., *J. Biol. Chem.*, 87, 435–439, 1930. C: Andrews, J. C., *J. Biol. Chem.*, 69, 209–217, 1926. D: Dev, S. and Koul, O., *Insecticides of Natural Origin*, Harwood Academic Publishers, Amsterdam, 1977. E: Kozma, D. and Fogassy, E., *Synthetic Commun.*, 29, 4315–4319, 1999. F: Pizarro, N., 10, 1085–1092, 2002. G: Stoll, A. and Hofmann A., assignors, Sandoz Ltd., Fribourg, Switzerland, U.S. Patent 2,438,259; Patented March 23, 1948. H: http://www.sigmaal drich.com/catalog/ProductDetail.do?lang=es&N4=86500|FLUKA&N5=SEARCH_CONCAT_ PNO|BRAND_KEY&F=SPEC. I: Sigma product no. E8875, *The Merck Index*, 12th ed., #3746, 1966. Free online encyclopedia (http://encyclopedia2.thefreedictionary.com/Estradiol). J: Snell, D. and Colby, J., *Enzyme Microbial Technol.*, 24, 160–163, 1999. K: Müller, T., Wiberg, K. B., and Vaccaro, P. H., *J. Phys. Chem.*, A104, 5959–5968, 2000. L: Skell, P. S., Allen, R. G., and Helmkamp, G. K., *J. Am. Chem. Soc.*, 82, 410–414, 1960. M: http://www.chem-online.org/food-ingredient/l-tartaric-acid.htm, http://www.sigmaaldrich.com/catalog/ProductDetail.do?lang=es& N4=95308|SIGMA&N5=SEARCH_CONCAT_PNO|BRAND_KEY&F=SPEC.

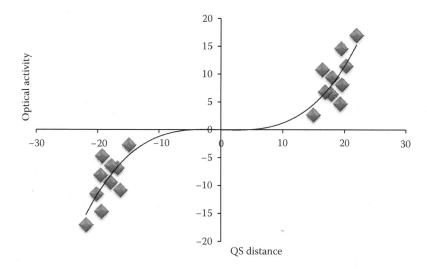

FIGURE 17.1 Plot of the specific optical rotation angle as a polynomial of the QS Euclidian distance for 10 amino acid molecules.

that appear at the top of the list. The obtained $[\alpha]_D \leftrightarrow D_{RS}$ results (as shown in Figure 17.1) present a clear nonlinear behavior as a third-order polynomial term gave a reasonable fit.

The graph of Figure 17.1 shows the obtained relation with a nonlinear term to improve the least squares fit, yielding $[\alpha]_D = 0.0015 D_{RS}^3 - 0.0304 D_{RS} (R^2 = 0.9146; R = 0.9563)$.

The resulting fitting polynomial was considered sufficiently interesting to attempt to study the same procedure but with a heterogeneous set of molecules.

17.5.4 QQSPR for Heterogeneous Optically Active Molecules

To test the ability of the QS theory in this way, a new collection of molecules has been chosen and added to the previous amino acid set. This additional set is made with several chiral natural insecticides, steroids, anti-inflammatory drugs, psychotropic substances, food additives, odorants, and simple alcohols (see Table 17.1 for more information). The $[\alpha]_D \leftrightarrow D_{RS}$ QQSPR in this larger molecular set, as shown in Figure 17.2, continues to possess a similar trend as the one previously found in the amino acid family. In this larger heterogeneous family of stereoisomers, the error is also larger than in the homogeneous amino acid set, though. In difference with the amino acid relation, which corresponds to some homogeneous molecular set with the specific rotation obtained in the same way, the heterogeneous set contains $[\alpha]_D$ results collected from a large variety of different solvents. Moreover, the literature sources are also far to be homogeneous; thus, experimental data could be submitted to different experimental variation errors. Also, some included molecules have several chiral centers, so this can bring problems at the moment in choosing and superposing the two compared chiral structures to compute the essential similarity measure Z_{RS}.

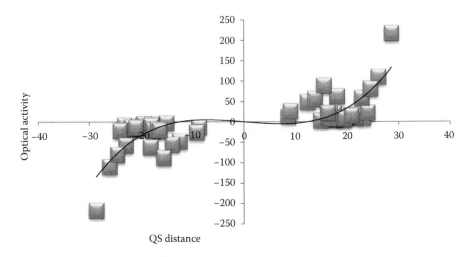

FIGURE 17.2   Plot of the specific optical rotation angle as a polynomial of the QS Euclidian distance for a set of 27 heterogeneous molecules.

The theoretical assessment of optical rotation is indeed not a simple problem.[29-31] The polynomial feature of the QQSPR plots can be caused by many factors that are too involved to be analyzed in depth here, with this relation being only a case example of the possibilities. Such a result must be taken as the basic QS technique that can be used to obtain what can be called QQSPR.[25,26,32-34] There is always a risk in obtaining a reduced precision when, from an initial homogeneous set, a truly heterogeneous one, as in the case shown above, is studied.

Nonetheless, the methodological potential of simple QQSPR techniques obviously appears in the present case, as no classic QSPR procedure can provide (1) a relationship for stereoisomer systems, able to offer a clear parameter, describing the distinction between R and S nature, as the one found in the form of a QS Euclidean distance and (2) a simple QQSPR for truly heterogeneous molecules. The origin of the nonlinear QQSPR structure found might be hidden in the approximate linearly defined structure of the QQSPR operator. As shown in Equation 17.9, the QQSPR operator is not completely appropriate to handle the R–S problem, as evidenced by the computed polynomial relations. Nevertheless, it can be said that there exists a linear relation of a type similar to Equation 17.10 for enantiomers, where the operator Ω needs the addition of a new nonlinear extra term correcting the expectation value. expression.

17.6   HYPERBOLIC SINE RELATION

For the 27 assorted heterogeneous molecules with a third-order polynomial, a concrete relation is obtained such as

$$[\alpha]_D = 0.007 D_{RS}^3 - 1.0769 D_{RS} \rightarrow R^2 = 0.6432 \rightarrow R = 0.8020. \qquad (17.11)$$

This polynomial relationship, which appears practically with a small second-order term, precludes that the function that can be employed to connect specific optical rotation and QS distance index perhaps can be associated to some hyperbolic sine, modified with some scaling variable a, that is, $h(D_{RS};a) = \frac{1}{2}(e^{+aD_{RS}} - e^{-aD_{RS}})$.

The above function can be approximated in turn by a Taylor series, whose first terms correspond to the structure of the polynomial Equation 17.11, as it can be written as $h(D_{RS};a) = \sum_{k=0}^{\infty} \frac{(aD_{RS})^{2k+1}}{(2k+1)!} = aD_{RS} + \frac{1}{6}(aD_{RS})^3 + O(5)$.

Then, one can try to obtain a relation such as $\pm[\alpha]_D = \lambda h(D_{RS};a) + \beta$, and taking into account the repeated positive and negative $\pm[\alpha]_D$ values into the data of the set property variable $\{\pi \equiv [\alpha]_D; D_{RS}\}$, a relation system can be written such as

$$\begin{pmatrix} |\pi\rangle \\ -|\pi\rangle \end{pmatrix} = \lambda \begin{pmatrix} |h\rangle \\ -|h\rangle \end{pmatrix} + \beta \begin{pmatrix} 1 \\ 1 \end{pmatrix}.$$

Such a system setup generates a quadratic error function that can be simply written as $\mathcal{E}^{(2)} = \||\pi\rangle - \lambda|h\rangle - \beta|1\rangle\|^2 = 2\langle\pi|\pi\rangle - 4\lambda\langle\pi|h\rangle + 2\lambda^2\langle h|h\rangle + 2\beta^2\langle 1|1\rangle$, providing an optimal fitting parameter pair $\{\lambda, \beta\}$, which can be expressed as $\beta = 0 \wedge -4\langle\pi|h\rangle + 4\lambda\langle h|h\rangle = 0 \rightarrow \lambda = \frac{\langle\pi|h\rangle}{\langle h|h\rangle}$.

Thus, in this case, a relation of the type $[\alpha]_D^{(*)} = \frac{\langle\pi|h\rangle}{\langle h|h\rangle} h(D_{RS};a)$ can always be written.

The hyperbolic sine scaling parameter has been optimized for the heterogeneous set of optically active molecules, providing a slight improvement of the polynomial Equation 17.11 results, such that $a[*] = 0.3000 \wedge \lambda = 0.0814225 \rightarrow R^2 = 0\ 779518 \rightarrow R = 0.882931$.

17.7 COMPUTATION DETAILS

The molecular structures employed here have been computed with the Spartan program[35] with a basis set 3-21G under HF procedure with full geometry optimization. The geometries and Mulliken populations obtained in this way as a final output have been entered in the MQSPS set of programs as described in reference 36 to obtain the involved QS measures and indices.

17.8 FINAL DISCUSSION

An updated presentation of the character and properties of QS has been carried out. Emphasis has been put on the nature and properties of the initially defined QS indices between two quantum DFs: CSI, a cosine of the subtended angle and dissimilarity indices, mainly the Euclidean distance. Both indices rely on the QS measure integrals, though. The set theoretical and geometrical characteristics of QS have also been discussed mainly in connection with centroid origin translation procedures, a simple DF manipulation overlooked in the past. Centroid shift of quantum point clouds allows one to have a possible general reference origin for any QOS collection, thus permitting one to consider in a general manner the study of heterogeneous molecular sets by means of QS. Two examples have been developed to illustrate the characteristics and useful capabilities of QS. First, the set of three p-type GTOs as a model of degenerate DF is presented, and the n-fold degenerate case has been discussed afterward. It is concluded that the QS geometrical picture of any set of n-fold degenerate DF corresponds from a QS point of view to an $(n - 1)$-dimensional simplex. A second example is also provided consisting in the QQSPR study of some optically active stereoisomers. A third-order polynomial model, relating the specific rotation angles with the QS Euclidean distance between the DF associated to the R–S forms, is found, which was first studied for several amino acids and then for a heterogeneous set of chiral molecules. Further model refinement promotes the possibility that such relation might be expressed as a scaled hyperbolic sine. In both examples, QS through the quantum mechanical DF structure permits one to topologically represent, on one hand, difficult wavefunction sets and, on the other hand, chiral sets; within classic QSPR, it is hard to find models with a sound parameter related to the specific optical rotation angle.

ACKNOWLEDGMENT

The author would like to thank Prof. A. Roglans (UdG) for disinterested help in obtaining specific rotation angle values for several molecular structures.

REFERENCES

1. Carbó, R., Leyda, L., and Arnau, M. *Int. J. Q. Chem.*, 17, 1185–1189, 1980.
2. Carbó-Dorca, R. and Mercado, L. D., *J. Comput. Chem.*, 31, 2195–2212, 2010.
3. Bultinck, P., Girones, X., and Carbó-Dorca, R., *Rev. Comput. Chem.*, Vol. 21, Lipkowitz, K. B., Larter R., and Cundari, T., eds., John Wiley & Sons, Inc., Hoboken, NJ, 2005, 127–207.
4. Carbó-Dorca, R. and Besalú, E., *J. Math. Chem.*, 50, 210–219, 2012.
5. Bultinck, P. and Carbó-Dorca, R. J., *Chem. Inf. Comput. Sci.*, 43, 170–177, 2003.
6. Hodgkin, E. E. and Richards, W. G., *Int. J. Q. Chem.*, 32S, 105–110, 1987.
7. Carbó-Dorca, R., *J. Math. Chem.*, 2011, DOI 10.1007/s10910-011-9920-6.
8. Carbó, R., Besalú, E., Calabuig, B., and Vera, L., *Adv. Q. Chem.*, 25, 253–313, 1994.
9. Carbó, R., Besalú, E., Amat, Ll., and Fradera, X., *J. Math. Chem.*, 19, 47–56, 1996.
10. Carbó-Dorca, R., *J. Math. Chem.*, 49, 2109–2115, 2011.
11. Burbea, J. and Rao, R., *IEEE Trans. Inf. Theory*, 28, 489–495, 1982.
12. Good, I. J., *Ann. Math. Stat.*, 34, 911–934, 1963.
13. Jeffreys, H., *Proc. R. Soc. London A*, 186, 453–461, 1946.
14. Kullback, S. and Leibler, R. A., *Ann. Math. Stat.*, 22, 79–86, 1951.
15. Mercado, L. D. and Carbó-Dorca, R., *J. Math. Chem.*, 49, 1558–1572, 2011.
16. Deza, M. M. and Deza, E., *Encyclopedia of Distances*, Springer-Verlag, Berlin, 2009.

17. Bultinck, P. and Carbó-Dorca, R., *J. Math. Chem.*, 36, 191–200, 2004.
18. Carbó-Dorca, R., *J. Math. Chem.*, 44, 628–636, 2008.
19. Carbó-Dorca, R., *J. Math. Chem.*, 49, 2109–2115, 2011.
20. Carbó-Dorca, R. and Besalú E., *J. Math. Chem.*, 2011, DOI 10.1007/s10910-011-9960-y.
21. Carbó-Dorca, R., *J. Math. Chem.*, 23, 353–364, 1998.
22. Carbó-Dorca, R., *J. Math. Chem.*, 22, 143–147, 1997.
23. Carbó-Dorca, R., *J. Math. Chem.*, 2011, DOI 10.1007/s10910-011-9921-5.
24. Carbó-Dorca, R., *J. Mol. Struct.*, 537, 41–54, 2001.
25. Carbó-Dorca, R., *SAR QSAR Environ. Res.*, 18, 265–284, 2007.
26. Carbó-Dorca, R. and Gallegos, A., Entry: 176, in *Encyclopedia of Complexity and Systems Science*, Meyers, R., ed., Springer, New York, 2009, 7422–7480.
27. Mezey, P. G., Ponec, R., Amat, L., and Carbó-Dorca, R., *Enantiomer*, 4, 371–378, 1999.
28. Boon, G., Van Alsenoy, Ch., De Proft, F., Bultinck, P., and Geerlings, P., *J. Phys. Chem. A*, 110, 5114–5120, 2006.
29. Atkins, P. W. and Woolley, R. G., *Proc. R. Soc. London A*, 314, 251–267, 1970.
30. Kondru, R. K., Wipf, P., and Beratan, D. N., *Science*, 282, 2247–2250, 1998.
31. Pedersen, T. B., Koch, H., Boman, L., and Sánchez de Merás, A. M., *J. Chem. Phys. Lett.*, 393, 319–326, 2004.
32. Carbó-Dorca, R. and Van Damme, S., *Int. J. Q. Chem.*, 108, 1721–1734, 2007.
33. Carbó-Dorca, R., Gallegos, A., and Sánchez, A. J., *J. Comput. Chem.*, 30, 1146–1159, 2008.
34. Carbó-Dorca, R. and Gallegos, A., *J. Comput. Chem.*, 30, 2099–2104, 2009.
35. *Spartan '10, Version 1.1.0*, Wavefunction, Inc., Irvine, CA, 2011.
36. Carbó-Dorca, R., Besalú, E., and Mercado, L. D., *J. Comput. Chem.*, 32, 582–599, 2011.

18 Electronic Excitation Energies of Molecular Systems from the Bethe–Salpeter Equation
Example of the H_2 Molecule

Elisa Rebolini, Julien Toulouse, and Andreas Savin

CONTENTS

18.1 INTRODUCTION

Time-dependent density functional theory (TDDFT)[1] within the linear response formalism[2–4] is nowadays the most widely used approach to the calculation of electronic excitation energies of molecules and solids. Applied within the adiabatic approximation and with the usual local or semilocal density functionals, TDDFT gives indeed in many cases excitation energies with reasonable accuracy and low computational cost. However, several serious limitations of these approximations are known, for example, for molecules, too low charge-transfer excitation energies,[5] lack of double excitations,[6] and wrong behavior of the excited-state surface along a bond-breaking coordinate (see, e.g., reference [7]). Several remedies to these problems are actively being explored, including long-range corrected TDDFT,[8, 9] which improves charge-transfer excitation energies; dressed TDDFT,[6, 10, 11] which includes double excitations; and time-dependent density-matrix functional theory (TDDMFT),[12–16] which tries to address all these problems.

In the condensed-matter physics community, the Bethe–Salpeter equation (BSE) applied within the GW approximation (see, e.g., references [17–19]) is often considered as the most successful approach to overcome the limitations of TDDFT. Although it has been often used to describe excitons (bound electron-hole pair) in periodic systems, it is also increasingly applied to calculations of excitation energies in finite molecular systems.[20–31] In particular, the BSE approach is believed to give accurate charge-transfer excitation energies in molecules,[29, 31] and when used with a frequency-dependent kernel, it is in principle capable of describing double excitations.[32, 33]

In this work, we examine the merits of the BSE approach for calculating excitation energies of the prototype system of quantum chemistry, the H_2 molecule. The paper is organized as follows. In Section 18.2, we give a review of Green's function many-body theory. In Section 18.3, we give the general working equations for a BSE calculation within the static GW approximation in a finite spin-orbital basis and the corresponding spin-adapted expressions for closed-shell systems. In Section 18.4, we apply the equations to the H_2 molecule in a minimal basis and discuss the possibility of obtaining correct spin-singlet and spin-triplet excited-state energy curves as a function of the internuclear distance. Section 18.5 contains our conclusions. Hartree atomic units are used throughout.

18.2   REVIEW OF GREEN'S FUNCTION MANY-BODY THEORY

We start by giving a brief review of Green's function many-body theory for calculating excitation energies. For more details, see references [17], [19], and [34].

18.2.1   ONE-PARTICLE GREEN'S FUNCTION

Let $|N\rangle$ be the normalized ground-state wavefunction for a system of N electrons described by the Hamiltonian $\hat{H}$. The time-ordered one-particle equilibrium Green's function is defined as

$$iG(1,2) = \langle N \mid \hat{T}[\hat{\Psi}(1)\hat{\Psi}^\dagger(2)] \mid N \rangle$$
$$= \theta(t_1 - t_2)\langle N \mid \hat{\Psi}(1)\hat{\Psi}^\dagger(2) \mid N \rangle - \theta(t_2 - t_1)\langle N \mid \hat{\Psi}^\dagger(2)\hat{\Psi}(1) \mid N \rangle. \quad (18.1)$$

Index 1 stands for space, spin, and time coordinates $(\mathbf{r}_1, \sigma_1, t_1) = (\mathbf{x}_1, t_1)$. $\hat{T}$ is the Wick time-ordering operator that orders the operators with larger times on the left, and θ is the Heaviside step function. The whole time dependence is contained in $\hat{\Psi}(1) = e^{i\hat{H}t_1}\hat{\Psi}(\mathbf{x}_1)e^{-i\hat{H}t_1}$ and $\hat{\Psi}^\dagger(2) = e^{i\hat{H}t_2}\hat{\Psi}^\dagger(\mathbf{x}_2)e^{-i\hat{H}t_2}$, which are the annihilation and creation field operators in the Heisenberg representation.

In the absence of external potential, the system is invariant under time translation; therefore, the Green's function depends only on $\tau = t_1 - t_2$. By introducing the closure relation for excited states with $N - 1$ or $N + 1$ particles, one can get

$$iG(\mathbf{x}_1, \mathbf{x}_2; \tau) = \theta(\tau) \sum_A \langle N | \hat{\psi}(\mathbf{x}_1) | N+1, A\rangle \langle N+1, A | \hat{\psi}^\dagger(\mathbf{x}_2) | N\rangle e^{-i(E_{N+1,A}-E_N)\tau}$$

$$-\theta(-\tau) \sum_I \langle N | \hat{\psi}^\dagger(\mathbf{x}_2) | N-1, I\rangle \langle N-1, I | \hat{\psi}(\mathbf{x}_1) | N\rangle e^{-i(E_N - E_{N-1,I})\tau},$$

$$(18.2)$$

where E_N, $E_{N+1,A}$, and $E_{N-1,I}$ are the energies of the ground state $|N\rangle$, the Ath excited state with $N + 1$ particles $|N + 1, A\rangle$, and the Ith excited state with $N - 1$ particles $|N - 1, I\rangle$, respectively. The Lehmann representation of the one-particle Green's function is obtained by Fourier transform

$$G(\mathbf{x}_1, \mathbf{x}_2; \omega) = \sum_A \frac{f_A(\mathbf{x}_1)f_A^*(\mathbf{x}_2)}{\omega - \mathcal{E}_A + i0^+} + \sum_I \frac{f_I(\mathbf{x}_1)f_I^*(\mathbf{x}_2)}{\omega - \mathcal{E}_I - i0^+},$$

$$(18.3)$$

where $f_A(\mathbf{x}) = \langle N | \hat{\psi}(\mathbf{x}) | N+1, A\rangle$ and $f_I(\mathbf{x}) = \langle N-1, I | \hat{\psi}(\mathbf{x}) | N\rangle$ are the Dyson orbitals, and $\mathcal{E}_A = E_{N+1,A} - E_N$ and $\mathcal{E}_I = E_N - E_{N-1,I}$ are minus the electron affinities and ionization energies, respectively.

18.2.2 Two-Particle Green's Function

The time-ordered two-particle Green's function is defined as

$$i^2 G_2(1,2;1',2') = \langle N | \hat{T}[\hat{\Psi}(1)\hat{\Psi}(2)\hat{\Psi}^\dagger(2')\hat{\Psi}^\dagger(1')] | N\rangle. \qquad (18.4)$$

Depending on the time ordering, it describes the propagation of a pair of holes, of electrons, or of a hole and an electron. In the case of optical absorption, one is only interested in the propagation of a hole-electron pair.

Let χ be the four-point polarizability,

$$\chi(1,2;1',2') = iG_2(1,2;1',2') - iG(1,1')G(2,2'). \qquad (18.5)$$

It describes the coupled motion of two particles minus the motion of the independent ones. When the times are appropriately ordered, the four-point polarizability reduces to the linear response function

$$\chi(\mathbf{x}_1, \mathbf{x}_2; \mathbf{x}_1', \mathbf{x}_2'; \tau) = \chi(\mathbf{x}_1, t_1, \mathbf{x}_2, t_2; \mathbf{x}_1', t_1^+, \mathbf{x}_2', t_2^+), \qquad (18.6)$$

where $t_1^+ = t_1 + 0^+$. The Lehmann representation of the response function explicitly gives the excitation energies as poles in ω,

$$\chi\left(\mathbf{x}_1, \mathbf{x}_2; \mathbf{x}_1', \mathbf{x}_2'; \omega\right) = \sum_{K \neq 0} \frac{\left\langle N \left| \hat{\Psi}^\dagger \left(\mathbf{x}_1'\right) \hat{\Psi}(\mathbf{x}_1) \right| N, K \right\rangle \left\langle N, K \left| \hat{\Psi}^\dagger \left(\mathbf{x}_2'\right) \hat{\Psi}(\mathbf{x}_2) \right| N \right\rangle}{\omega - (E_{N,K} - E_N) + i0^+}$$

$$- \sum_{K \neq 0} \frac{\left\langle N \left| \hat{\Psi}^\dagger \left(\mathbf{x}_2'\right) \hat{\Psi}(\mathbf{x}_2) \right| N, K \right\rangle \left\langle N, K \left| \hat{\Psi}^\dagger \left(\mathbf{x}_1'\right) \hat{\Psi}(\mathbf{x}_1) \right| N \right\rangle}{\omega + (E_{N,K} - E_N) - i0^+}, \tag{18.7}$$

where $|N,K\rangle$ is the Kth excited state with N particles of energy $E_{N,K}$. The ground state $|N,0\rangle = |N\rangle$ is excluded from the sum. It is also useful to define the independent-particle (IP) polarizability $\chi_{IP}(1,2;1',2') = -iG(1,2')G(2,1')$. Its Lehmann representation is easily obtained by calculating $\chi_{IP}\left(\mathbf{x}_1, \mathbf{x}_2; \mathbf{x}_1', \mathbf{x}_2'; \tau\right) = -iG\left(\mathbf{x}_1, \mathbf{x}_2'; \tau\right) G\left(\mathbf{x}_2, \mathbf{x}_1'; -\tau\right)$ with Equation 18.2 and taking the Fourier transform

$$\chi_{IP}(\mathbf{x}_1, \mathbf{x}_2; \mathbf{x}_1', \mathbf{x}_2'; \omega) = \sum_{IA} \frac{f_I^*(\mathbf{x}_1') f_A(\mathbf{x}_1) f_A^*(\mathbf{x}_2') f_I(\mathbf{x}_2)}{\omega - (\mathcal{E}_A - \mathcal{E}_I) + i0^+}$$

$$- \sum_{IA} \frac{f_I^*(\mathbf{x}_2') f_A(\mathbf{x}_2) f_A^*(\mathbf{x}_1') f_I(\mathbf{x}_1)}{\omega + (\mathcal{E}_A - \mathcal{E}_I) - i0^+}. \tag{18.8}$$

In practice, the one- and two-particle Green's function can be calculated with equations of motion.

18.2.3 Dyson Equation

To make easier the connection with expressions in a finite spin-orbital basis, we systematically use four-point indexes for all the two-electron quantities. The starting point is therefore a fully nonlocal time-dependent Hamiltonian

$$\hat{H}(t_1) = \int d\mathbf{x}_1 d1' \hat{\Psi}^\dagger(1) h(1,1') \hat{\Psi}(1')$$

$$+ \frac{1}{2} \int d\mathbf{x}_1 d2 d1' d2' \hat{\Psi}^\dagger(1) \hat{\Psi}^\dagger(2) v(1,2;1',2') \hat{\Psi}(1') \hat{\Psi}(2'), \tag{18.9}$$

where $v(1,2;1',2') = v_{ee}(|\mathbf{r}_1 - \mathbf{r}_2|)\delta(t_1,t_2)\delta(1,1')\delta(2,2')$ is the spin-independent instantaneous Coulomb electron–electron interaction, and $h(1,1')$ is the one-electron Hamiltonian that contains the electron kinetic operator and the nuclei–electron interaction V_{ne}

$$h(1,1') = -\delta(1,1')\frac{\nabla_1^2}{2} + \delta(1,1')V_{ne}(\mathbf{r}_1). \tag{18.10}$$

Using the equations of motion for the Heisenberg creation and annihilation operators in the expression of the derivative of G with respect to time,[17] one can obtain the following equation:

$$i\int d3\delta(1,3)\frac{\partial}{\partial t_1}G(3,2) - \int d3h(1,3)G(3,2)$$

$$+i\int d3d1'd3'v(1,3;1',3')G_2(1',3'^+;2,3^{++}) = \delta(1,2), \tag{18.11}$$

where $^{++}$ stands for $t_3^+ + 0^+$. A whole series of equations can be derived for the Green's functions, relating the one-particle Green's function to the two-particle Green's function, the two-particle Green's function to the three-particle Green's function, etc. However, solving this set of equations is not wanted.

To avoid this, one can use the Schwinger derivative technique. Introducing an external time-dependent potential $U(1,1') = U(\mathbf{x}_1, \mathbf{x}'_1, t_1)\delta(t_1, t'_1)$, one can express the two-particle Green's function in terms of the one-particle Green's function and its derivative with respect to U, evaluated at $U = 0$:

$$\frac{\delta G(1,2)}{\delta U(3,4)} = -G_2(1,4;2,3) + G(1,2)G(4,3). \tag{18.12}$$

Using this relation in Equation 18.11, one can get

$$\int d3\left[i\delta(1,3)\frac{\partial}{\partial t_1} - h(1,3)\right]G(3,2) - \int d3\Sigma_{Hxc}(1,3)G(3,2) = \delta(1,2), \tag{18.13}$$

where $\Sigma_{Hxc}(1,2)$ is the Hartree exchange-correlation self-energy that takes into account all the two-particle effects. It can be decomposed into a Hartree contribution

$$\Sigma_H(1,2) = -i\int d3d3'v(1,3;2,3')G(3'^+,3^{++}), \tag{18.14}$$

and an exchange-correlation one

$$\Sigma_{xc}(1,2) = i\int d3d1'd3'd4v(1,3;1',3')\frac{\delta G(1',4)}{\delta U(3^{++},3'^+)}G^{-1}(4,2). \tag{18.15}$$

One can define a Green's function G_h which shows no two-particle effects and therefore follows the equation of motion:

$$\int d3\left[i\delta(1,3)\frac{\partial}{\partial t_1} - h(1,3)\right]G_h(3,2) = \delta(1,2). \tag{18.16}$$

Using this relation in Equation 18.13, one finally gets the Dyson equation for the one-particle Green's function:

$$\int d3 \left[G_h^{-1}(1,3) - \Sigma_{Hxc}(1,3) \right] G(3,2) = \delta(1,2). \tag{18.17}$$

This equation is also often used under the form

$$G(1,2) = G_h(1,2) + \int d3d4 G_h(1,3)\Sigma_{Hxc}(3,4)G(4,2), \tag{18.18}$$

or

$$G^{-1}(1,2) = G_h^{-1}(1,2) - \Sigma_{Hxc}(1,2). \tag{18.19}$$

18.2.4 Bethe–Salpeter Equation

Starting from the Dyson equation (see Equation 18.19), and taking the derivative with respect to G, one can get the so-called BSE (see, e.g., reference [35]):

$$\chi^{-1}(1,2;1',2') = \chi_{IP}^{-1}(1,2;1',2') - \Xi_{Hxc}(1,2;1',2'), \tag{18.20}$$

or

$$\chi(1,2;1',2') = \chi_{IP}(1,2;1',2') + \int d3d4d5d6 \chi_{IP}(1,4;1',3)\Xi_{Hxc}(3,6;4,5)\chi(5,2;6,2'), \tag{18.21}$$

where Ξ_{Hxc} is the Hartree-exchange-correlation Bethe–Salpeter kernel defined as

$$\Xi_{Hxc}(3,6;4,5) = i \frac{\delta\Sigma_{Hxc}(3,4)}{\delta G(5,6)}. \tag{18.22}$$

18.2.5 Hedin's Equations

We now have equations of motion for the one- and two-particle Green's functions. They depend on the Hartree-exchange-correlation self-energy. Its Hartree part is trivial, but a practical way of calculating its exchange-correlation part is needed. Hedin [36] proposed a scheme that yields to a set of coupled equations and allows in principle for the calculation of the exact self-energy. This scheme can be seen as a perturbation theory in terms of the screened interaction W instead of the bare Coulomb interaction v. We show a generalization of this derivation for the case of a nonlocal potential.

Let $V(5,6) = U(5,6) - i \int d3d3' v(5,3;6,3')G(3',3^+)$ be the nonlocal classic potential. Using the chain rule in the exchange-correlation self-energy, we get

$$\Sigma_{xc}(1,2) = -i \int d3d1'd3'd4d5d6 v(1,3;1',3')G(1',4) \frac{\delta G^{-1}(4,2)}{\delta V(5,6)} \frac{\delta V(5,6)}{\delta U(3^{++},3'^+)}$$

$$= i \int d3d1'd3'd4d5d6 v(1,3;1',3')G(1',4)\tilde{\Gamma}(4,6;2,5)\epsilon^{-1}(5,3';6,3^+),$$

$$(18.23)$$

where the inverse dielectric function ϵ^{-1} that screens the bare Coulomb interaction v and the irreducible vertex function $\tilde{\Gamma}$ are defined by

$$\epsilon^{-1}(1,2;3,4) = \frac{\delta V(1,3)}{\delta U(4,2)} \text{ and } \tilde{\Gamma}(1,2;3,4) = -\frac{\delta G^{-1}(1,3)}{\delta V(4,2)}. \qquad (18.24)$$

We can therefore define a dynamically screened potential

$$W(1,2;1',2') = \int d3d3' \epsilon^{-1}(1,3;1',3'^+)v(2,3';2',3)$$

$$= \int d3d3' \epsilon^{-1}(1,3;1',3'^+)v(3',2;3,2'), \qquad (18.25)$$

where the symmetry of the Coulomb interaction v has been used, and we get the expression of the exchange-correlation self-energy:

$$\Sigma_{xc}(1,2) = i \int d1'd3d3'd4 G(1',4)\tilde{\Gamma}(4,3';2,3)W(3,1;3',1'). \qquad (18.26)$$

We still need to express the dielectric function and the irreducible vertex function without the use of V and U. To achieve this, we define the irreducible polarizability $\tilde{\chi}(1,2;3,4) = -i\delta G(1,3)/\delta V(4,2)$, which, with the properties of the inverse and the definition of the vertex correction, can be rewritten as

$$\tilde{\chi}(1,2;3,4) = -i \int d5d5' G(1,5)G(5',3)\tilde{\Gamma}(5,2;5',4). \qquad (18.27)$$

Using this relation, one can rewrite the dielectric function as

$$\epsilon(1,2;3,4) = \delta(1,4)\delta(2,3) - \int d5d5' v(1,5;3,5')\tilde{\chi}(5',2;5^+,4), \qquad (18.28)$$

and the irreducible vertex correction as

$$\tilde{\Gamma}(1,2;3,4) = \delta(1,4)\delta(2,3) - i \int d5d6 \frac{\delta \Sigma_{xc}(1,3)}{\delta G(5,6)} \tilde{\chi}(5,2;6,4). \qquad (18.29)$$

We now have a set of five coupled equations (see Equations 18.25 through 18.29) to calculate the self-energy. In practice, this set of equations is never solved exactly, and approximations are made.

18.2.6 Static GW Approximation

We discuss now the static GW approximation that is the most often used approximation in practice in the BSE approach.

In the GW approximation, one takes $\tilde{\Gamma}(1,2;3,4) = \delta(1,4)\delta(2,3)$. This greatly simplifies Hedin's equations. The irreducible polarizability becomes $\tilde{\chi}(1,2;3,4) = -iG(1,4)G(2,3) = \chi_{IP}(1,2;3,4)$, and the exchange-correlation self-energy becomes

$$\Sigma_{xc}(1,2) = i \int d1' d3 G(1',3) W(3,1;2,1').$$
(18.30)

If the derivative of W with respect to G is further neglected, as usually done, the corresponding Bethe–Salpeter kernel is then

$$\Xi_{Hxc}(1,2;1',2') = v(1,2;1',2') - W(2,1;1',2').$$
(18.31)

where W is obtained from Equation 18.25 and ϵ^{-1} with Equation 18.28 in which $\tilde{\chi}$ is replaced by χ_{IP}. The Coulomb interaction is instantaneous, and the one-particle Green's function depends only on the time difference; therefore, the time dependence of the screened interaction is

$$W(1,2;1',2') = W(\mathbf{x}_1,\mathbf{x}_2;\mathbf{x}_1',\mathbf{x}_2';\tau)\delta(t_1,t_1')\delta(t_2,t_2'),$$
(18.32)

where $\tau = t_1 - t_2$. If one considers the time dependence in W, the Fourier transform of the BSE is not straightforward.[32] We will only consider the usual approximation where the screened interaction is static, that is,

$$W(1,2;1',2') = W(\mathbf{x}_1,\mathbf{x}_2;\mathbf{x}_1',\mathbf{x}_2')\delta(t_1,t_1')\delta(t_2,t_2')\delta(t_1,t_2).$$
(18.33)

To summarize, the Fourier-space BSE in the static GW approximation writes

$$\chi^{-1}(\mathbf{x}_1,\mathbf{x}_2;\mathbf{x}_3,\mathbf{x}_4;\omega) = \chi_{IP}^{-1}(\mathbf{x}_1,\mathbf{x}_2;\mathbf{x}_3,\mathbf{x}_4;\omega) - \Xi_{Hxc}(\mathbf{x}_1,\mathbf{x}_2;\mathbf{x}_3,\mathbf{x}_4),$$
(18.34)

where the kernel $\Xi_{Hxc}(\mathbf{x}_1,\mathbf{x}_2;\mathbf{x}_3,\mathbf{x}_4) = v(\mathbf{x}_1,\mathbf{x}_2;\mathbf{x}_3,\mathbf{x}_4) - W(\mathbf{x}_2,\mathbf{x}_1;\mathbf{x}_3,\mathbf{x}_4)$ contains the static screened interaction W calculated from

$$W(\mathbf{x}_1,\mathbf{x}_2;\mathbf{x}_1',\mathbf{x}_2') = \int d\mathbf{x}_3 \, d\mathbf{x}_3' \epsilon^{-1}(\mathbf{x}_1,\mathbf{x}_3;\mathbf{x}_1',\mathbf{x}_3')v(\mathbf{x}_3',\mathbf{x}_2;\mathbf{x}_3,\mathbf{x}_2'),$$
(18.35)

and

$$\epsilon(\mathbf{x}_1, \mathbf{x}_2; \mathbf{x}_3, \mathbf{x}_4) = \delta(\mathbf{x}_1, \mathbf{x}_4)\delta(\mathbf{x}_2, \mathbf{x}_3) - \int d\mathbf{x}_5\, d\mathbf{x}_5'\, v(\mathbf{x}_1, \mathbf{x}_5; \mathbf{x}_3, \mathbf{x}_5')\chi_{IP}(\mathbf{x}_5', \mathbf{x}_2; \mathbf{x}_5, \mathbf{x}_4; \omega = 0).$$

(18.36)

We will refer to the approach of Equations 18.34 through 18.36 as the BSE-GW method. The one-particle Green's function G in $\chi_{IP} = -iGG$ is not yet specified. Different choices can be made. The simplest option is to use a noninteracting Green's function G_0 from a Hartree–Fock (HF) or Kohn–Sham (KS) calculation. In this case, $\chi_{IP} = -iG_0G_0 = \chi_0$ is just the noninteracting HF or KS response function. In the condensed-matter physics literature, the usual recipe is to use χ_0 in Equation 18.36 but an improved χ_{IP} in Equation 18.34 from a GW calculation. In the case of H_2 in a minimal basis, it is simple enough to use χ_{IP} constructed with the exact one-particle Green's function G. Finally, we note that the dielectric function of Equation 18.36 could be alternatively defined as including the HF exchange in addition to the Coulomb interaction, that is, $v(\mathbf{x}_1, \mathbf{x}_5; \mathbf{x}_3, \mathbf{x}_5') \rightarrow v(\mathbf{x}_1, \mathbf{x}_5; \mathbf{x}_3, \mathbf{x}_5') - v(\mathbf{x}_5, \mathbf{x}_1; \mathbf{x}_3, \mathbf{x}_5')$ (see, e.g., reference [37]), which removes the "self-screening error" for one-electron systems,[38] but we do not explore this possibility here.

18.3 EXPRESSIONS IN FINITE ORBITAL BASIS

18.3.1 Spin-Orbital Basis

To solve the BSE for finite systems, all the equations are projected onto an orthonormal spin-orbital basis $\{\phi_p\}$. As the equations are four-point equations relating two-particle quantities, they are in fact projected onto the basis of products of two spin orbitals. Each matrix element is thus indexed by two double indices.

We consider the simplest case for which $\chi_{IP} = \chi_0$. The Lehmann representation of χ_0 is

$$\chi_0(\mathbf{x}_1, \mathbf{x}_2; \mathbf{x}_1', \mathbf{x}_2'; \omega) = \sum_{ia} \frac{\phi_i^*(\mathbf{x}_1')\phi_a(\mathbf{x}_1)\phi_a^*(\mathbf{x}_2')\phi_i(\mathbf{x}_2)}{\omega - (\varepsilon_a - \varepsilon_i) + i0^+} - \frac{\phi_i^*(\mathbf{x}_2')\phi_a(\mathbf{x}_2)\phi_a^*(\mathbf{x}_1')\phi_i(\mathbf{x}_1)}{\omega + (\varepsilon_a - \varepsilon_i) - i0^+},$$

(18.37)

where ϕ_i is the ith occupied spin orbital of energy ε_i, and ϕ_a is the ath virtual spin orbital of energy ε_a. One can notice that χ_0 is expanded only on occupied-virtual (ov) and virtual-occupied (vo) products of spin orbitals. The matrix elements of χ_0 are given by

$$\left[\chi_0(\omega)\right]_{pq,rs} = \int d\mathbf{x}_1\, d\mathbf{x}_1'\, d\mathbf{x}_2\, d\mathbf{x}_2'\, \phi_p(\mathbf{x}_1')\phi_q^*(\mathbf{x}_1)\chi_0(\mathbf{x}_1, \mathbf{x}_2; \mathbf{x}_1', \mathbf{x}_2'; \omega)\phi_r^*(\mathbf{x}_2)\phi_s(\mathbf{x}_2').$$

(18.38)

The matrix representation of its inverse, in the (ov,vo) subspace, is

$$\chi_0^{-1}(\omega) = -\left[\begin{pmatrix} \Delta\varepsilon & 0 \\ 0 & \Delta\varepsilon \end{pmatrix} - \omega\begin{pmatrix} 1 & 0 \\ 0 & -1 \end{pmatrix}\right],$$

(18.39)

where $\Delta\varepsilon_{ia,jb} = \Delta\varepsilon_{ai,bj} = (\varepsilon_a - \varepsilon_i)\delta_{ij}\delta_{ab}$, where i,j refers to occupied spin orbitals, and a,b refers to virtual spin orbitals. The dimension of the matrix is thus $2M_oM_v \times 2M_oM_v$, where M_o and M_v are the numbers of occupied and virtual spin orbitals, respectively. To build the matrix χ^{-1}, one then needs to construct the matrix elements of the Bethe–Salpeter kernel Ξ_{Hxc}, which are given by

$$(\Xi_{Hxc})_{pq,rs} = v_{pq,rs} - W_{pr,qs}, \tag{18.40}$$

where $v_{pq,rs} = \langle qr|ps \rangle$ are the usual two-electron integrals, and the matrix elements of W can be obtained from Equation 18.35:

$$
\begin{aligned}
W_{pq,rs} &= \int d\mathbf{x}_1\, d\mathbf{x}_1'\, d\mathbf{x}_2\, d\mathbf{x}_2' \phi_p(\mathbf{x}_1')\phi_q^*(\mathbf{x}_1) \\
&\quad \times W(\mathbf{x}_1,\mathbf{x}_2;\mathbf{x}_1',\mathbf{x}_2')\phi_r^*(\mathbf{x}_2)\phi_s(\mathbf{x}_2') \\
&= \int d\mathbf{x}_1\, d\mathbf{x}_1'\, d\mathbf{x}_2\, d\mathbf{x}_2'\, d\mathbf{x}_3\, d\mathbf{x}_3' \phi_p(\mathbf{x}_1')\phi_q^*(\mathbf{x}_1) \\
&\quad \times \epsilon^{-1}(\mathbf{x}_1,\mathbf{x}_3;\mathbf{x}_1',\mathbf{x}_3')v(\mathbf{x}_3',\mathbf{x}_2;\mathbf{x}_3,\mathbf{x}_2')\phi_r^*(\mathbf{x}_2)\phi_s(\mathbf{x}_2').
\end{aligned}
\tag{18.41}
$$

To decouple the common coordinates in ϵ^{-1} and v, one can introduce two delta functions $\delta(\mathbf{x}_3,\mathbf{x}_4)$ and $\delta(\mathbf{x}_3',\mathbf{x}_4')$ and use the closure relations $\delta(\mathbf{x}_3,\mathbf{x}_4) = \sum_t \phi_t^*(\mathbf{x}_3)\phi_t(\mathbf{x}_4)$ and $\delta(\mathbf{x}_3',\mathbf{x}_4') = \sum_u \phi_u(\mathbf{x}_3')\phi_u^*(\mathbf{x}_4')$. By doing so, the matrix elements of v and ϵ^{-1} appear explicitly, and we get

$$W_{pq,rs} = \sum_{tu} \epsilon^{-1}_{pq,tu} v_{tu,rs}. \tag{18.42}$$

Similarly, for the dielectric function, we have

$$\epsilon_{pq,rs} = \delta_{pr}\delta_{qs} - \sum_{tu} v_{pq,tu}\left[\chi_0(\omega = 0)\right]_{tu,rs} = \delta_{pr}\delta_{qs} - v_{pq,rs}\left[\chi_0(\omega = 0)\right]_{rs,rs}, \tag{18.43}$$

where the last equality comes from the fact that χ_0 has only diagonal elements. It can be seen that the static screened interaction consists of an infinite-order perturbation expansion in the Coulomb interaction, namely, using matrix notations,

$$W = \epsilon^{-1} \cdot v = v + v \cdot \chi_0(\omega = 0) \cdot v + v \cdot \chi_0(\omega = 0) \cdot v \cdot \chi_0(\omega = 0) \cdot v + ..., \tag{18.44}$$

with the first term in this expansion corresponding to time-dependent HF (TDHF). The matrix representation of the inverse of the interacting response function, in the (ov,vo) subspace, is then

$$\chi^{-1}(\omega) = -\left[\begin{pmatrix} A & B \\ B^* & A^* \end{pmatrix} - \omega \begin{pmatrix} 1 & 0 \\ 0 & -1 \end{pmatrix}\right], \tag{18.45}$$

with the matrices

$$A_{ia,jb} = \Delta\varepsilon_{ia,jb} + v_{ia,jb} - W_{ij,ab},$$ (18.46)

$$B_{ia,jb} = v_{ia,bj} - W_{ib,aj}.$$ (18.47)

The block structure of Equation 18.45 is a consequence of the symmetry of the Coulomb interaction, $v_{qp,sr} = v_{pq,rs}^*$, and the static screened interaction, $W_{qs,pr} = W_{pr,qs}^*$. Moreover, the matrix A is Hermitian (because $v_{ia,jb} = v_{jb,ia}^*$ and $W_{ij,ab} = W_{ji,ba}^*$), and the matrix B is symmetric (because $v_{ia,bj} = v_{jb,ai}$ and $W_{ib,aj} = W_{ja,bi}$). The excitation energies ω_n are thus found by solving the usual linear response pseudo-Hermitian eigenvalue equation, just as in TDDFT,

$$\begin{pmatrix} A & B \\ B^* & A^* \end{pmatrix} \begin{pmatrix} X_n \\ Y_n \end{pmatrix} = \omega_n \begin{pmatrix} 1 & 0 \\ 0 & -1 \end{pmatrix} \begin{pmatrix} X_n \\ Y_n \end{pmatrix},$$ (18.48)

whose solutions come in pairs: excitation energies ω_n with eigenvectors (X_n, Y_n) and deexcitation energies $-\omega_n$ with eigenvectors $\left(Y_n^*, X_n^*\right)$. For real spin orbitals and if $A + B$ and $A - B$ are positive definite, the eigenvalues are guaranteed to be real numbers, and the pseudo-Hermitian eigenvalue equation (see Equation 18.48) can be transformed into a half-size symmetric eigenvalue equation (see Equation 18.3).

If, instead of starting from χ_0, one starts from $\chi_{IP} = -iGG$ with the exact one-particle Green's function G, the equations get more complicated because the matrix representation of χ_{IP} is generally not diagonal and has contributions not only in the (ov,vo) subspace of spin-orbital products but also in the occupied–occupied (oo) and virtual–virtual (vv) subspace of spin-orbital products. The dimension of the matrices thus becomes $M^2 \times M^2$, where M is the total number of (occupied and virtual) spin orbitals. In this case, the number of solutions of the response equations is generally higher than the number of single excitations, and in particular, double excitations might be obtained even without a frequency-dependent kernel. Spurious excitations are also found. This is similar to what happens in linear response TDDMFT.[12–15] We will show this later in the case of H_2 in a minimal basis.

18.3.2 Spin Adaptation

We give now the expressions for spin-restricted closed-shell calculations. For four fixed spatial orbitals referred to as p, q, r, and s, the Bethe–Salpeter kernel has the following spin structure:

$$\begin{pmatrix} \Xi_{p\uparrow q\uparrow,r\uparrow s\uparrow} & \Xi_{p\uparrow q\uparrow,r\downarrow s\downarrow} & 0 & 0 \\ \Xi_{p\downarrow q\downarrow,r\uparrow s\uparrow} & \Xi_{p\downarrow q\downarrow,r\downarrow s\downarrow} & 0 & 0 \\ 0 & 0 & \Xi_{p\uparrow q\downarrow,r\uparrow s\downarrow} & \Xi_{p\uparrow q\downarrow,r\downarrow s\uparrow} \\ 0 & 0 & \Xi_{p\downarrow q\uparrow,r\uparrow s\downarrow} & \Xi_{p\downarrow q\uparrow,r\downarrow s\uparrow} \end{pmatrix},$$ (18.49)

which can be brought to a diagonal form after rotation (see, e.g., refs. [35], [39], and [40]):

$$
\begin{pmatrix}
{}^1\Xi_{pq,rs} & 0 & 0 & 0 \\
0 & {}^3\Xi_{pq,rs} & 0 & 0 \\
0 & 0 & {}^3\Xi_{pq,rs} & 0 \\
0 & 0 & 0 & {}^3\Xi_{pq,rs}
\end{pmatrix},
\tag{18.50}
$$

with a spin-singlet term ${}^1\Xi_{pq,rs} = 2v_{pq,rs} - W_{pr,qs}$ and three degenerate spin-triplet terms ${}^3\Xi_{pq,rs} = -W_{pr,qs}$. It has been used that the Coulomb interaction v and the screened interaction W are spin independent: $v_{pq,rs} = v_{p\uparrow q\uparrow,r\uparrow s\uparrow} = v_{p\uparrow q\uparrow,r\downarrow s\downarrow} = v_{p\downarrow q\downarrow,r\uparrow s\uparrow} = v_{p\downarrow q\downarrow,r\downarrow s\downarrow}$ and $W_{pq,rs} = W_{p\uparrow q\uparrow,r\uparrow s\uparrow} = W_{p\uparrow q\uparrow,r\downarrow s\downarrow} = W_{p\downarrow q\downarrow,r\uparrow s\uparrow} = W_{p\downarrow q\downarrow,r\downarrow s\downarrow}$. The spin-adapted screened interaction is obtained by

$$
W_{pq,rs} = \sum_{tu} {}^1\epsilon^{-1}_{pq,tu} v_{tu,rs},
\tag{18.51}
$$

where t and u refer to spatial orbitals, and the singlet dielectric function ${}^1\epsilon_{pq,rs} = \epsilon_{p\uparrow q\uparrow,r\uparrow s\uparrow} + \epsilon_{p\uparrow q\uparrow,r\downarrow s\downarrow}$ is given by

$$
{}^1\epsilon_{pq,rs} = \delta_{pr}\delta_{qs} - 2v_{pq,rs}\left[\chi_0(\omega = 0)\right]_{rs,rs}.
\tag{18.52}
$$

The bottom line is that the linear response eigenvalue equation (see Equation 18.48) decouples into a singlet eigenvalue equation

$$
\begin{pmatrix}
{}^1A & {}^1B \\
{}^1B^* & {}^1A^*
\end{pmatrix}
\begin{pmatrix}
{}^1X_n \\
{}^1Y_n
\end{pmatrix}
= {}^1\omega_n
\begin{pmatrix}
1 & 0 \\
0 & -1
\end{pmatrix}
\begin{pmatrix}
{}^1X_n \\
{}^1Y_n
\end{pmatrix},
\tag{18.53}
$$

with the matrices

$$
{}^1A_{ia,jb} = \Delta\varepsilon_{ia,jb} + 2v_{ia,jb} - W_{ij,ab},
\tag{18.54}
$$

$$
{}^1B_{ia,jb} = 2v_{ia,bj} - W_{ib,aj},
\tag{18.55}
$$

and a triplet eigenvalue equation

$$
\begin{pmatrix}
{}^3A & {}^3B \\
{}^3B^* & {}^3A^*
\end{pmatrix}
\begin{pmatrix}
{}^3X_n \\
{}^3Y_n
\end{pmatrix}
= {}^3\omega_n
\begin{pmatrix}
1 & 0 \\
0 & -1
\end{pmatrix}
\begin{pmatrix}
{}^3X_n \\
{}^3Y_n
\end{pmatrix},
\tag{18.56}
$$

with the matrices

$$^3A_{ia,jb} = \Delta\varepsilon_{ia,jb} - W_{ij,ab},$$ (18.57)

$$^3B_{ia,jb} = -W_{ib,aj}.$$ (18.58)

18.4 EXAMPLE OF H₂ IN A MINIMAL BASIS

As a pedagogical example, we apply the BSE-GW method to the calculation of the excitation energies of H_2 in a minimal basis consisting of two Slater basis functions, φ_a and φ_b, centered on each hydrogen atom and with the same exponent $\zeta = 1$. This is a closed-shell molecule; therefore, all the calculations are done with spin adaptation in a spatial orbital basis. The molecular orbitals are $\psi_1 = (\varphi_a + \varphi_b)/\sqrt{2(1 + S_{ab})}$ (symmetry σ_g) and $\psi_2 = (\varphi_a - \varphi_b)/\sqrt{2(1 - S_{ab})}$ (symmetry σ_u), where S_{ab} is the overlap between φ_a and φ_b. The matrix representations of all two-electron quantities in the space of spatial–orbital products are of the following form:

$$P = \left(\begin{array}{cc|cc} P_{11,11} & P_{11,22} & P_{11,12} & P_{11,21} \\ P_{22,11} & P_{22,22} & P_{22,12} & P_{22,21} \\ \hline P_{12,11} & P_{12,22} & P_{12,12} & P_{12,21} \\ P_{21,11} & P_{21,22} & P_{21,12} & P_{21,21} \end{array} \right),$$ (18.59)

and we refer to the upper left block as the (oo,vv) block and to the bottom right block as the (ov,vo) block. All the values of the integrals as a function of the internuclear distance R can be found in reference [41]. Note that, in the condensed-matter physics literature, a simplified version of H_2 in a minimal basis with only on-site Coulomb interaction is often used under the name "half-filled two-site Hubbard model" (see, e.g., references [38] and [42])*.

18.4.1 BSE-GW METHOD USING THE NONINTERACTING GREEN'S FUNCTION

The simplest approximation in the BSE-GW method is to start from the noninteracting HF Green's function G_0, leading to the noninteracting HF linear response function $\chi_{IP} = -iG_0G_0 = \chi_0$ whose matrix representation reads

$$\chi_0(\omega) = \left(\begin{array}{cc|cc} 0 & 0 & 0 & 0 \\ 0 & 0 & 0 & 0 \\ \hline 0 & 0 & \dfrac{1}{\omega - \Delta\varepsilon} & 0 \\ 0 & 0 & 0 & \dfrac{-1}{\omega + \Delta\varepsilon} \end{array} \right),$$ (18.60)

---

* With the notations used here, the Hubbard model is obtained for $\Delta\varepsilon = 2t$ and $J_{11} = J_{22} = J_{12} = K_{12} = U/2$ where t is the hopping parameter and U is the on-site Coulomb interaction.

where $\Delta\varepsilon = \varepsilon_2 - \varepsilon_1$ is the difference between the energies of the molecular orbitals ψ_2 and ψ_1. The noninteracting linear response function has nonvanishing matrix elements only in the (ov,vo) block, but it will be necessary to consider the other blocks as well for the screened interaction W. The matrix of the Coulomb interaction is

$$
v = \left(
\begin{array}{cc|cc}
J_{11} & J_{12} & 0 & 0 \\
J_{12} & J_{22} & 0 & 0 \\
\hline
0 & 0 & K_{12} & K_{12} \\
0 & 0 & K_{12} & K_{12}
\end{array}
\right),
\tag{18.61}
$$

where $J_{pq} = \langle pq|pq \rangle$ and $K_{pq} = \langle pq|qp \rangle$ are the usual Coulomb and exchange two-electron integrals over the molecular orbitals ψ_1 and ψ_2. The off-diagonal blocks of v are zero by symmetry for H_2 in a minimal basis, but this is not the case in general. By matrix product and inversion, we get the static singlet dielectric matrix

$$
{}^1\epsilon = \left(
\begin{array}{cc|cc}
1 & 0 & 0 & 0 \\
0 & 1 & 0 & 0 \\
\hline
0 & 0 & 1+\dfrac{2K_{12}}{\Delta\varepsilon} & \dfrac{2K_{12}}{\Delta\varepsilon} \\
0 & 0 & \dfrac{2K_{12}}{\Delta\varepsilon} & 1+\dfrac{2K_{12}}{\Delta\varepsilon}
\end{array}
\right),
\tag{18.62}
$$

which, in this case, is block diagonal with the (oo,vv) block being the identity. By using its inverse, we finally get the static screened interaction matrix

$$
W = \left(
\begin{array}{cc|cc}
J_{11} & J_{12} & 0 & 0 \\
J_{12} & J_{22} & 0 & 0 \\
\hline
0 & 0 & \dfrac{K_{12}}{1+4K_{12}/\Delta\varepsilon} & \dfrac{K_{12}}{1+4K_{12}/\Delta\varepsilon} \\
0 & 0 & \dfrac{K_{12}}{1+4K_{12}/\Delta\varepsilon} & \dfrac{K_{12}}{1+4K_{12}/\Delta\varepsilon}
\end{array}
\right),
\tag{18.63}
$$

which is block diagonal and the (oo,vv) block is just the bare Coulomb interaction in the case of H_2 in a minimal basis, but this is not generally true. We have then everything to construct the ${}^1A$ and ${}^1B$ matrices of Equation 18.53 for singlet excitations, which in the present case are just one-dimensional

$$
{}^1A = \Delta\varepsilon + 2K_{12} - J_{12},
\tag{18.64}
$$

and

$$
{}^1B = 2K_{12} - \frac{K_{12}}{1+4K_{12}/\Delta\varepsilon},
\tag{18.65}
$$

and the ^3A and ^3B matrices of Equation 18.56 for triplet excitations

$$^3A = \Delta\varepsilon - J_{12}, \tag{18.66}$$

and

$$^3B = -\frac{K_{12}}{1 + 4K_{12}/\Delta\varepsilon}. \tag{18.67}$$

Solving then the response equations by the standard Casida approach,[3] we get the singlet excitation energy

$$^1\omega = \sqrt{\left(\Delta\varepsilon + 4K_{12} - J_{12} - \frac{K_{12}}{1 + 4K_{12}/\Delta\varepsilon}\right)\left(\Delta\varepsilon - J_{12} + \frac{K_{12}}{1 + 4K_{12}/\Delta\varepsilon}\right)}, \tag{18.68}$$

and the triplet excitation energy

$$^3\omega = \sqrt{\left(\Delta\varepsilon - J_{12} - \frac{K_{12}}{1 + 4K_{12}/\Delta\varepsilon}\right)\left(\Delta\varepsilon - J_{12} + \frac{K_{12}}{1 + 4K_{12}/\Delta\varepsilon}\right)}. \tag{18.69}$$

Note that, for this simple system, the A terms have the usual TDHF or configuration interaction single (CIS) forms, and the screening has an effect only on the B terms, decreasing the exchange integral K_{12} by a factor of $1 + 4K_{12}/\Delta\varepsilon$. Therefore, in the Tamm–Dancoff approximation,[43] which consists of neglecting B, the effect of screening would be lost and the method would be equivalent to CIS. It is interesting to analyze the effect of the screening as a function of the internuclear distance R. For small R, the orbital energy difference $\Delta\varepsilon$ is much greater than the exchange integral K_{12}, so the screening factor $1 + 4K_{12}/\Delta\varepsilon$ is close to 1 and TDHF excitation energies are recovered. For large R (dissociation limit), $\Delta\varepsilon$ goes to zero, so the screening factor diverges and the term $K_{12}/(1 + 4K_{12}/\Delta\varepsilon)$ vanishes.

The excitation energies from the ground state $^1\Sigma_g^+$ to the first singlet $^1\Sigma_u^+$ and triplet $^3\Sigma_u^+$ excited states are plotted as a function of R in Figure 18.1. The reference curves are from a full configuration-interaction (FCI) calculation giving the exact excitation energies on this basis. In a minimal basis, the singlet $^1\Sigma_u^+$ excited state is constrained to dissociate into the ionic configuration H^- ... H^+; so in the dissociation limit $R \to \infty$, the exact singlet excitation energy goes to a constant, $I - A \approx 0.625$ hartree, where I and A are the ionization energy and electron affinity of the hydrogen atom, respectively. The triplet $^3\Sigma_u^+$ dissociates into the neutral configuration $H^\bullet$... $H^\bullet$, as does the ground state, so the exact triplet excitation energy goes to zero in the dissociation limit. TDHF gives accurate excitation energies for small R but gives qualitatively wrong curves in the dissociation limit. For the singlet state, the TDHF excitation energy goes to zero, a wrong behavior inherited from the vanishing

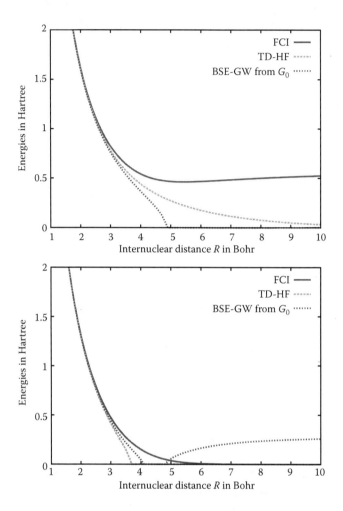

FIGURE 18.1   Excitation energies of the singlet $^1\Sigma_u^+$ (top) and triplet $^3\Sigma_u^+$ (bottom) states of H_2 in a minimal basis as a function of the internuclear distance R calculated by FCI, TDHF, and BSE-GW with the noninteracting HF Green's function G_0.

$\Delta\varepsilon$ in this limit. For the triplet state, the TDHF response equation suffers from a triplet instability for $R \geq 4$ Bohr and the excitation energy becomes imaginary. It is known that TDDFT with standard density functional approximations gives similarly incorrect energy curves.[7,42,44–46] The BSE-GW method using the noninteracting HF Green's function G_0 gives accurate excitation energies at small R but fails in the dissociation limit. The singlet excitation energy becomes imaginary for $R \geq 4.9$ Bohr. Indeed, in the dissociation limit, $\Delta\varepsilon$ goes to zero and Equation 18.68 leads to a negative term under the square root: $^1\omega \to \sqrt{(4K_{12} - J_{12})(-J_{12})}$. Similarly, the BSE-GW triplet excitation energy is imaginary between $R = 4.0$ and $R = 4.9$ Bohr and incorrectly tends to a nonzero value in the dissociation limit.

The BSE-GW method using the noninteracting HF Green's function G_0 thus badly fails for H_2 in the dissociation limit. As this method is based on a single-determinant reference, this should not come as a surprise. However, the BSE approach also allows one to start from an interacting Green's function G, taking into account the multiconfigurational character of stretched H_2. We will now test this alternative approach.

18.4.2 BSE-GW METHOD USING THE EXACT GREEN'S FUNCTION

18.4.2.1 Independent-Particle Response Function

We apply the BSE-GW equations (see Equations 18.34 through 18.36) with the IP response function $\chi_{IP} = -iGG$ constructed from the exact one-particle Green's function G, which can be calculated by the Lehmann formula 18.8 using the N-electron ground state and the $(N \pm 1)$-electron states. The states to consider for H_2 in a minimal basis are given in Figure 18.2. The ground state is composed of two Slater determinants, and its energy is $E_N = 2\varepsilon_1 - J_{11} + E_c$, where $E_c = \Delta - \sqrt{\Delta^2 + K_{12}^2}$ is the correlation energy with $2\Delta = 2\Delta\varepsilon + J_{11} + J_{22} - 4J_{12} + 2K_{12}$. The coefficients of the determinants are determined by $c_2 = c_1 K_{12}/(\Delta + \sqrt{K_{12}^2 + \Delta^2})$ and $c_1^2 + c_2^2 = 1$. The energies of the two $(N + 1)$-electron states are $E_{N+1,1} = 2\varepsilon_1 + \varepsilon_2 - J_{11}$ and $E_{N+1,2} = 2\varepsilon_2 + \varepsilon_1 - J_{11} + J_{22} - 2J_{12} + K_{12}$. The energies of the two $(N - 1)$-electron states are $E_{N-1,1} = \varepsilon_1 - J_{11}$ and $E_{N-1,2} = \varepsilon_2 - 2J_{12} + K_{12}$. We thus obtain four poles for the exact one-particle Green's function. Two of them correspond to minus the electron affinities:

$$\mathcal{E}_2 = E_{N+1,1} - E_N = \varepsilon_2 - E_c, \tag{18.70}$$

$$\mathcal{E}_2' = E_{N+1,2} - E_N = 2\varepsilon_2 - \varepsilon_1 + J_{22} - 2J_{12} + K_{12} - E_c, \tag{18.71}$$

and the other two correspond to minus the ionization energies:

$$\mathcal{E}_1 = E_N - E_{N-1,1} = \varepsilon_1 + E_c, \tag{18.72}$$

$$\mathcal{E}_1' = E_N - E_{N-1,2} = 2\varepsilon_1 - \varepsilon_2 - J_{11} + 2J_{12} - K_{12} + E_c. \tag{18.73}$$

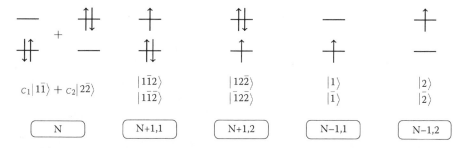

FIGURE 18.2 N-electron ground state and $(N \pm 1)$-electron states for H_2 in a minimal basis.

In condensed-matter physics, $\mathcal{E}_1$ and $\mathcal{E}_2$ are associated with "quasi-particle" peaks of photoelectron spectra, whereas $\mathcal{E}'_1$ and $\mathcal{E}'_2$ are associated with "satellites." The Dyson orbitals are also easily calculated, and we finally arrive at the matrix representation of χ_{IP} on the basis of the products of spatial orbitals

$$\boldsymbol{\chi}_{IP}(\omega) = \left(\begin{array}{cc|cc} \chi_{IP,11}(\omega) & 0 & 0 & 0 \\ 0 & \chi_{IP,22}(\omega) & 0 & 0 \\ \hline 0 & 0 & \chi_{IP,12}(\omega) & 0 \\ 0 & 0 & 0 & \chi_{IP,21}(\omega) \end{array} \right), \tag{18.74}$$

with the matrix elements

$$\chi_{IP,11}(\omega) = \frac{c_1^2 c_2^2}{\omega - (\mathcal{E}'_2 - \mathcal{E}_1)} - \frac{c_1^2 c_2^2}{\omega + (\mathcal{E}'_2 - \mathcal{E}_1)}, \tag{18.75}$$

$$\chi_{IP,22}(\omega) = \frac{c_1^2 c_2^2}{\omega - (\mathcal{E}_2 - \mathcal{E}'_1)} - \frac{c_1^2 c_2^2}{\omega + (\mathcal{E}_2 - \mathcal{E}'_1)}, \tag{18.76}$$

$$\chi_{IP,12}(\omega) = \frac{c_1^4}{\omega - (\mathcal{E}_2 - \mathcal{E}_1)} - \frac{c_2^4}{\omega + (\mathcal{E}'_2 - \mathcal{E}'_1)}, \tag{18.77}$$

$$\chi_{IP,21}(\omega) = \frac{c_2^4}{\omega - (\mathcal{E}'_2 - \mathcal{E}'_1)} - \frac{c_1^4}{\omega + (\mathcal{E}_2 - \mathcal{E}_1)}. \tag{18.78}$$

Therefore, whereas $\chi_0(\omega)$ has only one positive pole, $\chi_{IP}(\omega)$ has four distinct positive poles (and four symmetric negative poles). These poles are plotted in Figure 18.3. The lowest one, $\mathcal{E}_2 - \mathcal{E}_1$, called fundamental gap in the condensed-matter physics literature, can be considered as an approximation to a neutral single excitation energy because, in the limit of noninteracting particles, it equals the difference of the orbital eigenvalues $\Delta\varepsilon = \varepsilon_2 - \varepsilon_1$. The two intermediate poles, $\mathcal{E}'_2 - \mathcal{E}_1$ and $\mathcal{E}_2 - \mathcal{E}'_1$, can be interpreted as approximations to a double excitation energy because they reduce to $2\Delta\varepsilon$ in the limit of noninteracting particles. Surprisingly, the highest pole, $\mathcal{E}'_2 - \mathcal{E}'_1$, reduces to $3\Delta\varepsilon$ in this limit, and it is thus tempting to associate it with a triple excitation, although the system contains only two electrons. In the dissociation limit $R \to \infty$, the four poles tend to the same value, that is, $I - A \approx 0.625$ hartree, which is also minus twice the correlation energy $-2E_c$, showing that the nonvanishing fundamental gap in this limit is a correlation effect. Note that it has been shown[38] that the non-self-consistent GW approximation (G_0W_0) to the one-particle Green's function gives a fundamental gap that is too small by a factor of 2 in the dissociation limit, so we do not consider this approximation here.

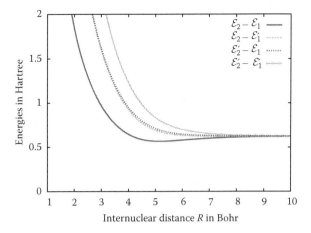

FIGURE 18.3 Positive poles of the IP linear response function as a function of the internuclear distance R.

18.4.2.2 Excitation Energies

Having calculated the IP response function, the next steps of the BSE-GW calculation of the excitation energies continue similarly as in Section 18.4.1, although the expressions get more complicated. From the matrix $\chi_{IP}(\omega = 0)$ and the Coulomb interaction matrix (Equation 18.61), we calculate the singlet dielectric matrix that is still block diagonal, but the upper left block is no longer the identity matrix. We calculate then the static screened interaction matrix that is still block diagonal, but the elements of its upper left block are now also affected by screening. We can then construct the corresponding singlet and triplet Bethe–Salpeter kernel $^1\Xi$ and $^3\Xi$. The response eigenvalue equations (see Equations 18.53 and 18.56) are no longer applicable, so the singlet excitation energies are found by searching the values of ω, giving vanishing eigenvalues of the inverse singlet linear response matrix $^1\chi(\omega)^{-1} = \chi_{IP}(\omega)^{-1} - {}^1\Xi$, and the triplet excitation energies are found by searching the values of ω, giving vanishing eigenvalues of the inverse triplet linear response matrix $^3\chi(\omega)^{-1} = \chi_{IP}(\omega)^{-1} - {}^3\Xi$. For H_2 in a minimal basis, $^1\chi(\omega)^{-1}$ and $^3\chi(\omega)^{-1}$ are 4×4 matrices that are block diagonal, the (oo,vv) block being uncoupled to the (ov,vo) block. For both singlet and triplet cases, the four positive poles of $\chi_{IP}(\omega)$ transform into four excitation energies (plus four symmetric deexcitation energies).

Between the two positive excitation energies coming from the (ov,vo) block of the matrix $^1\chi(\omega)^{-1}$, the lowest one is identified with the first singlet $^1\Sigma_u^+$ excitation energy, which is called the optical gap. It is plotted in Figure 18.4 and compared with the reference FCI excitation energy and also with the fundamental gap $\mathcal{E}_2 - \mathcal{E}_1$ to highlight the effect of the Bethe–Salpeter kernel. At small internuclear distance, $R \leq 3$ Bohr, the Bethe–Salpeter kernel brings the BSE-GW curve very close to the FCI curve. For large R, the BSE-GW excitation energy follows the curve of the fundamental gap, which slightly overestimates the excitation energy at $R = 10$ Bohr but eventually goes to the correct limit $I - A$ when $R \to \infty$. Thus, contrary to the BSE-GW method using the noninteracting Green's function, the obtained excitation energy curve has now a

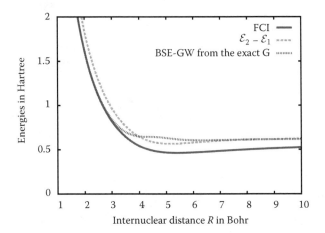

FIGURE 18.4 Excitation energy of the singlet $^1\Sigma_u^+$ state of H_2 in a minimal basis as a function of the internuclear distance R calculated by FCI and BSE-GW with the exact Green's function. The lowest pole of $\chi_{IP}(\omega)$, the fundamental gap $\mathcal{E}_2 - \mathcal{E}_1$, is also plotted for comparison.

correct shape. This relies on the fundamental gap being a good starting approximation to the optical gap. As regards the second excitation energy coming from the (ov,vo) block of the matrix $^1\chi(\omega)^{-1}$, which is connected to the highest pole $\mathcal{E}_2' - \mathcal{E}_1'$ of $\chi_{IP}(\omega)$, it is a spurious excitation due to the approximate Bethe–Salpeter kernel used.

The lowest positive excitation energy coming from the (oo,vv) block of the matrix $^1\chi(\omega)^{-1}$ is identified with the second singlet $^1\Sigma_g^+$ excited state that has a double excitation character. It is plotted in Figure 18.5 and compared with the FCI excitation

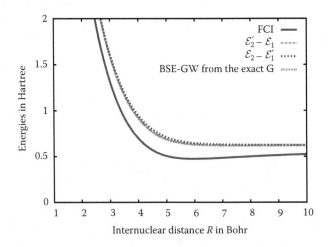

FIGURE 18.5 Excitation energy of the second singlet $^1\Sigma_g^+$ state of H_2 in a minimal basis as a function of the internuclear distance R calculated by FCI and BSE-GW with the exact Green's function. The poles $\mathcal{E}_2' - \mathcal{E}_1$ and $\mathcal{E}_2 - \mathcal{E}_1'$ of $\chi_{IP}(\omega)$ are also plotted for comparison.

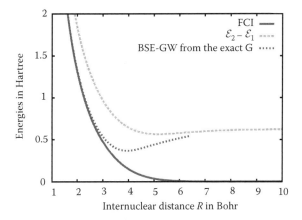

FIGURE 18.6   Excitation energy of the triplet $^3\Sigma_u^+$ state of H_2 in a minimal basis as a function of the internuclear distance R calculated by FCI and BSE-GW with the exact Green's function. The lowest pole of $\chi_{IP}(\omega)$, the fundamental gap $\mathcal{E}_2 - \mathcal{E}_1$, is also plotted for comparison.

energy for this state and with the poles $\mathcal{E}_2' - \mathcal{E}_1$ and $\mathcal{E}_2 - \mathcal{E}_1'$ of $\chi_{IP}(\omega)$. It is noteworthy that the BSE-GW method starting from $\chi_{IP}(\omega)$ instead of $\chi_0(\omega)$ but using a frequency-independent kernel does describe this double excitation state with an overall correct shape for the energy curve. However, the BSE-GW excitation energy is almost identical to the two poles $\mathcal{E}_2' - \mathcal{E}_1$ and $\mathcal{E}_2 - \mathcal{E}_1'$. The Bethe–Salpeter kernel in the static GW approximation thus brings virtually no improvement for this state over the starting poles of $\chi_{IP}(\omega)$. The (oo,vv) block of the matrix $^1\chi(\omega)^{-1}$ also gives a second higher excitation energy that is spurious.

We finally consider the triplet excited state $^3\Sigma_u^+$. The lowest positive excitation energy coming from the (ov,vo) block of the matrix $^3\chi(\omega)^{-1}$ should be identified with this state. It is plotted in Figure 18.6 and compared with the FCI excitation energy for this state and with the fundamental gap $\mathcal{E}_2 - \mathcal{E}_1$. For small internuclear distances, $R \leq 3$ Bohr, the BSE-GW method gives an accurate excitation energy, but for larger R, instead of going to zero, the BSE-GW excitation energy curve heads for the fundamental gap curve until the excitation energy becomes imaginary for $R \geq 6.5$ Bohr. The problem is that the poles of $\chi_{IP}(\omega)$ are the same for both singlet and triplet cases, and the fundamental gap $\mathcal{E}_2 - \mathcal{E}_1$ is not a good starting approximation to the triplet excitation energy in the dissociation limit. The Bethe–Salpeter kernel in the static GW approximation is not able to compensate for this bad starting point. In addition to this excitation energy, the BSE-GW method gives three other spurious triplet excitation energies.

18.5   CONCLUSION

We have applied the BSE approach in the static GW approximation for the calculation of the excitation energies on the toy model of H_2 in a minimal basis. We have tested two variants for the starting one-particle Green's function: the noninteracting HF one and the exact one. Around the equilibrium internuclear distance, both variants give accurate excitation energies to the first singlet $^1\Sigma_u^+$ and triplet $^3\Sigma_u^+$ excited states. In the dissociation limit, however, the two variants differ. The first variant, starting from the noninteracting one-particle Green's function, badly fails in this limit for both singlet and triplet states, giving imaginary excitation energies. The second variant, starting from the exact one-particle Green's function, gives a qualitatively correct energy curve for the singlet $^1\Sigma_u^+$ excited state up to the dissociation limit. This relies on the fact that the fundamental gap (given by the one-particle Green's function) is a good starting approximation to the first singlet excitation energy. However, the same variant gives an incorrect energy curve for the triplet $^3\Sigma_u^+$ excited state in the dissociation limit. In this case, the fundamental gap is a bad starting approximation to the first triplet excitation energy.

The second BSE variant using the exact one-particle Green's function gives more excitation energies than the first BSE variant. Most of them are spurious excitations due to the approximate Bethe–Salpeter kernel used. However, one of them can be identified with the excitation energy to the singlet $^1\Sigma_g^+$ excited state that has a double excitation character. It is remarkable that such a double excitation can be described at all without using a frequency-dependent kernel. However, the Bethe–Salpeter kernel in the static GW approximation is insufficient to describe accurately the energy curve of this state, even around the equilibrium distance.

ACKNOWLEDGMENTS

We thank L. Reining (École Polytechnique, Palaiseau, France) for suggesting to test the use of the exact one-particle Green's function in the BSE and F. Sottile (École Polytechnique, Palaiseau, France) and K. Pernal (Politechnika Łódzka, Łódź, Poland) for discussions.

REFERENCES

1. Runge, E. and Gross, E. K. U., *Phys. Rev. Lett.*, 52, 997, 1984.
2. Gross, E. K. U. and Kohn, W., *Phys. Rev. Lett.*, 55, 2850, 1985.
3. Casida, M. E., in *Recent Advances in Density Functional Methods, Part I*, Chong, D. P., ed., World Scientific, Singapore, 1995, 155.
4. Petersilka, M., Gossmann, U. J., and Gross, E. K. U., *Phys. Rev. Lett.*, 76, 1212, 1996.
5. Dreuw, A., Weisman, J. L., and Head-Gordon, M., *J. Chem. Phys.*, 119, 2943, 2003.
6. Maitra, N. T., Zhang, F., Cave, R. J., and Burke, K., *J. Chem. Phys.*, 120, 5932, 2004.
7. Gritsenko, O. V., van Gisbergen, S. J. A., Görling, A., and Baerends, E. J., *J. Chem. Phys.*, 113, 8478, 2000.
8. Tawada, Y., Tsuneda, T., Yanagisawa, S., Yanai, T., and Hirao, K., *J. Chem. Phys.*, 120, 8425, 2004.
9. Kronik, L., Stein, T., Refaely-Abramson, S., and Baer, R., *J. Chem. Theory Comput.*, 8, 1515, 2012.
10. Casida, M. E., *J. Chem. Phys.*, 122, 054111, 2005.
11. Huix-Rotllant, M. and Casida M. E., http://arxiv.org/abs/1008.1478.
12. Pernal, K., Gritsenko, O., and Baerends, E. J., *Phys. Rev. A*, 75, 012506, 2007.
13. Pernal, K., Giesbertz, K., Gritsenko, O., and Baerends, E. J., *J. Chem. Phys.*, 127, 214101, 2007.

14. Giesbertz, K. J. H., Baerends, E. J., and Gritsenko, O. V., *Phys. Rev. Lett.*, 101, 033004, 2008.
15. Giesbertz, K. J. H., Pernal, K., Gritsenko, O. V., and Baerends, E. J., *J. Chem. Phys.*, 130, 114104, 2009.
16. Pernal, K., *J. Chem. Phys.*, 136, 184105, 2012.
17. Strinati, G., *Riv. Nuovo Cim.*, 11, 1, 1988.
18. Rohlfing, M. and Louie, S. G., *Phys. Rev. B*, 62, 4927, 2000.
19. Onida, G., Reining, L., and Rubio, A., *Rev. Mod. Phys.*, 74, 601, 2002.
20. Rohlfing, M., *Int. J. Quantum Chem.*, 80, 807, 2000.
21. Grossman, J. C., Rohlfing, M., Mitas, L., Louie, S. G., and Cohen, M. L., *Phys. Rev. Lett.*, 86, 472, 2001.
22. Tiago, M. L. and Chelikowsky, J. R., *Solid State Commun.*, 136, 333, 2005.
23. Hahn, P. H., Schmidt, W. G., and Bechstedt, F., *Phys. Rev. B*, 72, 245425, 2005.
24. Tiago, M. L. and Chelikowsky, J. R., *Phys. Rev. B*, 73, 205334, 2006.
25. Tiago, M. L., Kent, P. R. C., Hood, R. Q., and Reboredo, F. A., *J. Chem. Phys.*, 129, 084311, 2008.
26. Grüning, M., Marini, A., and Gonze, X., *Nano Lett.*, 9, 2820, 2009.
27. Ma, Y., Rohlfing, M., and Molteni, C., *Phys. Rev. B*, 80, 241405, 2009.
28. Ma, Y., Rohlfing, M., and Molteni, C., *J. Chem. Theory Comput.*, 6, 257, 2010.
29. Rocca, D., Lu, D., and Galli, G., *J. Chem. Phys.*, 133, 164109, 2010.
30. Grüning, M., Marini, A., and Gonze, X., *Comput. Mater. Sci.*, 50, 2148, 2011.
31. Blase, X. and Attaccalite, C., *Appl. Phys. Lett.*, 99, 171909, 2011.
32. Romaniello, P., Sangalli, D., Berger, J. A., Sottile, F., Molinari, L. G., Reining, L., and Onida, G., *J. Chem. Phys.*, 130, 044108, 2009.
33. Sangalli, D., Romaniello, P., Onida, G., and Marini, A., *J. Chem. Phys.*, 134, 034115, 2011.
34. Bruneval, F., Ph.D. thesis, Ecole Polytechnique, 2005.
35. Toulouse, J., Zhu, W., Ángyán, J. G., and Savin, A., *Phys. Rev. A*, 82, 032502, 2010.
36. Hedin, L., *Phys. Rev.*, 139, 1965.
37. Shirley, E. L. and Martin, R. M., *Phys. Rev. B*, 47, 15404, 1993.
38. Romaniello, P., Guyot, S., and Reining, L., *J. Chem. Phys.*, 131, 154111, 2009.
39. Toulouse, J., Zhu, W., Savin, A., Jansen, G., and Ángyán, J. G., *J. Chem. Phys.*, 135, 084119, 2011.
40. Ángyán, J. G., Liu, R.-F., Toulouse, J., and Jansen, G., *J. Chem. Theory Comput.*, 7, 3116, 2011.
41. Dewar, M. J. S. and Kelemen, J., *J. Chem. Educ.*, 48, 494, 1971.
42. Aryasetiawan, F., Gunnarsson, O., and Rubio, A., *Europhys. Lett.*, 57, 683, 2002.
43. Hirata, S. and Head-Gordon, M., *Chem. Phys. Lett.*, 314, 291, 1999.
44. Cai, Z.-L. and Reimers, J. R., *J. Chem. Phys.*, 112, 527, 2000.
45. Casida, M. E., Gutierrez, F., Guan, J., Gadea, F.-X., Salahub, D., and Daudey, J.-P., *J. Chem. Phys.*, 113, 7062, 2000.
46. Giesbertz, K. J. H. and Baerends, E. J., *Chem. Phys. Lett.*, 461, 338, 2008.

19 Semiquantitative Aspects of Density-Based Descriptors and Molecular Interactions

A More Generalized Local Hard–Soft Acid–Base Principle

K. R. S. Chandrakumar,
Rahul Kar, and Sourav Pal

CONTENTS

19.1   INTRODUCTION

The challenging problems in chemistry are the understanding of the structure as well as reactivity of the systems at the molecular level and the prediction of a reaction pathway/mechanism.[1–6] This is primarily due to the existence of numerous varieties of chemical reactions and phenomena. Many experimental and theoretical groups have shown lots of interest and explained the nature of bonding and reactivity of the systems.[1–10] Most of the explanations were based on some intuitive ideas and empirical rules that were essentially derived from several experimental observations and many chemical facts. Among the many seminal works of different theoretical and experimental groups in correlating the structure and reactivity of molecular systems, the well-known and most celebrated principle is the hard–soft acid–base (HSAB) principle introduced by Pearson in 1963.[11–13] The principle states that there is an extra stabilization when the soft acid combines with soft base and the hard acid combines with hard base. Based on the large amount of chemical information from the experimental observations, this principle was proposed. The concepts of hardness and softness are, in general, related to the ionic (or electrostatic) and covalent (polarizing power or polarizability) types of interactions, respectively. The HSAB principle has been very successful in rationalizing most of the acid–base types of reactions at a qualitative level and is very popular among the chemist community because of its wider applicability and simplicity.[13–15] However, theoretical quantification of this qualitative principle is considered to be a great challenge. As the general concept of hardness and softness parameters lacks the mathematical definition or physical basis, the predictions and explanations made based on these descriptors remained at the qualitative level. Hence, the necessity for an intuitive and correct theoretical approach for these concepts was inevitable. The progress in this direction was realized through the density functional theory (DFT) applications, now known as conceptual DFT, which has been used to explain the hardness and softness parameters along with other useful concepts.[16–21]

There have been numerous works in terms of monographs and reviews on the conceptual DFT, bringing out the usefulness of these descriptors.[22–26] Hence, we will very briefly outline the reactivity descriptors without going into the details of their applications. In this review article, the objective is to highlight the quantitative features of the HSAB principle. More importantly, we will (1) illustrate the formal mathematical formulation of the HSAB principle and the local version of the HSAB principle, (2) illustrate the transformation of this qualitative principle into a quantitative formulation, especially in deriving the essential features of the chemical bonding and reaction energies in a generalized way, and (3) validate the applicability of these methods. The reactivity and stability of the DNA base-pair complexes in the presence of external field have been considered.

19.2   BRIEF OUTLINE OF GLOBAL AND LOCAL REACTIVITY DESCRIPTORS

According to the Hohenberg and Kohn theorem,[16] the ground state energy of a system is expressed as a unique functional of electron density:

$$E[\rho] = F_{HK}[\rho] + \int \rho(r)v(r)\,dr, \tag{19.1}$$

where $v(r)$ is the external potential and $F_{HK}[\rho]$ is the universal Hohenberg–Kohn functional, composed of kinetic energy, classic Coulomb energy, and exchange-correlation energy. The first and second partial derivatives of $E[\rho]$, with respect to the number of electrons N under the constant external potential $v(r)$, are defined as the chemical potential (μ) and the global hardness (η) of the system, respectively.[22,23] The inverse of the hardness is expressed as the global softness. Softness has been related to the overall reactivity of the system, and hardness can be related to the stability of the system. These parameters along with chemical potential are known as the global reactivity descriptors (GRDs). It has been customary to use a finite difference approximation for μ and η. Using the energies of N, $(N + 1)$, and $(N - 1)$ electron systems, we get the operational definition of μ and η as

$$\mu \approx -(IP + EA)/2 \text{ and } \eta \approx (IP - EA)/2, \tag{19.2}$$

where IP and EA are the first vertical ionization energy and the electron affinity of the chemical species, respectively. The operational definition of chemical potential is found to have direct correlation with the definition of electronegativity given by Pauling[22] and Mulliken[27], $\chi = \dfrac{1}{2}(I + A)$. Thus, the qualitative concept of electronegativity, which is a measure of the escaping tendency of electron from the interacting species, can directly be connected with the chemical potential within the framework of DFT. This was considered to be one of the successes of conceptual DFT in linking the qualitative concepts naturally within the framework of DFT.

As the chemical reactions are mostly driven by the functional groups or atomic centers in a molecule, the hardness or softness parameters cannot be used to analyze the reactivity of a particular system. It may be noted that GRD normally provides the information about the overall properties of the chemical systems and is useful for analyzing the thermodynamic aspects of chemical reactions. In general, the charge density fluctuations around any reactive centers will determine the reactivity of the system; hence, it is highly desirable to have a reactivity-oriented description of the molecule in terms of the local reactivity descriptor (LRD). Accordingly, we have two most useful descriptors, namely, local softness and Fukui function.[28] The local softness $s(r)$ of a chemical species or molecule is given by[29-31]

$$s(r) = \left(\frac{\partial \rho(r)}{\partial \mu} \right)_{v(r)} \tag{19.3}$$

such that

$$\int s(r)\,dr = S, \tag{19.4}$$

where

$$S = \left(\frac{\partial N}{\partial \mu} \right)_{v(r)}. \tag{19.5}$$

Further, local softness can also be written as

$$s(r) = \left(\frac{\partial \rho(r)}{\partial N} \right)_{v(r)} \left(\frac{\partial N}{\partial \mu} \right)_{v(r)} \tag{19.6}$$

$$= f(r)S, \tag{19.7}$$

where $f(r)$ is defined as the Fukui function.[32]

As it can be seen from the definition of the local softness and Fukui function, these descriptors clearly contain relative information about different regions in a given molecule. Parr and Yang introduced electronic Fukui function for the first time based on the molecular orbital (MO) concept.[33,34] The physical meaning of $f(r)$ is implied by its definition, that is, it measures the response of the chemical potential to the external potential at a particular point. The second definition implies how the electron density of the systems changes with respect to the small changes in the number of electrons. Using left and right derivatives with respect to the number of electrons gives electrophilic and nucleophilic Fukui function as shown below:

$$f^+(r) = \left(\frac{\partial \rho(r)}{\partial N} \right)^+_{v(r)} \quad \text{for a nucleophilic attack} \tag{19.8}$$

$$f^-(r) = \left(\frac{\partial \rho(r)}{\partial N} \right)^-_{v(r)} \quad \text{for an electrophilic attack.} \tag{19.9}$$

To describe the site selectivity and reactivity of an atom in a molecule, it is required to condense the values of $f(r)$ and $s(r)$ around each atomic site. Thus, for an atom k in a molecule, depending on the type of electron transfer, one can define three different types of Fukui functions as proposed by Yang and Mortier:[32]

$$f_k^+ \approx q_k^{N_o+1} - q_k^{N_o} \quad \text{nucleophilic attack}$$

$$f_k^- \approx q_k^{N_o} - q_k^{N_o-1} \quad \text{electrophilic attack} \tag{19.10}$$

$$f_K^o \approx \frac{1}{2} \left(q_K^{N_o+1} - q_K^{N_o-1} \right) \quad \text{radical attack}.$$

As depicted in Equation 19.10, the atom k having more f_k^+, f_k^- and f_k^o value will be more prone to nucleophilic, electrophilic, and radical attack, respectively.

The corresponding condensed local softness can be defined as

$$s_k^+ = f_k^+ S \tag{19.11}$$

$$s_k^- = f_k^- S. \tag{19.12}$$

Parr and Yang proposed that the larger the Fukui function value is for a particular atom, the center becomes more reactive.[29,31]

19.3 HSAB PRINCIPLE: DENSITY FUNCTIONAL VIEWPOINT

Having defined the physical basis and implications of GRD and LRD within the framework of DFT, the subsequent studies focused on providing the formal proof for the HSAB principle. These formulations essentially originated from DFT-based energy-density perturbation methods. One of the earliest proofs proposed by Parr and Pearson[19] was derived based on DFT-based energy-density perturbation methods. Herein, the energy of the system is perturbed with respect to the number of electrons.

Assuming that systems A and B are interacting systems, the energy expression for each system is expressed as[19]

$$E_A = E_A^0 + \mu_A \left(N_A - N_A^0 \right) + \eta_A \left(N_A - N_A^0 \right)^2 + \dots \tag{19.13}$$

$$E_B = E_B^0 + \mu_B \left(N_B - N_B^0 \right) + \eta_B \left(N_B - N_B^0 \right)^2 + \dots, \tag{19.14}$$

where μ and η are the chemical potential and hardness of the system, respectively, which are defined as the first- and second-order derivatives of the energy with respect to the number of electrons.

Truncating Equations 19.13 and 19.14 up to the second order, the total change in the energy will have the following form:

$$\Delta E = (\mu_A - \mu_B)\Delta N + (\eta_A - \eta_B)\Delta N^2 \tag{19.15}$$

where

$$\Delta N = N_B^0 - N_B = N_A - N_A^0. \tag{19.16}$$

The energy stabilization due to such a charge transfer is the second order in $(\mu_A - \mu_B)$. On minimizing $E_A + E_B$ with respect to ΔN, the result is

$$\mu_A = \mu_B, \tag{19.17}$$

where

$$\mu_A = \mu_A^0 + 2\eta_A \Delta N + \dots \tag{19.18}$$

and

$$\mu_B = \mu_B^0 + 2\eta_B \Delta N + \dots . \tag{19.19}$$

Consequently, to the first order,

$$\Delta N = \frac{(\mu_B - \mu_A)}{2(\eta_A + \eta_B)} . \tag{19.20}$$

On substituting the expression for ΔN, the interaction energy can be expressed as

$$\Delta E = \frac{(\mu_B - \mu_A)^2}{4(\eta_A + \eta_B)} . \tag{19.21}$$

It can be observed from Equation 19.21 that the favorable interaction can occur where there is an electron transfer process and differences in chemical potential. This process is assumed to take place continuously until the equilibrium is attained, and it is referred to as the chemical potential or electronegativity equalization process.[35] If we consider the systems soft acids and bases, $(\eta_A + \eta_B)$ is a small number, and for a reasonable difference in electronegativity, ΔE is substantial and stabilizing. This explains the HSAB principle in part: soft prefers soft. It does not explain, however, the hard–hard preference (large denominator). Following this work, Nalewajski[36] pointed out that the hard–hard interaction could be described by the inclusion of the first-order contribution due to the perturbing external potential [i.e., $E(N,Z)$]. The basic expression is written as

$$\Delta E_{AB} = \frac{(\mu_B^0 - \mu_A^0)^2}{2(\eta_A + \eta_B)} + \frac{(\mu_B^0 - \mu_A^0)(\alpha_B \Delta Z_B - \alpha_A \Delta Z_A)}{(\eta_A + \eta_B)} + (v_A^0 \Delta Z_A + v_B^0 \Delta Z_B), \tag{19.22}$$

where v is the electron-nuclear attraction per unit charge, $v = \dfrac{V_{ne}}{Z} < 0$, $\alpha = \dfrac{1}{2}\left(\dfrac{\partial \mu}{\partial Z}\right)_N = \dfrac{1}{2}\left(\dfrac{\partial V}{\partial N}\right)_Z < 0$, and $\beta = \dfrac{1}{2}\left(\dfrac{\partial V}{\partial Z}\right)_N < 0$.

In Equation 19.22, the second term is always small due to the cancellation of terms in the numerator for hard–hard and soft–soft cases. Even in the hard–soft/soft–hard cases also, the cancellation of terms in the numerator is substantial. From the first term, one can explain the soft–soft interaction, as discussed by Parr and Pearson.[19] For hard–hard combinations, the magnitude of the first term becomes small [large $(\eta_A + \eta_B)$], and the stability originates from the last term in Equation 19.22. For such a pair of hard–hard reactants, both ΔZ and $|v|$ are large, so that the favorable effect from the external field (predominantly ionic bond), due to relatively unshielded nucleus of the partner, becomes dominant. Thus, the physical basis for the HSAB principle and the model explanation of the hardness/softness were theoretically proposed.

Later on, Chattaraj, Lee, and Parr[37] have also derived the proof for the HSAB principle in an elegant way and established that, among potential partners of a given electronegativity, hard likes hard and soft likes soft. Gazquez[38] had derived an alternative proof for the HSAB principle in terms of the chemical potential μ and the hardness η,

$$E[\rho] = \mu N_\varepsilon - \frac{1}{2} N_\varepsilon^2 \eta + E_{core}[\rho], \qquad (19.23)$$

and he demonstrated that the interaction between the species whose softness is approximately equal to each other is energetically favored. Here, N_e is an effective number of valence electrons, and $E_{core}[\rho]$ represents the core contribution to the total energy. If one takes the derivative of the interaction energy expression with respect to S_B, keeping all the other variables constant, the optimum value of the softness of system B for a given softness of the system A can be found out. Following this idea, Gazquez has shown that the equality of the softness criterion, that is, $S_A = S_B$, emerges out naturally for specific values of other parameters.[38]

Besides the above proof, Li and Evans[39] have demonstrated the HSAB principle through the local descriptors and softness kernel in their proof. One of the impressive features of their study is that the Fukui function is shown to be one of the key concepts in relating the frontier MO theory and the HSAB principle. The final expression for the interaction energy, considering the interaction between two molecules i and j, takes the form

$$\Delta E_{ij} = (\mu_i - \mu_j)\Delta N + (\eta_i - \eta_i)^2 \Delta N^2 + \int [\rho_i(r)\Delta v_i(r) + \rho_j(r)\Delta v_j] dr$$

$$+ \Delta N \int [f_i(r)\Delta v_i(r) - f_i(r)\Delta v_i(r)] dr + \frac{\eta_i a}{2} \left(\int s_i(r)\Delta v_i(r) dr \right)^2 \qquad (19.24)$$

$$+ \frac{\eta_j a}{2} \left(\int s_j(r)\Delta v_j(r) dr \right)^2 - \frac{a}{2} \int [s_i(r)(\Delta v_i(r))^2 + s_j(r)(\Delta v_j(r))^2] dr.$$

Here, a and b are constants and μ, η, and s are the chemical potential, hardness, and local softness, respectively. Based on their proof, they proposed the following rule: for hard–hard interaction cases, the site of the minimum Fukui function is preferred; then again, the site with the maximum Fukui function is preferred in the soft–soft interaction cases. They have further shown that the global hardness indicates the chemical reactivity of a molecule as a whole, whereas the Fukui function determines chemical selectivity. From the above discussions, the energy-density perturbation method with the properly chosen perturbation variables, within the framework of DFT, indeed provides a formal proof for the HSAB principle. There have also been many numerical studies that have supported the HSAB principle.[40,41]

19.4 LOCAL HSAB PRINCIPLE AND INTERMOLECULAR INTERACTIONS

In what follows, we will now present a detailed description of the local HSAB principle, formulated by Gazquez and coworkers.[42–47] The local HSAB principle states that the interaction between two molecular systems is favorable when it occurs through those atoms whose local softnesses are approximately equal.[43,47] The proof is based on an approximate expression for the interaction energy between two chemical systems A and B, in terms of the chemical potentials and the softness of the isolated species, such that it allows one to establish the optimum value of the softness of system B for a given softness of system A.

For a general molecular interaction case, A + B $\rightarrow$ AB, the reaction is assumed to take place in two steps: in the first step, when A and B are far from each other, their chemical potentials μ_A and μ_B mutually equalize to the value of μ_{AB} at a constant external potential. In the second step, systems A and B evolve toward the equilibrium state through changes in the electronic density of the global system AB produced by changes in the external potential V_{AB} at constant chemical potential. Hence, the corresponding energy change during the interaction can be given as

$$\Delta E_{int} = \Delta E_v + \Delta E_\mu. \tag{19.25}$$

The expression for each term (ΔE_V and ΔE_μ) present in Equation 19.25 has been derived separately. This will together form the basis for the local HSAB principle. Herein, the atoms can be seen as open subsystems leading to a consideration of the grand potential Ω_A or Ω_B as natural quantity to describe the interactions in terms of atoms, in analogy with macroscopic thermodynamics. If the interaction between molecules A and B occurs between atoms x and y, one can get the expressions ΔE_{Ax} and ΔE_{By} in the same way as we have described above. The starting point for the above scheme is in terms of the Parr–Pearson's Equation 19.21:

$$\Delta E = -\frac{(\mu_B - \mu_A)^2}{4(\eta_A + \eta_B)} = -\frac{(\Delta\mu)^2}{2}\frac{S_A S_B}{S_A + S_B}. \tag{19.26}$$

Considering the grand potential, Ω_A and Ω_B for systems A and B, respectively, Gazquez and Mendez have shown that

$$\Delta\Omega_A + \Delta\Omega_B = \Delta E_A + \Delta E_B = \Delta\Omega_{AB} = \Delta E_{AB}, \tag{19.27}$$

where

$$\Delta\Omega_A + \Delta\Omega_B = -\frac{(\mu_B - \mu_A)^2}{2(S_A + S_B)^2}S_A S_B. \tag{19.28}$$

On minimizing the grand potential $\Delta\Omega_{Ak}$ with respect to S_A at fixed ($\mu_B - \mu_A$), S_{Bl} and f_{Ak} allow one to establish an optimum value of the softness of system B for a given softness of system A:

$$\frac{d(\Delta\Omega_{AK})}{dS_A} = -C\frac{d}{dS_A}\left(\frac{S_A}{(S_A + S_B)^2}\right) = C\left(\frac{S_B - S_A}{(S_A + S_B)^3}\right), \tag{19.29}$$

where $C = -\dfrac{(\mu_A - \mu_B)^2}{2} S_{Bl}^2 f_{Ak}.$

The minimization condition demands that S_A should be equal to S_B:

$$S_A = S_B \qquad (19.30)$$

The important outcome of the Gazquez and Mendez[47] formulation is that the grand potential of all the atoms in A and of all the atoms in B become minimum, when both species have approximately equal global softness. Because we have considered that the interaction between A and B occurs through the kth atom of A and the lth atom of B, one may assume that the most favorable situation corresponds to

$$(\Delta\Omega_{AK})_{min} \approx (\Delta\Omega_{Bl})min. \qquad (19.31)$$

It also implies that the interaction between A and B is favored when it occurs through those atoms whose softness becomes approximately equal. The important consequence of this statement is that although the softest atoms in a molecule A are, in general, the most reactive sites, there may be other sites, though not as soft, that may become the most reactive ones, depending on the softness of the reacting molecule B. This proves the local HSAB principle.[43,47] When f_{Ak} is equal to f_{Bl}, the total interaction energy expression at the minimum is given as

$$\Delta E_{Ak} + \Delta E_{Bl} = -\frac{(\mu_B - \mu_A)^2}{2(S_A + S_B)} S_A S_B f_{Ak}. \qquad (19.32)$$

From Equation 19.32, it can be seen that the greater the values of f_{Ak} and f_{Bl}, the greater the stabilization energy. If we assume that the Fukui functions of the reacting atoms are equal, then the interaction between A and B will not necessarily occur through the softer atoms but rather through those atoms whose Fukui functions are approximately equal.[47]

For the second expression, $\Delta E\mu$, which is the total energy change at constant chemical potential, it is assumed that the electron reshuffling process occurs with constant chemical potential and number of electrons during the early stage of the molecular interaction. Under such condition, Gazquez has formulated $\Delta E\mu$ as

$$\Delta E_\mu = -\frac{1}{2} N^2 \Delta\eta. \qquad (19.33)$$

The details of the expression are referred to elsewhere.[44]

From Equation 19.33, as $\eta > 0$, it implies that $\Delta E > 0$ when $\Delta\eta < 0$ and $\Delta E < 0$ when $\Delta\eta > 0$. Thus, it can be mentioned that as a system evolves toward a state of greater hardness ($\Delta\eta > 0$), its stability increases under the condition of constant chemical potential that is leading to the principle of maximum hardness.

Thus, the expression for the total interaction energy can be written from Equations 19.32 and 19.33[42–47]:

$$\Delta E_{int} = -\frac{(\mu_B - \mu_A)^2}{2(S_A + S_B)} S_A S_B f_{Ak} - \frac{1}{2} K \Delta \eta. \qquad (19.34)$$

The application of the local HSAB principle in the above form for the calculation of the bond energies of several polyatomic systems demonstrated that bond energies could be represented as a function of the chemical potential and the hardness of the system.[44–46] In addition, it has also been shown that the activation energy of a chemical reaction is dependent on the difference between the hardness of the initial state and the transition state.[46]

19.5   MORE GENERALIZED LOCAL HSAB PRINCIPLE FOR SINGLE AND MULTIPLE INTERMOLECULAR INTERACTION CASES

19.5.1   SINGLE INTERACTING SITES

According to Equation 19.25, the total interaction energy between A and B can be given as

$$\Delta E_{int} = -\frac{1}{2}\left(\frac{(\mu_B - \mu_A)^2}{\eta_A + \eta_A}\right)_v - \frac{1}{2} N_{AB}^2 \left(\eta_{AB} - \eta_{AB}^*\right)_\mu, \qquad (19.35)$$

and the interaction between systems A and B is assumed to take place in two steps: ΔE_v and ΔE_μ. In the first step, the interaction takes place at constant external potential through the equalization of chemical potential that is referred to as ΔE_v. In the second step, A and B evolve toward the equilibrium state through changes in the electron density of the global system produced at constant chemical potential, which is referred to as ΔE_μ. Equation 19.26 can be written in terms of softnesses as

$$\Delta E_v \approx -\frac{(\mu_B - \mu_A)^2}{2}\left(\frac{S_A S_B}{S_A + S_B}\right)_v. \qquad (19.36)$$

However, the corresponding changes in Equation 19.33 are not so simple, as they involve the total hardness of the system. One has to relate the total hardness of system AB in terms of the softnesses of the individual systems. In general, the total softness of system AB at equilibrium can be written as

$$S_{AB} = k(S_A + S_B), \qquad (19.37)$$

where k is the proportionality constant and S_A and S_B are the softness of the isolated systems A and B, respectively. As the total molecular softness is insensitive to the number of electrons, the difference in the softness $(S_{AB} - S_{AB}^*)$ can be approximated by

$$\Delta S = k'(S_A + S_B),\qquad(19.38)$$

where k' is another proportionality constant. By applying Equation 19.38 in Equation 19.33,

$$\Delta E_\mu = -\frac{1}{2} N_{AB}^2 K \left(\frac{1}{S_A + S_B} \right)_\mu = -\lambda \left[\frac{1}{4(S_A + S_B)} \right]_\mu,\qquad(19.39)$$

where N_{AB} is the total number of electrons of system AB. The product of the terms N_{AB}^2 and K is known as λ.

On substituting Equations 19.33 and 19.39 in Equation 19.35, one can get the global model in terms of the softness parameter of systems A and B:

$$\Delta E_{int} \approx -\frac{(\mu_B - \mu_A)^2}{2} \left(\frac{S_A S_B}{S_A + S_B} \right)_v - \frac{\lambda}{4} \left(\frac{1}{S_A + S_B} \right)_\mu.\qquad(19.40)$$

If the interaction between the systems occurs through atom x of A with the molecular system B, one can express the total interaction energy from the local point of view as

$$(\Delta E_{int})_{Ax} = -\frac{(\mu_A - \mu_B)^2}{2} \left(\frac{S_A f_{Ax} S_B}{S_A f_{Ax} + S_B} \right)_v - \frac{\lambda}{4} \left(\frac{1}{S_A f_{Ax} + S_B} \right)_\mu,\qquad(19.41)$$

where S_A and f_{Ax} are the global softness and condensed Fukui function (CFF) of atom x in a system A, respectively.

The constant λ cannot be computed rigorously only through the softness of the molecular complexes. In the literature, several values have been considered for this ad hoc parameter λ. In the initial study on the regioselectivity of enolate alkylation, Gazquez and Mendez[43] have used 0.5 as the value of λ. Geerlings and coworkers[48] used the values 0.5 and 1.0 for certain organic reactions. In the study by Mendez, Tamariz, and Geerlings[49] on 1,3-dipolar cycloaddition reactions, the dependence of the total interaction energy has been calculated at a local (dipole)–global (dipolarophile) level. The value of λ indicates that regioselectivity in the reactions between benzonitrile oxide and vinyl p-nitrobenzoate and 1-acetylvinyl p-nitrobenzoate is predicted correctly as long as $\lambda > 0.2$.

Pal and Chandrakumar[50] made an attempt to relate the parameter λ in terms of the change in the electron densities at the interacting site before and after the interaction process. This change will give the effective number of valence electrons that have participated in the interaction process. As the parameter λ contains several parameters including the number of electrons participating during the intermolecular interaction and other parameters related to the electron reshuffling process, it was logical to select the above approximate definition for the parameter λ. Thus, an expression for the term λ can be written as the difference of electron densities of system A before and after the interaction:

$$\lambda_A = \sum_{i=1}^{p} \rho_{A_i}^{eq} - \sum_{i=1}^{p} \rho_{A_i}^{0}. \tag{19.42}$$

Alternately, the term λ can be defined as the difference of electron densities for system B,

$$\lambda_B = \sum_{j=1}^{p} \rho_{B_j}^{eq} - \sum_{j=1}^{q} \rho_{B_j}^{0}, \tag{19.43}$$

where the first terms on the right-hand side of Equations 19.42 and 19.43 refer to the sum of the electron densities of each atom in A and B in the molecule AB at equilibrium, respectively, and the second terms in Equations 19.42 and 19.43 refer to electron densities of each atom in the isolated systems A and B, respectively. The indices p and q are the number of atoms of systems A and B, respectively.

Using the above scheme, Chandrakumar and Pal have made several studies to validate the working definition of the parameter λ and to obtain the energetic details of weak intermolecular interactions.[50–53] The first successful case was to have a quantitative estimate of the adsorption energy of small molecular systems with different alkali cations present at various reactive sites of zeolite-A.[50,51] This scheme results in interaction energies, which are found to be in very good agreement with experimental and other theoretical interaction energy values. It has also been shown that the arbitrary definition or complete neglect of the parameter λ can lead to erroneous results even at the qualitative level. This suggests the efficacy of the local HSAB principle in describing the weak intermolecular interaction and the validity of our quantitative definition of the parameter λ. To validate this quantitative approach for the purpose of studying chemical binding in a broad way, a detailed study of molecular interactions, including soft–soft, hard–hard, and hard–soft interactions, has been undertaken[52] to calculate the interaction energy of the molecular complexes with varying degrees of strength, especially weak-to-moderate type of interactions, ranging from the covalent, van der Waals, and other weak electrostatically held molecules. The results derived from different classes of systems (Lewis acid–base complexes) with the interaction energy ranging from about −41 to −1 kcal/mol are very encouraging, and the interaction energy values are found to be in good agreement with the experimental and theoretical results. It should be noted that the effectiveness and accuracy of these results heavily rely on the computation of the local descriptors, which are unfortunately dependent on the basis set and level of theory that are used in the calculation. Despite the arbitrary nature of the population analysis and the level of the theoretical methods, the final results are found to be very reliable and consistent with other conventional theoretical methods.

The expression of interaction energy (Equation 19.41) is derived based on the fact that only one specific atom in the molecule is interacting with the other molecule. If a system has several interacting or reactive sites, the above expression will no longer be valid. In the next subsection, an important generalization of the local HSAB principle has been formulated, which is applicable to a more general class of interactions taking place through a group of cooperative atoms or individual reactive atoms based on multiple sites.

19.5.2 MULTIPLE INTERACTING SITES: LOCALIZED REACTIVE MODEL AND SMEARED REACTIVE MODEL

The study of multiple interaction sites within the framework of the local HSAB principle requires modifications of the working equation (see Equation 19.41); accordingly, Chandrakumar and Pal[54,55] made an attempt to generalize the local HSAB principle using the group softness proposal as one of the important concepts. Essentially, it has been found to have two limiting cases: the so-called localized reactive model (LRM) and global (smeared) reactive model (SRM). Their origin stems from the nature and location of the reactive sites present in the molecular system. The feasibility of these models has been tested by selecting some prototype intermolecular hydrogen-bonded systems where the multiple-site interactions are important.

Herein, systems A and B are assumed to interact with each other through several reactive atoms or functional groups, viz., the locally identified distinctive reactive sites of A and B, which are designated as x, y, z, etc., and k, l, m, etc., respectively. These reactive atoms can be present in the system together or can be located far from each other, and based on the nature of these reactive atoms, the systems can interact in different ways. For simplicity, we can assume that the interaction between the different molecular systems is taking place through the individual reactive atoms that are located at the different part of the system. In such cases, one reactive part has no cooperative effect or influence on another atom/region. Hence, it can be safely formulated that although the reaction proceeds simultaneously through many reaction centers, interaction energy may be calculated in a decoupled manner. This model is known as the LRM. The total interaction energy for the complex AB may be represented as the sum of interactions arising from each part of the interacting atoms of A and B (x–k, y–l, z–m, etc.), and it can be obtained as a logical extension of the single-site local HSAB principle to multiple sites by adding the interaction energies of the individual sites in a decoupled manner.

According to this model, the interaction energy expression is given as

$$\Delta E_{int} \approx -\frac{(\mu_A - \mu_B)^2}{2}\left(\frac{S_A S_B f_{AX} f_{BK}}{S_A f_{AX} + S_B f_{BK}} + \frac{S_A S_B f_{Ay} f_{Bl}}{S_A f_{Ay} + S_B f_{Bl}} + \frac{S_A S_B f_{Az} f_{Bm}}{S_A f_{Az} + S_B f_{Bm}} + ...\right)_v$$

$$-\frac{\lambda}{4}\left(\frac{1}{S_A f_{AX} + S_B f_{BK}} + \frac{1}{S_A f_{Ay} + S_B f_{Bl}} + \frac{1}{S_A f_{AZ} + S_B f_{Bm}} + ...\right)_\mu. \quad (19.44)$$

Collecting the expression for the interaction between A_x and B_k, A_y and B_l, etc., from the first and second terms of Equation 19.44, one can write

$$\Delta E_{\text{int}} = \Delta E_{\text{A}x-\text{B}k} + \Delta E_{\text{A}y-\text{B}l} + \Delta E_{\text{A}z-\text{B}m}, \tag{19.45}$$

where $\Delta E_{\text{A}x-\text{B}k}$ defines the interaction energy derived from sites A_x and B_y.

In the second model, it has been assumed that the reacting sites consist of a group of atoms that can arise due to the participation of neighboring atoms at the reaction site or the proximity of two or more reacting sites. It can be safely assumed that these reactive atoms all together constitute one reacting site. In such cases, cooperative effects due to all the reactive atoms are strong. In other words, the individual reactivity of a particular site would be lost or smeared out due to the influence of other reactive sites; hence, this model is known as the SRM. This model is at the extreme limitation, or if all the atoms are reactive, the corresponding model should be reduced to the global HSAB principle. Let us denote the group of atoms of reacting sites x and y as x_1, x_2, x_3, ..., etc., and y_1, y_2, y_3, ..., etc., respectively. Similarly, for system B, one can denote the localized sets of reacting atoms as k, l, m, ..., etc. In such cases, Equation 19.44 can be written in terms of the softness of all the atoms, and it should be added to define the total softness of the reacting site. This can be called group softness,[56] and using the group softness of the cooperating atoms in a site, each term $\Delta E_{\text{A}x-\text{B}k}$ can generally be written as

$$\Delta E_{\text{A}x-\text{B}k} \approx -\frac{(\mu_A - \mu_B)^2}{2}\left[\frac{\left(\sum_{i=1}^{n} s_{\text{A}x_i}\right)\left(\sum_{j=1}^{m} s_{\text{B}k_j}\right)}{\left(\sum_{i=1}^{n} s_{\text{A}x_i} + \sum_{j=1}^{m} s_{\text{B}k_j}\right)}\right]_v - \frac{\lambda}{4}\left[\frac{1}{\left(\sum_{i=1}^{n} s_{\text{A}x_i} + \sum_{j=1}^{m} s_{\text{B}k_j}\right)}\right]_\mu, \tag{19.46}$$

where there are n participating atoms x_1, x_2, x_3, ..., x_n, etc., in site A_x, and similarly, there are m atoms k_1, k_2, k_3, ..., k_m, etc., in site B_k. ΔE_{int} is the sum of all such site interactions $\Delta E_{\text{A}x-\text{B}k}$, $\Delta E_{\text{A}y-\text{B}l}$, $\Delta E_{\text{A}z-\text{B}m}$, etc. Depending on the number of sites and group of atoms in a site, one can define different interaction patterns between two systems A and B. In the limit that each site contains only one atom, this model reduces to the previous model LRM. On the other hand, the other limit is the global HSAB, where there is only one site in each of the systems A and B and all atoms are cooperative. In such a limit, there is only one term in Equation 19.44, and this term now involves the group softness of all atoms, which is the global softness of the systems. SRM actually defines all other intermediate interactions between the limit of fully local (LRM) and the global model.

The applicability of the models LRM and SRM has been tested by selecting some prototype molecular systems, such as the complexes of formamide, acetamide, and acrylamide, with the formamide molecule, which represents a simple nucleic acid–base model.[54] The complexes have multiple hydrogen bonding, namely, carbonyl group interacting with NH group and amide, and acid OH group interacting with -NC group. In all these complexes, the reactive group is not directly bonded to another reactive group, and they are at least separated by two atoms. Hence, LRM has been applied to evaluate the interaction energy of these complexes, and the results are found to be in agreement and consistent with the available data. Similarly, for the case of SRM, the complexes between acetylene and ethylene molecules with alkali cations have also been tested wherein the reactive groups are located adjacent to each other. In this case, the two carbon atoms connected by double bond (the π-electron density is significant) are considered to be the reactive atoms; hence, the measure of the reactivity of these atoms was performed through the use of SRM. More importantly, we have also applied LRM successfully in determining the magnitude and nature of the hydrogen bonding interactions of biomolecules, which is responsible for the important unique properties of complementary Watson–Crick base pairs in DNA involving (1) adenine (Ade) with thymine (Thy) or uracil and (2) guanine (Gua) with cytosine (Cyt).[55] In general, the energies and geometry of these multiple molecular interactions are particularly important in understanding the molecular recognition processes. Many groups have made an attempt to calculate the nature and the stabilization of the DNA base-pair interactions using the different types of models (e.g., electrostatic potential, molecular similarity indexes, and electric field) and ab initio calculations. For these types of complexes, the interaction energy of the base pairs ranges from −21.3 to −8.09 kcal/mol and the Gua–Cyt base pair is the most stable base pair for which the LRM has predicted that the interaction energy including three hydrogen bonds (H-bonds) is −21.25 kcal/mol. Similarly, for the case of Ade–uracil, the interaction energy predicted by LRM is −11.8 kcal/mol. On comparing the values predicted by LRM with the conventional HF/MP2 methods, it has been found that the interaction energy values are in very good agreement with each other. For instance, MP2 values for the above two cases are −25.8 and −12.5 kcal/mol, respectively. It is gratifying to note that, despite the empirical nature of these models and calculations of these reactivity descriptors, the proposed models have been found to predict the nature of the molecular interactions in a systematic and consistent way. Nevertheless, these models have demonstrated that the interaction between molecular systems can be analyzed through the conceptual density descriptors in a quantitative manner. In what follows, we will further extend the applicability of these descriptors and reactivity models for the case of DNA bases and base pairs in the presence of external fields.

19.6 REACTIVITY AND STABILITY OF DNA BASES AND THEIR BASE PAIRS IN PRESENCE OF ELECTRIC FIELD

Chemically, DNA is a long polymer formed by units called nucleotides, with a backbone made of sugars and phosphate groups. Each sugar is attached to one of four types of bases, namely, Ade, Gua, Cyt, and Thy. It is the sequence of these four bases along the backbone that encodes information. The complementary Watson–Crick base pairs in DNA involve Ade with Thy as well as Gua with Cyt, which are held by H-bonds. In particular, the strength of the H-bond determines the magnitude and nature of interaction of the base pairs and is consequently responsible for the unique properties of DNA. As discussed above, owing to the importance of the DNA bases and the base-pair interactions, many theoretical studies have been devoted to understanding their reactivity and stability. Herein, we are interested in investigating the reactivity and stability of the DNA bases in the presence of external electric field with the aid of proposed semiquantitative reactivity models such as LRM.[54] We have successfully demonstrated the interaction between small prototype molecular systems in the presence of electric field through the descriptors.[57]

To calculate the reactivity descriptors of all the DNA bases (Ade, Gua, Cyt, and Thy) and their base pairs (Ade–Thy and Gua–Cyt), the corresponding geometries were optimized at the DFT level using 6-31++G (d,p) basis set with B3LYP hybrid functional at zero field. The external electric field was applied toward the atoms forming H-bonds, that is, in a direction parallel to the H-bonds (we designate it as $E_\parallel$), to calculate the energy of the neutral, cationic, and anionic systems for a field value increasing up to 0.006 au (1 au electric field = 51.4 V/Å), with an increment of 0.001 au at each step. Hence, the global and local parameters were calculated at each field value. For the calculation of chemical potential and hardness, we have used the values of ionization potential and electron affinity. The local parameters such as CFF were calculated using the Lowdin-based method of population analysis. Additionally, the electric field was applied in the two perpendicular directions, that is, perpendicular to the H-bonds and in the plane of the molecule ($E\perp$) and in the other direction perpendicular to the molecular plane. For calculating the interaction energy of the base pairs (Gua–Cyt and Ade–Thy), we have employed the LRM (Equation 19.44). All the calculations were performed by GAMESS software.[58] The quantum chemical (QC) method of calculating the interaction energy is the difference between the energy of the complex AB and the sum of the energy of monomers A and B: $\Delta E = E_{AB} - (E_A + E_B)$.

Let us first discuss the response of the descriptors when the field is applied parallel and toward the atoms forming the H-bonds, that is, in the $E_\parallel$ direction. It can be seen from Figures 19.1 through 19.4 that the chemical potential and hardness of all DNA bases decrease with the increase in field strength along this direction. However, the CFF for electrophilic attack (nucleophilicity) for more electronegative atom increases, whereas the electrophilicity of the H atoms decreases with the increase in the field values. For instance, the nucleophilicity of O atom in Gua and Thy (Figures 19.1b and 19.2b), N atom in Ade (Figure 19.3b), and both O and N atoms in Cyt (Figure 19.4b), all forming H-bonds, increase with increasing field values. On the other hand, the electrophilicity of H atoms, forming H-bonds, in all the above bases decreases as the field value is increased (Figures 19.1b through 19.4b).

When the field is applied in the opposite direction, the GRD increases up to some value of the electric field and then starts decreasing with the increasing field values in all bases (Figures 19.1a, 19.3a, and 19.4a), except in the case of Gua, where μ and η decrease with increasing field (Figure 19.2a). At this point, it should be mentioned that this is in contrast with the analyses performed earlier, where, as the field direction was reversed, the behavior of the descriptors was found to be altered in the opposite manner. However, the behavior of the LRD gets reversed as the field direction is reversed in all cases (Figures 19.1b through 19.4b). This anomaly may be attributed to the cooperative effect of all the atoms present in these molecules. Moreover, the complexity of the molecule has to be also taken into account.

If the field is applied perpendicular to the H-bonds and in the plane of the molecule ($E\perp$), the field is directed toward the H atom in all the bases. For Thy and Gua,

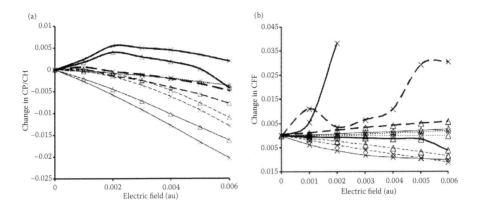

FIGURE 19.1 (a) Variation of change in chemical potential (×) and hardness (Δ) (in au) for Thy against electric field (in au). (b) Variation of change in CFF for Thy against electric field (in au). Δ represents CFF for electrophilic attack for O atom, whereas × represents CFF for nucleophilic attack for H atom. Solid line means that the field is parallel and toward the atoms forming H-bonds (E$_\parallel$ direction). Half dashed line means that the field is in perpendicular direction to the hydrogen bonds but in the plane of the molecule (E⊥). Fully dashed lines are for the field perpendicular to the plane of the molecule. Dark lines represent the values of the parameter for the fields in opposite direction.

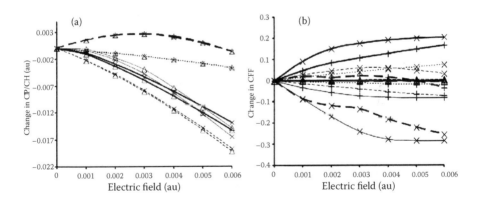

FIGURE 19.2 (a) Variation of change in chemical potential (Δ) and hardness (×) (in au) for Gua against electric field (in au). (b) Variation of change in CFF for Gua against electric field (in au). Δ represents CFF for electrophilic attack for O atom, whereas × represents CFF for nucleophilic attack for amide H atom and + represents CFF for nucleophilic attack for the other H atom. Solid line means that the field is parallel and toward the atoms forming H-bonds (E$_\parallel$ direction). Half dashed line means that the field is in perpendicular direction to the hydrogen bonds but in the plane of the molecule (E⊥). Fully dashed lines are for the field perpendicular to the plane of the molecule. Dark lines represent the values of the parameter for the fields in opposite direction.

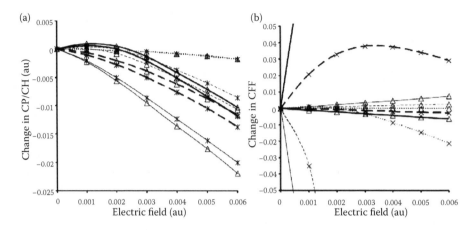

FIGURE 19.3 (a) Variation of change in chemical potential (×) and hardness (Δ) (in au) for Ade against electric field (in au). (b) Variation of change in CFF for Ade against electric field (in au). Δ represents CFF for electrophilic attack for N atom, whereas × represents CFF for nucleophilic attack for H atom. Solid line means that the field is parallel and toward the atoms forming H-bonds ($E_\parallel$ direction). Half dashed line means that the field is in perpendicular direction to the hydrogen bonds but in the plane of the molecule ($E\perp$). Fully dashed lines are for the field perpendicular to the plane of the molecule. Dark lines represent the values of the parameter for the fields in opposite direction.

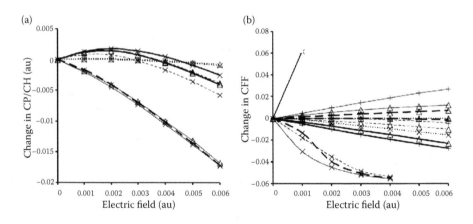

FIGURE 19.4 (a) Variation of change in chemical potential (Δ) and hardness (×) (in au) for Cyt against electric field (in au). (b) Variation of change in CFF for Cyt against electric field (in au). Δ represents CFF for electrophilic attack for O atom, whereas × represents CFF for nucleophilic attack for H atom and + represents change in CFF for electrophilic attack for N atom. Solid line means that the field is parallel and toward the atoms forming H-bonds ($E_\parallel$ direction). Half dashed line means that the field is in perpendicular direction to the hydrogen bonds but in the plane of the molecule ($E\perp$). Fully dashed lines are for the field perpendicular to the plane of the molecule. Dark lines represent the values of the parameter for the fields in opposite direction.

μ and η decrease (Figures 19.1a and 19.2a). However, for the other two (Cyt and Ade), they first increase and then start decreasing (Figures 19.3a and 19.4a). On the other hand, the nucleophilicity of the O and N atom(s) in Gua, Thy, and Cyt forming the H-bond decreases with the increasing field. Moreover, the electrophilicity of the reactive H atom (forming H-bonds) in these bases increases. Conversely, when the field is directed toward the reactive hydrogen atom in Ade, its electrophilicity decreases, whereas the nucleophilicity of the reactive N atom increases. This may be due to the nonplanarity of the amide H-atoms.

Similarly, when the field direction is reversed, the behavior of GRD and LRD does not reverse. Incidentally, μ and η decrease with the increase in field values, except in Thy where they increase and then begin to decrease (Figures 19.2a through 19.4a). Although, in the case of Thy, the behavior of LRD reverses as the field direction is reversed, the behavior of the reactive atoms in the other bases is different. As it can be seen from Figures 19.2b through 19.4b, the nucleophilicity of the O atom in Cyt and Gua reverses along with the N atom in Ade, whereas the N atom in Cyt remains constant and then starts decreasing at higher fields. Furthermore, the electrophilicity of the reactive amide H atom in Gua and Cyt decreases, but the other reactive H atom in Gua and the amide H atom in Ade increase and then decrease. This difference may also be attributed to the nonplanarity of the amide H atoms.

Now, we would confer with the qualitative results when the field was applied perpendicular to the plane of the molecule. It is interesting to note that μ and η decrease with the increasing field for all the bases (Figures 19.1a through 19.4a). On the other hand, the nucleophilicity of the more electronegative reactive atom(s) remains almost constant. For example, the nucleophilicity of O and/or N atom(s) in Ade, Thy, Gua, and Cyt remains almost constant with the increase in the field values (Figures 19.1b through 19.4b). However, the electrophilicity of the amide H atom in Cyt and Gua decreases with increasing field, whereas that of Ade first increases slightly and then decreases (Figures 19.2b through 19.4b). Similarly, the electrophilicity of the other H atom in Gua and Thy increases with increasing field values (Figures 19.1b and 19.2b).

Let us now discuss the effect of electric field on the stabilization of the base pairs by the local HSAB principle and compare our results with the QC method. The equilibrium structures of the base pairs Gua–Cyt and Ade–Thy are shown in Figure 19.5a and b, respectively. The interaction energy of these pairs is shown in Figure 19.6a and b.

When the field is applied along the H-bond toward the Ade molecule in Ade–Thy or Cyt molecule in Gua–Cyt base pair, the complex gets stabilized with the increasing field value. However, in the case of the Gua–Cyt base pair, the variation is almost linear in this field range. Moreover, they seem to be in well accordance with the variation in QC methods. It should, however, be noticed that the variation in QC methods is almost linear in all cases. This nonlinearity in the model interaction energies may be closely observed and attributed to the variation of the FF of the reactive atoms in the molecules. On the other hand, when the field is applied toward the Thy or Gua molecule, the base pair gets destabilized with the increasing field. This is an important conclusion that has to play a significant part when the base pairs are under some external perturbation.

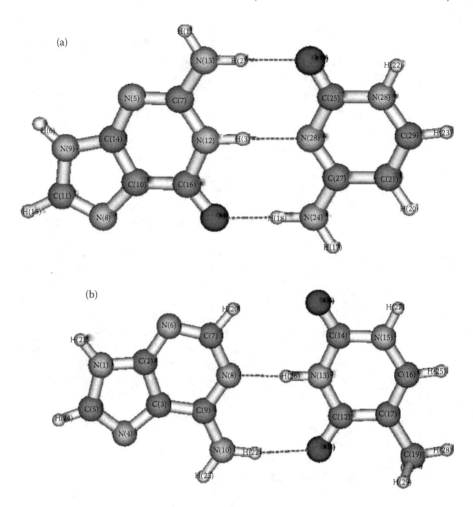

FIGURE 19.5 Optimized geometries of the DNA base pair (a) Gua–Cyt and (b) Ade–Thy at DFT level using 6-31++G (d,p) basis set with B3LYP hybrid functional.

Then again, when the field is in the perpendicular direction (in the plane of the base pairs) and when it is toward the amide H atom, the interaction energy of the Gua–Cyt and Ade–Cyt base pair decreases. However, the Gua–Cyt base pair gets stabilized irrespective of the field direction, and the stability increases when the field direction is toward the amide H atom. It should, however, be noted that the QC interaction energy of the complexes in this field direction is reverse. This may be due to the CFF values of the reactive atoms, which play a key role in deciding the stability of a complex.

When the field is in a perpendicular direction to the plane of the molecule, the interaction energy of the base pair Gua–Cyt increases slightly and then gets stabilized with the increasing field strength. However, there is a linear decrease in the QC interaction energy of the base pair (Figure 19.6b). Contrary to this, the Ade–Thy base pair gets destabilized with the increasing field values. Moreover, this is in complete

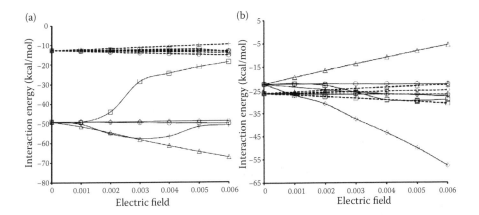

FIGURE 19.6 Variation of interaction energy (IE), both QC and local HSAB (in kcal/mol), for (a) Ade–Thy and (b) Gua–Cyt base pairs against electric field (in au); local HSAB and QC IE are denoted by ☐ when the field is applied parallel to the H-bonds and directed toward Thy (Cyt) but denoted by Δ when the field is directed toward Ade (Gua); perpendicular to the H-bonds (in the plane of the molecule) is denoted by + when the field is directed toward the amide H-atom but denoted by ◯ when the field is in opposite direction; perpendicular to the molecular plane is denoted by ◊. Dashed line represents QC IE, whereas solid line represents local HSAB IE.

disagreement with the QC interaction energy (Figure 19.6a). Close analysis of the anomaly reveals that the local HSAB interaction energy is dependent on both the charge transfer term and FF of the reactive atoms, and these factors decide the interaction energy, hence causing such variations.

In summary, the variation in the reactivity of the DNA bases in the presence of a weak external electric field indicates that when the field is applied parallel and toward the atoms forming the H-bonds, in all the DNA bases, the GRDs (μ and η) decrease with the increase in electric field strength, and the variation with the field is not always linear. The nonlinearity in the behavior of the reactivity descriptors in such external electric field may be attributed to the complexity in the molecule.

19.7 CONCLUSIONS

In conclusion, we have made an attempt to explain the theoretical basis for the empirical hardness/softness concepts to address the reactivity and stability of molecular complexes in a quantitative way through the energy expression derived from the energy-density perturbation theory and the local HSAB principle. To describe the different kinds of interactions in a broader perspective, due to the presence of single and multiple reactive sites, we have extended the local HSAB principle into a more generalized principle. The proposed semiquantitative models (i.e., LRM and SRM) are found to be applicable in describing the molecular interactions in certain distinguishable ways. We have also demonstrated that the reactivity and stability of molecular interactions occurring between simple prototype molecular systems as well as the DNA Watson–Crick base pairs can be described properly. In most cases, the results are consistent with the conventional QC values.

Although the proposed models have been successful in describing several weak to moderate types of molecular interactions, it is important to note the definition of ad hoc parameter λ. The computation of this parameter is rather difficult; hence, further study should be made on the evaluation of the parameter λ. Another critical aspect is that because the descriptors of isolated reactants are employed in the energy expression, these models are expected to be applicable only for the cases where the influence of one monomer reactant on another molecule is comparatively less. Having said the limitations of the proposed models, these models, nevertheless, can rationalize the relative influence of the hardness/softness parameters in determining the nature of different types of interactions and stabilization of the molecular complexes, thus transforming the once thought qualitative HSAB principle into a quantitative one.

ACKNOWLEDGMENTS

We thank Mr. Susanta Das and Ms. Deepti Mishra for their help in preparation and critical proofreading of this manuscript and Ms. Asha Shinde for providing secretarial assistance.

REFERENCES

1. Fukui, K., *Theory of Orientation and Stereo Selection*, Springer-Verlag, New York, 1975.
2. Pauling, L., *The Nature of Chemical Bond and Structure of Molecule and Crystals. The Nature of Chemical Bond and Structure of Molecule and Crystals*, Oxford & IBH, New Delhi, 1967.
3. *Theoretical Models of Chemical Bonding: The Concept of the Chemical Bond*, Maksic, Z. B., ed., Springer-Verlag, Berlin, 1990.
4. *The Force Concept in Chemistry*, Deb, B. M., ed., Van Nostrand-Reinhold, New York, 1981. Hanna, M. W. and Lippert, J. L., in *Molecular Complexes*, Foester, R., ed., Eleck, London, 1973, volume 1. *Molecular Interactions: From van der Waals to Strongly Bound Complexes*, Scheiner, S., ed., John-Wiley & Sons, New York, 1997.
5. Cohen, M. H., *Theory of Chemical Reactivity, Topics in Current Chemistry*, Nalewajski, R. F., ed., Springer-Verlag, Berlin, 1996, volume 183, 143.
6. Fukui, K., *Science*, 218, 747, 1982. Morokuma, K., *Acc. Chem. Res.*, 10, 294, 1977. Klopman, G., *J. Am. Chem. Soc.*, 90, 223, 1968. Coulson, C. A., *Disc. Faraday Soc.*, 2, 9, 1947. Wheland, G. W., *J. Am. Chem. Soc.*, 64, 900, 1942.
7. Dewar, M. J. S., *The Molecular Orbital Theory of Organic Chemistry*, McGraw-Hill, New York, 1969.
8. Fukui, K., Yonezawa, T., and Shingu, H., *J. Chem. Phys.*, 20, 722, 1952. Fukui, K., Yonezawa, T., Nagata, C., and Shingu, H., *J. Chem. Phys.*, 22, 1433, 1954. Nagakura, S. and Tanaka, J., *J. Chem. Soc. Jpn.*, 75, 993, 1954. Fukui, K., *Molecular Orbitals in Chemistry, Physics and Biology*, Lowdin, P. O. and Pullman, B., eds., Academic Press, New York, 1964, 513.
9. Lewis, G. N., *Valence and the Structure of Atoms and Molecules*, The Chemical Catalog Company, Reinhold Publication Corporation, New York, 1923.
10. Woodward, R. B. and Hoffmann, R., *J. Am. Chem. Soc.*, 87, 395, 1965. Woodward, R. B. and Hoffmann, R., *Acc. Chem. Res.*, 1, 17, 1968. Woodward, R. B. and Hoffmann, R., *The Conservation of Orbital Symmetry*, Academic Press, New York, 1989. Geerlings, G., Ayers, P. W., Toro-Labbe, A., Chattaraj, P. K., and De Proft, F., *Acc. Chem. Res.*, 45, 683, 2012.

11. Pearson, R. G., Recent advances in the concept of hard and soft acids and bases, *J. Chem. Educ.*, 64, 561, 1987.

12. Pearson, R. G., *Acc. Chem. Res.*, 26, 250, 1993.

13. Pearson, R. G., Hard and soft acids and bases, *J. Am. Chem. Soc.*, 85, 3533–3539, 1963.

14. Pearson, R. G., *Science*, 151, 172, 1966.

15. Pearson, R. G., *Coord. Chem. Rev.*, 100, 403, 1990.

16. Hohenberg, P. and Kohn, W., *Phys. Rev. B*, 136, 864, 1964.

17. Parr, R. G., Donnelly, R. A., Levy, M., and Palke, W. E., *J. Chem. Phys.*, 68, 3801, 1978.

18. Iczkowsksi, R. P. and Margrave, J. L., *J. Am. Chem. Soc.*, 83, 3547, 1961.

19. Parr, R. G. and Pearson, R. G., *J. Am. Chem. Soc.*, 105, 7512, 1983.

20. Pearson, R. G., *Proc. Natl. Acad. Sci. USA*, 83, 8440, 1986.

21. Parr, R. G. and Yang, W., *Annu. Rev. Phys. Chem.*, 46, 701, 1995.

22. *Electronegativity, Structure and Bonding, Volume 66*, Sen, K. D. and Jørgensen, C. K., eds., Springer-Verlag, Berlin, 1987.

23. *Chemical Hardness, Structure and Bonding, Volume 80*, Sen, K. D., ed., Springer-Verlag, Berlin, 1993.

24. Chattaraj, P. K., Sarkar, U., and Roy, D. R., *Chem. Rev.*, 106, 2065, 2006. Geerlings, P., De Proft, F., and Langenaeker, W., *Chem. Rev.*, 103, 1793, 2003.

25. *Density Functional Theory, Topics in Current Chemistry*, Nalewajski, R. F., ed., Springer, Berlin, 1996, volumes 1–4.

26. Pearson, R. G., *Hard and Soft Acids and Bases*, Dowden, Hutchinson, and Ross, Stroudsburg, PA, 1973. Pearson, R. G., *Chemical Hardness: Applications from Molecules to Solids*, Wiley-VCH Verlag GmbH, Weinheim, 1997.

27. Mulliken, R. S., *J. Chem. Phys.*, 2, 782, 1934.

28. Parr, R. G. and Yang, W., *Annu. Rev. Phys. Chem.*, 46, 701, 1995.

29. Yang, W. and Parr, R. G., *Proc. Natl. Acad. Sci.*, 82, 6723, 1985.

30. Ghosh, S. K. and Berkowitz, M., *J. Chem. Phys.*, 83, 2976, 1985.

31. Parr, R. G. and Yang, W., *J. Am. Chem. Soc.*, 106, 4049, 1984.

32. Yang, W. and Mortier, W., *J. Am. Chem. Soc.*, 108, 5708, 1986.

33. Ayers, P. W. and Levy, M., *Theor. Chim. Acta*, 103, 353, 2000.

34. Perdew, J. P., Parr, R. G., Levy, M., and Balduz, J. L. Jr., *Phys. Rev. Lett.*, 49, 1691, 1982.

35. Mortier, W. J., Ghosh, S. K., and Shankar, S., *J. Am. Chem. Soc.*, 108, 4315, 1986.

36. Nalewajski, R. F., *J. Am. Chem. Soc.*, 106, 944, 1984.

37. Chattaraj, P. K., Lee, H., and Parr, R. G., *J. Am. Chem. Soc.*, 113, 1855, 1991. Chattaraj, P. K. and Ayers, P. W., *J. Chem. Phys.*, 123, 086101, 2005.

38. Gazquez, J. L., *J. Phys. Chem. A*, 101, 4657, 1997.

39. Li, Y. and Evans, N. S., *J. Am. Chem. Soc.*, 117, 7756, 1995.

40. Chattaraj, P. K. and Schleyer, P. v. R., *J. Am. Chem. Soc.*, 116, 1067, 1994. Chattaraj, P. K., *J. Phys. Chem. A*, 105, 511, 2001.

41. Chattaraj, P. K.; Cedillo, A.; Parr, R. G., *Int. J. Quantum Chem.*, 103, 10621, 1995.

42. Mendez, F. and Gazquez, J. L., *Proc. Indian. Acad. Sci.*, 106, 183, 1994. Gazquez, J. L., *Structure and Bonding, Volume 80*, Sen, K. D., ed., Springer- Verlag, Berlin, 1993, 27.

43. Mendez, F. and Gazquez, J. L., *J. Am. Chem. Soc.*, 116, 9298, 1994.

44. Gazquez, J. L., *J. Phys. Chem. A*, 101, 9464, 1997.

45. Gazquez, J. L., *J. Phys. Chem. A*, 101, 8967, 1997.

46. Gazquez. J. L., Martinez, A., and Mendez, F., *J. Phys. Chem.*, 97, 4059, 1993.

47. Gazquez. J. L. and Mendez, F., *J. Phys. Chem.*, 98, 4591, 1994.

48. Damoun, S., Van de Woude, G., Choho, K., and Geerlings, P., *J. Phys. Chem. A*, 103, 7861, 1999.

49. Mendez, F., Tamariz, J., and Geerlings, P., *J. Phys. Chem. A*, 102, 6292, 1998.

50. Pal, S. and Chandrakumar, K. R. S., *J. Am. Chem. Soc.*, 122, 4145, 2000.

51. Chandrakumar, K. R. S. and Pal, S., *Colloids Surf. A*, 205, 127, 2002.

52. Chandrakumar, K. R. S. and Pal, S., *J. Phys. Chem. A*, 106, 11775, 2002.
53. Chandrakumar, K. R. S. and Pal, S., *Int. J. Mol. Sci.*, 3, 324, 2002.
54. Chandrakumar, K. R. S. and Pal, S., *J. Phys. Chem. A*, 105, 4541, 2001.
55. Chandrakumar, K. R. S. and Pal, S., *J. Phys. Chem. B*, 106, 5737, 2002.
56. Krishnamurty, S. and Pal, S., *J. Phys. Chem. A*, 104, 7639, 2000.
57. Kar, R., Chandrakumar, K. R. S., and Pal, S., *J. Phys. Chem. A*, 111, 375, 2007.
58. Schmidt, M. W., Baldridge, K. K., Boatz, J. A., Elbert, S. T., Gordon, M. S., Jensen, J. H., Koseki, S., Matsunga, N., Nguyen, K. A., Su, S. J., Windus, T. L., Dupuis, M., and Montgomery, J. A. (Department of Chemistry, North Dakota State University and Ames Laboratory, Iowa State University), GAMESS, General Atomic and Molecular Electronic Structure System, *J. Comput. Chem.*, 14, 1347, 1993.

20 First-Principles Design of Complex Chemical Hydrides as Hydrogen Storage Materials

S. Bhattacharya and G. P. Das

CONTENTS

20.1 RENEWABLE ENERGY

Energy is undoubtedly the most critical issue being faced by humanity for their very survival on Earth. Although there are various sources of energy that are available, starting from the Sun, which is a very large source, humans tend to use mostly fossil fuels to satisfy their need as well as their greed. However, the fossil fuel reserve on Earth is finite, and with our increasing population and materialistic needs, the intense use of fossil hydrocarbons causes some disproportionate increase of CO_2 in the atmosphere, thereby causing global warming with disastrous environmental consequences. So no matter how long our fossil fuel reserve is going to last, it is a forgone conclusion that we do need to switch over to renewable energy sources, such as solar, hydroelectric, geothermal, tidal wave, wind, and bioenergy, and thereby reduce our dependence on fossil fuels [1]. However, it is easier said than done because of the simple reason that the energy contents per kilogram or liter of fossil fuels are typically more than the other sources; in addition, it is easier to simply dig out coal or petroleum that Mother Nature has taken millions of years to generate. This makes fossil fuel the most natural choice for domestic, commercial, automobile, aviation, and other sectors. Also, there are other issues that have got to do with global politics rather than science and technology. To exploit the solar radiation incident on Earth (~85,000 TW), which is several orders of magnitude more than the entire energy need of the world (~15 TW), one needs to have suitable material that can store the energy and make it available for usage as and when needed. Extensive efforts are on to develop suitable materials for solar photovoltaic and solar thermal systems that can maximize efficiency as well as cost effectiveness [2].

Energy production, energy storage, and energy transportation are three issues that require multidisciplinary efforts. Hydrogen, which is the lightest element in the periodic table and the most abundant element in the universe, exhibits the highest energy density per unit mass (heating value) of all chemical fuels [3,4]. Furthermore, hydrogen can be produced by electrolysis from renewable energy sources and is therefore regenerative. It produces only water when burnt and is therefore environment-friendly. Hydrogen is not a "source" of energy but a "carrier" of energy, and it does not occur naturally as a gas on Earth. The primitive phase diagram of hydrogen shows that the critical point lies at a very low temperature of 33 K, implying that free H_2 molecules experience strong repulsive interaction; for hydrogen storage, one needs to somehow reduce this via interaction with another suitable material that has affinity for it. In fact, there are some metal hydrides that can store hydrogen with a density higher than that of liquid hydrogen. It is exciting as well as challenging to probe the possibility of storing hydrogen in a more compact and safer way compared with pressurized gas and cryogenic liquid [5]. Efficient hydrogen storage is a challenge faced by the materials scientists who have been trying out various kinds of materials, from bulk materials to nanomaterials, that show high volumetric density of H-atoms present in the host lattice. High storage capacity, satisfactory kinetics, and optimal thermodynamics are some of the essential criteria for a potential hydrogen storage material.

20.2 MATERIALS FOR HYDROGEN STORAGE

There are various ways for efficient storage of molecular hydrogen in solid state [6,7], for which a fundamental understanding of how hydrogen interacts with materials is of utmost importance [4]. Most of the metals in the periodic table, their alloys, or intermetallic compounds react with hydrogen to form metal hydrides. Hydrogen tends to go into the metal lattice as an octahedral/tetrahedral interstitial and form $M + (x/2)H_2 \leftrightarrow MH_x$ (x can be integral or nonintegral for stoichiometric and off-stoichiometric hydrides) by hybridizing with the metal band. The bonding between hydrogen and the metal can range from very covalent to very ionic as well as multicentered bonds and metallic bonding. Some elemental metals, such as Mg, Al, Ti, and Pd, show special affinity for hydrogen, whereas some, such as Pt, Ru, and Ni, act as good hydrogenation catalysts. A classic textbook example is palladium hydride (PdH_x) that can retain a substantial quantity of hydrogen within its crystal lattice. At room temperature and atmospheric pressure, palladium can adsorb up to 900 times its own volume of hydrogen in a reversible process. However, Pd is a heavy metal and hence does not yield good gravimetric efficiency, apart from the fact that it is quite costly. Intermetallic hydrides AB_xH_n are formed with A, an alkaline earth (AE) or a rare earth (RE) metal acting as a H-adsorber, and B, a transition metal (TM) acting as a H-activator. Appropriate A–B combination allows tailoring of hydride properties with hydrogen landing in the interstitial space as $H°$ neutral (e.g., $FeTiH_x$ and

$LaNi_5H_6$). Perovskite hydrides of ABH_3 structure (A is a monovalent alkali metal like K, Sr, Cs, and Rb, whereas B is a divalent alkaline earth metal like Ca and Mg), particularly Mg-based compounds, receive particular attention because of their lightweight characteristic and low-cost production. There are plenty of open areas of research that are yet to be explored both experimentally and theoretically for perovskite hydrides and their possible application in hydrogen storage [8].

Now, we come to complex hydrides formed by a combination of metals or metalloids (e.g., imides, alanates, borates, borohydrides, and aminoboranes) of low-Z elements, where the basic interaction tends to have partially ionic, partially covalent character that can be tuned. The hydrogen atoms are bonded covalently to a metal or metalloid atom to form a complex anion such as $(NH_2)^-$, $(BH_4)^-$, and $(AlH_4)^-$. This anion is then bonded ionically to the M-cation present to form a complex metal hydride [9]. In general, complex metal hydrides have the formula $A_xM_yH_z$, where A is an alkali metal or alkaline earth metal cation or cation complex and M is a metal or metalloid. Well-known examples feature anions of hydrogenated group 3 elements, particularly boron and aluminum. Compounds such as $LiBH_4$ (lithium borohydride) and $NaAlH_4$ (sodium alanate) are among the most widely studied. The variety in complex metal hydrides is very large. The possibility of forming complex metal hydrides using lightweight elements opens a promising route to achieve very high hydrogen content by weight; for example, $LiBH_4$ contains 18 wt.% hydrogen. Accordingly, there is an increasing interest to explore complex metal hydride systems and their subsequent optimization for practical use. Combining several complex hydrides into one storage system might improve the storage characteristics, but the complexity of reaction mechanisms requires further fundamental research on such materials.

Apart from complex hydrides, there are other kinds of novel materials that have been investigated, for example, carbon-based materials activated with nanocatalysts [10], clathrate hydrates [11], metal-organic complexes [12], and more recently, nanostructured cages, viz., fullerenes and nanotubes, decorated with simple or transition metals that serve to attract hydrogen in molecular form [13–17]. Nanostructure materials built from lightweight elements, such as boron, carbon, and nitrogen, have several attractive features, viz., large surface area, low density, and high structural stability, which can be exploited for efficient storage of hydrogen. The storage takes place as hydrogen molecules are adsorbed on the surface of the solids. The possibility of storing hydrogen in molecular form is advantageous over chemical storage in atomic form, which requires the dissociation of the hydrogen bond and the formation of a hydride. To understand and exploit these materials for H-storage, it is crucial to know the way hydrogen interacts with the surface or the bulk. There are mainly three ways in which hydrogen can be adsorbed on a material [4]:

1. Physisorption, where hydrogen remains in the molecular state (H_2) and gets bound on the surface rather weakly (BE approximately 10–100 meV)
2. Chemisorption, where H_2 dissociates into H atoms that migrate and gets strongly bound to the material (BE approximately 2–4 eV range)
3. Molecular chemisorption, where H–H bonding gets weakened but not broken (still H_2 molecular state is retained) and the strength of the binding is intermediate between physisorption and chemisorption

It is this third form of the quasimolecular bonding that is most suitable for optimal absorption and desorption of hydrogen. The basic quasimolecular interaction and bonding of hydrogen can be explained via what is known as Kubas interaction [18], that is, donation of charge from H_2 molecule to the unfilled d-orbitals of the TM atoms and back-donation from the TM atom to the antibonding orbital of H_2 molecule. Kubas interaction has been exploited for designing transition metal decorated nanomaterials [19,20], metal organic frameworks (MOFs) [21,22], spillover catalysts [23], and other kinds of materials for H-storage. However, it has limitations in explaining the bonding in alkaline earth metal complexes and alkali metal-doped nanostructures [24]. In the case of metal clusters containing a few atoms, the way hydrogen interacts is fundamentally different from bulk, and the reactivity and adsorption behavior change drastically with the addition and subtraction of a few metal atoms [25,26].

20.3 COMPUTATIONAL APPROACH

First-principles computational approach plays a crucial role in predicting the H_2 adsorption and desorption processes in complex hydrides and also their decomposition pathways. Calculations are performed within the so-called density functional theory (DFT) [27,28], where the exchange correlation potential is treated via some mean field approximation, and the problem of solving an inhomogeneous many-electron system is reduced to that of solving an effective one-electron Schrödinger equation with an effective potential. Such an effective single-particle approach has been embraced by materials scientists mainly because it provides a reliable computational tool yielding material-specific quantitative results with desirable accuracy for the ground state (cohesive, electronic, magnetic, etc.) properties of a large variety of systems. First-principles DFT calculations, based on local density approximation (LDA) and its improved variants such as generalized gradient approximation (GGA), have reached an unprecedented level of accuracy and reliability such that one not only can explain but also can predict material properties and phenomena [29].

The most widely used first-principles electronic structure method for materials with fixed geometry is based on either plane wave–based methods or linear methods or localized basis set methods. From the ground state total energy, one can estimate the force acting on the atoms, which is essential to do *ab initio* molecular dynamics simulations à la recipe proposed by Car and Parrinello within the DFT framework. Such dynamical simulation enables one to determine the so-called "energy landscape," that is, how the energy of a system evolves with the position of the atoms, to monitor the making and breaking of chemical bonds, for example, desorption of hydrogen molecule from a nanohost material as a function of temperature.

For the present studies, we have used state-of-the-art DFT-based methods with plane wave basis set, viz., VASP [30] with PAW potentials [31] for extended systems, and with localized atomic orbital or Gaussian basis set, viz., DMol3 [32] or GAUSSIAN03 [33] for molecular or cluster systems. In all our calculations, the ions are steadily relaxed toward equilibrium until the Hellmann–Feynman forces are converged to less than 10^{-3} eV/Å. Available experimental structural data have been used as input for some of the hydrides whenever they are available.

The calculations that we have carried out for studying the above-mentioned materials and phenomena can be broadly classified into the following three categories: (1) ground state geometry, electronic structure, and activation barrier estimation using the so-called nudge elastic band method [34] for different possible configurations; (2) transition state calculations and reaction pathways; and (3) *ab initio* molecular dynamics with Nose thermostat for estimating the desorption kinetics. There are a number of detailed investigations reported in the literature [35–37] where different kinds of complex hydrides have been treated using DFT at various levels of sophistication.

20.4   COMPLEX HYDRIDE: CASE STUDY OF LITHIUM IMIDE

Complex hydrides involving light metals show impressive gravimetric efficiencies, but the desorption temperature of H_2 is rather high. For example, amides and imides of low-Z alkali metals such as Li and Na are prospective candidates for hydrogen storage with ~7% to 10% gravimetric efficiency. Here, the dehydriding reaction takes place in one or more steps at varying desorption temperatures depending on the kinetic reaction barrier. It was demonstrated by Chen et al. [38] how lithium amide ($LiNH_2$) reacts with lithium hydride (LiH) to yield lithium imide (Li_2NH) or lithium nitride (Li_3N) and molecular hydrogen. The forward reaction results in desorption, whereas the reverse reaction results in absorption. The total reaction is a two-step reaction process (Figure 20.1) as follows:

Step 1: $Li_3N + H_2 \leftrightarrow Li_2NH + LiH$ (20.1)

Step 2: $Li_2NH + H_2 \leftrightarrow LiNH_2 + LiH$ (20.2)

Total reaction: $Li_3N + 2H_2 \leftrightarrow Li_2NH + LiH + H_2 \leftrightarrow LiNH_2 + 2LiH.$ (20.3)

The reaction is exothermic with $\Delta H \sim -96$ kJ/mol H_2, whereas the gravimetric efficiency turns out to be ~10 wt.%. This dehydrogenation reaction leading to

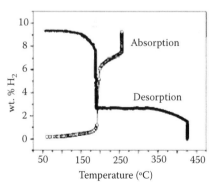

TG results of a Li_3N sample

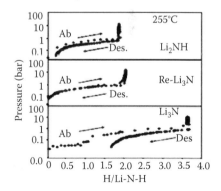

P-C-T curves of Li_3N and Li_2NH

FIGURE 20.1 Absorption and desorption of Li amide and Li imide undergoing reversible reaction to produce H_2. (Reproduced from Chen, P. et al., *Nature*, 420, 302, 2002. With permission.)

release of hydrogen is reversible, which is an additional attractive feature, although this reversibility is not always guaranteed, as will be discussed in this article.

Lithium atoms are ionized as Li^+ cations, whereas $[NH_2]^-$ forms a complex anion, and it is the strength of the interaction between Li^+ and $[NH_2]^-$ that dictates the enthalpy of reactions and hence the desorption kinetics of H_2. One way to do this is to alloy the binary hydride with some divalent alkaline earth metal such as Ca or Mg (Figure 20.2a) [9], and these ternary complex hydrides have been synthesized in recent years [39]. For pure lithium imide, the orthorhombic *Ima2* structure (Figure 20.2a) is found to have lower ground state energy (by ~0.6 eV) than that of the cubic structure (F-43m). The Ca-doped system, viz., $Li_2Ca(NH)_2$, has a trigonal structure space group *P-3m1*, with H occupying different possible 2d or 6i positions (Table 20.1) and three possible orientations of the N–H bond [40] The Li atoms in the 2d site occupy the tetrahedral hole created by the N-lattice (Figure 20.2b), whereas Ca atoms in the 1b site occupy the trigonal prismatic hole created by the N-lattice (Figure 20.2c). Li–Ca separation (≈3.077 Å) is ≈25% elongated as compared with the Li–Li bond length, which is very similar to the asymmetric Li–Li bond lengths in Li_3N. The most crucial, however, is the N–H bond where hydrogen can partially or randomly occupy the three possible positions with one-third probability of occupancy at any

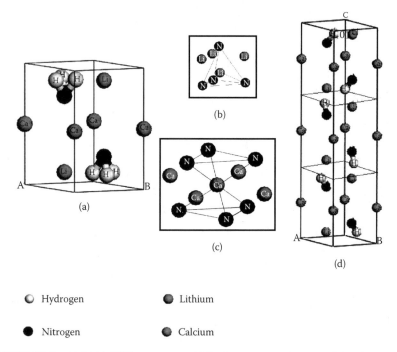

○ Hydrogen   ● Lithium

● Nitrogen   ● Calcium

FIGURE 20.2   (a) $Li_2Ca(NH)_2$ structure with hydrogen occupying any of the three equivalent positions above and below the respective nitrogen atoms. (b) Tetrahedral hole created by the N-lattice that is occupied by Li → 2d site (⅓, ⅔, 0.8841). d(Li–N) = 2.223 Å ×3, d(Li–Li) = 2.479 Å. (c) Octahedral hole created by the N-lattice that is occupied by Ca → 1b site (0, 0, ½). d(Ca–N) = 2.518 Å ×6, d(Ca–Ca) = 3.566 Å. (d) Supercell constructed by repeating the unit cell along the c-axis with different possible N–H bond orientations.

TABLE 20.1

Estimated Bond Lengths, Reaction Enthalpies, and Hydrogen Removal Energies for Undoped and Doped Li Imides

System	Structure Space Group (Formula Unit)	Chemical Reaction	Reaction Enthalpy, ΔH (kJ/ mol H$_2$)	ΔE_H (kJ/ mol H$_2$)	Average N-H Bond Length (Å)
LiNH$_2$	Tetragonal I-4 (4)	LiNH$_2$ + LiH $\leftrightarrows$ Li$_2$NH + H$_2$	68.9	268	1.03
Li$_2$NH	Orthorhombic $Ima2$ (8)	Li$_2$NH + LiH $\leftrightarrows$ Li$_3$N + H$_2$	108.8	192	1.04
Li$_2$Ca(NH)$_2$	Trigonal P-$3m1$ (3)	3Li$_2$Ca(NH)$_2$ + 2LiH $\leftrightarrows$ 4Li$_2$NH + Ca$_3$N$_2$ + 2H$_2$	102.6	181	1.05
Li$_2$Mg(NH)$_2$	Orthorhombic $Iba2$ (16)	3Li$_2$Mg(NH)$_2$ + 2LiH $\leftrightarrows$ 4Li$_2$NH + Mg$_3$N$_2$ + 2H$_2$	82.8	183	1.05

Source: Bhattacharya, S. et al., *J. Phys. Chem. B*, 112, 11381, 2008. With permission.

instant of time. To mimic this quantum delocalization effect, we have prescribed a supercell by repeating the unit cell three times along the c-axis and considered all these three possible H-positions built in the first, second, and third one-third of the super cell (Figure 20.2d). We have estimated the total energy for all the three different configurations designated as 6i (set-1), 6i (set-2), and 6i (set-3) and found that configuration 6i (set-2) is energetically most favorable.

The electronic structures of each of the constituents of Equations 20.1 through 20.3 and their Ca-doped counterparts as well as the heats of reactions (exothermic) have been estimated from first-principles density functional calculations [40]. The resulting total as well as partial densities of states (Figure 20.3) clearly show a semiconducting behavior with a GGA band gap of ~2.3 eV and a two-humped structure of the occupied part of electronic density of states (DOS) (Figure 20.3). N-2p bands predominantly contribute to the occupied DOS, as expected, whereas H-s character prevails in the lowest occupied band. Li behaves like a cation transferring its electron to [NH]$^{2-}$. The parent compound Li$_2$CaN$_2$ of the ternary hydride Li$_2$Ca(NH)$_2$ turns out to be metallic with the Fermi level lying near the antibonding peak. However, on introduction of hydrogen in this anti-La$_2$O$_3$ structure, this peak is pushed down below the Fermi level, thereby opening a band gap of ~2.3 eV, which is very similar to that of the pure imide. The Ca-4s band has a dominant contribution to the DOS and also affects the bonding between the Li$^+$ cation and the [NH]$^{2-}$ anion. The lower bonding peak arises because of strong hybridization between H-s and N-p orbitals. The upper bonding peaks arise out of Li–N interactions.

The average N–H bond lengths, hydrogen removal energies, and the enthalpy of formation of Li$_2$Ca(NH)$_2$ and Li$_2$Mg(NH)$_2$ have been estimated, and the results have been compared with the same quantities estimated for the pure Li imides and amides. The enthalpy of formation is the most fundamental and important quantity for hydrogen storage materials, which can be estimated from the difference between

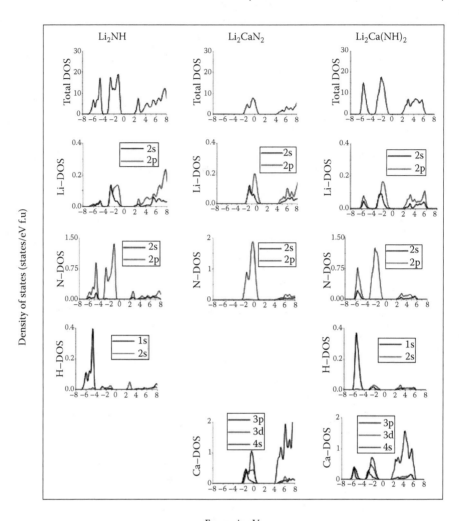

Energy in eV

FIGURE 20.3   Total and partial electronic densities of states calculated for Li_2NH, Li_2CaN_2, and $Li_2Ca(NH)_2$.

the energies before and after hydriding reaction (Equation 20.4). The enthalpy change in a reaction at 0 K was calculated using

$$\Delta H = \sum E_{products} - \sum E_{reactants} \qquad (20.4)$$

where E is the total energy of one of the bulk structures of interest as calculated by DFT. We investigate the thermodynamics of hydrogen release from the mixture of $Li_2Ca(NH)_2$ and LiH, which allows us to draw comparisons with the thermodynamics of hydrogen release from the other Li–N–based compounds, viz., parent imides and amides, along with $Li_2Ca(NH)_2$ with LiH. Table 20.1 summarizes our results for the specific exothermic chemical reactions that take place for H_2 desorption in different binary and ternary hydrides. We observe that ΔH decreases from 108.8 kJ/mol (1.13 eV) H_2 in Li-imide to 102.6 kJ/mol (1.06 eV) H_2 and to 82.8 kJ/mol (0.86 eV) H_2 for Ca and Mg ternary imides, respectively (Figures 20.4 and 20.5). The corresponding ΔH value estimated by Araujo et al. [41] is 118 kJ/mol H_2 for Li_2NH assuming *Pnma* space group and is 84 kJ/mol H_2 for $Li_2Mg(NH)_2$. It is interesting to note that the N–H bond lengths increase on ternary addition, indicating the weakening of the N–H bonds. Hydrogen removal energy ΔE_H for $Li_2Ca(NH)_2$ has been calculated using the relation

$$\Delta E_H[Li_2Ca(NH)_2] = E_T[Li_6Ca_3N_6H_5] + \frac{1}{2}E_T[H_2] - E_T[Li_6Ca_3N_6H_6], \quad (20.5)$$

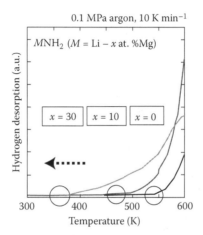

FIGURE 20.4 Hydrogen desorption reaction of lithium imide as a function of ternary addition ($x\%$ Mg). For $x = 30$, the desorption starts at around 370 K. (Reproduced from Orimo, S.-I. et al., *Chem. Rev.*, 107, 4111, 2007. With permission.)

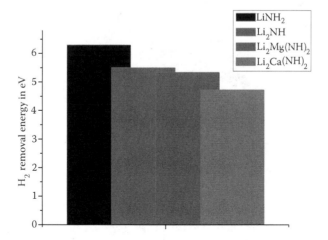

FIGURE 20.5 Estimated hydrogen removal energies for undoped and doped Li imides.

where $E_T[Li_6Ca_3N_6H_5]$, $E_T[Li_6Ca_3N_6H_6]$, and $E_T[H_2]$ are the ground state total energies of $Li_6Ca_3N_6H_5$, $Li_6Ca_3N_6H_6$ cell, and H_2 molecule in the gas phase, respectively. For $Li_2Ca(NH)_2$, we have used the optimized structure shown in Figure 20.2. It is interesting to note that ΔE_H reduces by ~5.5% for the ternary Ca imide. The enthalpy of reaction $\Delta H = T\Delta S$ for pure lithium imide decreases on ternary addition. Assuming the entropy change ΔS to remain more or less constant during the reactions, the dehydrogenation temperature T is expected to come down to a desirable range. The H removal energy correspondingly decreases by about 5.5% with a concomitant increase in the N–H bond length by about 0.01 Å for the ternary Ca imide system.

20.5 CHEMICAL HYDRIDE: CASE STUDY OF MONOAMMONIATED LITHIUM AMIDOBORANE

Ammonia borane (NH_3BH_3), AB for short, complexes have emerged as attractive candidates for solid-state hydrogen-storage materials because of their high percentage of available hydrogen (19.6 wt.%). However, relatively poor kinetics and high temperature of dehydrogenation as well as release of volatile contaminants, such as borazine, are posing big challenges for practical application of AB [42,43]. When one H atom in AB is replaced by an alkali or alkaline earth metal (M), a new class of materials called metal amidoboranes (MABs) is formed, which in turn can be used for efficient storage of molecular hydrogen [44,45]. These materials were highlighted as some of the best potential hydrogen storage materials in the 2008 Department of Energy (DOE) hydrogen program annual progress report. For example, LiAB and NaAB provide high storage capacity of 10.9 and 7.5 wt.%, respectively [46,47] at easily accessible temperatures without the unwanted release of borazine. $LiNH_2BH_3$ is environmentally harmless and stable in solid state at ambient temperature and normal pressure. The bonds get distorted as compared with those in pristine ammonia borane, as can be seen from the ball-and-stick model optimized geometries and corresponding bond lengths (Figure 20.6). However, to improve the operating properties of these materials, such as rapid H_2

FIGURE 20.6 Optimized structure of $LiNH_2BH_3$. Equilibrium bond distances estimated from first-principles calculations are B–N = 1.61 Å, Li–N = 1.85 Å, Li–B = 2.09 Å, B–H = 1.25 Å (av), and N–H = 1.01 Å (av).

release near room temperature, it is vital to understand the underlying mechanism for the release of H_2.

Recent experimental and computational studies have shown that NH_3 reacts with LiAB to yield H_2, and the dehydrogenation takes place in three different stages [48], each time resulting in an intermediate metastable product (adduct). The steps involved in the first dehydrogenation reactions along with the transition states and the intermediate products are shown in Figure 20.7 (similar things have been determined for the second and third dehydrogenation processes, which are not shown), whereas the estimated values from our first-principles calculations are summarized in Table 20.2. The first dehydrogenation from monomer occurs with an activation barrier of 78 kJ/mol (1 eV ≈ 100 kJ/mol) followed by H_2 removal energy of 0.16 eV/ H_2, leaving a metastable product $Li(NH_2)NH_2BH_2$. We have found a transition state (Figure 20.8) where the hydric B–H bond in the $[NH_2BH_3]$ unit interacts with the protic N–H bond of NH_3, which in turn leads to H_2 release from the system as a first dehydrogenation process. The reaction pathway having the minimum activation barrier has been estimated using transition state calculations and is shown schematically in Figure 20.8. Similarly, we have determined using our first-principles approach the second and third dehydrogenation processes that result in relatively high activation barriers and H_2 removal energies (shown in Table 20.2), whereas the metastable products left behind are $Li(NH)NH_2BH$ and $Li(NH)NBH$, respectively, with the final product matching with the available experimental results.

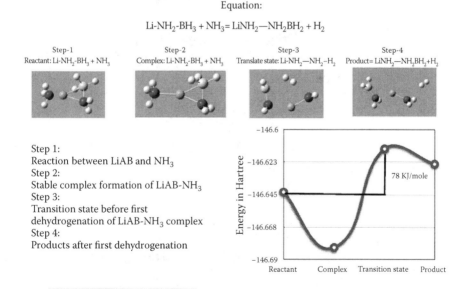

FIGURE 20.7 First dehydrogenation mechanism from LiAB monomer and NH_3 interaction and the corresponding reaction barrier.

TABLE 20.2

Activation Barriers and Hydrogen Removal Energies for First, Second, and Third Dehydrogenation Energies of Monoammoniated LiAB

Dehydrogenation	Activation Barrier (kJ/mol H_2)	H_2 Removal Energy (eV/H_2)
	Monomer of LiAB + NH_3	
First	78	0.16
Second	105	0.27
Third	353	1.3
	$[LiNH_2–BH_2–NH2]_3$	
Second [1]	No barrier	0.14
Second [2]	No barrier	0.20 (average ~ 0.25)
Second [3]	No barrier	0.40
Third [1]	230	0.14
Third [2]	236 (average ~ 236)	0.20 (average ~ 0.67)
Third [3]	243	0.40

Source: Bhattacharya, S. et al., *J. Phys. Chem. C*, 116, 8859, 2012. With permission.

The reactions are, however, not reversible. The H_2 removal energy defined as

$$E_{H_2\text{-removal}} = E_{total}[LiNH_2BH_2NH_2] - E_{total}[LiNH_2BHNH] - E_{total}[H_2] \quad (20.6)$$

is found to be 0.16, 0.27, and 1.30 eV/H_2 for the first, second, and third dehydrogenation processes, respectively. This increasing trend, especially the high activation barrier for the third dehydrogenation, rules out, in principle, the possibility of a spontaneous evolution of molecular hydrogen under normal conditions.

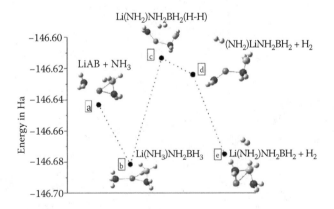

FIGURE 20.8 Three-stage dehydrogenation reaction path of NH_3 reacting with LiAB, with monomer. (From Bhattacharya, S. et al. *J. Phys. Chem. C*, 116, 8859, 2012. With permission.)

Based on these observations, we argued [48] that the LiAB dehydrogenation in presence of ammonia does not occur through single-stage reaction but possibly undergoes a combined reaction mechanism. We explored the possibility of forming a higher-order cluster, especially after the first dehydrogenation when the metastable product $LiNH_2$—BH_2NH_2 is reached. For the subsequent (i.e., after the first) dehydrogenation reactions, we have studied the stability of $[LiNH_2-BH_2-NH_2]_n$ clusters with n varying from 2 to 6. The stability of these clusters can be defined as

$$S = \frac{E_{total}[LiNH_2BH_2NH_2]_n - n \times E_{total}[LiNH_2-BH_2NH_2]_{metastable}}{n}. \quad (20.7)$$

The results (Figure 20.9) reveal that as the cluster size goes up from monomer to dimer to trimer ($n = 3$), the relative stability keeps on increasing (i.e., more negative) and tends to saturate for $n \geq 6$. The estimated activation barriers for three consecutive H_2 releases for the second dehydrogenation from the residual complex cluster ($n = 3$) are shown in Figure 20.10. The same trend has been observed for $n = 1 - 3$ in the case of the third dehydrogenation. The detailed reactions and their pathways are given in reference 48. It is this reduction in the activation barrier as a function of

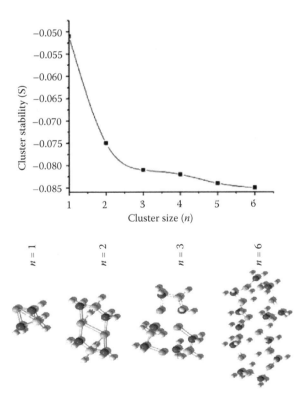

FIGURE 20.9   Relative stability of $[Li(NH_2)-NH_2BH_3]_n$ as a function of cluster size (n).

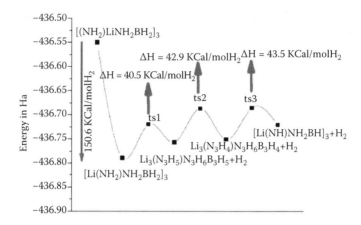

FIGURE 20.10 Reaction path for second dehydrogenation from the $[Li(NH_2)-NH_2BH_3]_3$ cluster.

increasing cluster size that provides an explanation for the dehydrogenation mechanism in the monoammoniated LiAB system.

20.6 CONCLUDING REMARKS

The first-principles density functional approach has been used to design efficient hydrogen storage materials, such as complex hydrides, viz., lithium imides, and chemical hydrides, viz., lithium amidoboranes, with improved dehydrogenation behavior at near ambient conditions. The enthalpy of reaction $\Delta H = T\Delta S$ for pure lithium imide decreases on ternary addition. Assuming the entropy change ΔS to remain more or less constant during the reactions, the dehydrogenation temperature T is expected to come down to a desirable range. The H removal energy correspondingly decreases by about 5.5% with a concomitant increase in the N–H bond length for the ternary Ca imide system. Another promising material for chemical storage of hydrogen is ammonia borane (NH_3–BH_3). Lithium amidoborane LiAB, in particular, provides high storage capacity ~10.9 wt.% of hydrogen at easily accessible temperature, without the release of any unwanted borazine. Our first-principles result suggests that this reaction is a three-step process, and each stage is combined with the evolution of hydrogen from the system. However, the first dehydrogenation occurs between the interactions of LiAB monomer and NH_3 molecule, and the second and third dehydrogenations are multicluster interactions.

ACKNOWLEDGMENTS

The authors have been collaborating on hydrogen energy with different groups in India and abroad. They would like to thank Chiranjib Majumder, Prasenjit Sen, Yuan Ping Feng, Ping Chen, Puru Jena, and Yoshiyuki Kawazoe for many useful discussions in the course of the work included in this article and Sonali Barman and Amrita Bhattacharya for their contributions.

REFERENCES

1. Smalley, R. E., *MRS Bull.*, 30, 412, 2005.
2. Abbott, D., *Proc. IEEE*, 98, 42, 2010.
3. Crabtree, G. W., Dresselhaus, M. S., and Buchanan, M. V., *Physics Today*, 57, 39, 2004. Crabtree, G. W. and Dresselhaus, M. S., *MRS Bull.*, 33, 421, 2008.
4. Jena, P., *J. Phys. Chem. Lett.*, 2, 206, 2011.
5. Schlapbach, L. and Zuttel, A., *Nature*, 414, 353, 2001. Zuttel, A., *Mater. Today*, 6, 24, 2003.
6. *Solid State Hydrogen Storage: Materials and Chemistry*, G. Walker, ed., Woodhead Publishing Ltd., United Kingdom, 2008.
7. Mandal, T. K. and Gregory, D. H., *Annu. Rep. Prog. Chem. A*, 105, 21, 2009.
8. Zuettel, A., *Naturwissenschaften*, 91, 157, 2004.
9. Orimo, S.-I., Nakamori, Y., Eliso, J. R., Zuettel, A., and Jensen, C. M., *Chem. Rev.*, 107, 4111, 2007.
10. Struzhkin, V. V., Militzer, B., Mao, W. K., Mao, H.-K., and Henley, R. J., *Chem. Rev.*, 107, 4133, 2007.
11. Panella, M. B., *Scripta Mater.*, 56, 809, 2007.
12. Dillon, A. C. and Heben, M. J., *Appl. Phys*, A72, 133, 2001.
13. Barman, S., Sen, P., and Das, G. P., *J. Phys. Chem. C*, 112, 19953, 2008.
14. Bhattacharya, S., Majumder, C., and Das, G. P., *J. Phys. Chem. C*, 112, 17487, 2008.
15. Bhattacharya, S., Majumder, C., and Das, G. P., *J. Phys. Chem. C*, 113, 15783, 2009.
16. Bhattacharya, S., Majumder, C., and Das, G. P., *Bull. Mater. Sci.*, 32, 353, 2009.
17. Bhattacharya, A., Bhattacharya, S., Majumder, C., and Das, G. P., *J. Phys. Chem. C*, 114, 10297, 2010.
18. Kubas, G. J., *Acc. Chem. Res.* 1988, 21, 120; *J. Organomet. Chem.*, 694, 2648, 2009.
19. Sun, Q., Wang, Q., Jena, P., and Kawazoe, Y., *J. Am. Chem. Soc.*, 127, 14582, 2005.
20. Chandrakumar, K. R. S. and Ghosh, S. K., *Nano Lett.*, 8, 13, 2008.
21. Hoang, T. K. A. and Antonelli, D. M., *Adv. Mater.*, 21, 1787, 2009.
22. Dixit, M., Maark, T. A., and Pal, S., *Int. J. Hydrogen Energy*, 36, 10816, 2011.
23. Singh, A. K., Ribas, M. A., and Yakobson, B. I., *ACS Nano*, 3, 1657, 2009.
24. Bhattacharya, S., Bhattacharya, A., and Das, G. P., *J. Phys. Chem. C*, 116, 3840, 2012.
25. Niu, J., Rao, B. K., and Jena, P., *Phys. Rev. Lett.*, 68, 2277, 1992.
26. Giri, S., Chakraborty, A., and Chattaraj, P. K., *J. Mol. Model*, 17, 777, 2011.
27. Hohenberg, P. and Kohn, W., *Phys. Rev.*, 136, B864, 1964. Kohn, W. and Sham, L. J., *Phys. Rev.*, 140, A1133, 1965.
28. Jones, R. O. and Gunnarson, O., *Rev. Mod. Phys.*, 61, 689, 1989.
29. Das, G. P., *Materials Research: Current Scenario and Future Projections*, Chidambaram, R. and Banerjee, S., eds., Allied Publisher, New Delhi, 2002, 634–669 and articles therein.
30. Hafner, J., *Comp. Phys. Commun.*, 177, 6, 2007. Kresse, G. and Hafner, J., *Phys. Rev. B*, 47, R6726, 1993. Kresse, G. and Furthmuller, J., *J. Comput. Mater. Sci.*, 6, 15, 1996.
31. Blöchl, P. E., *Phys. Rev. B*, 50, 17953, 1994. Kresse, G. and Joubert, J., *Phys. Rev. B*, 59, 1758, 1999.
32. Delley, B., *J. Chem. Phys.*, 113, 7756, 2000.
33. Frisch, M. J. et al., *GAUSSIAN03, Revision C.02*, Gaussian, Inc., Pittsburgh, PA, 2004.
34. Sheppard, D., Terrell, R., and Henkelman, G., *J. Chem. Phys.*, 128, 134106, 2008.
35. Hector, L. G. Jr. and Herbst, J. F., *J. Phys. Condens. Matter*, 20, 0642289, 2008.

36. Wolverton, C., Siegel, D. J., Akbarzadeh, A. R., and Ozolins, V., *J. Phys. Condens. Matter*, 20, 064228, 2008.
37. Vajeeston, P., Ravindran, P., and Fjellvag, H., *Materials*, 2, 2296, 2009.
38. Chen, P., Xiong, Z. T., Luo, J. Z. et al., *Nature*, 420, 302, 2002.
39. Wu, G., Xiong, Z., Liu, T., Liu, Y., Ju, J., Chen, P., Feng, Y., and Wee, A. T. S., *Inorg. Chem.*, 46, 517, 2007.
40. Bhattacharya, S., Wu, G., Chen, P., Feng, Y. P., and Das, G. P., *J. Phys. Chem. B*, 112, 11381, 2008.
41. Araujo, C. M., Scheicher, R. H., Jena, P., and Ahuja, R., *Appl. Phys. Lett.*, 91, 091924, 2007.
42. Staubitz, A., Robertson, A. P. M., and Manners, I., *Chem. Rev.*, 110, 4079, 2010.
43. Chua, Y. S., Chen, P., Wu, G., and Xiong, Z., *Chem. Commum.*, 47, 5116, 2011.
44. Wu, H., Zhou, W., and Yildirim, T., *J. Am. Chem. Soc.*, 130, 14834, 2008.
45. Shevlin, S. A., Kerkeni, B., and Guo, Z. X., *Phys. Chem. Chem. Phys.*, 13, 7649, 2011.
46. Xiong, Z., Yong, C. K., Wu, G., Chen, P., Shaw, W., Karmakar, A., Autrey, T., Jones, M. O., Johnson, S. R., Edwards, P. P., and David, W. I. F., *Nat. Mater.*, 7, 138, 2008.
47. Wu, C., Wu, G., Xiong, Z., Han, X., Chu, H., He, T., and Chen, P., *Chem. Mater.*, 22, 3, 2010.
48. Bhattacharya, S., Xiong, Z., Wu, G., Chen, P., Geng, Y. P., Majumder, C., and Das, G. P., *J. Phys. Chem. C*, 116, 8859, 2012.

21 | The Parameter I – A in Electronic Structure Theory*

Robert G. Parr and Rudolph Pariser

CONTENTS

PRELIMINARY REMARKS: Sixty years ago, Rudy Pariser and Bob Parr founded a new π-electron theory. It was ultimately called the Pariser–Parr–Pople (PPP) theory, in which there were parameters, including the difference between the ionization potential and electron affinity, $I – A$. Much followed.[1] When the editors of the Deb book invited Parr to contribute, he thought $I – A$ would make an interesting subject, and he secured the coauthorship of Pariser. He also shared his idea with one of the editors of this volume, Swapan Ghosh. To our surprise, Ghosh very soon produced a lucid and elegant first draft. This draft, in its entirety, with only minor changes by us, is reproduced as the full text below. This is an editor's major gift of authorship, of course, and we thank Ghosh profusely.

21.1 ON THE ROLE OF I – A IN CHEMISTRY

Chemistry is all about the transformation of molecular systems determined essentially by preferential holding/bonding together of a set of atoms/nuclei by an electron

---

* We are highly pleased to dedicate this work to Professor Deb. We would like to congratulate him for his career in teaching and his achievements, and would like to thank him for his kindnesses to many. Bob and Jane Parr have a treasured memory of a hillside picnic with the Deb family in which their young daughter delicately burst into an ethereal song.

glue. Thus, it basically involves a redistribution of electrons among a set of atomic centers, thus giving rise to the formation of new molecules dictated by the preferential tendency of electrons to be associated near selected regions/atoms.

The spatial distribution of the electron cloud in a molecular system is investigated quantitatively through the single-particle electron density,[2,3] which has served as a basic variable in the so-called density functional theory (DFT),[4–6] an approach that bypasses the many-electron wavefunction, which is the usual vehicle in conventional quantum chemistry or electronic structure theory (for a review of modern developments in quantum chemistry, see reference 7). The theoretical framework of DFT is well known for the associated conceptual simplicity as well as for the computational economy it offers. Another equally important aspect of DFT is its ability to rationalize the existing concepts in chemistry as well as to give birth to newer concepts, which has led to the important field of conceptual DFT.[5,6,8]

The simplicity of the DFT framework can be further enhanced by invoking a localized coarse-grained description, where the system is discussed in terms of partial atomic charges (and dipoles), for example. During chemical reactions, there may be some charge transfer between the reactant molecules (or between different parts of a reactant molecule, in case of intramolecular rearrangement reactions), which eventually again gets redistributed among the constituent atoms. Chemical reactivity deals essentially with the site-specific preference of accumulation or depletion of charges ultimately dictated by energetic considerations.

Electron transfer is governed, in global terms, by two important parameters of atomic or molecular systems, the concepts of which have been introduced, modified, redefined, and extended from time to time. These concepts of absolute electronegativity[9–11] and chemical hardness[12] form only a subset of the larger set of concepts that chemists have introduced over the years to understand, interpret, systematize, and rationalize chemical knowledge.

Electron transfer or redistribution is governed by the chemical potential of the electron cloud in different parts of a molecular system when reactants are brought close together or a molecule is subjected to an external field arising, for example, from interaction with another system. Although the chemical potentials dictate the direction of electron flow, the extent of charge transfer is controlled by another parameter, the chemical hardness. Both these concepts have been quantified through the theoretical framework of DFT.[13,14]

Focusing on an isolated single atom, the atomic parameter linked with the energetics of removal of an electron is the ionization potential (I), whereas that for the addition of an electron is the electron affinity (A). Therefore, these two parameters (either separately or jointly as a combination) should be instrumental in deciding the charge distribution/redistribution in molecular systems, during molecule formation and hence chemical binding, reactions, as well as reactivity. Although these two parameters are related to the fundamental charge transfer processes in molecular systems, it is interesting to note that the quantities obtained by their simple linear combinations, viz., $(I + A)$ and $(I - A)$, have played much higher important roles by serving the purpose of descriptors for chemical binding and reactivity as well as many other aspects of chemistry. The primary objective of this work is to highlight and discuss the role played by the particular combination, $(I - A)$, in various areas of the chemical world. This is, in fact, hardness.

21.2 REPRESENTING ENERGY DERIVATIVES

The concept of electronegativity has played a major role in the conceptual development of chemistry. From simple energetic considerations for the transfer of an electron between two dissimilar atoms X and Y, one can conclude that the energy associated with an electron transfer from X to Y and the reverse process Y to X will be equal if the condition $(I_X - A_Y) = (I_Y - A_X)$, and hence, $(I_X + A_X) = (I_Y + A_Y)$ is satisfied. This prompted the identification[10] of the electronegativity parameter χ, as $\chi = (I + A)/2$, which subsequently was translated[11] into the language of representing (within a finite difference approximation) the energy derivative with respect to the electron number, N, viz., $\chi = -(\partial E/\partial N)$. The same quantity has subsequently been interpreted[13] as the chemical potential μ, the Lagrange multiplier in an energy minimization procedure in DFT, viz.

$$\chi = -\left(\frac{\partial E}{\partial N}\right) = -\int d\mathbf{r}\, \frac{\delta E[\rho]}{\delta \rho(\mathbf{r})}\left(\frac{\partial \rho(\mathbf{r})}{\partial N}\right) = -\mu \int d\mathbf{r}\left(\frac{\partial \rho(\mathbf{r})}{\partial N}\right) = -\mu, \qquad (21.1)$$

where use has been made of the definition of the chemical potential μ as

$$\mu = \frac{\delta E[\rho]}{\delta \rho(\mathbf{r})} = v_{ext}(\mathbf{r}) + \frac{\delta F[\rho]}{\delta \rho(\mathbf{r})}. \qquad (21.2)$$

Here, the energy density functional $E[\rho]$ for a many-electron system characterized by an external potential $v_{ext}(\mathbf{r})$ has been expressed[6] as

$$E[\rho] = \int d\mathbf{r}\, v(\mathbf{r})\rho(\mathbf{r}) + F[\rho], \qquad (21.3)$$

consisting of the contributions from the external potential and the universal energy density functional $F[\rho]$, consisting of the kinetic, Coulomb, and exchange-correlation functionals.

In the same spirit, considering the transfer of an electron from an atom X to another identical atom X, the associated energy quantity is the energy difference, $(I_X - A_X)$, which should be a measure of the ease or difficulty of charge fluctuation between two identical atoms. This quantity, therefore, is interpreted as the measure of hardness/softness of an atomic species. The quantitative definition of the chemical hardness parameter η has thus been made $\eta = (I - A)/2$, which is interpreted[14] to represent (again, within a finite difference approximation) the second derivative of the energy quantity with respect to the electron number N, viz., $\eta = (1/2)(\partial^2 E/\partial N^2)$.

The definitions of χ and η, expressed as the first and second derivatives of energy, respectively, essentially correspond to the assumption of parabolic dependence of the energy $E(N)$ on N and are valid not only for atoms but also for other species, including molecules, clusters, and ions. Although the quantity $(I - A)$ can be obtained from evaluation or determination of I and A separately for a species, it is interesting to note that $(I - A)$ can also be evaluated directly by using the electron density, within a DFT framework, through its link with the concept of chemical hardness.

The density functional definition of the hardness parameter can in fact be obtained by introducing[15] the concept of local hardness, $\eta(\mathbf{r})$, through

$$\eta = \frac{1}{2}\left(\frac{\partial^2 E}{\partial N^2}\right) = \frac{1}{2}\int d\mathbf{r}\,\eta(\mathbf{r}) = \frac{1}{2}\int d\mathbf{r}\int d\mathbf{r}'\,\eta(\mathbf{r},\mathbf{r}')\left(\frac{\partial\rho(\mathbf{r}')}{\partial N}\right), \qquad (21.4)$$

where the hardness kernel $\eta(\mathbf{r},\mathbf{r}')$ is given[15] by

$$\eta(\mathbf{r},\mathbf{r}') = \frac{\delta^2 F[\rho]}{\delta\rho(\mathbf{r})\delta\rho(\mathbf{r}')}. \qquad (21.5)$$

The concept of softness has also been introduced, and the softness parameter S has been defined as the reciprocal of hardness, viz., $S = \eta^{-1} = 2(I - A)^{-1}$. The DFT-based definition for the softness parameter is also through the concept of local softness,[16] as given by

$$S = \int s(\mathbf{r})\,d\mathbf{r}; \quad s(\mathbf{r}) = \frac{\partial\rho(\mathbf{r})}{\partial\mu} = \frac{\partial\rho(\mathbf{r})}{\partial N}\frac{\partial N}{\partial\mu} = \frac{1}{S}f(\mathbf{r}), \qquad (21.6)$$

where the quantity $f(\mathbf{r}) = (\partial\rho(\mathbf{r})/\partial N)$ is the Fukui function, first introduced by Parr and Yang.[17] It is obvious that the local forms of hardness and softness are not reciprocal of each other; rather, the corresponding kernels $\eta(\mathbf{r},\mathbf{r}')$ and $s(\mathbf{r},\mathbf{r}')$ obey the relation[18]

$$\int \eta(\mathbf{r},\mathbf{r}')s(\mathbf{r}',\mathbf{r}'')\,d\mathbf{r}' = \delta(\mathbf{r} - \mathbf{r}''). \qquad (21.7)$$

These concepts of hardness and softness, defined originally in terms of the quantity $(I - A)$, have been extensively used in the theory of chemical binding and reactivity. The maximum hardness principle, as proven by Parr and Chattaraj[19] and also conjectured by Pearson, states that "molecules will arrange themselves to be as hard as possible." This is another triumph/victory of the quantity $(I - A)$ in chemistry. This principle has inspired a large number of investigations on the behavior of hardness along the reaction coordinate, during bond stretching, geometry variations, etc. Various correlations of the quantity $(I - A)$ with the size, bond length, polarizability, and other response properties have also been investigated. Other extensions through the Legendre transforms in DFT have also been proposed.[20] The chemical potential and the hardness have also been extended to represent the bond region[21,22] and generalized spin polarized version,[23-25] leading to the definition of spin-dependent reactivity parameters and a theory of covalent binding.[24,25]

Within the accepted approximations, the ionization potential is the energy of the highest occupied molecular orbital (HOMO), whereas the electron affinity is the energy of the lowest unoccupied molecular orbital (LUMO). Thus, by using Koopmans theorem, one has $\eta = (1/2)[\varepsilon_{HOMO} - \varepsilon_{LUMO}]$. The quantity $(I - A)$ thus represents the HOMO–LUMO gap or the band gap for extended systems and has been widely used in this context, thus enriching the area of solid-state chemistry and physics.

Among the other derivatives, mention may be made of the concept of Fukui function,[17] denoting the derivative of the electron density, viz., $f(\mathbf{r}) = (\partial\rho(\mathbf{r})/\partial N)$, which, within the finite difference approximation, represents the frontier orbitals—the HOMO and the LUMO. Nucleophilic and electrophilic attacks have thereby been rationalized in terms of these two approximations of the Fukui function.[17]

21.3 REPRESENTING MEASURE OF ELECTRON REPULSION

The earliest place where the quantity $(I - A)$ has made its appearance and played the role of a prominent character is the semiempirical electronic structure theory, which is known for its richness in scientific content, particularly for the physical basis in choosing the parameters and using the approximations. This theory provides[26] one of the simplest and transparent models through the semiempirical Hamiltonian as proposed by the Pariser–Parr–Pople (PPP) theory to approximate the total energy of a molecular system. In one of its simplest forms, the PPP Hamiltonian can be expressed[26] as

$$E = \frac{1}{2}\sum_{i,j} p_{ij}(H_{ij} + F_{ij})$$

$$F_{ii} = H_{ii}^{core} + \frac{1}{2}p_{ii}(ii,ii) + \sum_{i,j} p_{jj}(ii,jj), \qquad (21.8)$$

$$F_{ij} = H_{ij}^{core} - \frac{1}{2}p_{ij}(ii,jj)$$

where the symbols have their usual significance.

The PPP theory[27-30] has been widely applied to various systems of interest, successfully leading to fairly accurate description of electronic properties and the exact Hamiltonian of planar π-electron systems. One of the major ingredients of the PPP theory is the representation of the electron repulsion integral, which was first proposed to be approximated by Pariser and Parr[27-29] as the quantity $(I - A)$ and applied to the conjugated π-electron systems. Particularly significant applications of PPP theory are Pariser's calculations on polyacenes[31] and the single molecule azulene,[32] and his and Parr's treatment of ethylene-like molecules.[33]

The many-electron theory (e.g., Hartree–Fock–Roothan) of molecular systems within the LCAO approximation involves[26] the electron repulsion integrals of the type (ij,kl), which represent

$$(ij, kl) \equiv \left\langle \phi_i \phi_j \left| \frac{1}{r_{12}} \right| \phi_k \phi_l \right\rangle = \int d\mathbf{r}_1 \int d\mathbf{r}_2 \; \phi_i^*(1)\phi_j(1)\frac{1}{r_{12}}\phi_k^*(2)\phi_l(2), \qquad (21.9)$$

where $\phi_i(1)$ denotes the basis function centered at the ith atom and is a function of the coordinate of electron 1. Although, in the most general case, this integral is a four-center one, two-center integrals of the type (ii,kk) as well as one-center integrals (ii,ii) are often sufficient within suitable approximations. Although the evaluation using suitable forms of the basis function is a more elaborate and accurate method, here we are concerned with the semiempirical approach, in which the one-center integrals (ii,ii) have been estimated by an elegant procedure first proposed by Pariser and Parr.[27,28]

For two-carbon atoms separated at a distance, their π-electron energies can be assumed to be given by $2W_p$, where W_p is the ionization energy of the $2p_\pi$ electron. Now, considering the process of formation of C^+ and C^- ions through disproportionation, $2C \rightarrow C^+ + C^-$, that is, transfer of one electron from one carbon to the other carbon atom, one can easily see that the energy associated with the ions can be approximated as $2W_p + (ii,ii)$. The energy of formation of the ion pair is thus estimated to be given by $[2W_p + (ii,ii)] - 2W_p] = (ii,ii)$. If, on the other hand, one considers the same process to be the ionization of one carbon atom followed by the placement of the electron in the second carbon atom, the associated energy change is given by $(I - A)$. This provides an estimate of the intra-atomic electron repulsion integral as

$$(ii, ii) = I - A. \qquad (21.10)$$

Various estimates of the two quantities I and A were shown to provide a reasonable representation of the electron repulsion integral in a series of papers in the 1950s.

21.4 REPRESENTING MEASURE OF AROMATICITY

For conjugated π-electron systems, the quantity $(I - A)$ has been shown to be a measure of the electron repulsion integral as well as the band gap of the system. For cyclic conjugated π-electron systems, an important reactivity index is provided by the concept of aromaticity. Higher aromaticity is associated with higher stability, and as already discussed, the quantity $(I - A)$ is also associated with stability—the higher the gap, the higher the stability. Thus, $(I - A)$ has rightly been considered to be a measure of aromaticity. Zhou et al.[34] have in fact concluded that the HOMO–LUMO gap is a good indicator of the properties usually associated with aromatic character.

A review on aromaticity and conceptual DFT[35] has discussed various aspects of this issue.

Since $(I - A)$ is a measure of hardness according to the maximum hardness principle,[19] the stability of a system or the favorable direction of a physicochemical process is often dictated by this quantity. Because aromatic systems are much less reactive, especially toward addition reactions, $I - A$ may be considered to be a proper diagnostic of aromaticity. Moreover, $(I - A)$ has been used in different other contexts, such as stability of magic clusters, chemical periodicity, molecular vibrations and internal rotations, chemical reactions, electronic excitations, confinement, solvation, dynamics in the presence of external field, atomic and molecular collisions, toxicity and biological activity, chaotic ionization, and Woodward–Hoffmann rules.[36] The concept of absolute hardness as a unifying concept for identifying shells and subshells in nuclei, atoms, molecules, and metallic clusters has also been discussed by Parr and Zhou.[37]

21.5 REPRESENTING MEASURE OF HUBBARD *U* PARAMETER

An important member of the class of model Hamiltonian approaches is the Hubbard model, first introduced by Hubbard[38] to explain the physics of strongly correlated d-electron systems. It is the simplest among different models describing correlated electrons, and in its simplest form, the Hubbard Hamiltonian (one band model) can be represented[39] as

$$H = \sum_{i,j,\sigma} t_{ij} c_{i,\sigma}^{\dagger} c_{j,\sigma} + U \sum_{i} n_{i,\uparrow}^{\dagger} n_{i,\downarrow}, \qquad (21.11)$$

where t_{ij} and U are the two parameters of the model and $n_{i,\sigma} = c_{i,\sigma}^{\dagger} c_{i,\sigma}$ is the number operator in the second quantized notation in terms of the creation and destruction operators. Further simplification is possible from assuming the hopping matrix elements between the nearest neighbors to have a single value (t) and neglecting the others and also using strongly localized Wannier functions to evaluate the parameters through an integral in terms of these functions. The major concern here is about the Hubbard parameter U, representing the on-site Coulomb repulsion (correlation), expressed in a form analogous to Equation 21.9. Denoting the Coulomb-energy cost to place two electrons at the same site, it is usually obtained parametrically in terms of the quantity $(I - A)$. The relative magnitudes of the parameters t and U determine the interplay between the delocalization and the localization of the electrons.

21.6 REPRESENTING *U* PARAMETER IN LDA + *U* METHOD

In recent years, the DFT method involving the Kohn–Sham equation[4,6] with LDA (or better gradient corrected) forms for the exchange-correlation functional as given by the one-electron equation

$$\left[-\frac{1}{2}\nabla^2 + V_{\text{LDA}}(\mathbf{r};[\rho])\right]\phi_i(\mathbf{r}) = \varepsilon_{i,\text{LDA}}\phi_i(\mathbf{r});$$

$$\rho(\mathbf{r}) = \sum_i n_i \phi_i^*(\mathbf{r})\phi_i(\mathbf{r}); \qquad\qquad (21.12)$$

$$E_{\text{LDA}} = \sum_i \varepsilon_{i,\text{LDA}} - \frac{1}{2}\iint d\mathbf{r}\, d\mathbf{r}' \frac{1}{|\mathbf{r}-\mathbf{r}'|}\rho(\mathbf{r})\rho(\mathbf{r}') + E_{XC}[\rho] - \int d\mathbf{r}\rho(\mathbf{r})\frac{\delta E_{XC}[\rho]}{\delta\rho(\mathbf{r})}$$

represents the standard method for calculating the band structure of solids. Its inadequacy is, however, most prominent for strongly correlated many-electron systems, for which a hybrid of the one-electron Kohn–Sham–type Hamiltonian is obtained by augmenting it with a correction term representing correlation effects along the lines of the Hubbard U parameter. The motivation is provided by the important work of Perdew et al.,[40] demonstrating a discontinuity in the potential or orbital energy as the electron number is varied around an integer value.

The LDA energy is corrected along the lines of the model Hamiltonian approach by modifying the interaction energy between the electrons where Coulomb correlation is dominant. One expresses[41,42] the corrected energy to be given by

$$E = E_{\text{LDA}} - U\frac{N(N-1)}{2} + \frac{U}{2}\sum_{i\neq j} n_i n_j, \qquad\qquad (21.13)$$

where the LDA energy has been corrected by first subtracting the second term on the right, which approximately represents the overall Coulomb energy of the d-electrons, as determined by their number $N = \sum_i n_i$, and then adding the third term representing the Hubbard-like contribution determined by the d-orbital occupancies, n_i.

The orbital energies ε_i as obtained through the energy derivative are given by

$$\varepsilon_i = \frac{\partial E}{\partial n_i} = \varepsilon_{\text{LDA}} + U\left(\frac{1}{2} - n_i\right), \qquad\qquad (21.14)$$

from which it is clear that the LDA orbital energy is shifted by $-(U/2)$ for an occupied orbital ($n_i = 1$) and by $(U/2)$ for an unoccupied orbital ($n_i = 0$), thus opening a gap as a consequence of Coulomb correlation. Similarly, the LDA one-electron effective potential is corrected as

$$V_i(\mathbf{r}) = \frac{\delta E}{\delta\rho_i(\mathbf{r})} = V_{\text{LDA}}(\mathbf{r}) + U\left(\frac{1}{2} - n_i\right), \qquad\qquad (21.15)$$

where the discontinuity in the potential is quite evident. Thus, it is clear that the Coulomb term U, which essentially can be obtained from $(I - A)$, plays an important role in the modern LDA + U method[42] of electronic structure of solids.

21.7 CONCLUDING REMARKS

From the above, we see that the quantity $(I - A)$ represents a unique combination of I and A, which, although very simple in form, is rich in its scientific content and is useful for application in a number of areas in chemistry as well as physics. Although this number made its first appearance on stage through its role in electronic structure theory of molecular systems, providing a measure of the interatomic electron repulsion integral, as used in semiempirical molecular orbital theory, in particular, the well-known PPP theory, its reappearance among the solid-state community through the well-known U parameter of the Hubbard model has also been well received. It also acts in the LDA + U method for strongly correlated many-electron systems.

The representation of $(I - A)$ as the second derivative of the total energy with respect to the number of electrons and its identification as the hardness parameter in chemistry have led to another breakthrough. A rigorous calculation of the quantity through the DFT derivatives has also been possible. The horizon of $(I - A)$ has been further broadened through its link with other aspects of chemistry, such as aromaticity and various reactivity parameters. PPP, as it evolved, changed direction in the course of time under the influence of Ralph Pearson.

For molecules, I is very rarely equal to A, and that is what this extraordinary story of development within theoretical chemistry has been all about.

REFERENCES

1. Lowdin, P. O., Pariser, R., Parr, R. G., and Pople, J. A., *Int. J. Quantum Chem.*, 37, Number 4 (whole issue), 1990.
2. *The Single-Particle Density in Physics and Chemistry*, March, N. H. and Deb, B. M., eds., Academic Press, New York, 1987.
3. Bamzai, A. S. and Deb, B. M., *Rev. Mod. Phys.*, 53, 96, 593, 1981.
4. Hohenberg, P. and Kohn, W., *Phys. Rev. B*, 136, 864, 1964. Kohn, W. and Sham, L. J., *Phys. Rev. A*, 140, 1133, 1965.
5. Parr, R. G., *Annu. Rev. Phys. Chem.*, 34, 631, 1983. Kohn, W., Becke, A. D., and Parr, R. G., *J. Phys. Chem.*, 100, 12974, 1996.
6. Parr, R. G. and Yang, W. *Density-Functional Theory of Atoms and Molecules*, Oxford University Press, New York (Clarendon Press, Oxford), 1989.
7. *Reviews of Modern Quantum Chemistry: A Celebration of the Contributions of Robert G. Parr, Vol. 1 & 2*, Sen, K. D., ed., World Scientific, New Jersey, 2002.
8. Geerlings, P., DeProft, F., and Langenaeker, W., *Chem. Rev.*, 103, 1793–1873, 2003.
9. Pauling, L., *The Nature of the Chemical Bond*, 3rd ed., Cornell University Press, Ithaca, NY, 1960.
10. Mulliken, R. S., *J. Chem. Phys.*, 2, 782, 1934.
11. Iczkowski, R. P. and Margrave, J. L., *J. Am. Chem. Soc.*, 83, 3547, 1961.
12. Pearson, R. G., *J. Am. Chem. Soc.*, 85, 3533, 1963. Pearson, R. G., *Hard and Soft Acids and Bases*, Dowden, Hutchinson & Ross, Stroudsburg, PA, 1973. Pearson, R. G., *Chemical Hardness*, Wiley, New York, 1997.
13. Parr, R. G., Donnelly, R. A., Levy, M., and Palke, W. E., *J. Chem. Phys.*, 68, 3801, 1978.
14. Parr, R. G. and Pearson, R. G., *J. Am. Chem. Soc.*, 105, 7512, 1983.
15. Berkowitz, M., Ghosh, S. K., and Parr, R. G., *J. Am. Chem. Soc.*, 107, 6811, 1985. Ghosh, S. K. and Berkowitz, M., *J. Chem. Phys.*, 83, 2976, 1985. Ghosh, S. K., *Chem. Phys. Lett.*, 172, 77, 1990.

16. Yang, W. and Parr, R. G., *Proc. Natl. Acad. Sci. USA*, 82, 6723, 1985.
17. Parr, R. G. and Yang, W., *J. Am. Chem. Soc.*, 106, 4049, 1984. Parr, R. G. and Yang, W., *Annu. Rev. Phys. Chem.*, 46, 701, 1995.
18. Berkowitz, M. and Parr, R. G., *J. Chem. Phys.*, 88, 2554, 1988.
19. Parr, R. G. and Chattaraj, P. K., *J. Am. Chem. Soc.*, 113, 1854, 1991.
20. Nalewajski, R. F., *Int. J. Quantum Chem.*, 78, 168, 2000.
21. Ghosh, S. K. and Parr, R. G., *Theor. Chim. Acta*, 72, 379, 1987.
22. Ghanty, T. K. and Ghosh, S. K., *J. Phys. Chem.*, 95, 6512, 1991.
23. Galvan, M. and Vargas, R., *J. Phys. Chem.*, 96, 1625, 1992.
24. Ghosh, S. K., *Int. J. Quantum Chem.*, 49, 239, 1994.
25. Ghanty, T. K. and Ghosh, S. K., *J. Am. Chem. Soc.*, 116, 3943, 1994.
26. Parr, R. G., *The Quantum Theory of Molecular Electronic Structure*, W.A. Benjamin, Inc., New York, 1963.
27. Parr, R. G., *J. Chem. Phys.*, 20, 1499, 1952.
28. Pariser, R., *J. Chem. Phys.*, 21, 568, 1953.
29. Pariser, R. and Parr, R. G., *J. Chem. Phys.*, 21, 466, 1953.
30. Pople, J. A., *Trans. Faraday Soc.*, 49, 1375, 1953.
31. Pariser, R., *J. Chem. Phys.*, 24, 250, 1956.
32. Pariser, R., *J. Chem. Phys.*, 25, 1112, 1956.
33. Parr, R. G. and Pariser, R., *J. Chem. Phys.*, 23, 711, 1955.
34. Zhou, Z., Parr, R. G., and Garst, J. F., *Tetrahedron Lett.*, 29, 4843, 1988.
35. Chattaraj, P. K., Das, R., Duley, S., and Giri, S., *Chem. Modell.*, 8, 45–98, 2011.
36. Geerlings, P., Ayers, P. W., Toro-Labbé, A., Chattaraj, P. K., and De Proft, F., *Acc. Chem. Res.*, 45, 683, 2012.
37. Parr, R. G. and Zhou, Z., *Acc. Chem. Res.*, 26, 256, 1993.
38. Hubbard, J., *Proc. R. Soc. Lond. A*, 276, 238, 1963.
39. Albers, R. C., Christensen, N. E., and Svane, A., *J. Phys. Condens. Matter*, 21, 343201, 2009.
40. Perdew, J. P., Parr, R. G., Levy, M., and Balduz, J. L. Jr., *Phys. Rev. Lett.*, 49, 1691, 1982.
41. Anisimov, V. I., Solovyev, I. V., Korotin, M. A., Czyzyk, M. T., and Sawatzky, G. A., *Phys. Rev. B*, 48, 16929, 1993.
42. Anisimovy, V. I., Aryasetiawanz, F., and Lichtenstein, A. I., *J. Phys. Condens. Matter*, 9, 767, 1997.

22 Uncertainty and Entropy Properties for Coulomb and Simple Harmonic Potentials Modified by $\dfrac{ar^2}{1+br^2}$

S. H. Patil and K. D. Sen

CONTENTS

22.1 INTRODUCTION

Uncertainty relations are the basic properties of quantum mechanics; in particular, we have the Heisenberg uncertainty principle[1] for the product of the uncertainties in position and momentum

$$\sigma_x \sigma_p \geq \frac{1}{2}\hbar, \quad \sigma_x^2 = \left\langle \left(x - \langle x \rangle\right)^2 \right\rangle, \quad \sigma_p^2 = \left\langle \left(p_x - \langle p_x \rangle\right)^2 \right\rangle, \tag{22.1}$$

in terms of Planck's constant. The uncertainty product has many interesting properties for different potentials; for example, the product for bound states in homogeneous, power potentials is independent of the strength of the potentials.[2] There are many other interesting related properties such as entropy and information. Here, we will consider some general properties for the bound states in Coulomb and simple harmonic potentials with an additional nonpolynomial term. The modified harmonic potentials are commonly known as the nonpolynomial oscillator potentials (rational potentials) with applications in a variety of branches of physics.[3] It is observed[4] that the dimensionality and scaling properties lead to interesting properties of the uncertainty product and densities with implications for entropies and information.

22.2   HEISENBERG UNCERTAINTY RELATIONS

Here, we analyze some dimensionality properties and their implications for the uncertainty relations for the bound states in a power potential with an additional nonpolynomial term.

22.2.1   SUPERPOSITIONS OF POWER POTENTIALS

Consider a potential of the form

$$V(r) = Zr^n + \frac{ar^2}{1 + br^2} \tag{22.2}$$

where Z, a, b, and n are parameters (n may not be an integer) in which there are bound states for a particle of mass M. Specifically, we have

$$V_1(r) = -\frac{Z}{r} + \frac{ar^2}{1 + br^2} \tag{22.3}$$

for a modified Coulomb potential and

$$V_2(r) = Zr^2 + \frac{ar^2}{1 + br^2} \tag{22.4}$$

for a modified simple harmonic oscillator. The Schrödinger equation for the potential in Equation 22.2 is

$$-\frac{\hbar^2}{2M}\nabla^2\psi + \left[Zr^n + \frac{ar^2}{1 + br^2} \right]\psi = E\psi. \tag{22.5}$$

22.2.2 DIMENSIONALITY AND UNCERTAINTY RELATIONS

The basic dimensional parameters in our Schrödinger equation are $\hbar^2/M$, Z, a, and b. Of these,

$$s_1 = \frac{aM}{\hbar^2}\left(\frac{\hbar^2}{MZ}\right)^{4/(n+2)}, \quad s_2 = b\left(\frac{\hbar^2}{MZ}\right)^{2/(n+2)} \tag{22.6}$$

are the dimensionless parameters. Now, we consider the deviations

$$\sigma_{\vec{r}}^2 = \left\langle \left(\vec{r} - \langle\vec{r}\rangle\right)^2 \right\rangle, \quad \sigma_{\vec{p}}^2 = \left\langle \left(\vec{p} - \langle\vec{p}\rangle\right)^2 \right\rangle. \tag{22.7}$$

For our potential in Equation 22.5, the dimensionality properties imply that the deviations are of the form

$$\sigma_{\vec{r}} = \left(\hbar^2/MZ\right)^{1/(n+2)} g_1(s_1,s_2), \quad \sigma_{\vec{p}} = \hbar\left(MZ/\hbar^2\right)^{1/(n+2)} g_2(s_1,s_2), \tag{22.8}$$

so that the uncertainty product is

$$\sigma_{\vec{r}}\,\sigma_{\vec{p}} = \hbar g_1(s_1,s_2)g_2(s_1,s_2),$$

$$s_1 = \frac{aM}{\hbar^2}\left(\frac{\hbar^2}{MZ}\right)^{4/(n+2)}, \quad s_2 = b\left(\frac{\hbar^2}{MZ}\right)^{2/(n+2)}. \tag{22.9}$$

This implies that the uncertainty product depends only on the dimensionless parameters s_1 and s_2. Specifically, for the modified Coulomb potential $V_1(r)$ in Equation 22.3, it depends only on

$$s_1 = \frac{aM}{\hbar^2}\left(\frac{\hbar^2}{MZ}\right)^4, \quad s_2 = b\left(\frac{\hbar^2}{MZ}\right)^2, \tag{22.10}$$

and for the modified s.h.o. potential $V_2(r)$ in Equation 22.4, it depends only on

$$s_1 = a/Z, \quad s_2 = b\left(\frac{\hbar^2}{MZ}\right)^{1/2}. \tag{22.11}$$

It may also be noted that the bound state energies are of the form

$$E = (\hbar^2/M)^{n/(n+2)}Z^{2/(n+2)} g_3(s_1, s_2). \tag{22.12}$$

These results follow from just the dimensionality properties of the parameters.

22.3 SCALING PROPERTIES AND ENTROPIES

We will now consider some scaling properties for bound states in a power potential with an additional nonpolynomial term and their implications for a representative set of information measures.

22.3.1 SCALING PROPERTIES

For the Schrödinger equation in Equation 22.5, the energy E and eigenfunction ψ are functions of the form

$$E: E(\hbar^2/M, Z, a, b), \quad \psi: \psi(\hbar^2/M, Z, a, b). \tag{22.13}$$

Multiplying Equation 22.5 by $M/\hbar^2$ and introducing a scale transformation

$$\vec{r} = \lambda \vec{r}', \tag{22.14}$$

one gets

$$-\frac{1}{2}\nabla'^2\psi + (M/\hbar^2)[Z\lambda^{n+2}r'^n + a\lambda^4 r'^2(1 + b\lambda^2 r'^2)]\psi = (M/\hbar^2)\lambda^2 E\psi. \tag{22.15}$$

Taking

$$\lambda = \left(\frac{\hbar^2}{MZ}\right)^{1/(n+2)}, \tag{22.16}$$

it leads to

$$-\frac{1}{2}\nabla'^2\psi + \left[r'^n + \frac{Ma}{\hbar^2}\left(\frac{\hbar^2}{MZ}\right)^{4/(n+2)}\frac{r'^2}{1 + (\hbar^2/MZ)^{2/(n+2)}br'^2}\right]\psi = \frac{M}{\hbar^2}\left(\frac{\hbar^2}{MZ}\right)^{2/(n+2)}E\psi. \tag{22.17}$$

Comparing this with Equation 22.5, we obtain

$$E\left(\frac{\hbar^2}{M}, Z, a, b\right) = (\hbar^2/M)\lambda^{-2}E(1, 1, s_1, s_2), \quad \lambda = (\hbar^2/MZ)^{1/(n+2)}$$

$$\psi\left(\frac{\hbar^2}{M}, Z, a, b, r\right) = A\psi(1, 1, s_1, s_2, r'), \quad r = \lambda r', \tag{22.18}$$

$$s_1 = \frac{aM}{\hbar^2}\left(\frac{\hbar^2}{MZ}\right)^{4/(n+2)}, \quad s_2 = b\left(\frac{\hbar^2}{MZ}\right)^{2/(n+2)}.$$

Taking $\psi(1,1,s_1,s_2,r')$ to be normalized, the normalization of the wavefunction $\psi(\hbar^2/M,Z,a,b,r)$ leads to

$$1 = A^2 \int |\psi(1,1,s_1,s_2,r')|^2 \, d^3r,$$

$$= A^2 \lambda^3 \implies A = \lambda^{-3/2} = (MZ/\hbar^2)^{\frac{3}{2(n+2)}}, \qquad (22.19)$$

so that

$$\psi\left(\frac{\hbar^2}{M}, Z, a, b, r\right) = \lambda^{-3/2} \psi(1,1,s_1,s_2,r'), \quad r' = r/\lambda,$$

$$(22.20)$$

$$\lambda = (\hbar^2/MZ)^{1/(n+2)}, \quad s_1 = \frac{aM}{\hbar^2}\left(\frac{\hbar^2}{MZ}\right)^{4/(n+2)}, \quad s_2 = b\left(\frac{\hbar^2}{MZ}\right)^{2/(n+2)}.$$

For obtaining the wavefunction in the momentum space, we take the Fourier transform of the wavefunction in Equation 22.20, leading to

$$f(\hbar^2/M, Z, a, b, p) = \frac{1}{(2\pi\hbar)^{3/2}} \int d^3r \, e^{-\vec{p}\cdot\vec{r}/\hbar} \psi(\hbar^2/M, Z, a, b, r). \qquad (22.21)$$

Using the relation in Equation 22.20 and changing the integration variable to r', we get

$$f(\hbar^2/M, Z, a, b, p) = \lambda^{3/2} f(1,1,s_1,s_2,p'), \quad \lambda = (\hbar^2/MZ)^{1/(n+2)},$$

$$p' = \lambda p, \quad s_1 = \frac{aM}{\hbar^2}\left(\frac{\hbar^2}{MZ}\right)^{4/(n+2)}, \quad s_2 = b\left(\frac{\hbar^2}{MZ}\right)^{2/(n+2)}. \qquad (22.22)$$

From the relations in Equations 22.20 and 22.22, for the corresponding position and momentum densities, one obtains the following:

$$\rho(\hbar^2/M, Z, a, b, r) = \lambda^{-3}\rho(1,1,s_1,s_2,r'), \quad r' = r/\lambda,$$

$$\gamma(\hbar^2/M, Z, a, b, p) = \lambda^3 \gamma(1,1,s_1,s_2,p'), \quad p' = \lambda p, \qquad (22.23)$$

$$\lambda = (\hbar^2/MZ)^{1/(n+2)}, \quad s_1 = \frac{aM}{\hbar^2}\left(\frac{\hbar^2}{MZ}\right)^{4/(n+2)}, \quad s_2 = b\left(\frac{\hbar^2}{MZ}\right)^{2/(n+2)},$$

with s_1 and s_2 being the scaled parameters.

22.3.2 SHANNON ENTROPY SUM

The Shannon entropies[5-8] in the position space and momentum space are

$$S_r = -\int \rho(r)[\ln\rho(r)]d^3r, \quad S_p = -\int \gamma(p)[\ln\gamma(p)]d^3p. \tag{22.24}$$

Using the relations in Equation 22.23, for these entropies, we obtain

$$S_r(\hbar^2/M, Z, a, b) = 3\ln\lambda + S_r(1,1,s_1,s_2),$$

$$S_p(\hbar^2/M, Z, a, b) = -3\ln\lambda + S_p(1,1,s_1,s_2), \tag{22.25}$$

which imply that the Shannon entropy sum $S_T = S_r + S_p$ satisfies the relation

$$S_T(\hbar^2/M, Z, a, b) = S_T(1,1,s_1,s_2),$$

$$s_1 = \frac{aM}{\hbar^2}\left(\frac{\hbar^2}{MZ}\right)^{4/(n+2)}, \quad s_2 = b\left(\frac{\hbar^2}{MZ}\right)^{2/(n+2)}. \tag{22.26}$$

Therefore, for given values of the parameters Z, a, and b, the Shannon entropy sum depends only on the ratios $a/Z^{4/(n+2)}$ and $b/Z^{2/(n+2)}$. Specifically, for the potential $V_1(r)$ in Equation 22.2, it depends only on a/Z^4 and b/Z^2, and for the potential $V_2(r)$ in Equation 22.4, it depends only on a/Z and $b/Z^{1/2}$.

22.3.3 FISHER INFORMATION

The Fisher information measures[9-12] for position and momentum are

$$I_r = \int \frac{[\vec{\nabla}\rho(r)]^2}{\rho(r)}d^3r, \quad I_p = \int \frac{[\vec{\nabla}\gamma(p)]^2}{\gamma(p)}d^3p. \tag{22.27}$$

Using the relations in Equation 22.23, one obtains

$$I_r(\hbar^2/M, Z, a, b) = \frac{1}{\lambda^2}I_r(1,1,s_1,s_2), \quad I_p(\hbar^2/M, Z, a, b) = \lambda^2 I_p(1,1,s_1,s_2), \tag{22.28}$$

which together imply that the Fisher information product $I_r I_p$ satisfies the relation

$$I_{rp}(\hbar^2/M, Z, a, b) = I_{rp}(1,1,s_1,s_2), \quad I_{rp} = I_r I_p,$$

$$s_1 = \frac{aM}{\hbar^2}\left(\frac{\hbar^2}{MZ}\right)^{4/(n+2)}, \quad s_2 = b\left(\frac{\hbar^2}{MZ}\right)^{2/(n+2)}. \tag{22.29}$$

Here, for given values of the parameters Z, a, and b, the Fisher information product depends only on the ratios $a/Z^{4/(n+2)}$ and $b/Z^{2/(n+2)}$. Specifically, for the potential $V_1(r)$ in Equation 22.3, it depends only on a/Z^4 and b/Z^2, and for the potential $V_2(r)$ in Equation 22.4, it depends only on a/Z and $b/Z^{1/2}$.

22.3.4 RÉNYI ENTROPY

The Rényi entropies[13,14] in position and momentum spaces are

$$H_\alpha^{(r)} = \frac{1}{1-\alpha}\ln\int[\rho(r)]^\alpha d^3r, \quad H_\alpha^{(p)} = \frac{1}{1-\alpha}\ln\int[\gamma(p)]^\alpha d^3p. \tag{22.30}$$

With the relations in Equation 22.23, for these entropies, we obtain

$$
\begin{aligned}
H_\alpha^{(r)}(\hbar^2/M,Z,a,b) &= 3\ln\lambda + H_\alpha^{(r)}(1,1,s_1,s_2), \\
H_\alpha^{(p)}(\hbar^2/M,Z,a,b) &= -3\ln\lambda + H_\alpha^{(p)}(1,1,s_1,s_2),
\end{aligned}
\tag{22.31}
$$

which imply that the Rényi entropy sum $H_\alpha^{(T)} = H_\alpha^{(r)} + H_\alpha^{(p)}$ satisfies the relation

$$H_\alpha^{(T)}\left(\frac{\hbar^2}{M},Z,a,b\right) = H_\alpha^{(T)}(1,1,s_1,s_2),$$

$$s_1 = \frac{aM}{\hbar^2}\left(\frac{\hbar^2}{MZ}\right)^{4/(n+2)}, \quad s_2 = b\left(\frac{\hbar^2}{MZ}\right)^{2/(n+2)}. \tag{22.32}$$

Therefore, as in other cases, for given values of the parameters Z, a, and b, the Rényi entropy sum depends only on the ratios $a/Z^{4/(n+2)}$ and $b/Z^{2/(n+2)}$. In particular, for the potential $V_1(r)$ in Equation 22.3, it depends only on a/Z^4 and b/Z^2, and for the potential $V_2(r)$ in Equation 22.4, it depends only on a/Z and $b/Z^{1/2}$.

22.3.5 ONICESCU ENERGIES

The Onicescu energies[15] in position and momentum spaces are

$$E_r = \int[\rho(r)]^2 d^3r, \quad E_p = \int[\gamma(p)]^2 d^3p. \tag{22.33}$$

Using the relations in Equation 22.23, we get

$$E_r(\hbar^2/M,Z,a,b) = \frac{1}{\lambda^3}E_r(1,1,s_1,s_2), \quad E_p(\hbar^2/M,Z,a,b) = \lambda^3 E_p(1,1,s_1,s_2),$$

$$\tag{22.34}$$

which imply that the Onicescu energy product $E_{rp} = E_r E_p$ satisfies the relation

$$E_{rp}(\hbar^2/M,Z,a,b) = E_{rp}(1,1,s_1,s_2),$$

$$s_1 = \frac{aM}{\hbar^2}\left(\frac{\hbar^2}{MZ}\right)^{4/(n+2)}, \quad s_2 = b\left(\frac{\hbar^2}{MZ}\right)^{2/(n+2)}. \tag{22.35}$$

In this case also, for given values of the parameters Z, a, and b, the Onicescu energy product depends only on the ratios $a/Z^{4/(n+2)}$ and $b/Z^{2/(n+2)}$. In particular, for the potential $V_1(r)$ in Equation 22.3, it depends only on a/Z^4 and b/Z^2, and for the potential $V_2(r)$ in Equation 22.4, it depends only on a/Z and $b/Z^{1/2}$.

22.3.6 TSALLIS ENTROPY

The Tsallis entropies[16,17] in position and momentum spaces are

$$T_r = \frac{1}{q-1}\left[1 - \int [\rho(r)]^q d^3r\right], \quad T_p = \frac{1}{m-1}\left[1 - \int [\gamma(p)]^m d^3p\right], \quad \frac{1}{q} + \frac{1}{m} = 2.$$
$$\tag{22.36}$$

We consider the integral terms

$$J_r(\hbar^2/M,Z,a,b) = \int [\rho(r)]^q d^3r, \quad J_p(\hbar^2/M,Z,a,b) = \int [\gamma(p)]^m d^3p. \tag{22.37}$$

Using the relations in Equation 22.23, we get

$$J_r = (\hbar^2/M,z,a,b) = \lambda^{3-3q}J_r(1,1,s_1,s_2), \quad J_p = (\hbar^2/M,z,a,b) = \lambda^{3m-3}J_p(1,1,s_1,s_2). \tag{22.38}$$

Then, one obtains for the ratio

$$J_{p/r}(\hbar^2/M,Z,a,b) = J_{p/r}(1,1,s_1,s_2), \quad J_{p/r} = \frac{J_p^{1/2m}}{J_r^{1/2q}}, \quad \frac{1}{m} + \frac{1}{q} = 2,$$
$$\tag{22.39}$$

$$s_1 = \frac{aM}{\hbar^2}\left(\frac{\hbar^2}{MZ}\right)^{4/(n+2)}, \quad s_2 = b\left(\frac{\hbar^2}{MZ}\right)^{2/(n+2)}.$$

Therefore, in this case also, for given values of the parameters Z, a, and b, the ratio of Tsallis entropies depends only on the ratios $a/Z^{4/(n+2)}$ and $b/Z^{2/(n+2)}$. In particular, for the potential $V_1(r)$ in Equation 22.3, it depends only on a/Z^4 and b/Z^2, and for the potential $V_2(r)$ in Equation 22.4, it depends only on a/Z and $b/Z^{1/2}$.

22.4 STATISTICAL COMPLEXITY MEASURES

Finally, we consider the statistical complexity measure; that is, the LMC complexity measure is defined*[18,19] as

$$C_{\text{LMC}} = H_r.D_r, \tag{22.40}$$

where $H_r = -\int \rho(\vec{r})\log\rho(\vec{r})d\vec{r}$ and $D_r = \int \rho^2(\vec{r})d\vec{r}$ for the disequilibrium. We used the "exponential power Shannon entropy"[20] to work out the scaling properties of LMC statistical complexity:

$$H_r = \frac{1}{2\pi e}e^{\frac{2}{3}S_r}, \tag{22.41}$$

where S_r denotes the Shannon information entropy in the position space. For the applications of C_{LMC}, we refer to the published work in the literature.[21] Using Equations 22.25 and 22.34, it is clear that the LMC complexity measure in the position space itself obeys the scaling property obtained for the sum of Shannon entropy, product of Fisher measure, and other composite information theoretical measures. Similar property holds for the LMC complexity defined in the momentum space. In other words, for given values of the parameters Z, a, and b, the ratio of LMC complexity measure in either position or momentum space depends only on the ratios $a/Z^{4/(n+2)}$ and $b/Z^{2/(n+2)}$.

ACKNOWLEDGMENTS

It is a unique pleasure to join others to celebrate the contributions of Professor B. M. Deb, who introduced the density functional theory to the theoretical chemistry community in India. S. H. Patil acknowledges support from AICTE as an emeritus fellow. K. D. Sen is grateful to the Department of Science and Technology, New Delhi, for the award of the J. C. Bose National Fellowship.

REFERENCES

1. Heisenberg, W., *Z. Phys.*, 43, 172, 1927. Kennard, E. H., *Z. Phys.*, 44, 326, 1927.
2. Sen, K. D. and Katriel, J., *J. Chem. Phys.*, 125, 074117, 2006.
3. Saad, N., Hall, R. L., and Ciftci, H., *J. Phys. A Math. Gen.*, 39, 7745, 2006. Barakat, T., *J. Phys. A Math. Gen.*, 41, 015301, 2008. Roy, A. K., Jalbout, A. F., and Proynov, E. L., *Int. J. Quantum Chem.*, 108, 827, 2008.
4. Patil, S. H. and Sen, K. D., *Phys. Lett. A*, 362, 109, 2007. Patil, S. H. and Sen, K. D., *Int. J. Quantum Chem.*, 107, 1864, 2007.

---

* The structural entropy S_{str} introduced here as a localization quantity characteristic of the decay of the distribution function is related to the shape complexity as $\ln C_{\text{LMC}}$.

5. Shannon, C. E., *Bell Syst. Tech.*, 27, 379, 1948. Shannon, C. E., *Bell Syst. Tech.*, 27, 623, 1948.
6. Bialynicki-Birula, I. and Mycielski, J., *Commun. Math. Phys.*, 44, 129, 1975.
7. Gadre, S. R., Sears, S. B., Chakravorty, S. J., and Bendale, R. D., *Phys. Rev. A*, 32, 2602, 1985. Gadre, S. R., *Phys. Rev. A*, 30, 620, 1984. Gadre, S. R. and Bendale, R. D., *Int. J. Quantum Chem.*, 28, 311, 1985. Gadre, S. R., Kulkarni, S. A., and Shrivastava, I. H., *Chem. Phys. Lett.*, 16, 445, 1990. Gadre, S. R., Bendale, R. D., and Gejji, S. P., *Chem. Phys. Lett.*, 117, 138, 1985. Gadre, S. R., and Bendale, R. D., *Curr. Sci.*, 54, 970, 1985. Gadre, S. R., in *Reviews of Modern Quantum Chemistry*, Sen, K. D., ed., World Scientific, Singapore, 2002, 108.
8. Tripathi, A. N., Smith, V. H. Jr., Sagar, R .P., and Esquivel, R. O., *Phys. Rev. A*, 54, 1877, 1996. Ho, M., Weaver, D. F., Smith, V. H. Jr., Sagar, R. P., and Esquivel, R. O., *Phys. Rev. A*, 57, 4512, 1998. Ho, M., Smith, V. H. Jr., Weaver, D. F., Gatti, C., Sagar, R. P., and Esquivel, R. O., *J. Chem. Phys.*, 108, 5469, 1998. Ramirez, J. C., Perez, J. M. H., Sagar, R. P., Esquivel, R. O., Ho, M., and Smith, V. H. Jr., *Phys. Rev. A*, 58, 3507, 1998. Ho, M., Weaver, D. F., Smith, V. H. Jr., Sagar, R. P., Esquivel, R. O., and Yamamoto, S., *J. Chem. Phys.*, 109, 10620, 1998. Sagar, R. P., Ramirez, J. C., Esquivel, R. O., Ho, M., and Smith, V. H. Jr., *Phys. Rev. A*, 63, 022509, 2001. Guevara, N. L., Sagar, R. P., and Esquivel, R. O., *J. Chem. Phys.*, 119, 7030, 2003. Guevara, N. L., Sagar, R. P., and Esquivel, R. O., *J. Chem. Phys.*, 122, 084101, 2005. Ghosh, A. and Chaudhuri, P., *Int. J. Theor. Phys.*, 39, 2423, 2000. Shi, Q. and Kais, S., *J. Chem. Phys.*, 121, 5611, 2004. Shi, Q. and Kais, S., *J. Chem. Phys.*, 309, 127, 2005. Chatzisavvas, K. Ch., Moustakidis, Ch. C., and Panos, C. P., *J. Chem. Phys.*, 123, 174111, 2005.
9. Fisher, R. A., *Proc. Camb. Phil. Sec.*, 22, 700, 1925.
10. Frieden, B. R., *Science from Fisher Information*, Cambridge University Press, England, 2004.
11. Rao, C. R., *Linear Statistical Interference and its Applications*, Wiley, New York, 1965. Stam, A., *Inf. Control*, 2, 101, 1959.
12. Romera, E., Sanchez-Moreno, P., and Dehesa, J. S., *Chem. Phys. Lett.*, 414, 468, 2005. Dehesa, J. S., Lopez-Rosa, S., Olmos, B., and Yanez, R. J., *J. Math. Phys.*, 47, 052104, 2006. Romera, E., Sanchez-Moreno, P., and Dehesa, J. S., *J. Math. Phys.*, 47, 103504, 2006. Sanchez-Moreno, P., Gonzales-Ferez, R., and Dehesa J. S., *J. Phys. A*, 8, 1, 2006. Dehesa, J. S., Gonzales-Ferez, R., and Sanchez-Moreno, P., *J. Phys. A*, 40, 1845, 2007.
13. Rényi, A., Some fundamental questions of information theory, *MTA III Oszt. Közl.*, 10, 251, 1960; On measures of information and entropy, in *Proceedings of the Fourth Berkeley Symposium on Mathematics, Statistics and Probability*, Berkeley University Press, Berkeley, CA, 1960, 547; *Probability Theory*, North Holland, Amsterdam, 1970.
14. Bialynicki-Birula, I., *Phys. Rev. A*, 74, 052101, 2006.
15. Onicescu, O., *CR Acad. Sci. Paris A*, 263, 25, 1966.
16. Tsallis, C., *J. Stat. Phys.*, 52, 479, 1988.
17. Rajagopal, A. K., *Phys. Lett. A*, 205, 32, 1995. Ghosh, A. and Chaudhuri, P., *Int. J. Theor. Phys.*, 39, 2423, 2000.
18. López-Ruiz, R., Mancini, H. L., and Calbet, X., *Phys. Lett. A*, 209, 321, 1995.
19. Pipek, J., Varga, I., and Nagy, T., *Int. J. Quantum Chem.*, 37, 529, 1990. Pipek, J. and Varga, I., *Phys. Rev. A*, 46, 3148, 1992. Pipek, J. and Varga, I., *Phys. Rev. E*, 68, 026202, 2002.
20. Catalan, R. G., Garay, J., and López-Ruiz, R., *Phys. Rev. E*, 66, 011102, 2002. López-Ruiz, R., Nagy, A., Romera, E., and Sanudo, J., *J. Math. Phys.*, 50, 123528, 2009.
21. Sen, K. D., *Statistical Complexity: Applications in Electronic Structure*, Springer UK, 2011.

Index

Page numbers followed by *f* and *t* indicate figures and tables, respectively.